Dictyostelium discoideum Protocols

378. **Monoclonal Antibodies:** *Methods and Protocols,* edited by *Maher Albitar, 2007*

377. **Microarray Data Analysis:** *Methods and Applications,* edited by *Michael J. Korenberg, 2007*

376. **Linkage Disequilibrium and Association Mapping:** *Analysis and Application,* edited by *Andrew R. Collins, 2007*

375. **In Vitro Transcription and Translation Protocols:** *Second Edition,* edited by *Guido Grandi, 2007*

374. **Quantum Dots:** *Methods and Protocols,* edited by *Charles Z. Hotz and Marcel Bruchez, 2007*

373. **Pyrosequencing® Protocols,** edited by *Sharon Marsh, 2007*

372. **Mitochondrial Genomics and Proteomics Protocols,** edited by *Dario Leister and Johannes Herrmann, 2007*

371. **Biological Aging:** *Methods and Protocols,* edited by *Trygve O. Tollefsbol, 2007*

370. **Adhesion Protein Protocols,** *Second Edition,* edited by *Amanda S. Coutts, 2007*

369. **Electron Microscopy***: Methods and Protocols, Second Edition,* edited by *John Kuo, 2007*

368. **Cryopreservation and Freeze-Drying Protocols,** *Second Edition,* edited by *John G. Day and Glyn Stacey, 2007*

367. **Mass Spectrometry Data Analysis in Proteomics,** edited by *Rune Mattiesen, 2007*

366. **Cardiac Gene Expression:** *Methods and Protocols,* edited by *Jun Zhang and Gregg Rokosh, 2007*

365. **Protein Phosphatase Protocols:** edited by *Greg Moorhead, 2007*

364. **Macromolecular Crystallography Protocols:** *Volume 2, Structure Determination,* edited by *Sylvie Doublié, 2007*

363. **Macromolecular Crystallography Protocols:** *Volume 1, Preparation and Crystallization of Macromolecules,* edited by *Sylvie Doublié, 2007*

362. **Circadian Rhythms:** *Methods and Protocols,* edited by *Ezio Rosato, 2007*

361. **Target Discovery and Validation Reviews and Protocols:** *Emerging Molecular Targets and Treatment Options, Volume 2,* edited by *Mouldy Sioud, 2007*

360. **Target Discovery and Validation Reviews and Protocols:** *Emerging Strategies for Targets and Biomarker Discovery, Volume 1,* edited by *Mouldy Sioud, 2007*

359. **Quantitative Proteomics,** edited by *Salvatore Sechi, 2007*

358. **Metabolomics:** *Methods and Protocols,* edited by *Wolfram Weckwerth, 2007*

357. **Cardiovascular Proteomics:** *Methods and Protocols,* edited by *Fernando Vivanco, 2006*

356. **High-Content Screening:** *A Powerful Approach to Systems Cell Biology and Drug Discovery,* edited by *Ken Guiliano, D. Lansing Taylor, and Jeffrey Haskins, 2006*

355. **Plant Proteomics:** *Methods and Protocols,* edited by *Hervé Thiellement, Michel Zivy, Catherine Damerval, and Valerie Mechin, 2006*

354. **Plant–Pathogen Interactions:** *Methods and Protocols,* edited by *Pamela C. Ronald, 2006*

353. **DNA Analysis by Nonradioactive Probes:** *Methods and Protocols,* edited by *Elena Hilario and John. F. MacKay, 2006*

352. **Protein Engineering Protocols,** edited by *Kristian Müller and Katja Arndt, 2006*

351. *C. elegans:* *Methods and Applications,* edited by *Kevin Strange, 2006*

350. **Protein Folding Protocols,** edited by *Yawen Bai and Ruth Nussinov 2006*

349. **YAC Protocols,** *Second Edition,* edited by *Alasdair MacKenzie, 2006*

348. **Nuclear Transfer Protocols:** *Cell Reprogramming and Transgenesis,* edited by *Paul J. Verma and Alan Trounson, 2006*

347. **Glycobiology Protocols,** edited by *Inka Brockhausen-Schutzbach, 2006*

346. *Dictyostelium discoideum* **Protocols,** edited by *Ludwig Eichinger and Francisco Rivero, 2006*

345. **Diagnostic Bacteriology Protocols,** *Second Edition,* edited by *Louise O'Connor, 2006*

344. **Agrobacterium Protocols,** *Second Edition: Volume 2,* edited by *Kan Wang, 2006*

343. **Agrobacterium Protocols,** *Second Edition: Volume 1,* edited by *Kan Wang, 2006*

342. **MicroRNA Protocols,** edited by *Shao-Yao Ying, 2006*

341. **Cell–Cell Interactions:** *Methods and Protocols,* edited by *Sean P. Colgan, 2006*

340. **Protein Design:** *Methods and Applications,* edited by *Raphael Guerois and Manuela López de la Paz, 2006*

339. **Microchip Capillary Electrophoresis:** *Methods and Protocols,* edited by *Charles S. Henry, 2006*

338. **Gene Mapping, Discovery, and Expression:** *Methods and Protocols,* edited by *M. Bina, 2006*

337. **Ion Channels:** *Methods and Protocols,* edited by *James D. Stockand and Mark S. Shapiro, 2006*

336. **Clinical Applications of PCR,** *Second Edition,* edited by *Y. M. Dennis Lo, Rossa W. K. Chiu, and K. C. Allen Chan, 2006*

335. **Fluorescent Energy Transfer Nucleic Acid Probes:** *Designs and Protocols,* edited by *Vladimir V. Didenko, 2006*

334. **PRINS and *In Situ* PCR Protocols,** *Second Edition,* edited by *Franck Pellestor, 2006*

METHODS IN MOLECULAR BIOLOGY™

Dictyostelium discoideum Protocols

Edited by

Ludwig Eichinger
Francisco Rivero

Center for Biochemistry and Center for Molecular Medicine Cologne
University of Cologne, Cologne, Germany

HUMANA PRESS ✳ TOTOWA, NEW JERSEY

Preface

Dictyostelium discoideum is a simple but fascinating eukaryotic microorganism, whose natural habitat is deciduous forest soil and decaying leaves, where the amoebae feed on bacteria and grow as independent single cells. Exhaustion of the bacterial food source triggers a developmental program, in which up to 100,000 cells aggregate by chemotaxis towards cAMP. Morphogenesis and cell differentiation then culminate in the production of spores enabling the organism to survive unfavorable conditions. *Dictyostelium* offers unique advantages for studying fundamental cellular processes with the aid of powerful molecular genetic, biochemical, and cell biological tools. These processes include signal transduction, chemotaxis, cell motility, cytokinesis, phagocytosis, and aspects of development such as cell sorting, pattern formation and cell type differentiation. Recently, *Dictyostelium* was also described as a suitable host for pathogenic bacteria in which one can conveniently study the process of infection. In addition, *Dictyostelium* has many of the experimental conveniences of *Saccharomyces cerevisiae* and is probably the best experimentally manipulatable protozoan, providing insight into this diverse group of organisms, which includes some of the most dangerous human parasites.

The recent completion of the *Dictyostelium* genome sequencing project strengthens the position of *D. discoideum* as a model organism. The completed genome sequence and other valuable community resources constitute the source for basic biological and biomedical research and for genome-wide analyses. Together with a powerful armory of molecular genetic techniques that have been continuously expanded over the years, it further enhances the experimental attractiveness of *D. discoideum* and positions the organism on the same level as other fully sequenced model organisms like *S. cerevisiae*, *Caenorhabditis elegans*, or *Drosophila melanogaster*.

This book is divided into four major parts. It provides in the first part for the uninitiated an introduction to the organism, to important community resources and to genome-wide approaches. The second part describes basic methods and available molecular genetic techniques. The third part is dedicated to imaging and localization methods. The chapters in the fourth part emphasize the unique advantages of *Dictyostelium* as a model system. Throughout the book leading *Dictyostelium* scientists present their most useful and innovative techniques for studying fundamental biological processes in this attractive model organism.

Ludwig Eichinger
Francisco Rivero

Contents

Preface v

Contributors xi

**PART I THE ORGANISM, COMMUNITY RESOURCES,
AND GENOME-WIDE STUDIES**

1 The Secret Lives of *Dictyostelium*
   Richard H. Kessin 3

2 The Genome of *Dictyostelium discoideum*
   Adam Kuspa and William F. Loomis 15

3 The cDNA Sequencing Project
   Hideko Urushihara, Takahiro Morio, and Yoshimasa Tanaka 31

4 dictyBase and the Dicty Stock Center
   **Petra Fey, Pascale Gaudet, Karen E. Pilcher,
   Jakob Franke, and Rex L. Chisholm** 51

5 Analysis of Gene Expression Using cDNA Microarrays
   Marcel Kaul and Ludwig Eichinger 75

6 Proteomic Analysis of *Dictyostelium discoideum*
   Udo Roth, Stefan Müller, and Franz-Georg Hanisch 95

PART II BASIC METHODS AND MOLECULAR GENETIC TECHNIQUES

7 Cultivation, Spore Production, and Mating
   Hideko Urushihara 113

8 Parasexual Genetics Using Axenic Cells
   Jason King and Robert Insall 125

9 Slug Phototaxis, Thermotaxis, and Spontaneous Turning Behavior
   Paul R. Fisher and Sarah J. Annesley 137

10 Purification Techniques of Subcellular Compartments
   for Analytical and Preparative Purposes
   Laurence Aubry and Gérard Klein 171

11 Generation of Multiple Knockout Mutants Using
   the Cre-*loxP* System
   Alan R. Kimmel and Jan Faix 187

12 Restriction Enzyme-Mediated Integration (REMI) Mutagenesis
   Adam Kuspa 201

13 RNA Interference and Antisense-Mediated Gene Silencing
in *Dictyostelium*
Markus Kuhlmann, Blagovesta Popova, and Wolfgang Nellen 211

PART III IMAGING AND LOCALIZATION METHODS

14 Application of Fluorescent Protein Tags as Reporters in Live-Cell
Imaging Studies
Annette Müller-Taubenberger 229

15 Investigating Gene Expression: In Situ *Hybridization
and Reporter Genes*
Ricardo Escalante and Leandro Sastre 247

16 Application of 2D and 3D DIAS to Motion Analysis of Live Cells
in Transmission and Confocal Microscopy Imaging
Deborah Wessels, Spencer Kuhl, and David R. Soll 261

17 Using Quantitative Fluorescence Microscopy and FRET Imaging
to Measure Spatiotemporal Signaling Events
in Single Living Cells
Xuehua Xu, Joseph A. Brzostowski, and Tian Jin 281

18 Visualizing Signaling and Cell Movement During
the Multicellular Stages of *Dictyostelium* Development
Dirk Dormann and Cornelis J. Weijer 297

19 Under-Agarose Chemotaxis of *Dictyostelium discoideum*
David Woznica and David A. Knecht 311

20 Optimized Fixation and Immunofluorescence Staining Methods
for *Dictyostelium* Cells
Monica Hagedorn, Eva M. Neuhaus, and Thierry Soldati 327

21 Cryofixation Methods for Ultrastructural Studies
of *Dictyostelium discoideum*
Mark J. Grimson and Richard L. Blanton 339

PART IV DICTYOSTELIUM AS MODEL ORGANISM

22 Analysis of Signal Transduction: *Formation of cAMP, cGMP,
and Ins(1,4,5)P₃ In Vivo and In Vitro*
Peter J. M. Van Haastert 369

23 Assaying Chemotaxis of *Dictyostelium* Cells
Michelle C. Mendoza and Richard A. Firtel 393

24 Characterization of Cross-Linked Actin Filament Gels and Bundles
Using Birefringence and Polarized Light Scattering
Ruth Furukawa and Marcus Fechheimer 407

25 Quantitative and Microscopic Methods for Studying
the Endocytic Pathway
Francisco Rivero and Markus Maniak 423

26 Preparation of Intact, Highly Purified Phagosomes
from *Dictyostelium*
**Daniel Gotthardt, Régis Dieckmann, Vincent Blancheteau,
Claudia Kistler, Frank Reichardt, and Thierry Soldati** 439

27 Assaying Cell–Cell Adhesion
Salvatore Bozzaro 449

28 Periodic Activation of ERK2 and Partial Involvement of G Protein
in ERK2 Activation by cAMP in *Dictyostelium* Cells
Mineko Maeda 469

29 An Improved Method for *Dictyostelium* Centrosome Isolation
**Irene Schulz, Yvonne Reinders, Albert Sickmann,
and Ralph Gräf** 479

30 Epigenetics in *Dictyostelium*
Markus Kaller, Wolfgang Nellen, and Jonathan R. Chubb 491

31 *Dictyostelium discoideum* as a Model to Study Host–Pathogen
Interactions
Can Ünal and Michael Steinert 507

32 Pharmacogenetics: *Defining the Genetic Basis of Drug Action
and Inositol Trisphosphate Analysis*
Kathryn E. Adley, Melanie Keim, and Robin S. B. Williams 517

33 How to Assess and Study Cell Death in *Dictyostelium discoideum*
**Artemis Kosta, Catherine Laporte, David Lam, Emilie Tresse,
Marie-Françoise Luciani, and Pierre Golstein** 535

Index 551

Contributors

KATHRYN E. ADLEY • *Department of Biology and Wolfson Institute for Biomedical Research, University College London, WC1 E6BT, UK*

SARAH J. ANNESLEY • *Department of Microbiology, La Trobe University, Bundoora, VIC 3086, Australia*

LAURENCE AUBRY • *Laboratoire de Biochimie et Biophysique des Systèmes Intégrés (UMR5092), Département Réponse et Dynamique Cellulaires, CNRS-CEA-UJF, CEA-Grenoble, 17, rue des Martyrs, 38054 Grenoble Cedex 09, France*

VINCENT BLANCHETEAU • *Department of Biological Sciences, Sir Alexander Fleming Building, Imperial College London, South Kensington Campus, London SW7 2AZ, UK*

RICHARD L. BLANTON • *Department of Botany and University Honors Program, North Carolina State University, Raleigh, NC, USA*

SALVATORE BOZZARO • *Department of Clinical and Biological Sciences, University of Turin, Ospedale S. Luigi, 10043 Orbassano, Torino, Italy*

Joseph A. Brzostowski • *Chemotaxis Signal Section, National Institutes of Health, NIAID/LIG/CSS, Rockville, MD, USA*

REX L. CHISHOLM • *dictyBase, Center for Genetic Medicine, Lurie 7-125, Northwestern University, Chicago, IL, USA*

JONATHAN R. CHUBB • *Department of Anatomy and Structural Biology, Albert Einstein College of Medicine, The Bronx, New York, USA (Current address: University of Dundee, WTB/MSI Complex, Division of Cell and Developmental Biology, Dow St., Dundee, UK)*

RÉGIS DIECKMANN • *Department of Biochemistry, University of Geneva, Sciences II, 30 quai Ernest Ansermet, CH-1211 Geneva, Switzerland*

DIRK DORMANN • *Division of Cell and Developmental Biology, Welcome Trust Biocentre, University of Dundee, Dundee DD1 5EH, UK*

LUDWIG EICHINGER • *Center for Biochemistry and Center for Molecular Medicine Cologne, University of Cologne, 50931 Cologne, Germany*

RICARDO ESCALANTE • *Instituto de Investigaciones Biomédicas, CSIC/UAM, Arturo Duperier, 4, 28029-Madrid, Spain*

JAN FAIX • *Institute for Biophysical Chemistry, Hannover Medical School D-30623 Hannover, Germany*

MARCUS FECHHEIMER • *Department of Cellular Biology, University of Georgia, Athens, GA, USA*

PETRA FEY • *dictyBase, Center for Genetic Medicine, Lurie 7-125, Northwestern University, Chicago, IL, USA*

RICHARD A. FIRTEL • *Section of Cell and Developmental Biology, Division of Biological Sciences and Center for Molecular Genetics, University of California, San Diego, La Jolla, CA, USA*

PAUL R. FISHER • *Department of Microbiology, La Trobe University, Bundoora, VIC 3086, Australia*

JAKOB FRANKE • *Dicty Stock Center, Columbia University, P&S 12-442, New York, NY, USA*

RUTH FURUKAWA • *Department of Cellular Biology, University of Georgia, Athens, GA, USA*

PASCALE GAUDET • *dictyBase, Center for Genetic Medicine, Lurie 7-125, Northwestern University, Chicago, IL, USA*

PIERRE GOLSTEIN • *Centre d'Immunologie INSERM-CNRS-Univ.Medit. de Marseille-Luminy, Case 906, 13288 Marseille cedex 9, France.*

DANIEL GOTTHARDT • *Department of Molecular Cell Research, Max-Planck-Institute for Medical Research, Jahnstrasse 29, D-69120 Heidelberg, Germany (Current address: Department of Internal Medicine IV, University Hospital of Heidelberg, Im Neuenheimer Feld 410, 69120 Heidelberg, Germany)*

RALPH GRÄF • *Adolf-Butenandt-Institut für Zellbiologie, Ludwig-Maximilians-Universität München, Schillerstrasse 42, D-80336 München, Germany*

MARK J. GRIMSON • *Department of Biological Sciences, Texas Tech University, Lubbock, TX, USA*

MONICA HAGEDORN • *Department of Biochemistry, University of Geneva, Sciences II, 30 quai Ernest Ansermet, CH-1211 Geneva, Switzerland*

FRANZ-GEORG HANISCH • *Central Bioanalytics, Center for Molecular Medicine Cologne (CMMC), and Center for Biochemistry , University of Cologne, 50931 Cologne, Germany*

ROBERT H. INSALL • *School of Biosciences, The University of Birmingham, Edgbaston, Birmingham, B15 2TT, UK*

TIAN JIN • *Chemotaxis Signal Section, National Institutes of Health, NIAID/LIG/CSS, Rockville, MD, USA*

MARKUS KALLER • *Abt. Genetik, Kassel University, Heinrich-Plett-Str. 40, 34132 Kassel, Germany*

MARCEL KAUL • *Center for Biochemistry and Center for Molecular Medicine Cologne, University of Cologne, 50931 Cologne, Germany*

MELANIE KEIM • *Department of Biology and Wolfson Institute for Biomedical Research, University College London, WC1 E6BT, UK*

RICHARD H. KESSIN • *Department of Anatomy, Cell Biology, and Pathology Columbia University, New York, NY, USA*

ALAN R. KIMMEL • *Laboratory of Cellular and Developmental Biology, NIDDK National Institutes of Health, Bethesda, MD, USA*

JASON KING • *Cardiff University, Museaum Avenue, Cardiff CF10 3US, UK*

CLAUDIA KISTLER • *Department of Molecular Cell Research, Max-Planck-Institute for Medical Research, Jahnstrasse 29, D-69120 Heidelberg, Germany (Current address: Abteilung D070, DKFZ, German Cancer Research Centre, Im Neuenheimer Feld 280, 69120, Heidelberg, Germany)*

GÉRARD KLEIN • *Laboratoire de Biochimie et Biophysique des Systèmes Intégrés (UMR5092), Département Réponse et Dynamique Cellulaires, CNRS-CEA-UJF, CEA-Grenoble, 17, rue des Martyrs, 38054 Grenoble Cedex 09, France*

DAVID A. KNECHT • *Department of Molecular and Cell Biology, University of Connecticut, Storrs, CT, USA*

SPENCER KUHL • *W. M. Keck Dynamic Image Analysis Facility, Department of Biological Sciences, The University of Iowa, Iowa City, IA, USA*

MARKUS KUHLMANN • *Abt. Genetik, Kassel University, Heinrich-Plett-Str. 40, 34132 Kassel, Germany*

ADAM KUSPA • *Verna and Marrs McLean Department of Biochemistry and Molecular Biology, Baylor College of Medicine, Houston TX, USA*

ARTEMIS KOSTA • *Centre d'Immunologie INSERM-CNRS-Univ.Medit. de Marseille-Luminy, Case 906, 13288 Marseille cedex 9, France*

DAVID LAM • *Centre d'Immunologie INSERM-CNRS-Univ.Medit. de Marseille-Luminy, Case 906, 13288 Marseille cedex 9, France*

CATHERINE LAPORTE • *Centre d'Immunologie INSERM-CNRS-Univ.Medit. de Marseille-Luminy, Case 906, 13288 Marseille cedex 9, France*

WILLIAM F. LOOMIS • *Cell and Developmental Biology, Division of Biology, University of California San Diego, La Jolla, CA, USA*

MARIE-FRANÇOISE LUCIANI • *Centre d'Immunologie INSERM-CNRS-Univ.Medit. de Marseille-Luminy, Case 906, 13288 Marseille cedex 9, France*

MINEKO MAEDA • *Biological Science, Graduate School of Science, Osaka University, Machikaneyama-cho, Toyonaka, Osaka 560-0043, Japan*

MARKUS MANIAK • *Department of Cell Biology, University of Kassel. Heinrich-Plett-Str. 40, 34132 Kassel, Germany*

MICHELLE C. MENDOZA • *Section of Cell and Developmental Biology, Division of Biological Sciences and Center for Molecular Genetics, University of California, San Diego, La Jolla, CA, USA*

TAKAHIRO MORIO • *Graduate School of Life and Environmental Sciences, University of Tsukuba, 1-1-1 Tennoudai, Tsukuba-shi, Ibaraki-ken 305-8572, Japan*

STEFAN MÜLLER • *Central Bioanalytics, Center for Molecular Medicine Cologne (CMMC), 50931 Cologne, Germany*

ANNETTE MÜLLER-TAUBENBERGER • *Ludwig-Maximilians-Universität München, Institut für Zellbiologie (ABI), Schillerstr. 42, 80336 München, Germany*

WOLFGANG NELLEN • *Abt. Genetik, Kassel University, Heinrich-Plett-Str. 40, 34132 Kassel, Germany*

EVA M. NEUHAUS • *Lehrstuhl für Zellphysiologie, Ruhr-Universität Bochum, ND 4-132, Universitätsstrasse 150, 44780 Bochum, Germany*

KAREN E. PILCHER • *dictyBase, Center for Genetic Medicine, Lurie 7-125, Northwestern University, Chicago, IL, USA*

BLAGOVESTA POPOVA • *Abt. Genetik, Kassel University, Heinrich-Plett-Str. 40, 34132 Kassel, Germany*

FRANK REICHARDT • *Department of Molecular Cell Research, Max-Planck-Institute for Medical Research, Jahnstrasse 29, D-69120 Heidelberg, Germany*

YVONNE REINDERS • *Rudolf Virchow Zentrum/Protein Mass Spectometry Versbacher Strasse 9, D-97078 Würzburg, Germany*

FRANCISCO RIVERO • *Center for Biochemistry and Center for Molecular Medicine Cologne, University of Cologne, 50931 Cologne, Germany*

UDO ROTH • *Central Bioanalytics, Center for Molecular Medicine Cologne (CMMC), 50931 Cologne, Germany*

LEANDRO SASTRE • *Instituto de Investigaciones Biomédicas, CSIC/UAM. Arturo Duperier, 4, 28029-Madrid, Spain*

IRENE SCHULZ • *Adolf-Butenandt-Institut für Zellbiologie, Ludwig-Maximilians-Universität München, Schillerstrasse 42, D-80336 München, Germany*

ALBERT SICKMANN • *Rudolf Virchow Zentrum/Protein Mass Spectometry, Versbacher Strasse 9, D-97078 Würzburg, Germany*

THIERRY SOLDATI • *Department of Biochemistry, University of Geneva, Sciences II, 30 quai Ernest Ansermet, CH-1211 Geneva, Switzerland and Department of Biological Sciences, Sir Alexander Fleming Building, Imperial College London, South Kensington Campus, London SW7 2AZ, UK*

DAVID R. SOLL • *W. M. Keck Dynamic Image Analysis Facility, Department of Biological Sciences, The University of Iowa, Iowa City, IA, USA*

MICHAEL STEINERT • *Institut für Molekulare Infektionsbiologie, Universität Würzburg, Röntgenring 11, 97070 Würzburg, Germany*

YOSHIMASA TANAKA • *Graduate School of Life and Environmental Sciences, University of Tsukuba, 1-1-1 Tennoudai, Tsukuba-shi, Ibaraki-ken 305-8572, Japan*

EMILIE TRESSE • *Centre d'Immunologie INSERM-CNRS-Univ.Medit. de Marseille-Luminy, Case 906, 13288 Marseille cedex 9, France*

CAN ÜNAL • *Institut für Molekulare Infektionsbiologie, Universität Würzburg, Röntgenring 11, 97070 Würzburg, Germany*

HIDEKO URUSHIHARA • *Graduate School of Life and Environmental Sciences, University of Tsukuba, 1-1-1 Tennoudai, Tsukuba-shi, Ibaraki-ken 305-8572, Japan*

PETER J. M. VAN HAASTERT • *Department of Molecular Cell Biology, University of Groningen, Kerklaan 30, 9751NN Haren, The Netherlands*

CORNELIS J. WEIJER • *Division of Cell and Developmental Biology, Welcome Trust Biocentre, University of Dundee, Dundee DD1 5EH, UK*

DEBORAH WESSELS • *W. M. Keck Dynamic Image Analysis Facility, Department of Biological Sciences, The University of Iowa, Iowa City, IA, USA*

ROBIN S. B. WILLIAMS • *Department of Biology and Wolfson Institute for Biomedical Research, University College London, WC1 E6BT, UK*

DAVID WOZNICA • *Department of Molecular and Cell Biology, University of Connecticut, Storrs, CT, USA*

XUEHUA XU • *Chemotaxis Signal Section, National Institutes of Health, NIAID/LIG/CSS, Rockville, MD, USA*

I

The Organism, Community Resources, and Genome-Wide Studies

1

The Secret Lives of *Dictyostelium*

Richard H. Kessin

Summary

Research on *Dictyostelium* has a long history, the trend of which has been to add cell biology, natural products, and evolutionary biology research to the traditional studies of development. The methods presented here and the recent publication of the genome present vast new opportunities.

Key Words: Cellular slime mold; natural products; soil ameba; amoeboid genome.

1. Introduction

In 1991, the archives of the Soviet Union and its satellites began to open and to inform us in the West about events that we had perceived only dimly. This history was meticulously stored in endless files, and as these came under scrutiny, we came to realize that truth does not just spring forth. It requires sifting, close reading, and interpretation by linguists and historians. Was so and so a spy? What actually happened during the Cuban Missile Crisis? How extensive was the Soviet biological warfare effort? Slowly, new perceptions formed and we came to understand events differently or in greater detail. You may think that this is a strange way to introduce a chapter on an ameba, but we will see how far this metaphor, the opening of an archive, carries us. The opening of an archive and the sequencing of a genome are similar in the sense that at one moment you do not know something and then, within a very short time, you do. They are parallel in that dramas of the past, some completely unsuspected— the give-and-take of spy and counter-spy or of pathogen and host, for example—are revealed. They are parallel in the sense that whether you are a historian or a scientist, you can be overwhelmed by the volume of new information, and new ways to handle such volume must be invented.

From: *Methods in Molecular Biology, vol. 346:* Dictyostelium discoideum *Protocols*
Edited by: L. Eichinger and F. Rivero © Humana Press Inc., Totowa, NJ

We biologists have been teasing out phenomena (chemotaxis, motility, development, signal transduction) that are the products of long evolutionary histories. We have been asking how a particular mechanism works, but also, as historians do, we ask how it evolved. Until the genome became available, we were like historians who can only see a few books or, in our case, genes. We once thought that getting a single gene to study was a great triumph— and how hard it was to get them, whether the cAMP phosphodiesterase, the cAMP receptor, or a legion of others! Some of us remember endless time in the cold room purifying our favorite proteins. Now, that is not so common. With the completion of the genome, we find genes more easily and we have something in common with historians who have been given access to a complete archive. It will take time, but with this new access we will discover genes related to those that we know and, more stunning perhaps, we will find new families—large families, even, that have quietly eluded our attention. These little amoeboid secrets will be revealed by careful reading, and analysis and will draw us into new ventures.

The seminal event, just past, in our own research history is the publication of the genome sequence of *Dictyostelium discoideum* *(1)*. This is a historical document perhaps not unique in science— there are other sequences, after all— but it is the personal history of the species that we know to do extraordinary things and, frankly, to be just plain beautiful. It is our archive, and we owe a debt to the many people who participated in throwing it open and analyzing it. They faced many difficulties, and the result is a triumph. It is described in the next chapter by Adam Kuspa and Bill Loomis and in the superbly written *Nature* article.

Others will detail the rich collection of techniques and methods that people have developed for the study of *Dictyostelium*, and this is timely because the last such book was published in 1987 *(2)*. My charge is to describe the earlier pivotal moments in the history of research on this organism, taking a note from one of my favorite scientists, André Lwoff (of lysogeny and *lac* operon fame), who said, "it is dangerous to parachute young scientists into a field whose history they do not know."

2. A History of *Dictyostelium* Research

The first person to be amazed by these organisms, as far as we know, was the mycologist Oskar Brefeld in 1869. Brefeld isolated the amebae from dung and carefully noted that they aggregate *(3)*. There was no way to photograph the structures, so Brefeld made beautiful drawings, some of which I have included in **Fig. 1**. Nonetheless, he got things wrong. Initially, he thought that the slugs of *D. mucoroides* were syncytia, as in *Mucor*, but later he corrected his error and learned that they could be disaggregated into individual cells *(4)*.

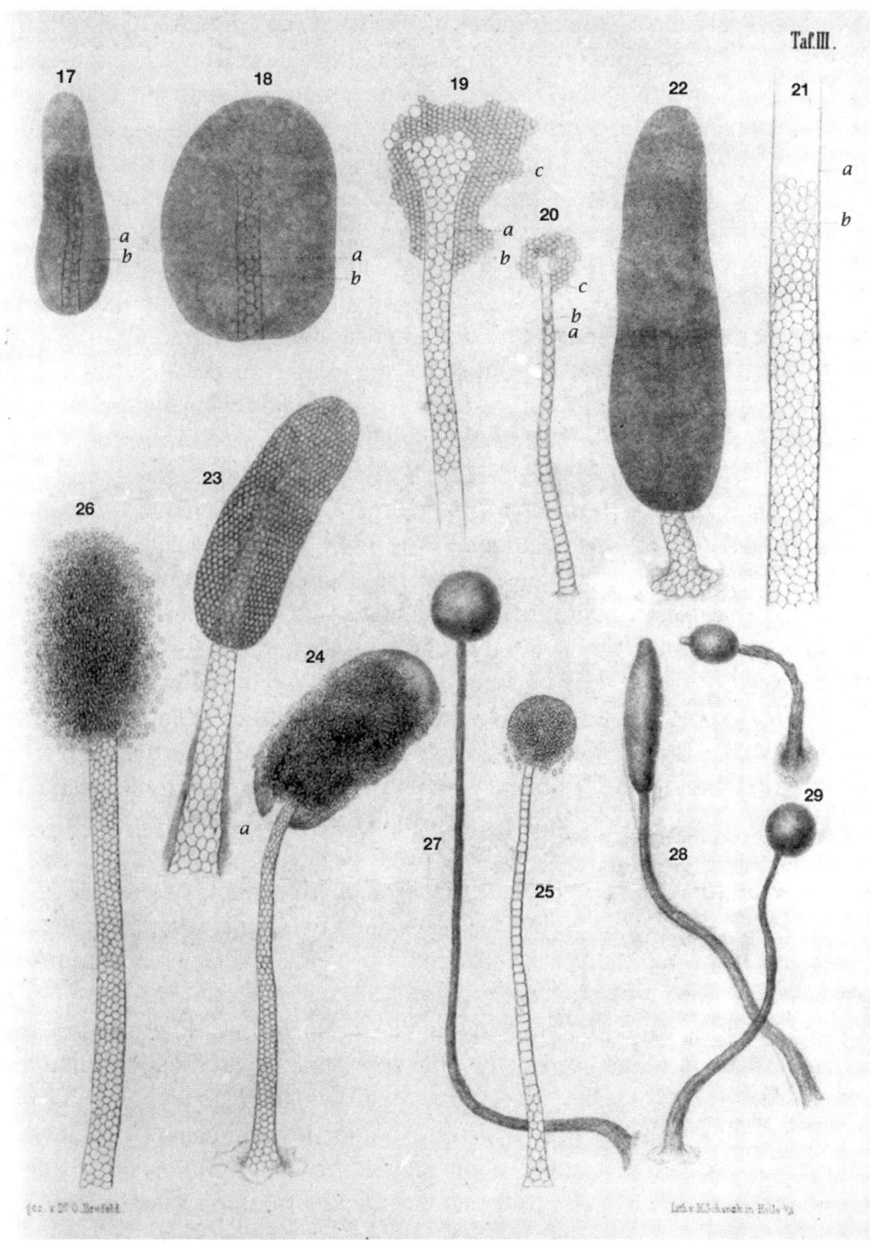

Fig. 1. The original drawings by Brefeld *(3)* show the culmination and fruiting body stages of *Dictyostelium mucoroides*. These specimens were recovered from horse dung. The name *Dicty* means "net-like," and presumably describes the aggregation stage, while *-stelium* means "tower." The word *mucoroides* was meant to denote a similarity to the fungus Mucor.

He was probably the first to note the co-operativity of *Dictyostelium* cells during development, but essentially his interests were descriptive.

Until about 1903, to grow *Dictyostelium* meant to inoculate dung, and it was not clear what the amebae were eating until Potts, in 1902, developed a medium that was more sophisticated than dung (anything is) and realized that the growth of the amebae depends on the presence of bacteria *(5)*. The next breakthrough, although only a one-page note, was by Vuillemin *(6)*, who realized that the amebae fed by phagocytosis, a process that had been discovered by Elie Metchnikoff, who was also at the Pasteur Institute. The idea that the amebae captured their food and digested it internally was another step away from thinking of the Dictyostelia as fungi, which, having a cell wall, secrete enzymes and digest their food externally before uptake. Many years later, it became clear what a large number of different bacterial species *D. discoideum* could digest *(7)*. One hundred and three years after Vuillemin and Potts, the genome will surely help reveal an extraordinary number of hydrolytic enzymes capable of recognizing and digesting the large variety of bacterial surface structures. Phagocytosis itself remains a poorly understood and complex process to which *Dictyostelium* workers continue to contribute.

In 1933, Kenneth Raper isolated *D. discoideum*, which is still the species we use. Raper was then a graduate student, and he took his girlfriend (later wife) on a camping trip to Little Butt's Gap near Ashville, North Carolina to collect samples. Many years ago Mrs. Raper told me it was a little unusual to go camping unchaperoned in 1933, but they had a good time. The samples that came back are the parents of the ones we work on now, and were first described in 1935 *(8)*. Before embarking on his classical studies of proportioning and development, Raper revisited phagocytosis using the giant bacterium *Bacillus megatherium*, and identified the digestive and contractile vesicles of the cell. Made at a time when there were no digital cameras and no green fluorescent protein (GFP), his *camera lucida* drawings are magnificent *(9)*.

Raper and his mentor Thom then published their studies of proportionality, and the first thing they showed was that the proportions of a slug were constant, no matter what its size. They were the first to appreciate that the proportionality problem is at the heart of *Dictyostelium* development *(10)*. Raper and Thom created what is essentially a fate map of the *Dictyostelium* slug by determining that the rear gave rise to the spores and the front to the stalk. The front of the slug, they showed, controlled the movement of the rest. Not having GFP, they used *Serratia marcesens*, a red bacterium, to pigment the cells. By this time, the early 1940s, the fundamental developmental biology was established. There are more details to be found on the early history of *Dictyostelium* research in two books *(11,12)*. For the historically minded, the *Dictyostelium* strain repository is ready to send copies of early papers.

A book could be written on chemotaxis in *Dictyostelium,* and it would start with the experiments of John Tyler Bonner and extend through many workers, including the neglected Brian Shaffer, who probably understood the nature of adaptation and the refractory period for chemotaxis better than anyone at the time *(13)*. This book would continue through the discovery of cAMP as the chemotactic agent by Theo Konijn and others in John Bonner's lab *(14)*. It would include the suspended cell pulsation experiments of Guenther Gerisch *(15)*. It would feature Peter Devreotes' initial studies, which brought a neurobiologist's subtlety to the study of the cAMP receptors *(16,17)*. With many students and collaborators, he has continued the molecular dissection of chemotaxis for the last 25 yr. We are now at the point where we know more about chemotaxis and motility in *Dictyostelium* than in any other cell, and the results have proved revealing for higher organisms.

Although it might once have been possible to study chemotaxis without considering the cytoskeleton, that was long ago. Now we have made fundamental contributions to cell biology, both regarding the cytoskeleton and its regulation during chemotaxis. The many workers who have studied the *Dictyostelium* cytoskeleton have made fundamental observations. If there was ever a problem that *Dictyostelium* was meant to solve, it is the connection between chemotaxis and the mobilization of the cytoskeleton, as has been done so elegantly by Peter Devreotes, Rick Firtel, Guenther Gerisch, and many others.

The chapters of this book will detail the methods of modern molecular biology, including the introduction of transformation, restriction enzyme-mediated integration, the astonishing effect of specific promoters driving GFP or other markers, and microarray technology. For the record, and because I believe that, once in awhile, old papers should be resurrected, let me say that the idea of a molecular genetics of *Dictyostelium* belongs to Maurice Sussman *(18)*, who was the first to create mutants and use them for a variety of purposes, including the development of a parasexual system. The first look at the genome we owe to Rick Firtel and Raquel Sussman, who discovered its size and AT richness, the extrachromosomal rRNA palindrome, and the large amount of mitochondrial DNA *(19,20)*. In those days, there was not even a thought that we would eventually know the full sequence.

In the 1960s, it was known that bacteria induced genes and that there was a subsequent synthesis of protein, but it was not clear that genes could be induced in eukaryotes. Maurice Sussman realized that the spore coat and other structures of the developing *Dictyostelium* would require basic enzymes of polysaccharide synthesis, and he and his colleagues worked out assays for these enzymes, which rose in activity during development. They also showed that the amount of protein rose. The countervailing school suggested that enzymes were not induced and that everything occurred through substrate fluxes, not requiring any genetic induction. I did not understand this then and I do not now, but it is safe to say that this school lacked clarity and that it lost.

3. *Dictyostelium* Lab Strains

There has been evolution of the strains that most *Dictyostelium* workers use. Maurice and Raquel Sussman may have been the first to realize that it would be a great convenience to be able to grow *Dictyostelium* in a broth medium rather than on live or dead bacteria, and reported the first isolation of an axenically growing strain, Ax-1, by subculturing their lab strain DdB (NC-4) *(21)*. The original medium had all manner of nutrients, including liver extract and fetal calf serum. The paper does not mention the use of mutagens. DdB is an NC-4 derivative selected by the continuous subculturing of colonies that showed more synchronous development and less spreading colony morphology.

Both Schwalb and Roth *(22)* and Watts and Ashworth *(23)* eliminated the serum and liver extract from the medium by prolonged subculturing of Ax-1. Again, there is no mention of mutagens. Watts and Ashworth call the isolate, which can grow in the simplified medium, Ax-2 (AX2), and it is the progenitor of many commonly used strains. Ax-1 has been lost. Cocucci and Sussman *(24)* used the term HL-5 (HL5) medium to describe the simplified axenic medium. At about this time, Bill Loomis *(25)* isolated an independent axenic strain with the use of *N*-methyl- *N'*-nitro- *N*-nitrosoguanidine, and this strain was called A3 (AX3). AX3 gave rise to a number of derivatives and contains a large duplication that is not present in AX2. Both strains grow well in the defined minimal medium *(26)*.

The origins and variations of the AX2 and AX3 strains have been the subject of a protracted correspondence, which can be found on the dictybase.org website. There has been much debate as to whether to use AX2 or AX3 derivatives and as to their individual origins. New laboratories making this decision should consult the correspondence. One thing is clear—always get a well defined culture and store a great deal of it frozen. Every month or 6 wk, go back to the frozen culture and grow new stocks, otherwise the culture will evolve unpredictably. The Kessin lab's AX3 strain, originally from the Loomis laboratory, gave rise to AX3-K and AX4 *(27)*, although we call it AX3-1. It has been stored without change since 1979. It, AX2 and AX4, and their many derivatives are available through the *Dictyostelium* stock center. AX4 was used in the sequencing project.

Finally, something is known about the genetics of axenic growth. The control of axenic growth is attributed to three genes, which probably control the rate of macropinocytosis and thus of nutrient uptake. Williams et al. *(28)* established that at least two recessive mutations (*axeA* on linkage group II and *axeB* on linkage group III) are involved in axenic growth. North and Williams established that a third locus, *axeC* on linkage group II, is involved in the axenic growth of strains AX2 and AX3 *(29)*.

4. A Short Walk Along Chromosome 5

When we search the genome, we usually do not wander. We request a particular gene and land there. Maybe we will look at the gene to the left or the right of it, but usually not. Similarly, most people going to the library find the book that they want online and go almost straight to it, although they might see what is to the left or right of the wanted volume, but not me. Sometimes I wander through the stacks or browse through a particular section, and I have found many of interesting things this way. I came across a volume from the International Bacteriology Conference in London in 1881, at which Louis Pasteur and Robert Koch fumed at each other. Koch did not think that Pasteur's cultures were pure (true), or that his anthrax vaccine worked (not true). It was all very nasty—one of the great hate–hate relationships in the history of science. Yet it was instructive to read about how, over many years, each developed his science, and how, for not entirely scientific reasons, they came into conflict. Alas, such is progress that my university has decided that modernism does not require these old books, and they are all being sent to a disused limestone mine in New Jersey. We are going to have a coffee shop in their place. I am told there will be lots of places to hook up my laptop. *Au revoir,* Louis. *Auf Wiedersehen,* Robert.

The habits of a lifetime are not easily erased, and although I am sorry to see the books go, I am looking for a new place to browse. There is no sense in regretting what is past. Happily, I have discovered that I can jump into the genomic equivalent of the stacks online. Maybe there are unimagined genes or genes that I previously thought boring.

This is not quite as satisfying as a real library, but there is less dust.

So here goes: choosing a site at random, we land in . . . chromosome 5 at position 331051; that would be a gene *BC5V2_0_00143* (DDB0187493). It looks like something to do with RNA transport in the nuclear pore. This is not interesting to me, but I make a note to call a friend who might be interested. Moving to the right on chromosome 5, there is *BC5V2_0_00146* (DDB0187496). It codes for *pksE*, which resembles a short-chain alcohol dehydrogenase, an acyl-carrier-protein, and an *S*-malonyltransferase with 3-oxoacyl-(acyl-carrier-protein) synthase I domains. Quite a mouthful, that is. I must have slept through that part of biochemistry. We could ignore it, but let us scroll down for a second. Maybe there is something comprehensible. Aha! This is similar to *Clostridium acetobutylicum* polyketide synthase. I remember that Rob Kay's Differentiation Inducing Factor is a polyketide; then I recall that so is erythromycin. I search for "polyketide synthase" online and find a nice little review. I look again in the *Nature* article, whose authors have pointed out that *D. discoideum* has 47 of these genes, more than any other known eukaryote. How odd.

Polyketide synthases are highly modular proteins that resemble fatty acid synthases and polymerize 2 and 4 carbon fragments into an amazing variety of biologically active compounds. Could *Dictyostelium* amebae have a secret chemical life, one that we have almost completely ignored? The presence of all of these polyketide synthases and also of a great deal of ABC transporters suggests that there is opportunity, but is there motive? Well, yes: the soil is a noxious place, populated by those other champions of antibiotics and secondary metabolites, the *Streptomycetes* and the fungi. *Dictyostelium* must compete with all of them. And do not forget those marauding nematodes, which we know that *D. purpureum* can repel *(30)*. Why should *Dictyostelium* not have a chemical repertoire of its own to aid in its competition with the thousands of other denizens of the soil? It probably does. Could we have predicted that this would be the case before the sequencing of the genome? Perhaps, but now the possibility cannot be ignored.

5. New Directions

As the site of the *Dictyostelium* strain repository, our laboratory tries to predict developments so that we can make sure that the resources are collected. This is like predicting the stock market, but certain trends that have occurred over the last few years have confirmed the first principle of curatorship—you cannot predict. For example, for many years we were concerned with only a few strains—AX2 or AX3 or their immediate derivatives—and it is still true that these occupy a large proportion of the items in the strain collection. A description of the Strain Repository, currently directed by Jakob Franke, is included in this volume (*see* Chapter 4 by Fey et al.). We worried about whether it was worthwhile to collect strains from the wild. Yet we now collect wild strains, which can be employed in experiments that do not require axenic growth. What are these experiments?

First, in the last 6 or 7 yr, it has become clear that *Dictyostelium* has an unusual evolutionary biology. Not only did it diverge long ago, but unlike a multicellular organism arising from a single cell, its cells may face competition from other *Dictyostelium* amebae. This has attracted great interest from evolutionary biologists *(31)*. Using new array technologies, certain problems having to do with competition can be addressed here better than in any other organism. These experiments are not limited to the axenic strains, and so there is every rationale for collecting a variety of wild strains.

Second, the metabolites discussed previously are not going to be the same in every isolate from the wild. *D. discoideum* populates vast territory in the

Appalachian Mountains in the Southern United States. If we believe in Darwin at all, we should know that there is diversity in the wild and that this would extend to the synthesis of interesting compounds. Although we know more about *D. discoideum* than any other species, there is no reason to believe that other species are not also good at making novel compounds. Hence, we have stored a sample of them for the future. The National Science Foundation has also funded a program to collect a wide range of Dictyostelids, Acrasids, and Protostelids. It is led by Dr. Fred Spiegel of the University of Arkansas.

A third area that is beginning to grow is that of bacterial pathogenesis. A number of pathogens, including *Legionella pneumophila* and various myco-bacterial species, infect *Dictyostelium* amebae. In the case of *L. pneumophila*, it is clear that the pathogenic capabilities of the organism evolved to survive its natural hosts, the aquatic and soil amebae. Thus, it grows in many amebae including *Acanthameba, Hartmanella,* and *Dictyostelium.* That *Legionella* infects the macrophages of the human lung and causes pneumonia is an acci-dent. Of the amebae, the only one with genetic capacity and a sequenced genome is *Dictyostelium.* How the bacterium replicates and how the amebae fight back will soon be clearer. This work has gone somewhat slowly, but given the importance of these pathogens, it should soon accelerate.

The chapters that follow will detail the extraordinary variety of techniques that have been developed to study important biological questions. Some devel-oping areas of investigation have been suggested, but there are others that will arise. These problems are more diverse than they were when the last tech-niques book was written in 1987. This is a testimony to the progress we have made in understanding the events of chemotaxis, motility, autophagy, and many other processes. We should be delighted at the entry of pathobiologists and evolutionary biologists into our field. The current volume is one of a number of efforts to make *Dictyostelium* useful for newcomers and veterans and reflects the culture of the field, which is usually one of genuine cooperation and good will.

References

1. Eichinger, L., Pachebat, J. A., Glockner, G., et al. (2005) The genome of the social ameba *Dictyostelium discoideum. Nature* **435,** 43–57.
2. Spudich, J. A. (1987) *Methods in Cell Biology:* Dictyostelium discoideum: *Molecular Approaches to Cell Biology.* Academic, Orlando.
3. Brefeld, O. (1869) *Dictyostelium mucoroides.* Ein neuer Organismus aus der Verwandtschaft der Myxomyceten. *Abhandlungen der Senckenbergischen Naturforschenden Gesellschaft Frankfurt* **7,** 85–107.
4. Brefeld, O. (1884) *Polysphondylium violaceum* und *Dictyostelium mucoroides* nebst Bemerkungen zur Systematik der Schleimpilze. *Untersuchungen aus dem Gesamtgebiete der Mykologie* **6,** 1–34.

5. Potts, G. (1902) Zur Physiologie des *Dictyostelium mucoroides*. *Flora (Jena)* **91**, 281–347.

6. Vuillemin, P. (1903) Une Acrasiee bacteriophage. *C.R. Acad. Sc. Paris* **137**, 387–389.

7. Depraitere, C. and Darmon, M. (1978) Croissance de l'amibe sociale *Dictyostelium discoideum* sur differentes especes bacteriennes. *Ann. Microbiol. (Institute Pasteur)* **129B**, 451–461.

8. Raper, K. B. (1935) *Dictyostelium discoideum*, a new species of slime mold from decaying forest leaves. *J. Agr. Res.* **50**, 135–147.

9. Raper, K. B. (1937) Growth and development of *Dictyostelium discoideum* with different bacterial associates. *J. Agr. Res.* **55**, 289–316.

10. Raper, K. B. (1940) Pseudoplasmodium formation and organization in *Dictyostelium discoideum*. *J. Elisha Mitchell Sci. Soc.* **56**, 241–282.

11. Raper, K. B. (1984) The Dictyostelids. Princeton University Press, Princeton, NJ.

12. Kessin, R. H. (2001) Dictyostelium—*Evolution, Cell Biology, and the Development of Multicellularity*. Cambridge University Press, Cambridge, UK.

13. Shaffer, B. M. (1957) Apects of aggregation in cellular slime moulds. I. Orientation and chemotaxis. *Am. Naturalist* **91**, 19–35.

14. Konijn, T. M., Barkley, D. S., Chang, Y. Y., and Bonner, J. T. (1968) Cyclic AMP: a naturally occurring acrasin in the cellular slime molds. *Am. Naturalist* **102**, 225–233.

15. Gerisch, G., Hulser, D., Malchow, D., and Wick, U. (1975) Cell communication by periodic cyclic-AMP pulses. *Phil. Trans. R. Soc. London. B.* **272**, 181–192.

16. Devreotes, P. N., Derstine, P. L., and Steck, T. L. (1979) Cyclic 3',5'-AMP relay in *Dictyostelium discoideum*. I. A technique to monitor responses to controlled stimuli. *J. Cell Biol.* **80**, 291–299.

17. Devreotes, P. N. and Steck, T. L. (1979) cAMP relay in *Dictyostelium discoideum*. II. Requirements for the initiation and termination of the response. *J. Cell Biol.* **80**, 300–309.

18. Sussman, M. (1954) Synergistic and antagonistic interactions between morphogenetically deficient variants of the slime mould *Dictyostelium discoideum*. *J. Gen. Microbiol.* **10**, 110–120.

19. Firtel, R. and Bonner, J. (1972) Characterization of the genome of the cellular slime mold *Dictyostelium discoideum*. *J. Mol. Biol.* **66**, 339–361.

20. Sussman, R. and Rayner, E. P. (1971) Physical characterization of deoxyribonucleic acids in *Dictyostelium discoideum*. *Arch. Biochem. Biophys.* **144**, 127–137.

21. Sussman, R. and Sussman, M. (1967) Cultivation of *Dictyostelium discoideum* in axenic culture. *Biochem. Biophys. Res. Commun.* **29**, 53–55.

22. Schwalb, M. and Roth, R. (1970) Axenic growth and development of the cellular slime mould *Dictyostelium discoideum*. *J. Gen. Microbiol.* **60**, 283–286.

23. Watts, D. J. and Ashworth, J. M. (1970) Growth of myxamebae of the cellular slime mould *Dictyostelium discoideum* in axenic culture. *Biochem. J.* **119**, 171–174.

24. Cocucci, S. M. and Sussman, M. (1970) RNA in cytoplasmic and nuclear fractions of cellular slime mold amebas. *J. Cell Biol.* **45**, 399–407.

25. Loomis, W. F. (1971) Sensitivity of *Dictyostelium discoideum* to nucleic acid analogues. *Exp. Cell Res.* **64,** 484–486.
26. Franke, J. and Kessin, R. (1977) A defined minimal medium for axenic strains of *Dictyostelium discoideum. Proc. Natl. Acad. Sci. USA* **74,** 2157–2161.
27. Knecht, D. A., Cohen, S. M., Loomis, W. F., and Lodish, H. F. (1986) Developmental regulation of *Dictyostelium discoideum* actin gene fusions carried on low-copy and high-copy transformation vectors. *Mol. Cell. Biol.* **6,** 3973–3983.
28. Williams, K. L., Kessin, R. H., and Newell, P. C. (1974) Parasexual genetics in *Dictyostelium discoideum*: Mitotic analysis of acriflavin resistance and growth in axenic medium. *J. Gen. Microbiol.* **84,** 59–69.
29. North, M. J. and Williams, K. L. (1978) Relationship between the axenic phenotype and sensitivity to ω-aminocarboxilic acids in *Dictyostelium discoideum. J. Gen. Microbiol.* **107,** 223–230.
30. Kessin, R. H., Gundersen, G. G., Zaydfudim, V., Grimson, M., and Blanton, R. L. (1996) How cellular slime molds evade nematodes. *Proc. Natl. Acad. Sci. USA* **93,** 4857–4861.
31. Strassmann, J. E., Zhu, Y., and Queller, D. C. (2000) Altruism and social cheating in the social ameba *Dictyostelium discoideum. Nature* **408,** 965–967.

2

The Genome of *Dictyostelium discoideum*

Adam Kuspa and William F. Loomis

Summary

The *Dictyostelium discoideum* genome has been sequenced, assembled and annotated to a high degree of reliability. The parts-list of proteins and RNA encoded by the six chromosomes can now be accessed and analyzed. One of the initial surprises was the remarkably large number of genes that are shared with plants, animals, and fungi that must have been present in their common progenitor over a billion years ago. The genome encodes a total of about 10,300 proteins including protein families involved in cytoskeletal control, posttranslational protein modification, detoxification, secondary metabolism, cell adhesion, and signal transduction. The genome has a higher proportion of homopolymeric tracts and simple sequence repeats, such as $[CAA]_n$, than most other genomes. Triplet repeats in translated regions produce the highest known proportion of polyglutamine tracts in any known proteome. Phylogenetic analyses based on complete proteomes confirm that the amoebozoa are a sister group to the animals and fungi, distinct from plants and early diverging species such as *Leishmania*, *Plasmodium*, or *Giardia*. The completed *Dictyostelium* sequence opens the door to large-scale functional exploration of its genome.

Key Words: DNA sequence; proteome; amoebozoa; phylogeny.

1. Introduction

The advantages to knowing the complete genome sequence for efficient and productive functional analyses in any organism are becoming increasingly apparent. Model systems for which the genome has not yet been sequenced must present unique attributes to be the subject of continued investigations. The genome defines the information content available to the organism and allows one to predict potential physiological processes. It is the starting point for molecular manipulations to test those predictions.

From: *Methods in Molecular Biology, vol. 346:* Dictyostelium discoideum *Protocols*
Edited by: L. Eichinger and F. Rivero © Humana Press Inc., Totowa, NJ

In 1998, an international consortium began sequencing the *Dictyostelium discoideum* genome and completed the entire sequence early last year *(1)*. High-throughput shotgun sequencing of DNA enriched from the individual chromosomes allowed contigs to be assembled on the basis of overlapping sequences. When combined with high-resolution physical maps, the sequence of each of the six chromosomes could be assembled from one end to the other. The finished sequence conformed well to earlier low-resolution physical maps that defined the chromosomes and provided landmarks along them *(2,3)*. Fewer than 300 gaps remain, and many of these are known to consist entirely of complex repetitive elements such as retrotransposons. Coverage is estimated to include at least 99% of the protein coding information. As one measure of the completeness of the genome sequence, 966 of 967 previously well characterized *Dictyostelium* genes were identified in the assembled sequence. The lengthy and challenging task of sequencing the genome has been worth the effort because it can now be used to characterize the structure of the chromosomes and the genetic information they carry. All the sequence information is publicly available in a convenient and attractive form at dictyBase.org *(4)*.

2. The Chromosomes

The nuclear genome of *Dictyostelium* consists of six chromosomes totaling 34 Mb and approx 100 copies of a linear 88-kb palindrome that carries genes for the ribosomal RNAs and no other functional genes *(1,5)*. Each cell has several hundred mitochondria, each of which carries a 57-kb genome that comprises about 30% of the total cellular DNA *(6)*. All of these genetic elements have been sequenced and annotated; however, the major interest lies in the chromosomes.

Early studies on the number of chromosomes in *Dictyostelium* based on Giemsa and Hoechst staining reported that there were seven *(7,8)*. It now turns out that one of the stained structures is an aggregate of the 100 or so copies of the 88-kb palindrome that together hold 9 Mb of DNA, slightly more than any of the individual chromosomes *(5)*. Such aggregates may normally function in the segregation of the palindrome copies at cytokinesis, but they cannot be considered a chromosome. Physical separation of the chromosomes on pulsed-field gels and long-range physical mapping showed that there are only six chromosomes *(2,3,9)*. Moreover, repetitive *Dictyostelium* inverted repeat sequence (DIRS) elements were shown to form six complex clusters that mapped to one end of each of the chromosomes *(2,3)*. Previous cytological evidence had suggested that the *Dictyostelium* chromosomes are telocentric, and *in situ* hybridization with DIRS showed six strongly stained regions, each at the end of a chromosome *(1)*. These DIRS clusters also localize near the nuclear membrane, which is consistent with the behavior of subtelomeric repeats in other organisms *(10)*.

The sequences of these putative centromeric regions presented a major challenge to assembly because of their complex repeated nature. Only the DIRS region of chromosome 1 was completely assembled, as the result of having above-average coverage *(1)*. The 187-kb terminal region of chromosome 1 contains 14 complete or near-complete DIRS elements as well as several complete and partial copies of eight other long terminal repeat (LTR), non-LTR, and DNA transposons. It is likely that centromeric functions are encoded within, or near, the DIRS elements. However, there is no functional evidence demonstrating the centromeric function of the DIRS clusters or its neighboring sequences.

Although the DIRS clusters are established features marking one subtelomeric end of each chromosome, the telomeres themselves remain somewhat of a mystery. Previous work had suggested that the chromosomes and palindromic elements terminate in AG_{1-8} repeats that could be extended by a telomerase *(2,5,11)*. Dense clusters of repeats at the ends of the chromosome assemblies made it impossible to distinguish one from another and determine the sequence to the very end of the chromosomes *(1)*. However, there were 12 "floating" contigs with complex repetitive elements on one end and specific short segments of the rDNA element on the other end that could be derived from the actual telomeres. Because there are two ends to each of the six linear chromosomes, there are just enough of these contigs to account for the telomeres. Differences in the repeat elements of some of these contigs allowed them to be physically mapped to the ends of individual chromosomes. Others could be assigned based on the prevalence of their composite reads among the reads from enriched chromosomes. In this way, each putative telomeric contig has been tentatively assigned to a chromosome end *(1)*. The presence of a short portion of the palindromic sequence at the distal end of each chromosome raises the possibility that these sequences act as signals for telomere addition to both the rDNA palindrome and the chromosomes.

Although the palindromic rDNA elements are thought to be autonomously replicating mini-chromosomes, they are ultimately encoded by a master copy embedded in chromosome 4 *(5,12)*. Further characterization of the master copy locus by the sequencing project revealed the likely junctions between the embedded element and the rest of the chromosome. The locus carries a complete half element and extends past the asymmetric center ending in a G/C-rich sequence that could snap back to form a hairpin primer/template for extrachromosomal replication *(1)*. Such a transcription-based replication process could explain the complete absence of sequence variation between the two halves of the element and between the complete elements.

The sizes of the six chromosomes range from 8.5 Mb (chromosome 2) to 3.5 Mb (chromosome 6). Otherwise, they have few differences in gene density, number of complex repeats, or gaps. There is a perfect inverted repeat of 1.5 Mb on chromosome 2, but it appears to have entered the genome of strain AX4, the one used for sequencing, when its progenitor strain AX3 was isolated 35 yr ago *(2,3,13)*.

A considerable number of duplications encompassing several kilobases of DNA have occurred relatively recently *(1)*. There are 269 pairs of genes encoding nearly identical proteins and 351 other gene families that contain 3–81 members. Most of the genes in the larger families are clustered, with the most similar family members closest to each other in physical distance along the chromosomes. These observations indicate that most duplication occurs in adjacent positions along the chromosomes. Twenty percent of the tRNA genes occur as closely linked pairs with nearly identical sequence, also suggesting a recent wave of duplications.

Each chromosome is studded with simple sequence repeats that can be generated by slippage of the lagging strand during replication and further extended and contracted by unequal crossing over *(1)*. About 10% of the genome consists of quite long homopolymers, as well as repeats of two, three, and six nucleotides. In intergenic regions the A+T content of these repeats is 99.2%, which is much higher than the average base composition for the same regions (85% A+T). In coding regions, the simple sequence repeats consist mainly of triplets that encode polyglutamine, polyasparagine, or polythreonine. There is one or more homopolymer tract in one-third of all predicted protein coding genes. In fact, a higher proportion of the *Dictyostelium* genome encodes polyglutamine and polyasparagine than has been observed in any other sequenced eukaryote.

3. Protein-Coding Genes

Several Hidden Markov Model programs designed to recognize protein-coding genes have been trained with manually annotated *Dictyostelium* genes and used to predict protein sequences in the 34-Mb of the *Dictyostelium* genome. Information from each of these automated predictions has been consolidated with the GFMerge program developed by the Pathogen Sequencing Unit at the Sanger Institute and then subjected to manual curation by the team at dictyBase and the rest of the consortium *(1)*. In an effort to include all potentially functional genes, the initial criteria were quite permissive. A total of 13,541 genes were predicted, but 2000 of these encoded proteins of less than 100 amino acids, many of which are unlikely to be functional. Using the simplifying assumption that half were mispredictions, the number of genes was estimated by the consortium to be about 12,500 *(1)*. However, the definition of a gene is a subject of debate. Olsen started with the 13,541 predicted genes and then subtracted genes encoding proteins with less than 50 amino acids (786),

Table 1
Predicted Protein-Coding Genes of *Dictyostelium discoideum*
Compared With Other Organisms

	Species				
Feature	*D.* *discoideum*	*S.* *cerevisiae*	*A.* *thaliana*	*D.* *melanogaster*	*H.* *sapiens*
Genome size (Mb)	34	13	125	180	2917
Number of genes	12,500	5538	25,498	13,676	31,400
Gene spacing (kbp/gene)	2.5	2.2	4.9	13.2	132.5
Mean coding size (amino acids)	518	475	437	538	447
% genes with introns	69	5	79	38	85
Mean intron size (bp)	146	ND	170	ND	3365
Mean no. of introns (in spliced genes)	1.9	1.0	5.4	4.0	8.1
Total a.a. encoded (thousands)	7021	2471	11,143	9267	9838

Modified from Eichinger et al. *(1)*.

recently duplicated (nearly identical) genes (355), apparent pseudogenes (1659), and genes from retrotransposons (434) to arrive at 10,307 protein-coding genes *(14)*. As a result of uncertainties in predicting transcriptional signals and protein stability, the total number of genes is likely to be in the range of 10,000–10,600. The complement of predicted protein coding genes identified 99% of the previously characterized genes and sequenced cDNAs *(15)*. Although such measures suggest that the current predicted proteome is nearly complete, continuing manual curation and experimental verification will improve the inventory.

On average, there is a gene in every 2.5 kb of sequence, a gene density similar to that of the yeasts (*see* **Table 1**). Compared with most other eukaryotes, *Dictyostelium* genes are smaller and have fewer introns, which are themselves shorter, but encode proteins of about the same average length (*see* **Table 1**). The exception is *Saccharomyces cerevisiae*, which has a more compact genome and smaller genes than *Dictyostelium*. Considering that *Dictyostelium* was long thought to be a relatively simple organism, it was surprising to find that it encodes twice as many proteins as *S. cerevisiae* and almost as many as *Drosophila* (*see* **Table 1**). The human genome only encodes about twice as many proteins as *Dictyostelium*.

4. The Proteome

The major protein families found in *Dictyostelium*, such as G protein-coupled receptors (GPCRs), protein kinases, and transcription factors, were discussed in the paper presenting the *Dictyostelium* genome and many have recently been further analyzed *(1,16)*. The most striking aspect of the proteome is the diversity of protein types among the broad classes of proteins and super-families *(17)*. For example, *Dictyostelium* has at least one member of each of the major subfamilies of ABC transporters that are found in mammals *(18)*. *Dictyostelium* also possesses a large number of Frizzled/smoothened and GABA$_B$ GPCRs that were previously thought to be specific to metazoa *(19)*. One of them, GrlE, has recently been shown to be sensitive to an antagonist specific to GABA$_B$ receptor and to be a functional GABA$_B$ receptor (Anjard and Loomis, submitted).

Global analyses of protein domains in the *Dictyostelium* genome also revealed some interesting insights and surprises. The presence or absence of Pfam domains within eukaryotic proteomes can be determined with increasing resolution as genome sequences accumulate. There are 53 Pfam domains found in *Dictyostelium*, animals, and fungi that are not present in any fully sequenced plant genome (*see* **Fig. 1**). These domains either arose soon after plants diverged and before *Dictyostelium* diverged from the line leading to animals or they were lost from all plants. The major classes of domains in this group of proteins include those involved in small and large G protein signaling (e.g., regulator of G protein signaling [RGS] proteins), cell cycle control, and domains involved in signaling. It also appears that glycogen storage and utilization arose (or was retained) as a metabolic strategy soon after the plant/animal divergence, because glycogen synthetase appears in this evolutionary interval.

Also particularly striking are the cases in which otherwise ubiquitous Pfam domains appear to be completely absent in one group or another. For instance, *Dictyostelium* appears to have lost the genes that encode collagen domains and basic helix-loop-helix (bHLH) transcription factors. Metazoa, on the other hand, appear to have lost receptor histidine kinases that are common to plants and fungi, whereas *Dicyostelium* has retained 14 of them. The current patterns of gene retention and loss in eukaryotes are likely to change as more genomes are sequenced, and it may turn out that lineage-specific gene loss may be more species-specific than it now appears. For instance, an animal may yet be discovered to have receptor histidine kinases, in the same way that *Ciona* was found to have a cellulose synthase gene similar to that of *Dictyostelium*'s and

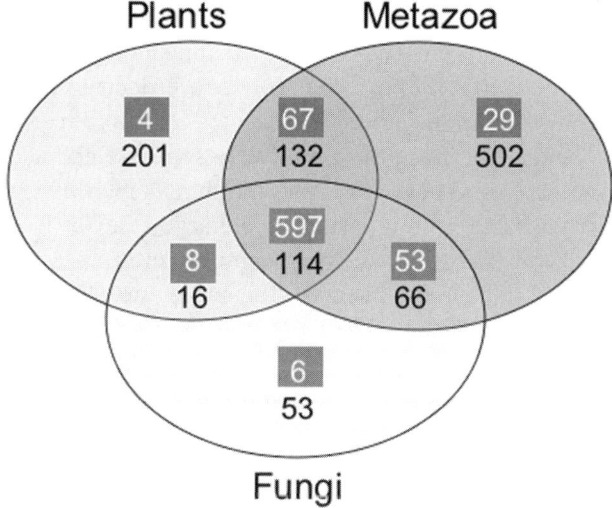

Fig. 1. Distribution of Pfam domains among the eukaryotes. The number of eukaryotic Pfam protein domains present in the major groups of organisms is shown. The numbers of domains present in *Dictyostelium* are boxed. The metazoa are represented by *Homo sapiens, Fugu rubripes, Caenorhabditis elegans, Drosophila melanogaster*; the fungi are represented by *Neurospora crassa, Aspergillus nidulans, Schizosaccharomyces pombe*, and *Saccharomyces cerevisiae*; and the plants are represented by *Arabidopsis thaliana, Oryza. Sativa*, and *Chlamydonomas reinhardtii*. Modified from **ref. 1**, in which a complete description of the analysis and a listing of the domains can be found.

plants were recently found to have "animal-specific" SH2 domain proteins *(1,20)*.

Bacterial orthologs can be recognized for at least 1450 genes that must have been in the common ancestor of plants and animals because they exist in at least one proteome within each of the major groups of eukaryotes (Olsen and Loomis, unpublished observations). About one-quarter of these genes are most similar to orthologs found among the archaebacteria, the likely progenitor of eukaryotes. However, about one-half of these genes are most similar to orthologs found in the proteobacteria that are likely to have entered the eukaryotic genome when a proteobacterium became an established endosymbiont and gave rise to mitochondria. Likewise, about one-fifth of these genes are most similar to orthologs in cyanobacteria that may have been the major food source as eukaryotes took up a predatory life style. The remaining genes appear to have gradually entered the eukaryotic genome from other bacterial types, chiefly the actinobacteria.

There have been more recent cases of gene transfer from bacteria to specific lineages among the eukaryotes. *Dictyostelium* appears to have benefited from horizontal gene transfer (HGT) of genes for such properties as a resistance to tellurite, which is so far unique among eukaryotes *(1)*. Moreover, *Dictyostelium* clearly lost the eukaryotic form of thymidylate synthase and acquired a completely unrelated, rare form of the enzyme found in a few bacteria *(21)*. Predictions of HGT from bacteria to a particular eukaryote suffer from incomplete sampling of eukaryotic genomes. As more genomes are completed, the tests for HGT become more stringent. For instance, 18 genes in the *Dictyostelium* genome were proposed as candidates for HGT *(1)*, but the recent release of the draft genome of *Entamoeba histolytica* showed that one of these was present in their common ancestor and was not recently acquired from a bacterial species *(22)*.

Looking at the protein repertoire of *Dictyostelium* and the other major phylogenetic groups, it becomes clear that the common ancestor had a broad array of proteins and that specific ones were amplified into superfamilies of more specialized proteins in particular lineages. Very few functioning proteins cannot be traced back to a gene that was present when plants and animals shared a common ancestor. Lineage-specific gene loss has turned out to be much more common than was previously supposed, and invoking multiple independent losses to explain the extant phylogenetic pattern is no longer thought to be implausible *(23)*.

5. Genome-Based Phylogeny

The phylogeny of *Dictyostelium* has been clarified as more and more genome sequence has accumulated over the past 15 years. Based on sequence comparisons of small ribosomal subunit RNAs (18S), *Dictyostelium* had been thought to be among the disparate group of early diverging eukaryotes that are quite distinct from the crown group of organisms *(24)*. However, Loomis and Smith realized that the unusually high A+T base composition of the *Dictyostelium* genome could easily skew phylogenetic interpretations made from rRNA sequences, and began to compare the available protein sequences *(25)*. These initial protein sequence comparisons told a very different story, and suggested that *Dictyostelium* proteins are actually more similar to mammalian proteins than are the fungal proteins *(25,26)*. This observation was confirmed and extended by an analysis of more than 100 protein sequences, predicted from the genome project, that indicated that the amoebozoa were monophelytic and a sister group to the animals and fungi *(27)*.

Olsen and Loomis extended the phylogenetic protein sequence comparisons of eukaryotes to include thousands of clusters of orthologs from organisms with complete or near completed genome sequences *(28)*. They examined the predicted proteomes of *Dictyostelium* and 22 other eukaryotes and assembled a

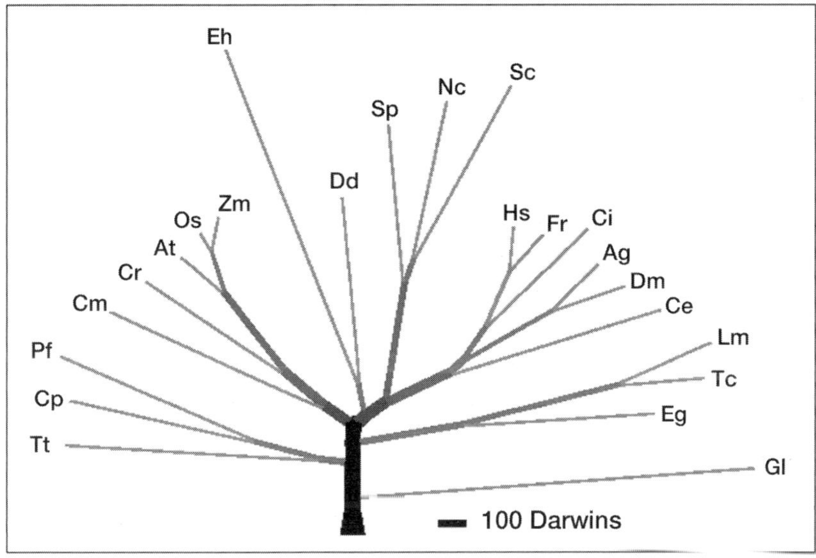

Fig. 2. Proteome based phylogeny of eukaryotes. A phylogenetic tree based on a set of 5908 eukaryotic clusters of orthologs (ECOs) shared by most eukaryotes and rooted on seven archaebacterial proteomes *(1,28)*. Modified from Song et al. *(22)*. One Darwin is equivalent to 1/2000 the divergence between *Saccharomyces cerevisiae* and *Homo sapiens*. From left to right the organisms shown are: *Tetrahymena thermophila, Cryptosporidium parvum, Plasmodium falciparum, Cyanidioschyzon merolae, Chlamydomonas rheinhardtii,, Arabidopsis thaliana, Oryza sativa, Zea mays, Entamoeba histolytica, Dictyostelium discoideum, Schizosaccharomyces pombe, Neurospora crassa, Saccharomyces cerevisiae, Homo sapiens, Fugu rubripes, Ciona intestinalis, Anopheles gambiae, Drosophila melanogaster, Caenorhabditis elegans, Leishmania major, Trypanosoma cruzi, Euglena gracilis,* and *Giardia lamblia.*

set of 5908 eukaryotic clusters of orthologs (ECOs) based on a new model of protein sequence divergence *(28)*. From this, they derived a phylogenetic tree of the eukaryotes rooted on a set of seven archaebacteria, with all of the available completed genomes, that confirms *Dictyostelium's* placement among the amoebozoa and within the crown group (*see* **Fig. 2**). It appears that *Dictyostelium* diverged after the plant/animal split, along the line leading to the animals. This tree also explains the higher similarity between *Dictyostelium* and animal proteins relative to fungal proteins. Higher rates of evolutionary change along the fungal lineage have lead to more highly divergent proteins.

ECOs that include members from early diverging organisms as well as at least one crown organism must have been present in the common ancestor of the crown group eukaryotes. Likewise, ECOs that include members from a plant as well as at least one from the branch leading to animals must have been present in this common ancestor. A total of 2258 such ancestral ECOs were found that could be used to determine patterns of gene loss in specific lineages. The greatest loss of these "ancient genes" occurred in the fungi, for which 40% cannot be recognized in the complete genomes of *Neurospora, Schizosaccharomyces*, or *Saccharomyces (29)*. *Dictyostelium* has lost 35% of these ancient genes, whereas *Drosophila* has lost only 17% and *Arabidopsis* has lost only 12% (reviewed in **ref.** *28*). The distribution of eukaryotic Pfam domains among eukaryotes revealed a similar pattern of gene retention and gene loss (*see* **Fig.** *1* and **ref.** *1*). The plants, metazoa, fungi, and *Dictyostelium* all share 32% of the eukaryotic Pfam domains. Consistent with the phylogeny, Fungi and metazoa share more Pfam domains (119) than do *Dictyostelium* and metazoa (82). Intriguingly, *Dictyostelium* carries a considerable number of Pfam domains that are uniquely found among the metazoa (29). Thus, valuable clues to the functions of proteins containing these domains may come from studies in *Dictyostelium*.

6. Comparisons With Another Amoebozoa

Whole-genome comparisons among related species have yielded dramatic insights, as illustrated by studies of yeasts, fruit flies, and mammals *(30–34)*. The amoebozoa lack the morphological traits needed for precise taxonomic categorization, so sequence comparisons are more critical for classification and genome characterization. Previous analysis of 100 genes has clustered *Dictyostelium* and *Entamoeba* as representative genera of the amoebozoa *(27)*. They represent the two major arms of the conosa lineage: the free-living mycetozoa and the amitochondrial archamoeba, respectively.

The *Dictyostelium* genome was the first of the amoebozoa to be completely sequenced and remains the only free-living amoeba sequence available. The genome of the human pathogen *Entamoeba histolytica* has been subjected to deep shotgun sampling and assembly into unordered scaffolds, so most of its coding capacity is known *(35)*. These two genomes have been compared with each other and with other eukaryotic genomes in an effort to identify ameba specific properties *(22)*. Of the 1500 orthologous gene families shared between the two amobae, most are also shared with plant, animal, and fungal genomes. Surprisingly, only 42 gene families could be defined as distinct to the ameba lineage. Among the ameba-specific proteins are a large number that contain repeats of the FNIP domain, the function of which is unknown. The transcription factor CudA was only known previously in *Dictyostelium*, but an ortholog is now known to exist in *Entamoeba (22,36)*. The amoebozoa-specific genes may prove to be useful for designing of diagnostics or novel therapies for amoebal pathogens such as *Entamoeba* or *Acanthamoeba*.

The small number of lineage-specific genes indicates an ancient split in the conosa lineage. When *Entamoeba* is included in the phylogenetic analysis of ECOs, the expanded tree indicates that the divergence of these two amoebae is greater than the divergence between the budding and fission yeasts and probably happened shortly after the amoebozoa split from the opisthokont lineage (*see* **Fig. 2**).

7. Prospects for Functional Studies

Achieving a meaningful understanding of a single eukaryotic cell will be an enormous task. Moreover, an understanding of the emergent properties of robustness and evolutionary adaptability inherent in all genomes will necessitate a thorough exploration of the genomic potential of a number of organisms *(37,38)*. History has demonstrated that this will come from the study of relatively simple systems such as *Dictyostelium*, to which powerful technologies can be brought to bear. The *Dictyostelium* genome sequence opens up enormous possibilities for functional studies. Groups from around the world have begun global investigations of gene function through directed knockout strategies and expression profiling of mutants using DNA microarrays (e.g., **ref. 39**). The "molecular anatomy" of *Dictyostelium* is being defined by *in situ* hybridization to establish the temporal and spatial patterns of gene expression throughout development *(40–42)*. Specialists in all areas of eukaryotic biology will be able to enrich the initial interpretations and make useful extrapolations to other species.

Homology comparisons between proteins remains the most reliable and efficient way of deriving functional predictions because they allow information from other species to be integrated and used to make testable hypotheses. Although making functional inferences from data obtained with other species has its limits, the steady accumulation of sequence and functional data offers the possibility of continuous refinement of the predictions. There are a significant number of predicted *Dictyostelium* proteins that have close homologs in other species but whose function in any species remains elusive. For example, there are numerous *Dictyostelium* orthologs to human genes implicated in various diseases that could be fruitfully studied *(1)*. Studies in *Dictyostelium* could provide information on the basic cellular function of these proteins that might be applicable to understanding human pathologies.

Additional, relatively unexplored areas of genome function in *Dictyostelium* remain. For example, the extent to which micro-RNAs (miRNA) regulate expression is an open question. The genome sequence indicates that many of the components needed for miRNA-mediated regulation are present, but bioinformatics analyses of the genome sequence and cDNA databases have so far failed to uncover potential miRNAs on the basis of cross species similarities

(43,44). Novel small noncoding RNAs have been identified, but it is not known whether any of them function as regulatory miRNAs *(45)*.

Determination of the *Dictyostelium* genome sequence has marked a turning point in functional analyses of this organism. Over the last few years, information in the Preliminary Directory of *Dictyostelium* Genes, which was based on the sequences of contigs several years before the complete assembly of the chromosomal sequences, has proven immensely useful to those working with *Dictyostelium*. Genes encoding novel cGMP binding proteins, transcription factors, lipid phosphatases and kinases, histidine kinases, and members of the GPCR and ABC superfamilies were recognized and used in molecular genetic studies that have begun to provide exciting new insights. More such studies can be expected in the years to come. Improvements in data structures for describing biological information will facilitate comparisons between systems. The mechanistic details of a biological process need not be identical in *Dictyostelium* for them to illuminate functions in other species.

References

1. Eichinger, L., Pachebat, J. A., Glockner, G., et al. (2005) The genome of the social amoeba *Dictyostelium discoideum. Nature* **435,** 43–57.
2. Loomis, W. F., Welker, D., Hughes, J., Maghakian, D., and Kuspa, A. (1995) Integrated maps of the chromosomes in *Dictyostelium discoideum. Genetics* **141,** 147–157.
3. Kuspa, A. and Loomis, W. F. (1996) Ordered yeast artificial chromosome clones representing the *Dictyostelium discoideum* genome. *Proc. Natl. Acad. Sci. USA* **93,** 5562–5566.
4. Kreppel, L., Fey, P., Gaudet, P., Just, E., Kibbe, W. A., Chisholm, R. L., and Kimmel, A. R. (2004) dictyBase: a new *Dictyostelium discoideum* genome database *Nucleic Acids Res.* **32(Database issue),** D332–D333.
5. Sucgang, R., Chen, G., Liu, W., et al. (2003) Sequence and structure of the extra-chromosomal palindrome encoding the ribosomal RNA genes in *Dictyostelium. Nucleic Acids Res.* **31,** 2361–2368.
6. Ogawa, S., Yoshino, R., Angata, K., et al. (2000) The mitochondrial DNA of *Dictyostelium discoideum*: complete sequence, gene content and genome organization. *Mol. Gen. Genet.* **263,** 514–519.
7. Robson, G. E. and Williams, K. L. (1977) The mitotic chromosomes of the cellular slime mould *Dictyostelium discoideum*: a karyotype based on Giemsa banding. *J. Gen. Microbiol.* **99,** 191–200.
8. Zada-Hames, I. M. (1977) Analysis of karyotype and ploidy of *Dictyostelium discoideum* using colchicine induced metaphase arrest. *J. Gen. Microbiol.* **99,** 201–208.
9. Cox, E. C., Vocke, C. D., Walter, S., Gregg, K. Y., and Bain, E. S. (1990) Electrophoretic karyotype for *Dictyostelium discoideum. Proc. Natl. Acad. Sci. USA* **87,** 8247–8251.

10. Louis, E. J. (2002) Are Drosophila telomeres an exception or the rule? *Genome Biol* **3**, REVIEWS0007.1–07.6.
11. Emery, H. S. and Weiner, A. M. (1981) An irregular satellite sequence is found at the termini of the linear extrachromosomal rDNA in *Dictyostelium discoideum*. *Cell* **26**, 411–419.
12. Welker, D. L., Hirth, K. P., and Williams, K. L. (1985) Inheritance of extrachromosomal ribosomal DNA during the asexual life cycle of *Dictyostelium discoideum*: examination by use of DNA polymorphisms. *Mol. Cell. Biol.* **5**, 273–280.
13. Loomis, W. F. (1971) Sensitivity of *Dictyostelium discoideum* to nucleic acid analogues. *Exp. Cell Res.* **64**, 484–486.
14. Olsen, R. M. (2005) How many protein coding genes does *Dictyostelium* have, in Dictyostelium *Genomics* (Loomis, W. F. and Kuspa, A., eds.), Horizon Scientific, Norwich, UK: pp. 265–278.
15. Urushihara, H., Morio, T., Saito, T., et al. (2004) Analyses of cDNAs from growth and slug stages of *Dictyostelium discoideum*. *Nucleic Acids Res.* **32**, 1647–1653.
16. Loomis, W. F. and Kuspa, A. (eds.) (2005) Dictyostelium *Genomics*. Horizon Scientific, Norwich, UK.
17. Anjard, C. (2005) Multigene Families of *Dictyostelium*, in Dictyostelium *Genomics* (Loomis, W. F. and Kuspa, A., eds.), Horizon Scientific, Norwich: pp. 59–82.
18. Anjard, C. and Loomis, W. F. (2002) Evolutionary analyses of ABC transporters of *Dictyostelium discoideum*. *Euk. Cell* **1**, 643–652.
19. Hereld, D. (2005) Signal transduction via G-protein-coupled receptors, trimeric G-proteins and RGS proteins, in Dictyostelium *Genomics* (Loomis, W. F. and Kuspa, A., eds.), Horizon Scientific, Norwich: pp. 103–124.
20. Williams, J. G. and Zvelebil, M. (2004) SH2 domains in plants imply new signaling scenarios. *Trends Plant Sci.* **9**, 161–163.
21. Myllykallio, H., Lipowski, G., Leduc, D., Filee, J., Forterre, P., and Liebl, U. (2002) An alternative flavin-dependent mechanism for thymidylate synthesis. *Science* **297**, 105–107.
22. Song, J., Xu, Q., Olsen, R., et al. (2005) Comparing the *Dictyostelium* and *Entamoeba* genomes reveals an ancient split in the Conosa lineage, *PLoS Comput Biol.* **1**(7):e71.
23. Bapteste, E. and Gribaldo, S. (2003) The genome reduction hypothesis and the phylogeny of eukaryotes. *Trends Genet.* **19**, 696–700.
24. McCarroll, R., Olson, G. J., Stahl, X. D., Woese, C. R., and Sogin, M. L. (1983) Nucleotide sequence of the *Dictyostelium discoideum* small-subunit ribosomal ribonucleic acid inferred from the gene sequence: evolutionary implications. *Biochemistry* **22**, 5858–5868.
25. Loomis, W. F. and Smith, D. W. (1990) Molecular phylogeny of *Dictyostelium discoideum* by protein sequence comparison. *Proc. Natl. Acad. Sci. USA* **87**, 9093–9097.
26. Loomis, W. F. and Smith, D. W. (1995) Consensus phylogeny of *Dictyostelium*. *Experientia* **51**, 1110–1115.

27. Bapteste, E., Brinkmann, H., Lee, J. A., et al. (2002) The analysis of 100 genes supports the grouping of three highly divergent amoebae: *Dictyostelium, Entamoeba,* and *Mastigamoeba. Proc. Natl. Acad. Sci. USA* **99,** 1414–1419.

28. Olsen, R. and Loomis, W. F. (2005) A collection of amino acid replacement matricies derrived from clusters of orthologs *J. Mol. Evol.* **61(5),** 659–665.

29. Kuspa, A. (2005) Whole-genome functional analyses in *Dictyostelium,* in Dictyostelium *Genomics* (Loomis, W. F. and Kuspa, A., eds.), Horizon Scientific, Norwich, UK: pp. 279–296.

30. Kellis, M., Patterson, N., Endrizzi, M., Birren, B., and Lander, E. S. (2003) Sequencing and comparison of yeast species to identify genes and regulatory elements. *Nature* **423,** 241–254.

31. Kellis, M., Patterson, N., Birren, B., Berger, B., and Lander, E. S. (2004) Methods in comparative genomics: genome correspondence, gene identification and regulatory motif discovery. *J. Comput. Biol.* **11,** 319–355.

32. Kellis, M., Birren, B. W., and Lander, E. S. (2004) Proof and evolutionary analysis of ancient genome duplication in the yeast *Saccharomyces cerevisiae. Nature* **428,** 617–624.

33. Gibbs, R. A., Weinstock, G. M., Metzker, M. L., et al. (2004) Genome sequence of the Brown Norway rat yields insights into mammalian evolution. *Nature* **428,** 493–521.

34. Richards, S., Liu, Y., Bettencourt, B. R., et al. (2005) Comparative genome sequencing of *Drosophila pseudoobscura*: Chromosomal, gene, and cis-element evolution. *Genome Res.* **15,** 1–18.

35. Loftus, B., Anderson, I., Davies, R., et al. (2005) The genome of the protist parasite *Entamoeba histolytica. Nature* **433,** 865–868.

36. Fukuzawa, M., Hopper, N., and Williams, J. (1997) CudA: a *Dictyostelium* gene with pleiotropic effects on cellular differentiation and slug behaviour. *Development* **124,** 2719–2728.

37. Alm, E. and Arkin, A. P. (2003) Biological networks. *Curr. Opin. Struct. Biol.* **13,** 193–202.

38. McAdams, H. H., Srinivasan, B., and Arkin, A. P. (2004) The evolution of genetic regulatory systems in bacteria. *Nat. Rev. Genet.* **5,** 169–178.

39. Van Driessche, N., Demsar, J., Booth, E. O., et al. (2005) Epistasis analysis with global transcriptional phenotypes. *Nature Genet.* **37,** 471–477.

40. Maeda, M., Kuwayama, H., Yokoyama, M., et al. (2000) Developmental changes in the spatial expression of genes involved in myosin function in *Dictyostelium. Dev. Biol.* **223,** 114–119.

41. Tsujioka, M., Yokoyama, M., Nishio, K., et al. (2001) Spatial expression patterns of genes involved in cyclic AMP responses in *Dictyostelium* development. *Devel. Growth Differ.* **43,** 275–283.

42. Maeda, M., Sakamoto, H., Iranfar, N., et al. (2003) Changing patterns of gene expression in *Dictyostelium* prestalk cell subtypes recognized by in situ hybridization with genes from microarray analyses. *Euk. Cell* **2,** 627–637.

43. Martens, H., Novotny, J., Oberstrass, J., Steck, T. L., Postlethwait, P., and Nellen, W. (2002) RNAi in *Dictyostelium*: the role of RNA-directed RNA polymerases and double-stranded RNase. *Mol. Biol. Cell* **13,** 445–453.

44. Graf, S., Borisova, B. E., Nellen, W., Steger, G., and Hammann, C. (2004) A database search for double-strand containing RNAs in *Dictyostelium discoideum*. *Biol. Chem.* **385,** 961–965.

45. Aspegren, A., Hinas, A., Larsson, P., Larsson, A., and Soderbom, F. (2004) Novel noncoding RNAs in *Dictyostelium discoideum* and their expression during development. *Nucleic Acids Res.* **32,** 4646–4656.

3

The cDNA Sequencing Project

Hideko Urushihara, Takahiro Morio, and Yoshimasa Tanaka

Summary

The *Dictyostelium discoideum* cDNA sequencing project started in 1995, preceding the genome sequencing project. Altogether, 14 cDNA libraries, including full-length ones, were constructed from five different stages of growth and asexual and sexual development, from which nearly 100,000 randomly chosen clones were sequenced to yield over 150,000 expressed sequence tags (ESTs). The data have been publicized online to facilitate clone distribution and collaboration using the whole clone set for microarray analyses. The EST reads were assembled to 6700 independent genes, which constitute about 55% of the total estimated *Dictyostelium* genes. Utilization of wet and dry resources have contributed to the understanding of the genetic system controlling the multicellular development in *Dictyostelium*.

Key Words: *Dictyostelium discoideum*; development; gene expression; EST analysis; DNA microarray; *in situ* hybridization; sequence database; gene annotation; promoter sequence.

1. History and Project Outline

The story of the Japanese cDNA project began in 1995 at the annual *Dictyostelium* meeting when Dr. William Loomis discussed with Dr. Ikuo Takeuchi the possibility of laboratories in Japan participating in sequencing the *Dictyostelium* genome. Dr. Takeuchi consulted with Dr. Mineko Maeda and Dr. Yoshimasa Tanaka, and we decided to organize a project team to explore the possibilities. After considerable discussion, our team decided that we should focus on sequencing a large number of cDNAs because we could carry out such an expressed sequence tag (EST) project more rapidly and at less cost than sequencing the 34 Mb genome with its high A/T content and multiple repetitive elements.

From: *Methods in Molecular Biology, vol. 346:* Dictyostelium discoideum *Protocols*
Edited by: L. Eichinger and F. Rivero © Humana Press Inc., Totowa, NJ

Our application for a Grant-in-Aid for Scientific Research on Priority Areas from the Ministry of Education, Culture, Sports, Science and Technology of Japan (MEXT) was approved in 1996 for 5 yr with Dr. Takeuchi as Director. Dr. Maeda later served as Director. 1996 was a pivotal year for Japanese basic science because the government set in place, as Research For The Future (RFTF), several top-down directives for big science including genome research through the Japan Society for the Promotion of Science (JSPS). JSPS and, later, MEXT funded our cDNA project (project leader: Dr. Tanaka; later, Dr. Hideko Urushihara) for 9 yr from 1996. At the onset of the project, we decided to make the sequence data immediately available to the public and to freely distribute clones upon request.

Using established procedures, we generated cDNA libraries from mRNA isolated from growing cells and cells at the slug stage. The most critical, time-consuming, and laborious step was the preparation of plasmids. Fortunately, we were able to use the newly developed automated 96-well plasmid preparation system, without which our work would have been severely delayed. We subsequently collaborated with Dr. Sumio Sugano at the University of Tokyo to generate cDNA libraries enriched in copies of full-length mRNA using the oligo-capping method *(1)*. Libraries were made from vegetative cells, aggregating cells, slugs, and culminants. We were able to collaborate with Dr. Yuji Kohara at National Institute of Genetics of Japan to carry out high-throughput sequencing, which brought us a large step toward the completion of the project.

Many specific cDNAs were distributed to the community throughout this project. Moreover, efforts were made to prepare a nonredundant set of several thousand cDNAs, which was provided to several laboratories for microarray studies (*see* **ref. 2** for an example). We collaborated with Dr. Gad Shaulsky and his colleagues in studies of genes regulated developmentally *(3)* or by dedifferentiation *(4)*. We also collaborated with Dr. Loomis and his colleagues on the spatial expression patterns of cell-type specific genes recognized on the microarrays *(5,6)* and with Dr. Eichinger and his colleagues on the differential gene expression in response to infection of *Dictyostelium* cells with *Legionella pneumophila (7)*.

Success of the *Dictyostelium* cDNA project in Japan was dependent on generous financial support, cooperation among the labs, and the timely appearance of new improved methods just when we needed them.

2. cDNA Libraries and Clone Resources

To obtain as many genes as possible, we generated 14 sets of the cDNA library from four developmental stages (vegetative, aggregation, migrating slug, and early culmination) and sexually maturated cells or gametes. Historically, we first generated conventional cDNA library sets, which we will refer to as EST libraries *(8,9)*. Second, we produced libraries enriched in cDNA clones containing whole sequences of mRNA (full-length enriched cDNA libraries) *(10)*. In addition, to collect genes specifically expressed in the macrocyst cycle efficiently, we generated a gamete-specific subtracted cDNA library *(11)*. In this section, we provide an overview of the construction of the cDNA libraries and sequencing analyses. The properties of the libraries are summarized in **Table 1**. Because the *Dictyostelium* genome is highly A/T-rich, one might have to consider the unusual rearrangement and/or deletion of the cDNA cloned in *Escherichia coli* cells. Selection of the vector, host bacterial strain, and/or conditions of the bacterial culture, such as the medium and temperature, can overcome the problems associated with a biased DNA composition. We tested some of the conditions mentioned previously and found that even a standard protocol of library construction is sufficient to produce good results.

2.1. Generation of the EST Library

The *Dictyostelium discoideum* strain Ax4 was used for the cDNA library of developmental stages *(8,9)*. Axenically grown cells were harvested at the exponential growth phase (3×10^6 cells/mL), washed to remove nutrient, spread on a nitrocellulose filter saturated with lipopolysaccharide (LPS) buffer (20 mM KCl, 0.24 mM MgCl$_2$, and 40 mM K$_2$HPO$_4$/KH$_2$PO$_4$, pH 6.4), and incubated at 22°C. Poly (A)$^+$ RNA was prepared from the cells harvested at the growth phase and the first finger stage (14–16 h of development). Using the SuperScript Plasmid System (GIBCO BRL, Life Technologies, Inc.), cDNA was synthesized by priming with an oligo (dT)$_{15}$ primer carrying a *Not* I adapter. For directional cloning, the cDNA has a *Sal*I-linker (5'-TCGACCCACGCGTCCG-3') at the 5' end and primer-derived (dA)$_{15}$ stretch and the *Not*I-adaptor sequences (5'-A$_{15}$GGGCGGCCGC-3') at the 3' end. The cDNA was size-fractionated to recover the fraction over about 0.4 kb (fraction S) and 1.0 kb (fraction L). The cDNA included in fraction S was ligated to *Sal*I-*Not*I sites of pBluescript II KS (GenBank/EMBL/DDBJ, accession number X52329) and introduced into *E. coli* DH5α by electroporation to generate libraries SS (slug stage) and VS (vegetative stage). The cDNA of fraction L were ligated into the corresponding sites of pSPORT1 (accession number U12390) and transformed into *E. coli* DL795 cells [K12 SH28 *Dhsd, mcrBC, mrr; e14 (mcrA)0; sbc201; recA*::Cm^r; *su*pE44] to obtain library SL (slug stage).

For preparation of the gamete cDNA library, strain KAx3 cells were used. To induce sexual maturation, the cells grown on SM agar plates with *Klebsiella aerogenes* were harvested and resuspended in Bonner's Salt Solution (10 mM NaCl, 10 mM KCl, 3 mM CaCl$_2$) *(12)* containing *K. aerogenes* and shaken at 22°C for 15 h in the dark. Poly (A)$^+$ RNA was isolated and cDNA was synthesized as described previously. The cDNA was size-fractionated to

Table 1
Specification of the cDNA Library Sets

Library	Stage	Plasmid vector	*E. coli* strain	Library size (cfu)	Mean insert size (kb)	Number of clones picked
VS	Vegetative	pBluescriptII KS-	DH5α	>1.0E6	0.8	8148
SS	Slug	pBluescriptII KS-	DH5α	>6.0E6	1.0	9984
SL	Slug	pSPORT1	DL795	>2.5E6	1.3	8448
FC	Gamete	pSPORT1	DH5α	>3.5E5	1.0	1104
FCL	Gamete	pSPORT1	DH5α	>1.0E4	1.6	96
VF	Vegetative	pME18SFL3	DH10B	>1.2E5	1.5	11,520
VH	Vegetative	pME18SFL3	DH10B	>4.8E4	2.1	12,672
AF	Aggregation	pME18SFL3	DH10B	>2.9E5	1.8	10,752
AH	Aggregation	pME18SFL3	DH10B	>3.6E4	2.6	13,440
SF	Slug	pME18SFL3	DH10B	>1.9E5	1.7	8832
SH	Slug	pME18SFL3	DH10B	>9.6E3	3	9216
CF	Culmination	pME18SFL3	DH10B	>3.6E5	1.5	8064
CH	Culmination	pME18SFL3	DH10B	>2.9E4	2.5	14,208
FC-IC	Gamete	pCR2.1	DH5α	>8.0E3	0.3	1786

recover shorter (library FC) and longer (library FCL) fractions and cloned into *Sal*I-*Not*I sites of the pSPORT1 plasmid vector.

2.2. Generation of the Full-Length Enriched cDNA Library

Total RNA was isolated from axenically growing Ax4 cells (designated as V series), cells developed on nitrocellulose filter at the aggregation (8 h; A series), slug (16 h; S series), and early culmination stage (20 h; C series). The quality of the RNA was assessed by comparing the amount of large and small rRNA subunits and Northern analyses probing with longer (>3 kb) genes. To enrich the cDNA containing whole sequences of mRNA, cDNA was synthesized using an oligo-capping method *(1,13)*. In principle, the method consists of the following four steps:

1. The total RNA was treated with bacterial alkaline phosphatase followed by tobacco acid pyrophosphatase to remove the phosphate group at the 5' end of the truncated mRNA and cleave the cap structure to expose the phosphate group at the 5' end of the capped (full-length) mRNA.
2. Oligo RNA was ligated to the 5' end of the mRNA. Because only full-length mRNA has the phosphate group exposed at the 5' end, the oligo RNA can be ligated only to full-length mRNA, not truncated mRNA.
3. Using the treated mRNA as a template, the first strand cDNA was synthesized by priming with the oligo-dT primer with the adapter sequence at its 5' end.
4. PCR was performed with primers complementary to the 5' oligo RNA and 3' adapter sequences. It is expected that only cDNA derived from capped mRNA was amplified.

The amplified cDNA was size-fractionated using agarose electrophoresis. One fraction (fraction F) contains cDNA longer than 0.8 kb and the other (fraction H) contains that longer than 2 kb. Each fraction was directionally ligated to the *Dra*III site of the plasmid vector pME18SFL3 (Acc. No. AB009864; *see* **Fig. 1A**). Because the vector was originally designed for expression in mammalian cells, the promoter, intron, and terminator may not work in *Dictyostelium* cells. The sequence around the cDNA insert is shown in **Fig. 1B**. The ligated cDNA was introduced into *E. coli* DH10B cells. Finally, we generated eight sets of cDNA library in combination with four developmental stages (V, A, S, and C series) and two fractions by size (fractions F and H). The names of the libraries are defined as the combination of two letters, each of which indicates the developmental stage and fraction. For example, library AF shows the one prepared from the >0.8 kb fraction of the aggregation stage cDNA.

2.3. Generation of the Gamete-Enriched Subtracted Library

Although sexually mature cells are clearly distinct from "immature" cells, these two cells share many features as growing cells. Therefore, it is difficult to

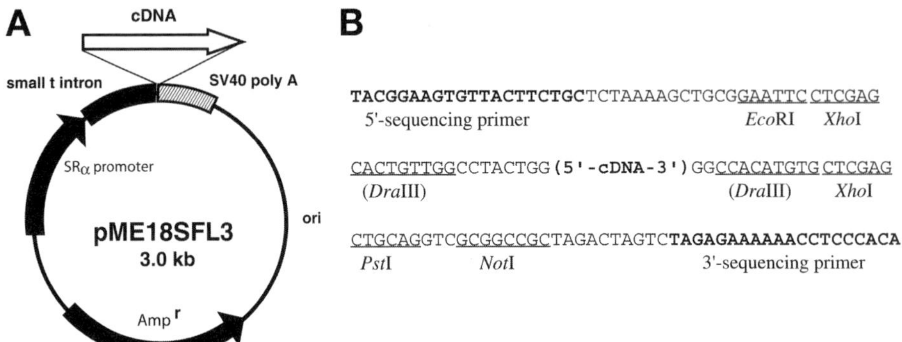

Fig. 1. Structure of a clone in the full-length enriched cDNA library. (**A**) Schematic map of vector plasmid pME18SFL3. The direction of the inserted cDNA is shown as the arrow. (**B**) Nucleotide sequences around the cDNA insertion site of the plasmid. Restriction enzyme sites available for excision of the cDNA insert are underlined. The *Dra*III sites used for cloning are no longer available. Primer sites used for sequencing from the ends are shown in bold face. It should be noted that they are not conventional M13 primer sequences.

find genes specific to sexual maturation by randomly sequencing cDNA clones of the FC and FCL libraries. To overcome this difficulty, a gamete-specific subtraction cDNA library (FC-IC library) was generated using the suppression subtractive hybridization technique (SSH) *(14)*. The tester cDNA was prepared from the sexually matured cells, whereas the driver cDNA was from the cells grown on SM plates fed with bacteria. Using a PCR-select cDNA subtraction kit (Clontech Laboratories), SSH was performed and the amplified cDNA fragments were cloned into a pCR2.1 plasmid vector (Invitrogen) *(11)*. It should be noted that the cDNA was cloned into the vector in a random orientation, unlike the other cDNA libraries.

2.4. Sequencing of cDNA Clones

The *E. coli* clones carrying the cDNA insert were randomly picked and transferred into 96-well plates and then plasmid DNA was isolated manually using a 96-well Alkaline Lysis Miniprep Kit (Edge BioSystems.) or automatically with a BIOMEK 2000 robot (Beckman) using Multiscreen™ NA and FB filter plates (Millipore) following the manufacturer's instructions. For the full-length enriched library sets, the cDNA insert was amplified using PCR. Sequence of the cDNA insert was read once from the 5' and 3' ends. Primers used for sequencing were conventional M13 forward (5'-GTTTTCCCAGTCACGACGTTGTA-3') and reverse (5'-CAGGAAACAGCATTGAC-3') primers. For full-length enriched cDNA clones, vector-specific primers were used (*see* **Fig. 1B**). The quality of the resulting sequences was examined automatically or by eye and reliable sequence data were deposited to the cDNA database (*Dicty*_cDB) (*see* **Subheading 4.**).

2.5. Availability of the cDNA Clone Resource

The cDNA clones were stored as frozen bacterial stock and/or plasmid DNA. Individual cDNA clones are available by sending us a request. The full-length enriched cDNA clones are provided as bacteria and other cDNA clones are sent as plasmid DNA. For detailed contact information, one can visit the cDNA project website (http://dictycdb.biol.tsukuba.ac.jp/cDNA/req_clones.html).

3. Utilization of the cDNA Resources for Functional Genomics

With the *Dictyostelium* cDNA project, a large collection of genes expressed in the organism became available. As well as a tool for discovering genes of interest, it provides a set of probes for monitoring the expression of thousands of genes. In this section, we introduce examples of utilization of the cDNA resources for genome-scale analyses.

3.1. Expression Profiling Using cDNA Microarray

DNA microarray technology enables us to monitor the behavior of thousands of genes simultaneously using solid supports, such as glass slides, on which DNA fragments corresponding to the genes are spotted. Our cDNA collection has been served as a source of the gene set for microarray experiments using *Dictyostelium*.

Because the results of microarray experiments are usually presented as a ratio of the transcription level of each gene between two samples, it is easy to imagine that the microarray is a useful tool for the screening of differentially expressed genes between two distinct cell types or genotypes. By comparing the transcription levels of the genes between prespore and prestalk cells, novel prespore- or prestalk-enriched genes were identified *(5,15,16)*. Downstream target genes of some transcription factors were identified by comparison between the wild-type and mutants of the transcription factor genes *(2,17,18)*. As well as a direct comparison between the two distinct cell populations, clustering analysis using multiple sets of expression profiles is also helpful to identify groups of genes showing the same expression pattern. As genes involved in a co-expressed gene group are expected to share a mechanism of transcriptional regulation, it provides a clue to the identification of regulatory promoter sequences and corresponding transcription factors.

Another use for expression profiling using microarray is as a tool for monitoring the physiological status of cells. This use is based on the idea that an expression profile reflects on the cell physiology. Because expression profiling is more sensitive and more comprehensive than other phenotypic markers, such as morphological and biochemical markers, it should be useful for describing even subtle phenotypes. By comparing transcription profiles along a time sequence of a biological event such as development, one can identify the critical stages when dramatic changes in cell physiology occur. Van Driessche et al. demonstrated that the largest change of transcription profile coincides with the transition from unicellular to multicellular development *(3)*. Katoh et al. *(4)* showed that the process of dedifferentiation consists of three distinct phases based on changes of transcription profile. On the other hand, comparison of the transcription profile between the wild-type and a mutant and/or between mutants helps us to predict the genetic network. For example, Van Driessche et al. reconstructed the protein kinase A (PKA) pathway using only the information of dissimilarities of the expression profiles between the relevant mutants *(19)*.

3.2. Spatial Expression Profiling With in situ Hybridization

Discovery and utilization of cell-type-specific marker genes provides a clue to understanding cell differentiation and pattern formation during *Dictyostelium* multicellular development. Marker genes of novel expression patterns will suggest new regulatory mechanisms of cell differentiation, whereas genes sharing their expression pattern will provide an opportunity to identify regulatory promoter elements by comparing with the upstream regions. As described previously, many novel cell-type enriched genes were identified using microarray analyses. As one of the research activities of our *Dictyostelium* cDNA project, Maeda and her colleagues investigated the spatial expression pattern of 54 prespore-specific and 104 prestalk-specific genes from the tipped aggregate to late culmination stage using *in situ* hybridization (ISH), as well as 10 and 7 genes involved in cAMP signalling and myosin function, respectively *(5,16,20,21)*. The resulting images of spatial expression were deposited in the ATLAS database (http://dictycdb.biol.tsukuba.ac.jp/~tools/bin/ISH/index.html) and can be retrieved using the database (*see* **Fig. 2**).

With the ISH analyses, many new genes specific to one of the prestalk subregions were identified. At the same time, they were divided into several subclasses according to their spatial and temporal expression pattern *(5)*. For example, mRNA recognized by cDNA SLF308 was expressed in the most anterior zone of the PstA region, whereas SSK861 mRNA was detected in the posterior half of the PstA region. As well as expression profiling with microarrays, comparison of the gene expression patterns with ISH is also useful for the elucidation of molecular mechanisms of development. For example, investigation of the spatial expression pattern of the PstO-specific genes in the *dmtA*⁻ mutant revealed that some genes were expressed in the PstO region of

SLA128

(DictybaseID:DDB0187890)

product:arabinofuranohydrolase weak homolog Accession number:AU060076

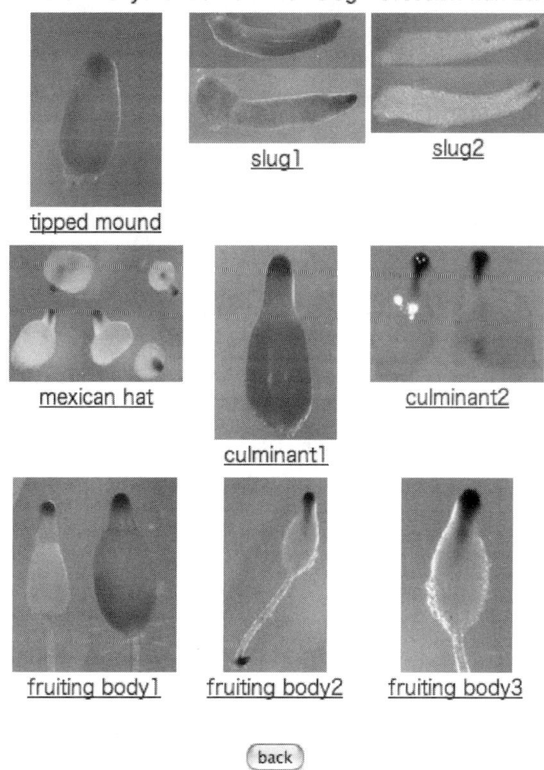

Fig. 2. Spatial expression pattern of SLA128 mRNA as shown in the ATLAS database. By setting expression pattern, cDNA clone name, gene name, gene product name and/or GenBank/EMBL/DDBJ accession number, expression pattern of gene of interest is shown.

the mutant slugs *(5)*. Because there was only one PstO marker (*ecmO* promoter element-reporter gene fusion) and it was DIF-1 dependent, it was the first instance. As the *dmtA⁻* mutant produces a very low level of DIF-1 *(22)*, it

Table 2
Major URLs for *Dictyostelium* cDNA Information

#	Title	URL
1	Dicty_cDB top	http://dictycdb.biol.tsukuba.ac.jp/cDNA/database.html
2	Clone Overview	http://dictycdb.biol.tsukuba.ac.jp/CSM/
3	Contig Overview	http://dictycdb.biol.tsukuba.ac.jp/CSM_Contig/
4	Catalog Index	http://dictycdb.biol.tsukuba.ac.jp/catalogue/Catalogue.html
5	Search_master	http://dictycdb.biol.tsukuba.ac.jp/~tools/html/ search_master.html
6	ftp site	http://dictycdb.biol.tsukuba.ac.jp/downloadFiles/

suggests that not only DIF-1 signaling, but also some other molecular mechanism may be involved in the differentiation of PstO cells.

4. *Dictyostelium* cDNA Database

The nucleotide sequences generated by the *Dictyostelium* cDNA Project in Japan have been gathered to construct the Dicty_cDB for their analysis, and made accessible via the internet for public use (URLs shown in **Table 2**). In this section, the outline and a brief summary of Dicty_cDB are described.

4.1. Clone-Based Sequence and Related Information

The major part of the database contains information on cDNA clones. As described previously, we experimentally determined the 5' end and/or 3' end sequences of the cDNA clones. These read pairs were merged into single sequences if possible. If not, the read pairs were artificially joined with an affixed gap of 10 'N's (shown by 10 "–"s in the web pages). Currently, there are, all together, 157,284 sequences derived from 95,884 clones from 14 cDNA libraries in Dicty_cDB (*see* **Table 3**).

The nucleotide sequences are used to collect biological information on the relevant clones by bioinformatics treatment or by internet searches. The abstracted information for each clone is shown in an "index" page of the clone, where the nucleotide (Seq) and deduced amino acid sequences (Translated), top 10s of homology search results against Dicty_cDB entries (CSM), nr-DNA database (DNA), nr-protein database (Protein), and *Dictyostelium* genome database (Genome) are shown. The detailed information for each can be obtained through the links "all frames" for amino acid sequence and "more info" for homology searches. The clone-index pages can be accessed via the

Table 3
Data Summary in Dicty_cDB

Stage	Directional library			Full-length library			Subtraction library			Total	
	Name	Clones	Reads	Name	Clones	Reads	Name	Clones	Reads	Clones	Reads
Vegetataive	VS	6733	10,850	VF, VH	21,272	38,718	*	*	*	28,005	49,568
Aggregating	*	*	*	AF, AH	19,768	33,022	*	*	*	19,768	33,022
Slug	SL,SS	14,423	19,759	SF, SH	14,099	22,806	*	*	*	28,522	42,565
Culminant	*	*	*	CF, CH	17,564	28,783	*	*	*	17,564	28,783
Gamete	FC, FCL	1122	1450	*	*	*	FC-IC	903	1896	2025	3346
Overall		22,278	32,059		72,703	123,329		903	1896	95,884	157,284

These reads were assembled by PHRAP program to yield 11,129 nonredundant sequences and further clustered into 8402 cDNA contigs by consideration of read pairs, which represent 6790 independent genes.

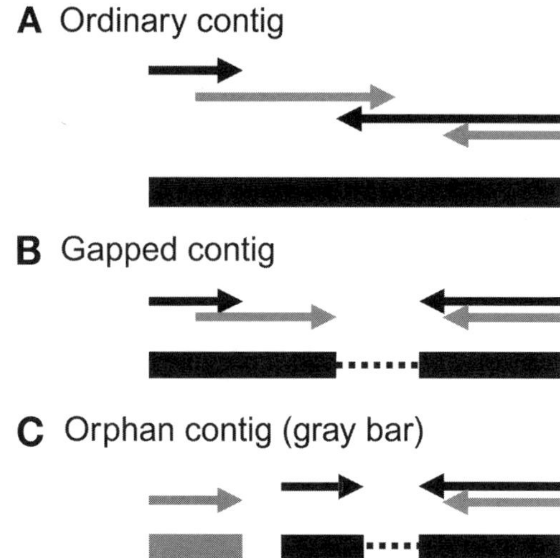

Fig. 3. Categories of cDNA contigs. The 5' and 3' reads shown by rightward and leftward arrows were assembled to generate cDNA contigs shown by thick bars below. The dotted line indicates a sequence gap. The gray bar shown in C represents an orphan contig. The reads shown by black and gray arrows each are derived from the same clone.

"Clone-overview" page (URL#3), or can be searched for using the Search-master page (URL#5). For the analysis of the sequence set, the FASTA files for nucleotide sequences and for deduced amino acid sequences are available from the ftp site (URL#6).

4.2. Clustering and Assembling ESTs to Generate Independent Gene Sequences

An extensive collection of cDNA sequences from nonnormalized libraries results in enormous redundancies of highly represented mRNAs. To obtain nonredundant and consensus sequences from EST data, we used the PHRAP assembly program *(23)*, used widely for the assembly of genome sequences. The sequences that possibly interfere with the assembly, such as the short sequences (<100 nt) and *dutA*-containing sequences, had been eliminated prior to the assembly. The cDNA copies of the dutA RNA, a noncoding poly-adenylated RNA sequence *(24)*, quite frequently appeared as portions of the inserts of otherwise unrelated clones. Short repeat sequences such as $(TAA)_n$ and $(TCC)_n$ were masked using the Repeat-masker program included in the PHRED-PHRAP package. The PHRAP conditions we used were: stringency 0.85, vector bound 0, and pre-assembly.

The assembly of sequences in the entire database resulted in 11,129 PHRAP contigs (including singletons). When the read pairs of clones were found in independent contigs, the relevant contigs were artificially joined in the same way as for the clone sequences, and are termed "Gapped-contigs" (*see* **Fig. 3B**). This insures against an overestimation of gene numbers. If more than two contigs were to be joined but their relative physical positions were unclear, the third and any additional contigs were left separate and are termed "Orphan-contigs" (unjoined contig in **Fig. 3C**). The singlets, which represent the sequences of unique cDNA clones, and the remainders of the PHRAP contigs are termed "Ordinary-contigs" (*see* **Fig. 3A**). The sum of the Ordinary-contigs and Gapped-contigs (6790) corresponds to the number of independent genes in Dicty_cDB. This number roughly corresponds to 55% of the predicted gene number in *D. discoideum* (**10**).

Where have the remaining 45% of the genes gone? First of all, there are genes expressed at very low frequencies and left unpicked by the random sequencing of the libraries. Second, we have not constructed cDNA libraries from all the stages of the *D. discoideum* life cycle. Nor have we analyzed non-standard or stressed conditions such as high or low osmotic pressures and temperatures. Third, mRNAs without poly-A tails, such as histone mRNAs, should have escaped from oligo-d(T) primed reverse transcription for library construction. Finally, there may be errors in PHRAP assembly. For example, we encountered the presence of unacceptably long cDNA contigs, which suggests the incidence of overassembly. In addition, very similar members of the gene families may have been clustered into a single gene. Therefore, close and careful examination of the cDNA contigs is still necessary. Sequence and related information of the cDNA contigs are publicized in the same way as for the clones described in **Subheading 4.1.**

4.3. Sequence Catalogs and Functional Classification

In order to find clones of interest or to comprehend the properties of the libraries or clone/gene sets, a sequence catalog with annotation is a convenient entrance. Basically, we refer to the descriptions of the highest score hits of BLASTX (with E value <e-4) for annotation, and classified the sequences into 15 categories by function (1: Metabolism, 2: Energy, 3: Transcription, 4: Translation, 5: Protein destination, 6: Cellular biogenesis and organization, 7: Transport facilitation, 8: Cell proliferation, 9: Movement, 10: Stress response and cell rescue, 11: Signal transduction, 12: Multicellular organization, 13: Retrotransposon and plasmid protein, 14: Classification uncertain, 15: Not classified). To assign the genes to appropriate functional categories, we first made a list of keywords extracted from descriptions in BLASTX results and assigned them to appropriate functional categories. We then searched the highest hit

BLASTX description of each sequence with keywords, and assigned it to the corresponding categories. Our classification correlates mainly to the cellular process of Gene Ontology (GO).

Recently, we modified the original process of catalog in the interest of greater reliability, because there were incidents where BLASTX replied that the query sequence was most homologous to, for example, a human gene with an E value of 1×10^{-30} or so, whereas BLASTN reported that it was identical to a well characterized *D. discoideum* gene. In general, BLASTX is better than BLASTN to find homologous sequences across species, but this is apparently not true in those cases. This sort of problem may occur when deletions or insertions occur during the sequencing process that result in artificial frameshifts. Therefore, we now compare the scores of BLASTN and BLASTX and use the one with the better score. The second problem is that about one-third of the sequences are classified into the category 14.1 (Classification uncertain). This category includes ambiguous descriptions such as "probable membrane protein" and "unknown sequence of" and does not contribute to the characterization of the clone. As a result of the great achievements of genome sequencing in various organisms, the chance of hitting to the predicted genes with those descriptions is increasing. To avoid the accumulation of functionally uncharacterized genes, we re-treat sequences in the category 14.1 to refer to the next highest hit of search results up to the 25th until they are classified into meaningful categories (categories 1 to 13).

5. Utilization of the cDNA Data for Genome Informatics

In this section, we describe how the collection of cDNA data can be used to analyze the genetic system in *D. discoideum*. Whereas the genome represents the complete genetic potential of the organism as a whole, transcriptomes represent the real activity of it in given physical and biological situations. The following three examples of analysis are based on the fact that the cDNA data reflect the actually transcribed genes.

5.1. Transcriptome Analysis

As can be seen in **Table 2**, the cDNA libraries were constructed without the intention of normalization except for the FC-IC library. Clone prevalence in a given library is, therefore, a good approximation of the relative levels of the relevant mRNAs. Strictly, there is a bias for a shorter cDNA to be cloned more efficiently. Especially in full-length libraries constructed with the oligo-capping method (*see* **Subheading 2.2.**), incomplete reverse transcripts, which were more frequent in longer mRNAs, had no chance of cloning. Even so, the stage-dependent redundancy of the genes will be useful information for their expression specificity. Thus, we newly identified the slug-stage or vegetative-stage specific genes by comparison of the SL, SS, and VS libraries *(9)*. By comparing the genes and clones that appeared in the full-length libraries, the drastic changes in the transcriptome properties were observed. Namely, genes in category 4 (translation) are actively transcribed at the growth phase but are quiescent during development, whereas those in categories 11 (signaling) and 12 (multicellular construction) are in reverse. In addition, these shifts are observed exclusively at the aggregation stage, which well agrees with the results of microarray analysis *(3)*, indicating that the formation of multicellular structures triggers extensive transcriptome alteration.

5.2. Transcript and Open Reading Frame Determination

The cDNA contigs containing the full-length cDNA clones as members are expected to represent the complete transcripts containing full-length of open reading frames (ORFs), further expanding up to transcription start sites (TSSs) and down to poly-adenylation sites. To analyze the gene function, the structure of the protein product is, no doubt, of major importance. We have about 4000 contigs meeting the above criterion. Even without full-length clones, we can predict the ORF using the genome sequences more accurately than without cDNA information, and many of the contigs without full-length clones possibly cover the entire ORF. This also means that we can obtain 5' and 3' untranslated regions believed to affect the efficiency of translation and processing and stability of mRNA *(25)*.

One unexpected observation was the variation in TSS among the full-length clones derived from the same gene. Although one specific TSS for the known gene is usually described in the literature and submitted to the DNA database, the TSS in *D. discoideum* varies either over a short range (e.g., <20 nt) or extensively (e.g., >500 nt). This is in accordance with the report of human TSS analysis *(26)*. As to the former variation, the nucleotide preference is primarily adenine in *D. discoideum*. TSS variation of this type seems to affect little the completeness of the ORF, because the start sites of all cDNA clones are upstream of the first methionine for 86% of the genes examined, and because the most frequent TSS are so for 96% of the genes. Here "ORFs" had been predicted as from the first ATG of the cDNA extending more than 150 nucleotides before the termination codons in frame.

5.3. Promoter Analysis In Silico

Once the complete sequences of mRNA are obtained, we can determine the exon–intron structures by aligning them to the genome sequences. In addition, upstream sequences of the TSS that correspond to the promoter regions important for regulation of gene expression can be obtained. We thus aligned our cDNA contigs to the genome sequences obtained from the ftp site of the dictyBase

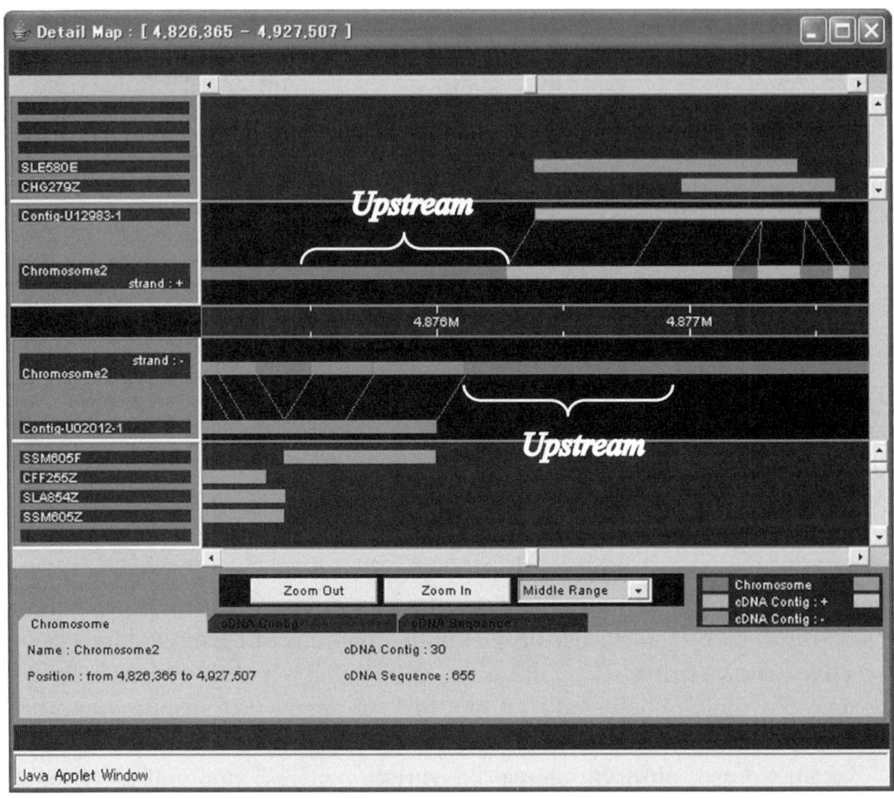

Fig. 4. Alignment of expressed sequence tag (EST) and cDNA contig sequences to the genome. Two cDNA contigs, Contig-U12931-1 and Contig-U02012-1 are mapped to + and—strands of chromosome 2, respectively. Their constituent ESTs are also shown. The 5' flanking region of each contig represents the upstream sequence.

(http://dictyBase.org/db/cgi-bin/dictyBase/download/blast_databases.pl) using the GeneSeqer program *(27)* (*see* **Fig. 4**). About 80% of the cDNA contigs were successfully aligned to six chromosomes, although some of the alignments may require revision. We obtained a genome sequence from –2000 to +100 nt of the TSS for each cDNA contig to construct the "*Dictyostelium* Upstream database." This database can be used to search for possible target genes of the transcription factors with known binding motifs. We used this database to identify possible *cis* elements commonly present in the gene groups with same stage-specificity determined by the cDNA prevalence in Dicty_cDB. We generated all possible sequences (5–10 nt) and aligned each with the upstream sequences of chromosome 2 with the DP-matching method, and statistically analyzed the results with respect to stage-specificity of the gene. We obtained two slug-stage-specific sequence elements, 5'-TTGSSCAA-3' and 5'-GRGTGTGAT-3', both of which are known development-specific *cis* elements, as well as vegetative-stage specific new elements, 5'-C/AATGS-3' and 5'-TCCTC-3'. The validity of these novel elements waits for experimental confirmation, but apparently this kind of approach is a combined fruit of cDNA and genome sequencing.

6. Prospects

In 1996, when the cDNA sequencing started, the lineup of clones carrying *D. discoideum* genes together with sequence information increasing daily in number were of enormous value themselves. The value of clones has not changed, but the implication of the cDNA sequences seems to have changed. Today, we can easily amplify the genes of interest with PCR using the genome sequences, and can clone them. Even the clone set for microarray analysis must be substituted by the PCR products based on the genome sequences and gene prediction. In this sense, the primary role of cDNA analysis seems to have finished.

Then what else, if anything, should be done? The next decade should see the complete understanding of the genetic system in *D. discoideum*. To step into this new era, one thing left undone must be achieved: complete identification of the gene repertoire in *D. discoideum*. We should once again examine the validity of the gene structures represented by the cDNA contigs. After that, we must examine the predicted genes without cDNA clones. Once this is accomplished, we can then step forward in the analysis of the genetic system in *D. discoideum* in combination with the approaches of bioinformatics and molecular biological techniques, in which a simple model organism may still be too complicated, but also a worthwhile challenge.

Acknowledgments

This work was supported by the Grant-in-Aid for Scientific Research on Priority Areas (A) "Genome Science" and (C) "Genome Biology" from the Ministry of Education, Science, Sports and Culture of Japan, Japan Society for the Promotion of Science (Research for the Future), and Grant-in-Aid for Publication of Scientific Research Result by the Japan Society for the Promotion of Sciences.

References

1. Suzuki, Y., Yoshimoto-Nakagawa, K., Maruyama, K., Suyama, A. and Sugano, S. (1997) Construction and characterization of a full length-enriched and a 5'-end-enriched cDNA library. *Gene* **200**, 149–156.

2. Araki, T., Tsujioka, M., Abe, T., et al. (2003) A STAT-regulated, stress-induced signalling pathway in *Dictyostelium*. *J. Cell Sci.* **116,** 2907–2915.
3. Van Driessche, N., Shaw, C., Katoh, M., et al. (2002) A transcriptional profile of multicellular development in *Dictyostelium discoideum*. *Development* **129,** 1543–1552.
4. Katoh, M., Shaw, C., Xu, Q., et al. (2004) An orderly retreat: Dedifferentiation is a regulated process. *Proc. Natl. Acad. Sci. USA* **101,** 7005–7010.
5. Maeda, M., Sakamoto, H., Iranfar, N., et al. (2003) Changing patterns of gene expression in prestalk cell subtypes of *Dictyostelium* recognized by *in situ* hybridization with genes from microarray analyses. *Eukaryot. Cell* **2,** 627–637.
6. Maruo, T., Sakamoto, H., Iranfar, N., et al. (2004) Control of cell type proportioning in *Dictyostelium discoideum* by Differentiation-Inducing Factor as determined by *in situ* hybridization. *Eukaryot. Cell* **3,** 1241–1248.
7. Farbrother, P., Wagner, C., Na, J., Tunggal, B., Morio, T., Urushihara, H., and Tanaka, Y. (2006) *Dictyostelium* transcriptional host cell response upon infection with *Legionella*. *Cell Microbiol.* **8,** 438–456.
8. Morio, T., Urushihara, H., Saito, T., et al. (1998) The *Dictyostelium* developmental cDNA project: generation and analysis of expressed sequence tags from the first-finger stage of development. *DNA Res.* **5,** 335–340.
9. Urushihara, H., Morio, T., Saito, T., et al. (2004) Analyses of cDNAs from growth and slug stages of *Dictyostelium discoideum*. *Nucleic Acids Res.* **32,** 1647–1653.
10. Eichinger, L., Pachebat, J. A., Glockner, G., et al. (2005) The genome of the social amoeba *Dictyostelium discoideum*. *Nature* **435,** 43–57.
11. Muramoto, T., Suzuki, K., Shimizu, H., et al. (2003) Construction of a gamete-enriched gene pool and RNAi-mediated functional analysis in *Dictyostelium discoideum*. *Mech. Dev.* **120,** 965–975.
12. Bonner, J. T. (1947) Evidence for the formation of the aggregates by chemotaxis in the development of the slime mold *Dictyostelium discoideum*. *J. Exp. Zool.* **106,** 1–26.
13. Maruyama, K. and Sugano, S. (1994) Oligo-capping: a simple method to replace the cap structure of eukaryotic mRNAs with oligoribonucleotides. *Gene* **138,** 171–174.
14. Diatchenko, L., Lau, Y.-F. C., Campbell, A. P., et al. (1996) Suppression subtractive hybridization: A method for generating differentially regulated or tissue-specific cDNA probes and libraries. *Proc. Natl. Acad. Sci. USA* **93,** 6025–6030.
15. Iranfar, N., Fuller, D., Sasik, R., Hwa, T., Laub, M., and Loomis, W. F. (2001) Expression patterns of cell-type-specific genes in *Dictyostelium*. *Mol. Biol. Cell* **12,** 2590–2600.
16. Maruo, T., Sakamoto, H., Iranfar, N., et al. (2004) Control of cell type proportioning in *Dictyostelium discoideum* by differentiation-inducing factor as determined by *in situ* hybridization. *Eukaryot. Cell* **3,** 1241–1248.
17. Zhukovskaya, N. V., Fukuzawa, M., Tsujioka, M., et al. (2004) Dd-STATb, a *Dictyostelium* STAT protein with a highly aberrant SH2 domain, functions as a regulator of gene expression during growth and early development. *Development* **131,** 447–458.

18. Escalante, R., Iranfar, N., Sastre, L., and Loomis, W. F. (2004) Identification of genes dependent on the MADS box transcription factor SrfA in *Dictyostelium discoideum* development. *Eukaryot. Cell* **3,** 564–566.
19. Van Driessche, N., Demsar, J., Booth, E. O., et al. (2005) Epistasis analysis with global transcriptional phenotypes. *Nat. Genet.* **37,** 471–477.
20. Maeda, M., Kuwayama, H., Yokoyama, M., et al. (2000) Developmental changes in the spatial expression of genes involved in myosin function in *Dictyostelium. Dev. Biol.* **223,** 114–119.
21. Tsujioka, M., Yokoyama, M., Nishio, K., et al. (2001) Spatial expression patterns of genes involved in cyclic AMP responses in *Dictyostelium discoideum* development. *Dev. Growth Differ.* **43,** 275–283.
22. Thompson, C. R. and Kay, R. R. (2000) The role of DIF-1 signaling in *Dictyostelium* development. *Mol. Cell* **6,** 1509–1514.
23. Gordon, D., Abajian, C., and Green, P. (1998) Consed: a graphical tool for sequence finishing. *Genome Res.* **8,** 195–202.
24. Yoshida, H., Kumimoto, II., and Okamoto, K. (1994) *dutA* RNA functions as an un-translatable RNA in the development of *Dictyostelium discoideum. Nucleic Acids Res.* **11,** 41–46.
25. Rivero, F. (2002) mRNA processing in *Dictyostelium*: sequence requirements for termination and splicing. *Protist* **153,** 169–176.
26. Suzuki, Y., Taira, H., Tsunoda, T., et al. (2001) Diverse transcriptional initiation revealed by fine, large-scale mapping of mRNA start sites. *EMBO Rep.* **2,** 388–393.
27. Usuka, J., Zhu, W., and Brendel, V. (2000) Optimal spliced alignment of homologous cDNA to a genomic DNA template. *Bioinformatics* **16,** 203–211.

4

dictyBase and the Dicty Stock Center

Petra Fey, Pascale Gaudet, Karen E. Pilcher,
Jakob Franke, and Rex L. Chisholm

Summary

dictyBase is the model organism database that houses all the sequence and associated data for *Dictyostelium discoideum*, including literature, researchers, and strains from the Dicty Stock Center. The database makes it possible to connect genes, proteins, and publications and is designed to address the needs of biologists and bioinformaticists alike. We provide tools for retrieving and analyzing data and strive to compile the highest quality, most up-to-date information. Here we will describe how to navigate the website and mine the extensive database to help users make optimal use of this invaluable resource.

Key Words: *Dictyostelium discoideum*; dictyBase; Dicty Stock Center; model organism database; genome sequence; Genome Browser; gene annotation; phenotype; Gene Ontology.

1. Introduction

In June of 2003, the prospect of a complete *D. discoideum* genome sequence gave rise to dictyBase, the comprehensive model organism database for researchers interested in the biology, genomics, and phylogenetics of *Dictyostelium (1,2)*. dictyBase (www.dictybase.org) provides a single access point to the most comprehensive, up-to-date, highly curated online information available for *Dictyostelium*. dictyBase has gone through several rounds of expansions and upgrades to accommodate the rapid increase in scientific data and the needs of users as the knowledge of the genome naturally drives research into new areas.

The *Dictyostelium* nuclear genome consists of approx 34 million base pairs organized in six chromosomes and believed to encode approx 12,500 proteins *(3)*. The gene products are the central focus of dictyBase: their function, their

From: *Methods in Molecular Biology, vol. 346:* Dictyostelium discoideum *Protocols*
Edited by: L. Eichinger and F. Rivero © Humana Press Inc., Totowa, NJ

interaction partners, the pathways to which they contribute, whether they are part of a wider gene family, and how they relate to proteins from other organisms. Automated processes are used as an initial attempt to assign gene function; however, the strength of dictyBase lies in manual verification of all data by curators who have extensive expertise in the field. Curators remove incorrect annotations, add information, and make associations that would be impossible to determine electronically.

Accuracy, usability, and service are the highest priorities. In view of this, there is detailed Help documentation on the dictyBase site, and the dictyBase staff promptly responds to questions and requests from users. dictyBase promotes the use of *Dictyostelium* as an experimental system as there is extensive information about the organism, a complete collection of the *Dictyostelium* literature, and a curated strain and plasmid Stock Center. In addition, we provide a medium for collaboration within the research community with an archived ListServ and a comprehensive colleague database with contact information. In this chapter, we intend to give the new and seasoned user insights and tips on how to find information in dictyBase.

2. Contents of dictyBase

2.1. Data and Annotations

dictyBase is the central repository for all information related to *Dictyostelium*. The data and annotations currently held in dictyBase are listed in **Table 1**. The data in dictyBase are continually updated with many datasets such as references, GenBank records, and changes to the genome sequence provided by the *Dictyostelium* Genome Sequencing Consortium.

2.2. The Curated Model

Each Automated Gene Prediction is reviewed by a curator who takes into account all available data including GenBank sequences, expressed sequence tags (ESTs), and similarity to known proteins. Reviewed gene models are called **Curated Models**, and represent the highest quality gene model in dictyBase. When there is evidence for alternative transcripts, two or more separate Curated Models are displayed. In rare cases in which the underlying sequence has a deletion or insertion that results in a frameshift, a Curated Model cannot be generated. In such cases, the responsible Sequencing Center is notified of the problem to determine if a resolution is possible, and a note is displayed explaining the discrepancy. When a gene model has been curated, a Curated Model appears as the track just below the Gene Track in the Genome Browser window (*see* **Subheading 4.**) and it is the first of the Associated Sequences on the Gene Page (*see* **Subheading 6.**).

Table 1
Contents of dictyBase as of June 2005

Data	Annotations

• 13,573 Automated Gene Predictions *(3)*
• 1380 GenBank records
• 155,032 expressed sequence tags *(4)*
• 5801 PubMed references
• External data:
  ○ 6801 Microarray expression profiles (Baylor College of Medicine [BCM] *[5]* and University of California, San Diego *[6]*)
  ○ *In situ* hybridization: 150 images (Tsukuba Atlas *[7,8]*)
• Insertional mutants: 817 links (BCM)
• 1131 Colleagues
• Dicty Stock Center:
  ○ 690 strains
  ○ 125 plasmids, 2334 Curated Models*a*
	• 8 Alternative transcripts
	• Gene products for 5147 genes
	• Brief Descriptions for 1538 genes
	• Mutant Phenotypes for 306 genes
	• Gene Ontology annotations for 5692 genes
	• Summary paragraphs for 494 genes

a See **Subheading 2.2.**

2.3. Other dictyBase Web Pages

In addition to the data and annotations, dictyBase contains a number of other resources such as a *Dictyostelium* tutorial Learn about *Dictyostelium*, pictures and videos, the dictyNews, the ListServ, techniques, nomenclature guidelines, and Help. These pages are accessible from the front page.

2.4. Downloading Data

Accessible from the side bar on the dictyBase homepage is the **Downloads** page, where the following types of information can be downloaded: *Dictyostelium* DNA and protein sequences

GFF3 files (full genome sequence, positions of Gene Predictions and Curated Models, annotations such as Gene Ontology [GO])
Gene annotations (gene names, synonyms, dictyBaseIDs, gene products)
Mutant phenotype information
Gene family information
GO annotations
Franke *Dictyostelium* Reference Library

These pages contain lists of information that users can view or download to perform their own analysis on their local computer. The date the file was last updated can be found next to each file. Because new data and annotations are continually added to the database, it is recommended to download these files frequently to have the most up-to-date version.

3. How to Find Data in dictyBase

The next sections of this chapter are aimed at helping users to successfully locate the data they seek in dictyBase. We will describe the text-based search tools, the Genome Browser (*see* **Subheading 4.**) and the BLAST Server (*see* **Subheading 5.**). Following the description of the text- and sequence-based search tools in dictyBase, we will discuss the Gene Page (*see* **Subheading 6.**). The Gene Page is a comprehensive source for gene information containing literature, phenotypes, sequences, GO annotations, links to expression data, links to insertional mutants, and researchers working on the gene. **Subheading 7.** will address searching the Stock Center.

> **Tip**: Each dictyBase page links to a **Help** page through the question mark on the right-hand side. The Help is context-sensitive and relates to the page currently being viewed. For example, clicking on the Help link on the Gene Page goes to the Gene Page Help; on the BLAST Server, the link goes to the BLAST Server Help. Help files and frequently asked questions (FAQs) are also accessible from any dictyBase page through the link on the top menu bar.

3.1. Searchable Items

dictyBase has a search box in the upper right-hand corner of every page. This search box searches the database portion of dictyBase. Gene Predictions, ESTs, and GenBank records can be found using either dictyBaseIDs or GenBank accession numbers. Gene names, synonyms, gene products, Gene Ontology terms, Colleagues, and Authors can also be searched using this search box.

> **Tip**: The difference between an **Author** and **Colleague** is that information about Authors is limited to the articles that they have published that mention *Dictyostelium*. Colleagues have filled out an online form to be part of our Colleague database. E-mail addresses and other information can be viewed by other users. Also, Colleagues can receive the dictyBase newsletter and they can be linked as Researchers to genes they are investigating. The Colleague Sign-Up/Update form is accessible through the Colleague link in the top menu bar on every page.

Table 2
Search Results for "cAMP"

Gene products	Associated genes
cAMP phosphodiesterase	*pdsA, regA*
cAMP receptor	*carA, carB, carC, carD*
cAMP receptor-like protein	*crlA, crlB, crlC*
...	

Gene Ontology terms	Associated genes
G protein signaling, coupled to **cAMP** nucleotide second messenger	*dagA, gpaB, gpbA, rasC, rasG*
cAMP biosynthesis	*acaA, acgA, acrA...*
cAMP receptor activity	*carA, carB, carC...*
...	

3.2. Search Box Output

When the search term is a unique gene name (for example, *abpA*) the search leads directly to the corresponding Gene Page. All other searches result in an intermediate page, the **Quick Search Result** page, which lists all searchable categories with links to the pages where results were found. For example, searching for "cAMP" results in 11 gene products, 40 GO terms, and 9 Authors (*see* **Table 2**). Following the link to gene products gives a list of gene products containing "camp" and their corresponding genes, such as cAMP receptor, cAMP phosphodiesterase, and cAMP-dependent protein kinase. Similarly, the link to Gene Ontology terms returns a list of GO terms and the associated genes, and clicking on the 9 Authors link leads to a list of authors of papers in dictyBase whose last name start with "Camp."

> **Tip**: The search is case-insensitive. Searching for cAMP, camp, or CAMP returns the same results.

> **Tip**: The number of items found by the search engine does not necessarily correspond with the number of genes associated with this item. For instance, searching for "serine/threonine kinase" returns two gene products. For these two gene products, *protein* serine/threonine kinase and *putative protein* serine/threonine kinase, there are 26 and 145 associated genes, respectively.

Table 3
Searching for "PKA" (Protein Kinase A) With and Without a Wildcard Character[a]

Search results for **PKA**	Search results for **PKA***	Search Results for ***PKA***
2 Gene names/synonyms:	3 Gene names/synonyms:	5 Gene names/synonyms:
• pkaC/**PKA**	• pka<u>C</u>	• pka<u>C</u>
• pkaR/**PKA**	• pka<u>D</u>	• pka<u>D</u>
	• pka<u>R</u>	• pka<u>R</u>
		• <u>p</u>pkA
		• <u>s</u>pkA

[a]The string found by the Search tool is shown in bold; the position of the wildcard in the results is underlined.

3.3. The Wildcard (*)

The **wildcard** (*) can be used to widen the possible results of a search. The wildcard can be added at any position of the query, as shown in **Table 3**.

3.4. Searching for Literature

There are two ways to find literature references. First, articles can be retrieved through the Search Box by searching for an Author using the author's last name, or the last name and first initial, separated by a comma. Second, the publications pertaining to specific genes are listed in the References section of each Gene Page (*see* **Subheading 6.6.**).

3.5. Expanded Search

For an expanded search, there is a Search dictyBase link on the left-hand side menu of the home page. From this page (*see* **Fig. 1**), users can search the dictyNews archive and all dictyBase web pages (*see* **Subheading 2.3.**) in addition to all of the previously mentioned database items. There are currently over 400 issues of the newsletter in the archive and more than 1000 separate pages on the dictyBase website.

4. Navigating the Genome and Finding Data With the Genome Browser

All sequence information in dictyBase is displayed graphically by the **Genome Browser**. The Genome Browser is an immensely versatile and customizable tool developed by the Generic Model Organism Database Project (www.gmod.org) *(9)*. The genomic position of genes and gene models, both automatically predicted and curated, is based on their chromosomal coordi-

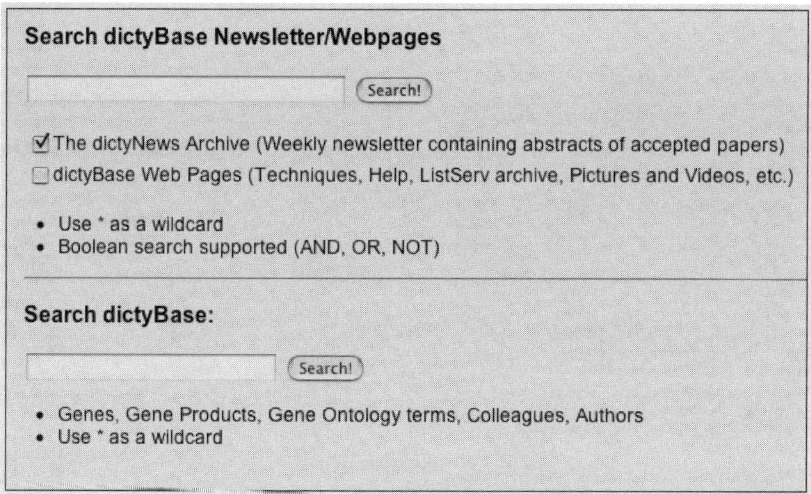

Fig. 1. Search options on the Search dictyBase page. The Search dictyBase Newsletter/Webpages search box (top) accepts any key word and searches the dictyNews Archive or all dictyBase web pages. The Search dictyBase search box (bottom) searches the database and performs the same function as the search box in the upper right-hand corner of every page in dictyBase. http://dictyBase.org/cgi-bin/search_news.cgi

nates on the genome sequence. Other sequences such as GenBank records and EST sequences are shown as BLAST alignments to the genome sequence.

4.1. The Genome Browser Display

4.1.1. Organization of Sequences in the Genome Browser

The Genome Browser displays a selected chromosomal region up to 100 kb. **Figure 2** shows chromosome 2, positions 6,393,671 to 6,413,670. The gray panel at the top of the image is the **Chromosome Overview**. The overview shows the whole chromosome, with the current location enclosed within red lines. Clicking on a different region of the gray bar changes the position of the Genome Browser to that region of the chromosome.

Each chromosomal feature is represented by a **Track** (*see* **Fig. 2A**). The uppermost track represents the **Gene**. The Gene Track shows the beginning to end as defined by the best available gene model. The complete list of tracks is shown in **Table 4**. All tracks are clickable and lead either to the **Gene Page** (*see* **Subheading 6.**) for the Gene Track or to the corresponding **Sequence Info Page** (*see* **Subheading 6.3.**) for all other tracks.

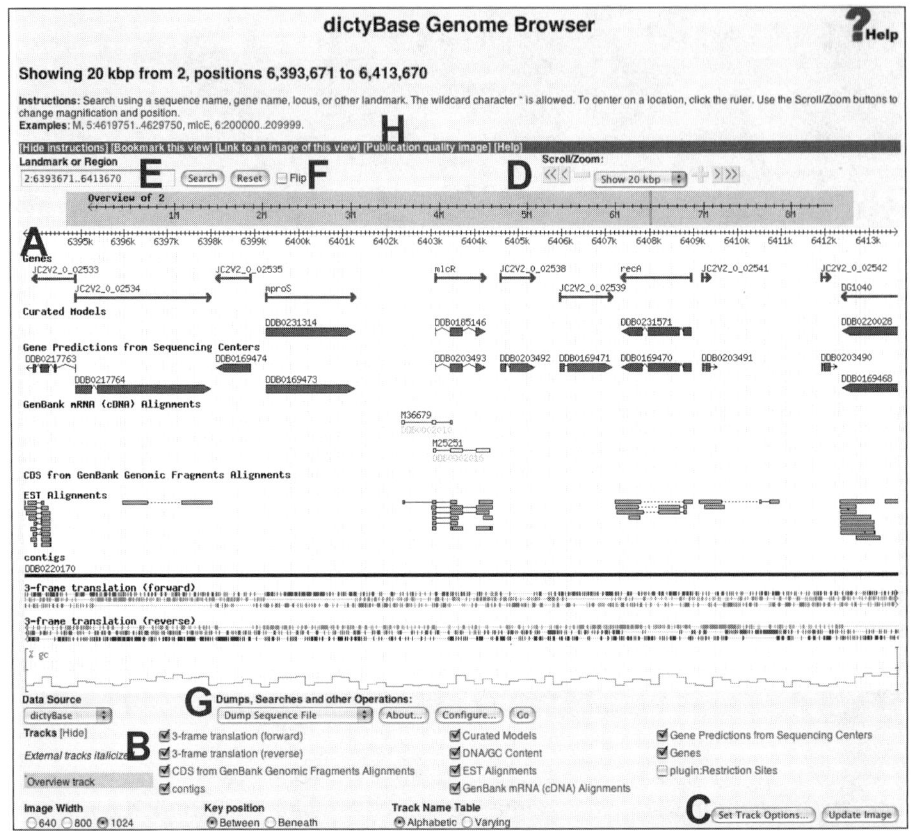

Fig. 2. View of 20 kb of chromosome 2 in the Genome Browser window, position 6,393,671 to 6,413,670. (**A**) Tracks, starting with the Gene Track; (**B**) check boxes to choose tracks; (**C**) Set Track Options button; (**D**) Scroll/Zoom tool; (**E**) Landmark or Region search box; (**F**) Flip image/sequence; (**G**) Dumps, Searches and Other Operations; (**H**) Publication quality image. Except for the restriction sites plug-in, all tracks are selected and shown here. http://dictyBase.org/db/cgi-bin/ggb/gbrowse/dictyBase? name=2:6393671..6413670

4.1.2. Configuring the Display of the Genome Browser

Below the Genome Browser image are check boxes to select the tracks displayed in the Genome Browser window (*see* **Fig. 2B**). The chosen tracks will appear once the **Update Image** button is clicked. To further customize the tracks, a page accessible by the **Set Track Options** button (*see* **Fig. 2C**) allows you to change the order of the tracks or the number of sequences to be displayed. All of these settings are saved by your browser. Note that the number

Table 4
Genome Browser Tracks

Track	Description
Genes	Region of best available gene model
Curated Models	Manually curated gene model
Gene Predictions from Sequencing Centers	Automated gene model *(3)*
GenBank mRNA (cDNA) Alignments	mRNA sequences from GenBank
CDS from GenBank Genomic Fragments Alignments	Genomic sequences from GenBank
EST Alignments	ESTs from Tsukuba expressed sequence tag project *(4)*
Contigs	Contigs from genome assembly
3-frame translation (forward)	Translation of sequence when zoom set at 100 bp; displays stop codons for Watson strand in a region >100 bp
3-frame translation (reverse)	Translation of sequence when zoom set at 100 bp; displays stop codons for Crick strand in a region >100 bp
DNA/GC Content	Graph of %GC in genomic sequence
plugin:Restriction sites	Selected restriction sites in genomic sequence

of ESTs to be displayed and the size of the region determine the speed of the browser window to load.

4.2. Navigating the Genome Browser

4.2.1. Scroll/Zoom

The **Scroll/Zoom** tool (*see* **Fig. 2D**) allows you to move left or right using the arrows or double arrows; the view shifts by either 50% or 100% of the currently displayed genome segment, respectively. The plus (+) and minus (–) buttons zoom in or out by 10%. The drop-down option displays preselected sizes between 100 bp and 100 kbp, remaining centered on the currently displayed sequence. In the web browser, the Scroll/Zoom tool is easily recognizable by its yellow buttons.

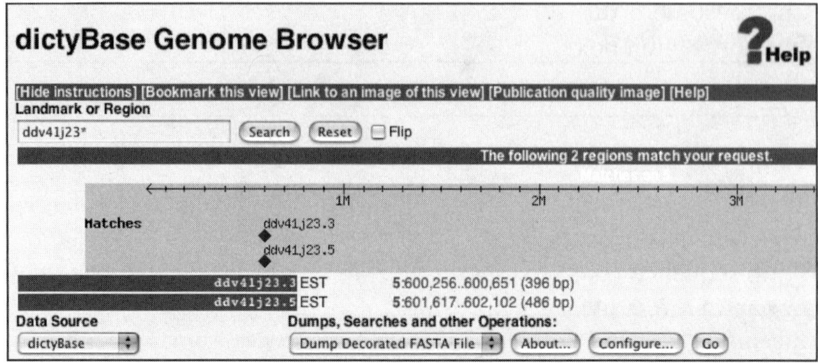

Fig. 3. Result for searching ddv41j23* using the Genome Browser's Landmark or Region box. The window shows that two clones were found, ddv41j23.3 and ddv41j23.5, in adjacent locations on chromosome 5. Clicking on the diamond next to the clone name or on the coordinates leads to exactly that region on the chromosome in the Genome Browser window. In a similar fashion, this tool can be used to search genes with similar name symbols such as gpa* or cpn*. http://dictyBase.org/db/cgi-bin/ggb/gbrowse/dictyBase?name=ddv41j23*.

4.2.2. Jumping to a New Location

The Genome Browser **Landmark or Region** box (*see* **Fig. 2E**) can be used to search for:

- *Chromosomal coordinates of up to 100 kbp.* Example: 5:10000..25000 goes to a 15 kbp region on chromosome 5 between the coordinates 10,000 and 25,000.
- *Gene names or EST clone names.* Examples: mlcE, dagA, cbp*, ddv41j23.3.
- *Identifiers such as dictyBaseIDs, GenBank accession numbers, and gi numbers.* Examples: DDB0185150, AY232265, 639923.

> **Tip**: Try using the **wildcard** (*) when searching **EST names** because many EST clones have more than one sequence. For example, searching for "ddv41j23" will produce no results. Searching for "ddv41j23*" results in an intermediate page that displays the chromosomal locations of ddv41j23.3 and ddv41j23.5 (*see* **Fig. 3**). Clicking on the small diamonds loads the full length of that sequence in the Genome Browser.

> **Tip**: You can search any **Landmark or Region**, regardless of your current location. For example, if the Genome Browser is centered on the *mlcE* gene on Chromosome 3, you can go to ddv41j23.3 on

Chromosome 5 directly just by entering ddv41j23.3 in the Land-mark or Region box and clicking on Search.

4.2.3. Flip Image/Sequence

The Flip function (*see* **Fig. 2F**) reverses the orientation of the image. To activate this feature, the Flip box must be checked and the genome searched again by clicking on **Search**. This function is persistent when dumping the underlying sequence (*see* **Subheading 4.3.**) and therefore returns the reverse complement of the sequence. This is especially useful for genes that are on the Crick strand.

4.3. Dumps, Searches and Other Operations

Genome Browser operations (*see* **Fig. 2G**) allow you to retrieve additional information from the genomic sequence in the current view. It is recommended that each operation first be configured. The **Configure...** and **Go** buttons will execute the respective commands on the operation selected in the pull-down menu. Currently, it is possible to choose between the following three operations.

4.3.1. Dump Sequence File

This tool simply retrieves the sequence in the region displayed in the Genome Browser. Different sequence formats, such as FASTA, GenBank, or Raw Sequence, can be chosen by clicking the **Configure...** button. The sequence can also be downloaded to your computer.

4.3.2. Dump Decorated FASTA File

This is a very useful tool for visualizing intron/exon boundaries, EST alignments, and more. There are many decoration options: upper/lowercase letters, font style, and colors as well as background colors, all of which first must be configured. **Figure 4** shows an example of how the Curated Model differs from the Sequencing Center Gene Prediction with support by the ESTs. Once this configuration is set, your browser will remember the settings.

4.3.3. Annotate Restriction Sites

The position of restriction sites of interest can be displayed in the Genome Browser image. Clicking on **Configure...** when **Annotate Restriction Sites** is chosen in the drop-down menu leads to a list of restriction sites from which any number can be selected. The configuration will be stored in your browser. To view the restriction sites, the **plugin:Restriction Sites** track (*see* **Fig. 2B**) must be selected.

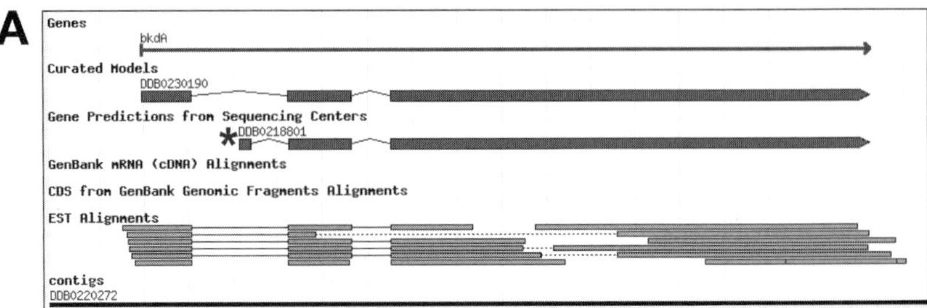

B

4:4250569,4252850

```
>4:4250569,4252850
gtaaaaaaaagtgtattttttttatttatttatttatttatttttatttatttatttatt
tatttatttatttatttatttattcgcttacttttttttttttaaaaaaaaaaatatatta
tttgtctgacatgtaaaatatcgtctttttgtttagaaaaaaaaaattataaaattaaaa
ttttttttttttttttttttttttttttttttttgaaatttcatattttagaatcacgcgag
tatattgtttttttttttttttttttcgttgttttttatcttttttttttttgaaacaat
aacataattttgtattttttattacgatgataagccaatcttacagaatttatcaagaa
tcagtagaataatgaattaaaaaaaacctttttaaccaatttaaactgtaaatcatcat
caccatcaattatcagagtatgttaataattttctctcgttaaaataaaacaaataaaga
aaataaataaataaataataattgataaatacaataaaagaccacaataaaaataaata
aataaATGATAATTTTAAAAAAAAATAACAAgtatcacaaaattctaaataaagaaatt
aaactatatctatctatatatatattattatttatatttgattgaaatcatagagTTTCT
GTAAAAAACAAAATTTAGATGAAAATTTTGAATATACAAACAAATTAGAAGTTCAAGAAC
TTAAACACTATATACCATGTTATACTATTATGGATCAAGAAGGTGTCGTTTCAAAACCTG
ACCAAGACCCAAATgtaagttttgcattattattataattattttttttatttttttatt
tttattattattattaatatttattttgtttatttatatagTTTAGTAAAGAAGAAGTT
ATTAAATGTATACAACAATGTTAACATTAAATGTTATGGATTCAATTCTTTATGATGTA
CAAAGACAAGGTAGAATTTCATTTTATATGACATCATTTGGTGAAGAAGCAATTCATATT
GGATCAGCAGCAGCATTAGAGATGAGTGATACAATTTTTGCACAATATAGAGAGACTGGT
GTATTCATGTGGAGAGGATTCACAATCAATGATATTATCAATCAATGTTGTACCAATGAA
CATGATTTAGGTAAAGGTAGACAAATGCCAATGCATTTCGGTTCAAGAAAGATTAATCTT
CAAACTATTTCATCACCATTAACAACTCAATTACCACAAGCAGTGGGTAGCTCATACGCT
CAAAAATTAGCAGGTGAAAAGAATTGTACCATCGTTTACTTTGGTGAGGGTGCTGCAAGT
GAGGGTGATTTCCATGCTGCTATGAATTTCGCCGCAGCATTGTCAACACCAACTATCTTT
TTCTGTCGTAACAATAAATGGGCAATCTCTACACCCTCAAAAGAGCAATACAAAGGTGAT
GGTATTGCTGGTCGTGGTCCAAATGGCTACGGTATGAAAACTATCAGAGTCGATGGTAAC
GATATTTGGGCCGTATACAATGTGACTAAATTAGCACGTAAAATCGCCGTCGAAGAACAA
GTTCCAGTTCTCATCGAAGCTATGACCTATCGTGTTGGTCATCATTCTACCTCTGATGAC
TCTTCTCGTTATCGTACTGTTGAAGAGATCAACGCTTGGAAAGAAGGTAAAAATCCAATT
AGTCGTTTAAGAAACTACATGAATCACAAAGGTTGGTGGAGTGATGCTCAAGAGAAGGAA
ACTATCGCAAACGCTCGTACTACCGTTCGTGAATCCTTAGTCAACGCTGAAAAACAATAT
AAACCATCAATCAATGAAATCTTTACCGATGTTTATGATAAACCAACTCCAAATCTCATC
GAACAACAAAAAGAATTAATTGAACATTTGAAATTATATCCAGATGAATATCCATTAAAT
CAATTTGCTGATTCAAAATTAATTTTAAAAGATTAAtttaataataataaaataaaaaa
taaaataaaaaataaaaattaaaattaaaatctaaaataaaataaaaaaaaaacatt
gttttattttctattttgggtaaaaaaaaaaaaaaaaaaaaaaaacctatgtaaaagtatca
```

Fig. 4. The Genome Browser: Dump Decorated FASTA File. (**A**) Genome Browser window centered on chromosome 4, position 4,250,569 to 4,252,850. (**B**) The chromosomal region shown in (**A**) as a decorated FASTA file; gray background: Curated Model; uppercase: Gene Prediction from Sequencing Center; underlined: expressed sequence tags. * Start of the gene prediction exon 1. http://dictyBase.org/db/cgi-bin/ggb/gbrowse/dictyBase?name=4:4250569..4252850.

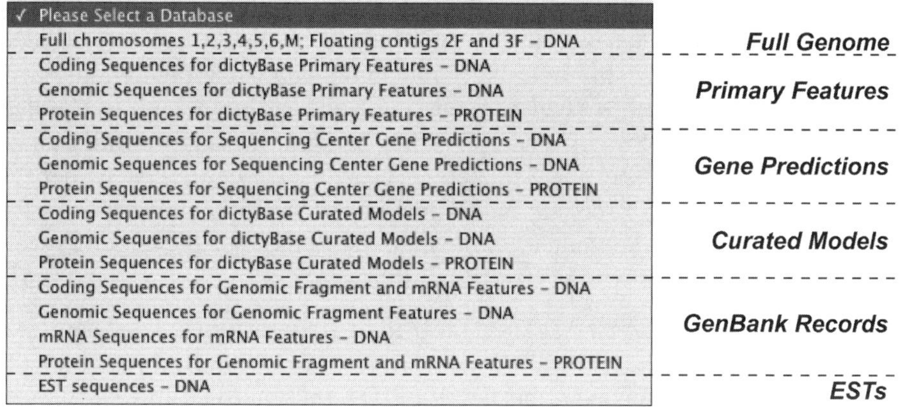

Fig. 5. The dictyBase BLAST databases. To blast against the best nonredundant dataset, use the Primary Features datasets. Accessible from http://dictyBase.org/db/cgi-bin/blast.pl.

4.4. Publication-Quality Image

The Genome Browser image can be saved as a high-quality scalable vector graphics (SVG) file (*see* **Fig. 2H**). The SVG image is resizable without any loss of resolution, and can be opened and edited in any vector graphics application such as Adobe Illustrator or Macromedia Freehand. If desired, from there it can easily be saved as a raster-based image (ESP, TIFF, JPEG).

5. The dictyBase BLAST Server

The dictyBase BLAST Server offers an additional gateway to the sequence data and Gene Pages (*see* **Subheading 6.**) and can be accessed from the top menu bar on every page. BLAST is also accessible from the **Sequence Info Page** (*see* **Subheading 6.3.**), where the selected sequence is pasted directly into the query window. The BLAST Server offers the choice of different BLAST programs, several different datasets, and configurable parameters. Results of BLAST searches show alignments and provide links to Gene Pages that contain additional information about the identified genes or gene products.

5.1. The BLAST Databases and Sequence Download

The BLAST Server offers several different *Dictyostelium* sequence datasets (*see* **Fig. 5**). The first is the full sequence of all six chromosomes plus the mitochondrion and floating contigs, the contigs that have not yet been assembled into a chromosome. Next, there is the **Primary Feature** set, which corresponds to the best available sequence for a given gene and includes the

Coding Sequence, Genomic Sequence (coding sequence +/– 1000 bp), and Protein Sequence for all genes in the genome. If a gene model has not yet been curated, the Primary Feature is usually the automated Gene Prediction; however, if a gene model has been curated (*see* **Subheading 2.2.**), the Curated Model becomes the Primary Feature. Other datasets include the fully automated prediction set from the Sequencing Centers, *Dictyostelium* sequences from GenBank, and EST sequences. All datasets can be downloaded as FASTA files, accessible from the same drop-down menu as the BLAST search as well as from the **Downloads** page (*see* **Subheading 2.4.**).

> **Tip**: If you "BLAST" against the **Full Chromosomes** database, make note of the chromosomal coordinates in addition to the dictyBaseID. In this case, the dictyBaseID represents the entire chromosome, and hence is not suitable for later referrals to a distinct region of the genome.

5.2. Optimizing BLAST Options

Better results can be obtained by optimizing the BLAST settings. Because the database is restricted to *Dictyostelium* sequences, the dictyBase BLAST Server returns results relatively fast. The dictyBase BLAST Server has options to change the **E Value**, the **Number of alignments to show**, turn **Gap alignments** and **Filtering** on or off, and five different **Matrices**. Detailed information about BLAST in dictyBase can be found in the Help documentation and in **ref. 10**.

> **Tip**: As *Dictyostelium* proteins often contain low complexity regions, it is sometimes useful to turn off the **Filtering** to get the most complete alignment. Note that this makes the search slower, so it is advisable to lower the E value and decrease the number of sequences output at the same time.

6. The dictyBase Gene Page

The dictyBase Gene Page serves as the central resource for all available *Dictyostelium* gene information. As shown in **Fig. 6**, the Gene Page contains an overview of the information regarding each gene. Numerous links provide

---

Fig. 6. *(opposite page)* A dictyBase Gene Page. **A–L** show some important links to in-depth information. (**A**) Genome Browser; (**B**) Back to Home Page (www.dictybase.org); (**C**) Help page; (**D**) Sequence Info pages; (**E**) dictyBase BLAST Server; (**F**) BLASTP at NCBI; (**G**) detailed Gene Ontology info; (**H**) links to external expression data; (**I**) detailed phenotype data; (**J**) external databases; (**K**) complete list of references; (**L**) detailed paper information. http://dictyBase.org/db/cgi-bin//gene_page.pl?gene_name=mhka.

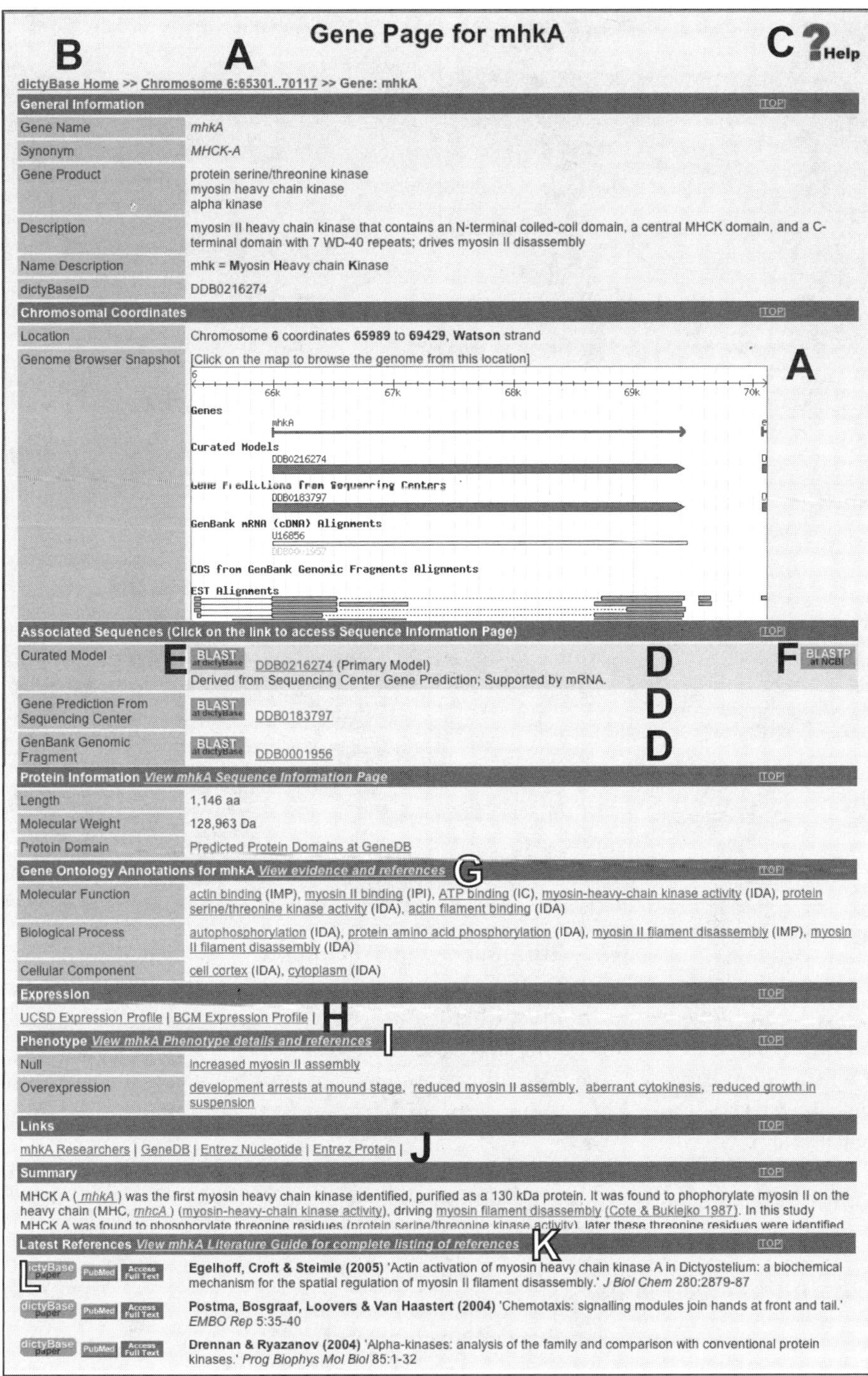

B   A   Gene Page for mhkA   C ? Help
dictyBase Home >> Chromosome 6:65301..70117 >> Gene: mhkA
General Information   [TOP]
Gene Name   mhkA
Synonym   MHCK-A
Gene Product   protein serine/threonine kinase
myosin heavy chain kinase
alpha kinase
Description   myosin II heavy chain kinase that contains an N-terminal coiled-coil domain, a central MHCK domain, and a C-terminal domain with 7 WD-40 repeats; drives myosin II disassembly
Name Description   mhk = Myosin Heavy chain Kinase
dictyBaseID   DDB0216274
Chromosomal Coordinates   [TOP]
Location   Chromosome 6 coordinates 65989 to 69429, Watson strand
Genome Browser Snapshot   [Click on the map to browse the genome from this location]
A
66k   67k   68k   69k   70k
Genes
mhkA
Curated Models
DDB0216274
Gene Predictions from Sequencing Centers
DDB0183797
GenBank mRNA (cDNA) Alignments
U16856
DDB0001956
CDS from GenBank Genomic Fragments Alignments
EST Alignments
Associated Sequences (Click on the link to access Sequence Information Page)   [TOP]
Curated Model   E BLAST at dictyBase   DDB0216274 (Primary Model)   D   F BLASTP at NCBI
Derived from Sequencing Center Gene Prediction; Supported by mRNA.
Gene Prediction From Sequencing Center   BLAST at dictyBase   DDB0183797   D
GenBank Genomic Fragment   BLAST at dictyBase   DDB0001956   D
Protein Information View mhkA Sequence Information Page   [TOP]
Length   1,146 aa
Molecular Weight   128,963 Da
Protein Domain   Predicted Protein Domains at GeneDB
Gene Ontology Annotations for mhkA View evidence and references G   [TOP]
Molecular Function   actin binding (IMP), myosin II binding (IPI), ATP binding (IC), myosin-heavy-chain kinase activity (IDA), protein serine/threonine kinase activity (IDA), actin filament binding (IDA)
Biological Process   autophosphorylation (IDA), protein amino acid phosphorylation (IDA), myosin II filament disassembly (IMP), myosin II filament disassembly (IDA)
Cellular Component   cell cortex (IDA), cytoplasm (IDA)
Expression   [TOP]
UCSD Expression Profile | BCM Expression Profile | H
Phenotype View mhkA Phenotype details and references I   [TOP]
Null   increased myosin II assembly
Overexpression   development arrests at mound stage, reduced myosin II assembly, aberrant cytokinesis, reduced growth in suspension
Links   [TOP]
mhkA Researchers | GeneDB | Entrez Nucleotide | Entrez Protein | J
Summary   [TOP]
MHCK A (mhkA) was the first myosin heavy chain kinase identified, purified as a 130 kDa protein. It was found to phophorylate myosin II on the heavy chain (MHC, mhcA) (myosin-heavy-chain kinase activity), driving myosin filament disassembly (Cote & Bukiejko 1987). In this study MHCK A was found to phosphorylate threonine residues (protein serine/threonine kinase activity), later these threonine residues were identified
Latest References View mhkA Literature Guide for complete listing of references K   [TOP]
L dictyBase paper PubMed Access Full Text   Egelhoff, Croft & Steimle (2005) 'Actin activation of myosin heavy chain kinase A in Dictyostelium: a biochemical mechanism for the spatial regulation of myosin II filament disassembly.' J Biol Chem 280:2879-87
dictyBase paper PubMed Access Full Text   Postma, Bosgraaf, Loovers & Van Haastert (2004) 'Chemotaxis: signalling modules join hands at front and tail.' EMBO Rep 5:35-40
dictyBase paper PubMed Access Full Text   Drennan & Ryazanov (2004) 'Alpha-kinases: analysis of the family and comparison with conventional protein kinases.' Prog Biophys Mol Biol 85:1-32

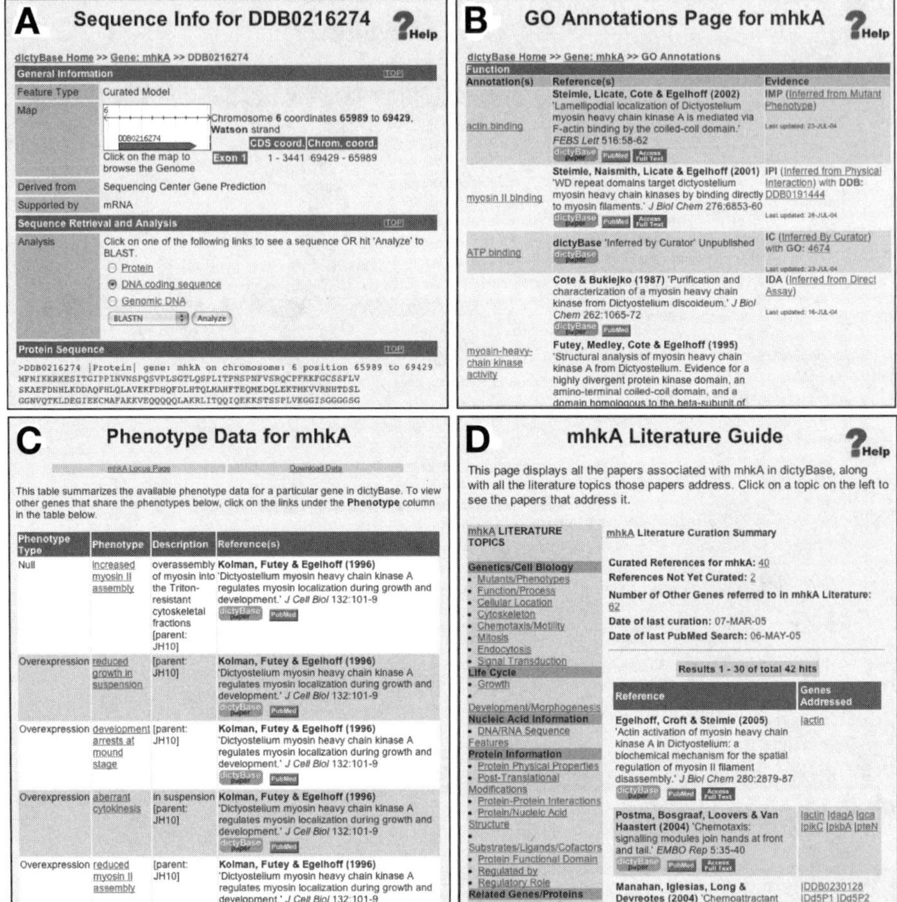

Fig. 7. dictyBase pages with in-depth information. (**A**) Sequence Info page; (**B**) complete Gene Ontology Info; (**C**) complete phenotype info; (**D**) complete literature info. (**A**) http://dictyBase.org/db/cgi-bin/feature_page.pl?dictybaseid=DDB0216274 (**B**) http://dictyBase.org/db/cgi-bin/dictyBase/GO/goAnnotation.pl?gene_name=mhkA (**C**) http://dictyBase.org/db/cgi-bin/dictyBase/phenotype/phenotype.pl?feat=mhkA& type=locus (**D**) http://dictybase.org/db/cgi-bin/dictyBase/reference/geneinfo.pl? locus=mhkA.

additional pages with more in-depth records as shown in **Fig. 7** and described in more detail as follows.

6.1. The Navigation Bar

At the top of every Gene Page, a **Navigation Bar** links to the chromosomal location of the gene in the Genome Browser (*see* **Fig. 6A**) and also back to the

Home Page (*see* **Fig. 6B**). There is also a link to the Gene Page **Help Page** (*see* **Fig. 6C**) on the right side.

6.2. Names and Identifiers

The **General Information** section contains the gene name, all synonyms, gene product(s), a short description, and the primary dictyBase identifier (ID). The dictyBaseID is a unique identifier consisting of the letters DDB followed by seven digits and corresponding to the ID of the Primary Feature of that gene (*see* **Subheading 5.1.**). This dictyBaseID should be used when referring to a gene in dictyBase.

> **Tip**: The synonyms field is very important, as it makes searches possible using any published name for a gene that the curators have identified in the literature. Use of a synonym in a search also takes the user to the correct Gene Page.

> **Tip**: To be sure that you will find the same page on a future visit to dictyBase, make a note of the dictyBaseID. In the rare case that a dictyBaseID becomes obsolete, you will be redirected to the new ID.

6.3. Chromosomal Coordinates and Associated Sequences

The next sections on the Gene Page show Chromosomal Coordinates and a graphical image of the chromosomal location, as well as the sequences that are associated with the gene. The Genome Browser Snapshot gives an instant overview of the region immediately surrounding the gene. Clicking on the image (*see* **Fig. 6A**) links to the **Genome Browser** (*see* **Subheading 4.**). Below the Genome Browser image, Associated Sequences are listed separately by category. The dictyBaseID (*see* **Fig. 6D**) of each sequence links to their respective Sequence Info Page (*see* **Fig. 7A**). This section also provides access to the dictyBase **BLAST Server** (*see* **Fig. 6E**) and BLASTP at NCBI (*see* **Fig. 6F**).

Every sequence in dictyBase has its own **Sequence Info Page** (*see* **Fig. 7A**) that displays all available information. For Curated Models and Gene Predictions, this information includes a clickable mini-map as well as the gene coordinates including intron/exon boundaries. This is followed by the protein sequence, the coding sequence, and genomic DNA sequence (including 1000 bp on either side). All Sequence Info Pages provide a gateway to the dictyBase BLAST Server, from which the selected sequence is automatically pasted into the BLAST search box.

6.4. Gene Ontology

The GO project (www.geneontology.org) *(11)* is an effort to produce a system for annotating gene products that can be applied across all organisms. GO is divided into three categories describing molecular functions, biological processes, and cellular compartments. GO annotations are displayed on the Gene Page. For most users, the GO provides a way to quickly get an overview of the cellular role of a gene. GO can also be used for analysis of high-throughput experiments such as microarrays.

The GO annotations are listed on the Gene Page and each term links to a page with a definition and a list of other *Dictyostelium* genes that are annotated with this term. The listed GO terms also include **evidence codes** indicating the type of supporting information for a given annotation.

A link at the top of the GO section on the Gene Page (*see* **Fig. 6G**) leads to a separate **GO Annotations Page** (*see* **Fig. 7B**) where all annotations for that gene are listed with their evidence and reference. The reference given for a GO annotation is the source of the annotation; this reference is often a published paper but can also be an unpublished method, for example, tools to assess sequence similarity.

> **Tip**: Annotations with the evidence code IEA (inferred from electronic annotation) are purely automated and, although they are often correct, may contain inaccurate information. Evidence codes other than IEA have been reviewed by a curator and therefore are indicative of higher-quality annotations.

6.5. Expression and Phenotype Data

The next two sections of the Gene Page provide information about expression profiles and phenotypes. The **Expression** field (*see* **Fig. 6H**) contains links to expression data provided by different groups. Currently, there are two datasets that were obtained by microarray analysis from the Baylor College of Medicine and the University of California at San Diego, respectively. Each is available as a graphic representation of the developmental gene expression *(5,6)*. The third dataset contains images of *in situ* hybridization patterns generated at Tsukuba University *(7,8)*.

The **Phenotype** section (*see* **Fig. 6I**) is manually curated from the literature. Similarly to the GO (*see* **Subheading 6.4.**), each phenotype links to another page where all *Dictyostelium* genes sharing this annotation are listed with references, and this list can be downloaded as an Excel® file. The phenotype section also links out to a separate **Phenotype Data** page (*see* **Fig. 7C**), which lists all phenotypes associated with a gene, and includes the type of mutant, a short description, parental strain information, and the reference used to make the annotation. In addition, a list of all mutants annotated in the database is available from the Research Tools drop-down in the top menu bar on every page as well as from the **Downloads** page (*see* **Subheading 2.4.**).

6.6. References and Summary

An important part of the data in dictyBase is the published literature, which is imported weekly from PubMed. The literature in dictyBase is restricted to *Dictyostelium discoideum*, which is extremely useful considering the exponential growth of publications. Another advantage to using dictyBase to access published literature is the power of both a relational database and manual curation: all genes addressed in a paper can be associated with that paper. In comparison, a PubMed search will only return gene names that are in the title, abstract, and key words of the paper.

Only the five most recent articles are listed on the Gene Page (*see* **Fig. 6K**, reference list truncated). The full list of references for a gene can be accessed from the Gene Page by clicking on **View Literature Guide** (*see* **Fig. 7D**). In the Literature Guide, the references are listed with title, a link to PubMed, a link to dictyBase Curated Paper (*see* next paragraph), and additional genes addressed in the paper. On the left side of the Literature Guide page is a list of **Literature Topics**. These Literature Topics are general categories such as Chemotaxis/ Motility, Adhesion, Development/Morphogenesis, and Mutants/Phenotypes and are aimed at providing the user with a fast overview of the focus of publications. For example, to view papers that discuss mutations in a certain gene, click on the Mutants/Phenotypes category in the Literature Guide of that gene.

The Literature Topics are assigned by curators to every gene discussed in an article. Therefore, each gene in a paper may have different combinations of topics. This information can be viewed on the **dictyBase Curated Paper** page, which is accessible by clicking on the **dictyBase paper** icon (*see* **Fig. 6L**). The Curated Paper page contains the abstract and a table displaying the genes addressed as well as their Literature Topics.

7. The Dicty Stock Center

Since its inception, the Dicty Stock Center collection has grown to more than 690 strains and more than 125 plasmids as of June 2005. The strain collection includes natural isolates, a large collection of axenic strains, motility mutants, labeled strains for cell biological studies, insertional (restriction enzyme-mediated integration [REMI]) mutants, null mutants obtained by homologous recombination, chemical mutants, and tester strains for asexual genetic analysis. Materials are acquired by scanning the current *Dictyostelium* publications and requesting strains and plasmids described in the literature. In addition, older literature is searched on a laboratory-by-laboratory basis and attempts are made to collect the strains and plasmids described there. The collection of strains and plasmids in a central repository ensures that they will always be available to the research community.

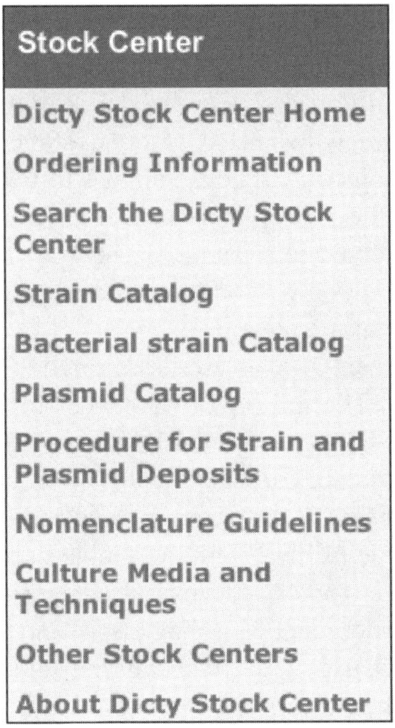

Fig. 8. The Stock Center drop-down menu accessible from the top menu bar on every page in dictyBase.

The Stock Center drop-down menu (*see* **Fig. 8**) is accessible from the top menu bar on every page in dictyBase. The pages and functions of this menu will be discussed below.

7.1. The Strain and Plasmid Catalogs

The catalogs provide alphabetical listings of all strains and plasmids stored in the Stock Center. Clicking on a name will bring up the **Strain Details** page or **Plasmid Details** page. As shown in **Fig. 9**, the Strain Details page displays genotypes and phenotypes in addition to information on how the strain was created and the publication in which the strain was described. When available, the Plasmid Details page shows vector maps as well as sequences, either present as text files or as link to the GenBank record. All detailed pages have an **Add to Cart** button at the bottom; clicking on this deposits the strain or plasmid in the Shopping Cart (*see* **Subheading 7.3.**).

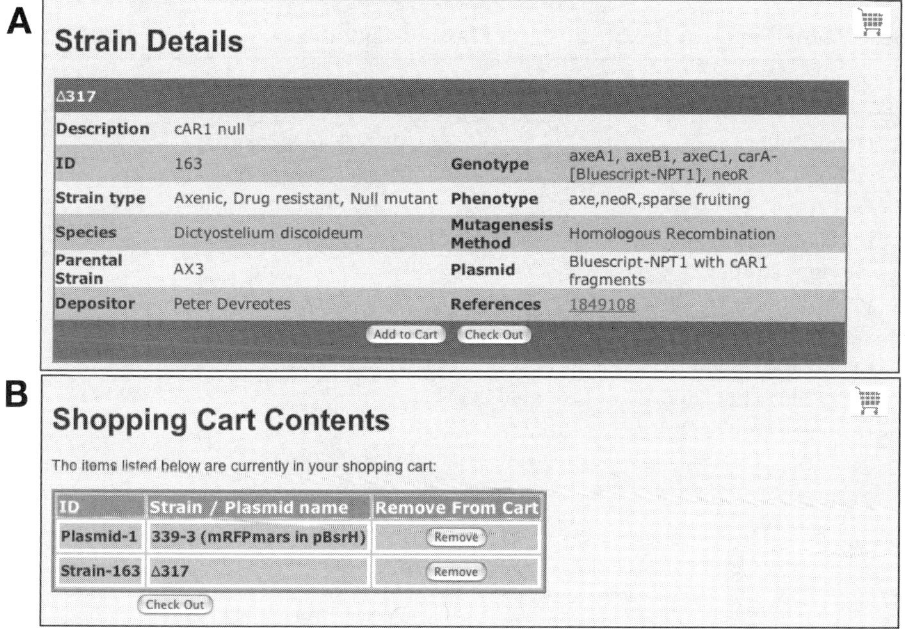

Fig. 9. (**A**) Example of a Strain Details Page of the Stock Center. A strain, or equally a plasmid, can be added to the Shopping Cart by clicking the Add to Cart button. Clicking on the Shopping Cart icon on top of the page links to the Shopping Cart Contents Page (**B**) to view the item(s) in the cart. Clicking on Check Out on either page leads to a form that needs to be filled out. http://dictyBase.org/db/cgi-bin/dictyBase/SC/strain_details.pl?id=163 http://dictyBase.org/db/cgi-bin/dictyBase/SC/manage_cart.pl

7.2. Searching the Stock Center

To search the contents of the Stock Center, you must first choose to search the **Strains** or the **Plasmids** database. The default "All function" searches all of the respective fields shown in **Table 5**, whereas choosing a specific field narrows the search. Key words are usually taken from the literature reference or the Franke *Dictyostelium* Reference Library. Note that searching the Stock Center does not allow for the use of the wildcard (*) character; rather, wildcards are added by default.

> **Tip**: The name by which a Strain or Plasmid is listed in the catalog may not always be the name by which it is best known. However, all known names exist as synonyms in the Stock Center and are fully searchable. For example, the strain *pi3k1-2-* is listed as *GMP1*

Table 5
Searchable Fields in the Strains and Plasmids Databases

Strains	Plasmids
• Depositor	• Depositor
• Genotype	• GenBank accession number
• Key word	• ID
• Mutagenesis method[a]	• Key word
• Parental strain	• Name
• Phenotype	
• Plasmid (transforming plasmid used to create the strain)	
• Reference (PubMed ID or the label from the Franke Reference Library)	
• Species	
• Strain ID	
• Strain name	
• Strain type[a]	

[a]A pop-up window appears when this field is selected, from which search options can be specified. In addition, a link at the bottom of the Dicty Stock Center search page achieves the same result.

(pi3k1/2-null) in the catalog, but can be found when searching for any of the following names: *pik1-/2-, deltaDdpik1 deltaDdpik2 null, pik1/pik2 double knockout, pi3k1-2-, pi3k1-/2-, pik1-/pik2-, Ddpi3k1/2-null, Ddpi3k1-/2-.*

7.3. Ordering and Depositing Strains and Plasmids

All pages of the Stock Center have a **Shopping Cart** link on the right side that allows you to order strains and plasmids online. After adding desired strains and plasmids to the Shopping Cart via the links on the Strain or Plasmid Details pages, click on the Cart icon to view the selected materials. If the items are correct, click on Check Out to process your order. A confirmation E-mail will notify you that the order has been placed. The strains and plasmids are free when ordered for research purposes, but shipping will be charged.

Tip: For prompt shipment, please include your full address, a telephone number, and the billing number.

The Stock Center accepts strains as axenic cultures, frozen cells, colonies on lawns of bacteria, lyophilized spores, or spores in silica gel. If strains are sent on plates, please identify the medium and the bacterial strain that were used. Plasmids can be deposited as either DNA or as a transformed bacterial culture. More information can be found at http://dictybase.org/StockCenter/Deposit.html.

7.4. Nomenclature Guidelines

It is highly recommended that each laboratory assigns itself a unique two- or three-letter code, as described in the **Nomenclature Guidelines**. Uniform naming conventions make for easier cataloging, both in the laboratory and in the Stock Center. The current laboratory strain designations are listed at http://dictybase.org/NomenclatureGuidelines.htm.

8. New Data and Future Directions

In addition to the continuous curation of the database, dictyBase strives to improve the database by adding new types of data and by making the database more user-friendly. The following list briefly describes some priorities for the expansion of dictyBase.

- *Curation of new sequence types.* We have implemented changes to the database that provide more flexibility for storing biological information. This will allow us to display sequences for non protein-coding genes, such as tRNAs and micro RNAs. We will also add 5' and 3' untranslated regions to the Curated Models and annotate repetitive elements in the genome.
- *Display of protein domains.* Every Gene Page will have a graphical display of conserved functional domains in the Protein Information section. The domains will link to the respective InterPro records *(12)* and, similarly to the GO and Phenotype pages, will link to a page listing all *Dictyostelium* proteins containing the domain.
- *dictyCyc.* Using the GMOD Pathway Tools (www.gmod.org/ptools.shtml) *(13)*, we will implement dictyCyc, a graphical and interactive display of metabolic pathways in *Dictyostelium*.
- *Related proteins.* Every Gene Page will have a new section where the proteins with highest similarity from several eukaryotes will be shown.
- *Additional expression data.* In the Expression section of the Gene Page, we plan to integrate data from the literature such as reverse-transcription PCR and Northern and Western blots.
- *Integration with AAAS Signal Transduction Knowledge Environment (STKE).* Reciprocal links will take users to the signal transduction pathways curated at STKE *(14)* and bring STKE users to dictyBase Gene Pages.

Acknowledgments

The authors wish to thank Eric M. Just, Sohel Merchant, and Warren A. Kibbe for comments on the manuscript and for developing such a valuable resource. dictyBase is supported by grants from the National Institutes of Health, GM64426 and HG02273.

References

1. Chisholm, R. L. (2005) dictyBase: Using the genome to organize *Dictyostelium* biology, in Dictyostelium *Genomics* (Loomis, W. F. and Kuspa, A., eds.), Horizon Scientific, Norwich, UK: pp. 23–40.
2. Kreppel, L., Fey, P., Gaudet, P., et al. (2004) dictyBase: using the genome to organize *Dictyostelium* biology. *Nucleic Acids Res.* **32,** D332–D333.
3. Eichinger, L., Pachebat, J. A., Glockner, G., et al. (2005) The genome of the social amoeba *Dictyostelium discoideum. Nature* **435,** 43–57.
4. Morio, T., Urushihara, H., Saito, T., et al. (1998) The *Dictyostelium* developmental cDNA project: generation and analysis of expressed sequence tags from the first-finger stage of development. *DNA Res.* **5,** 335–340.
5. Van Driessche, N., Shaw, C., Katoh, M., et al. (2002) A transcriptional profile of multicellular development in *Dictyostelium discoideum. Development.* **129,** 1543–1552.
6. Iranfar, N., Fuller, D., and Loomis, W. F. (2003) Genome-wide expression analyses of gene regulation during early development of *Dictyostelium discoideum. Euk. Cell.* **2,** 664–670.
7. Maeda, M., Sakamoto, H., Iranfar, N., et al. (2003) Changing patterns of gene expression in *Dictyostelium* prestalk cell subtypes recognized by in situ hybridization with genes from microarray analyses. *Euk. Cell.* **2,** 627–637.
8. Maruo, T., Sakamoto, H., Iranfar, N., et al. (2004) Control of cell type proportioning in *Dictyostelium discoideum* by differentiation-inducing factor as determined by in situ hybridization. *Euk. Cell.* **3,** 1241–1248.
9. Stein, L. D., Mungall, C., Shu, S., et al. (2002). The generic Genome Browser: a building block for a model organism system database. *Genome Res.* **10,** 1599–1610.
10. Altschul, S. F., Madden, T. L., Schaffer, A. A., et al. (1997) Gapped BLAST and PSI-BLAST: a new generation of protein database search programs. *Nucleic Acids Res.* **25,** 3389–3402.
11. Ashburner, M., Ball, C. A., Blake, J. A., et al. (2000) Gene ontology: tool for the unification of biology. The Gene Ontology Consortium. *Nat. Genet.* **1,** 25–29.
12. Mulder, N. J., Apweiler, R., Attwood, T. K., et al. (2005) InterPro, progress and status in 2005. *Nucleic Acids Res.* **33,** D201–D205.
13. Krieger, C. J., Zhang, P., Mueller, L. A., et al. (2004) MetaCyc: a multiorganism database of metabolic pathways and enzymes. *Nucleic Acids Res.* **32,** D438–442.
14. Kimmel, A. R. and Parent, C. A. (2003) The signal to move: *D. discoideum* go orienteering. *Science* **300,** 1525–1527.

5

Analysis of Gene Expression Using cDNA Microarrays

Marcel Kaul and Ludwig Eichinger

Summary

A DNA microarray consists of an orderly arrangement of DNA, cDNA, or oligo-nucleotide probes that represent individual genes of an organism. Today, microarrays are the most important tools used to analyze gene transcription on a large scale. Investigations reach from finding key genes in whole genomes to looking at overall patterns of gene expression. This provides a better understanding of the architecture of genetic regulatory networks and helps to identify those genes in the genome that are differentially expressed, e.g., during cellular programs, in tumors or as the result of a treatment. The use of different fluorescent dyes allows mRNAs from two different sources (tissues, strains, or differently treated cells) to be labeled in different colors, mixed, and hybridized to the same array, which results in competitive binding of the target to the arrayed sequences. After hybridization and washing, the slide is scanned using two different wavelengths corresponding to the dyes used, and the intensity of the same spot in both channels is compared. This results in a measurement of the ratio of transcript levels for each gene represented on the array. In order to find significantly regulated genes, the resulting data are analyzed by different bioinformatic tools such as R and significance analysis of microarrays (SAM).

Key Words: DNA-microarrays; *Dictyostelium discoideum*; bioconductor; R; SAM, gene expression.

1. Introduction

Analysis of gene expression as an answer to changes in the conditions of the environment is an important approach to understanding the workings of genes, cells, and organisms. Before the breakthrough and establishment of DNA microarrays, Northern blots were used to study the expression of genes. The problem with this method is its limitation to a few genes, making it a very cumbersome work if one wants to study a large number of genes. In contrast, microarrays are powerful tools to investigate differential expression of thousands

From: *Methods in Molecular Biology, vol. 346:* Dictyostelium discoideum *Protocols*
Edited by: L. Eichinger and F. Rivero © Humana Press Inc., Totowa, NJ

of genes in parallel *(1–4)*. A DNA microarray consists of an orderly arrange-ment of probes of cDNAs, genomic DNA, or synthesized oligonucleotides that represent individual genes of an organism *(1)*. cDNA microarrays are the most commonly used, and were first described in 1995 *(2)*. DNA microarrays of all types have been generated for several organisms and will be produced for more and more species as sequence data become available. Unlike printed arrays, Affymetrix slides are produced by photolithography and contain much shorter oligonucleotides *(5,6)*.

To determine a significant change in gene transcription, control and experi-ment RNAs are required. They are reverse transcribed and labeled with fluo-rescent dyes such as Cy3 and Cy5. Simultaneous hybridization with separate detection of signals provides the comparative analysis, allowing one to deter-mine the amounts of different transcripts present from control and experiment that are bound to every probe on the array *(7)*. For this purpose, the slide is scanned. The fluorescent dyes are excited by laser light of pertinent wave-length, and emission is detected by a photomultiplier, quantitated, and ana-lyzed by different bioinformatics tools (**Fig. 1**).

Microarrays often carry probes representing whole genomes or large parts of genomes. They can be used for a variety of experimental approaches. Expres-sion profiles for genes with known metabolic functions are useful for analyz-ing metabolic reprogramming, which occurs during applied treatments such as heat-shock, hypoxia, or osmotic stress *(8)*. Other possible applications are the analysis of processes such as cell cycle, the profiling of complex diseases such as cancer, and the discovery of novel disease-related genes *(9–12)*. Tumors can be classified by their individual transcriptional profile *(13)* and different mutants can be characterized by monitoring their expression in comparison with that of the wild-type *(14)*. Connection of expression profiles to positions of regulated genes on the genome led to the discovery of sections of chromo-somes that contain similarly regulated genes *(15)*. Also, co-expression of genes of known function with poorly characterized or novel genes may provide important clues to the functions of many uncharacterized genes, for which information is currently not available *(11,16)*.

The *Dictyostelium* genome and cDNA projects have paved the way for the generation of different *Dictyostelium* DNA microarrays. The microarrays at our institute (http://www.uni-koeln.de/med-fak/biochemie/transcriptomics/) and at the Baylor College of Medicine, Houston, TX (http://www.bcm.edu/db/db_fac-shaulsky.html) are based on nonredundant expressed sequence tags (ESTs) from the Japanese cDNA project *(17)*. Both arrays cover approximately half of the genes in the *Dictyostelium* genome. A new generation of microarray has been developed at the Wellcome Trust Sanger Institute, Cambridge, UK (http://www.sanger.ac.uk/PostGenomics/PathogenArrays/Dicty/). It is based

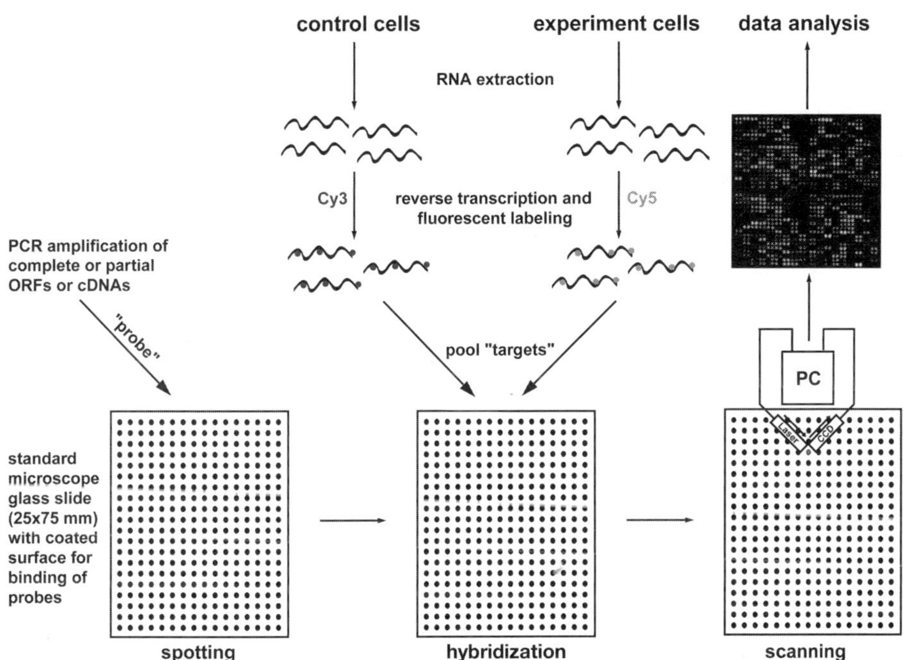

Fig. 1. A DNA microarray consists of an orderly arrangement of cDNA probes, oligonucleotides, or longer partial gene sequences that represent individual genes of an organism. Partial ORFs or cDNAs are amplified by PCR and spotted on the array. The targets of a microarray experiment are two RNA pools that are reverse transcribed and labeled with fluorescent dyes such as Cy3 and Cy5. The targets are hybridized to the microarray and subsequently scanned to quantify the amount of target bound to every probe on the array. Differentially expressed genes are then detected and further analyzed.

on gene predictions of the recently completed *Dictyostelium* genome *(18)* and currently represents 9000 of the approx 12,500 genes in the genome. Here, we describe the protocols that we use for the analysis of microarray data at our institute. The microarray contains partial sequences of 450 selected published genes, 5423 nonredundant ESTs from the *Dictyostelium* cDNA project, and appropriate positive and negative controls as well as the Spot Report-10 Array Validation System for sensitivity and ratio determination (**Table 1**; Gene Expression Omnibus [GEO] accession number GLP1972, http://www.ncbi. nlm.nih.gov/geo/). Positive controls consist of spots for 12 *Dictyostelium* genes with known expression profiles for different developmental stages as well as genomic DNA. As negative controls, we used *Bst* EIII-digested DNA of phage lambda and microarray spotting solution. All controls are spotted many times in different positions on the microarray and also serve to monitor spatial bias.

Table 1
Spots of the *Dictyostelium discoideum* Microarray

	Genes		Spots	
	Number	Proportion	Number	Proportion
Selected genes	450	7.61%	900	6.16%
cDNAs	5423	91.82%	10,846	74.19%
Controls	33	0.57%	2874	19.65%
Sum	5906		14,620	

The complete microarray dataset is available at the Gene Expression Omnibus (GEO; http://www.ncbi.nlm.nih.gov/geo/; accession number GPL1972).

We are using this microarray to study genome-wide gene expression patterns of *Dictyostelium* cells. The main projects pursued are the analysis of transcriptional regulation in response to environmental stresses, particularly under hyper-/hypoosmotic or low oxygen (hypoxia) conditions, and infection with pathogens such as *Legionella pneumophila* as well as the transcriptional regulation in a variety of knockout mutants in comparison to wild-type cells.

2. Materials

2.1. General

1. Ethanol 100%, 95%, 75%, 70%, DNase/RNase free.
2. 3 M sodium acetate, pH 4.5.
3. H_2O, DNase/RNase free.
4. β-Mercaptoethanol (β-ME).
5. Dimethylsulfoxide (DMSO).
6. 20X SSC: 3 M NaCl, 0.3 M sodium citrate.
7. 10% (w/v) sodium dodecyl sulfate (SDS) solution.
8. Uvette (sterile plastic cuvet, Eppendorf AG, Germany).

2.2. RNA Extraction

1. RNA-Mini Kit, Qiagen (components not listed here).
2. Buffer RLN: 50 mM Tris-HCl, pH 8.0, 140 mM NaCl, 1.5 mM MgCl$_2$, 0.5% (v/v) Nonidet P-40 (1.06 g/mL); autoclave and store at 4°C. Just before use, add 1000 U/mL RNase inhibitor (optional), 1 mM dithiothreitol (DTT) (optional).
3. β-ME must be added to Buffer RLT before use. Add 10 μL β-ME per 1 mL Buffer RLT. β-ME is toxic; dispense in a fume hood and wear appropriate protective clothing. Buffer RLT is stable for 1 mo after addition of β-ME.
4. Buffer RPE is supplied as a concentrate. Before using for the first time, add 4 volumes of ethanol (96–100%), as indicated on the bottle, to obtain a working solution.
5. Soerensen buffer: 15 mM KH$_2$PO$_4$, 2 mM Na$_2$HPO$_4$, pH 6.0.

2.3. Spike Mix

SpotReport—10 Array Validation System, Stratagene.

2.4. Labeling With the FairPlay® Kit

1. FairPlay® Microarray Labelling Kit, Stratagene(components not listed here).
2. 1 *M* NaOH.
3. 1 *M* HCl.
4. Cy5™ Mono-Reactive Dye Pack, Amersham Biosciences.
5. Cy3™ Mono-Reactive Dye Pack, Amersham Biosciences.
6. TE buffer: 10 m*M* Tris-HCl, pH 8.0, 1 m*M* ethylenediamine tetraacetic acid (EDTA).
7. 10 m*M* Tris-HCl, pH 8.5.
8. Uvette (Eppendorf AG, Germany).

2.5. Prehybridization

1. Hellendahl staining trough, Roth.
2. Formamide.
3. 10 mg/mL bovine serum albumin in H_2O.
4. Isopropanol.

2.6. Hybridization

1. UltraGAPS Microarray-Slides, Corning.
2. Fish sperm DNA, Roche.
3. Oligo dA 18mer, 100 µ*M* in H_2O.
4. 100X Denhardt's reagent: 2% Ficoll 400, 2% polyvinylpyrrolidone, 2% bovine serum albumin.
5. 1.2 *M* Phosphate buffer: 2 vol. 1.2 *M* Na_2HPO_4, 1 vol. 1.2 *M* NaH_2PO_4, pH 6.8.
6. Hybridization buffer: 0.12 *M* Phosphate buffer, pH 6.8, 2 m*M* EDTA, 50% formamide, 0.1% Na-laurylsarcosinate, 0.1% SDS, 4X Denhardt's reagent, 2X SSC.
7. Microarray hybridization chambers, Corning.

2.7. Washing

1. 2X SSC, 0.1% SDS.
2. 0.1X SSC, 0.1% SDS.
3. 0.1X SSC.
4. 0.01X SSC.

2.8. Scanning/Quantification

1. Microarray-Scanner: Scan-Array® 4000XL, PerkinElmer Life Sciences.
2. ScanArrayExpress 3.0, PerkinElmer Life Sciences.
3. *Dictyostelium* microarray GAL (GenePix Array List format, **.gal*) file (*see* **Note 1**).

2.8. Data Analysis

All programs/tools listed here are freely available.

1. ArrayTools: http://www.uni-koeln.de/medfak/biochemie/transcriptomics/download/arraytools.xla.
2. R Version 1.6.2: *The R Project for Statistical Computing*, http://www.r-project.org as well as modules from Bioconductor, http://www.bioconductor.org: marrayClasses, marrayInput, marrayNorm, marrayPlots, Biobase.
3. Significance Analysis of Microarrays (SAM) Version 1.21, Department of Statistics, Stanford University, http://www-stat.stanford.edu/~tibs/SAM/.
4. *Dictyostelium* microarray GAL file for R (*see* **Note 1**).

3. Methods

3.1. RNA Extraction

Total RNA is extracted from *D. discoideum* cultures with the Qiagen RNeasy Mini Kit according to the "Protocol for Isolation of Cytoplasmic RNA from Animal Cells" with some modifications, marked "***." After extraction, the RNA concentration is determined by measuring the OD_{260} (should be >500 µg/mL) and the RNA is examined on a denaturing agarose gel and should give two bands with sizes of 4.1 and 1.9 kb for 26S and 18S rRNA, respectively. For the hybridization of one slide, we usually use 20 µg of both control and experiment RNA.

1. Harvest cells grown in suspension (*see* **Note 2**): Pellet the appropriate number of cells for 3 min at ***250g in a centrifuge tube. Decant supernatant. ***Wash cells two times with Soerensen buffer, centrifuge after each step for 3 min at 250g; remove supernatant.
2. Carefully resuspend cells in 175 µL cold (4°C) Buffer RLN to lyse the plasma membrane and incubate on ice for 5 min (*see* **Note 3**).
3. ***Centrifuge lysate at 4°C for 10 min at 3300g. Transfer supernatant to a new centrifuge tube, and discard the pellet (*see* **Note 4**). After centrifuging, heat the centrifuge to 20–25°C if the same centrifuge is to be used in the following centrifugation steps of the protocol.
4. Add 600 µL of Buffer RLT to the supernatant. Mix thoroughly by vigorously vortexing.
5. Add 430 µL of ethanol (96–100%) to the homogenized lysate. Mix thoroughly by pipetting. Do not centrifuge.
6. Apply 700 µL of the sample, including any precipitate that may have formed, to an RNeasy mini column placed in a 2 mL collection tube. Close the tube gently and centrifuge for 15 s at ≥9000g. Discard the flow-through. Reuse the collection tube in **step 7**. If the volume exceeds 700 µL, load aliquots successively onto the RNeasy column, and centrifuge as above. Discard the flow-through after each centrifugation step.

7. Add 700 µL of Buffer RW1 to the RNeasy column. Close the tube gently and centrifuge for 15 s at ≥9000*g* to wash the column. Discard the flow-through and collection tube.

8. Transfer the RNeasy column into a new 2 mL collection tube. Pipet 500 µL of Buffer RPE onto the RNeasy column. Close the tube gently, and centrifuge for 15 s at ≥9000*g* to wash the column. Discard the flow-through. Reuse the collection tube in **step 9**.

9. Add another 500 µL of Buffer RPE to the RNeasy column. Close the tube gently, and centrifuge for 2 min at ≥9000*g* to dry the RNeasy silica-gel membrane.

10. ***Discard the flow-through, and centrifuge in a microcentrifuge at full speed for 1 min to eliminate any chance of possible Buffer RPE carryover.

11. To elute, transfer the RNeasy column to a new 1.5 mL collection tube. ***Pipet 50 µL of RNase-free water directly onto the RNeasy silica-gel membrane. Close the tube gently and centrifuge for 1 min at ≥9000*g* to elute.

12. ***Repeat **step 11** as described with a second volume (50 µL) of RNase-free water. Elute into the same collection tube.

3.2. Addition of Spike Mix and Precipitation of RNA

Quality control is an important issue in DNA microarray analysis. We use positive controls, negative controls, and the SpotReport Validation Kit, which consists of ten internal mRNA controls from *Arabidopsis thaliana* genes. These are added (spiked) as a mix with different known amounts of each mRNA to the *D. discoideum* RNA prior to cDNA generation and labeling. Two different mixes are used for the two labeling reactions of one microarray experiment (**Table 2**).

1. Add 10 µL of spike mix A to 20 µg of control-RNA and 10 µL of spike mix B to 20 µg of experiment RNA.

2. Precipitate the RNA mixes by adding 0.1 volumes of 3 *M* Na-acetate, pH 4.5, and 2.5 volumes 100% ethanol.

3. Store at –20°C for 2 h (probe can be stored overnight if desired) and centrifuge in a tabletop centrifuge at maximum speed for 30 min.

4. Remove ethanol by aspiration and wash pellet with 70 % ethanol.

5. Centrifuge 15 min at ≥9000*g*, aspirate the ethanol, and dry the pellet at room temperature.

6. Dissolve the pellet in 12 µL of RNase-free water.

3.3 Labeling With the FairPlay Kit

Various labeling protocols have emerged for the conversion of test and reference mRNA into labeled cDNA in a reverse-transcription (RT) reaction. One method is the direct incorporation of labeled nucleotides into the cDNA during the RT reaction. An inherent problem associated with this method is the biased incorporation of the different fluorescent dyes, which can result in uneven

Table 2
Spike Mix for the *Dictyostelium discoideum* Microarray

Gene	Spike A [pg/10 µL]	Spike B [pg/10 µL]	Quotient A/B	Total amount [pg]
Cab	300	150	2	450
RCA	250	250	1	500
rbcL	250	250	1	500
LTP4	5	5	1	10
LTP6	2	2	1	4
XCP2	1	1	1	2
RCP1	300	150	2	450
NAC1	400	40	10	440
TIM	375	75	5	450
PRKase	375	75	5	450

The SpotReport—10 Array Validation System consists of 10 internal mRNA controls from *Arabidopsis thaliana* genes. These are added (spiked) as a mix with different amounts of each mRNA to the *D. discoideum* RNA prior to cDNA generation and labeling.

distribution of fluorescence and/or overall low-level fluorescence in the resulting cDNA. The FairPlay microarray labeling kit solves the problems associated with direct incorporation through the use of indirect labeling and through the division of the labeling procedure into two parts: (1) preparation of amino-allyl modified cDNA and (2) chemical coupling of the fluorescent dye to the modified cDNA. The kit contains all components necessary to convert mRNA to amino-allyl modified cDNA ready to be coupled to any fluorescent dye containing an NHS- or STP-ester leaving group.

3.3.1. cDNA Generation

Prepare separate cDNA labeling reactions for each fluorescent dye being used. This protocol produces a sufficient amount of labeled cDNA per reaction to hybridize to a total microarray surface area of approx 10 cm^2.

1. Add 1 µL of 500 ng/µL oligonucleotide d(T)12–18 to the dissolved RNA from **Subheading 3.2., step 6**. Incubate at 70°C for 10 min. Cool on ice until ready for use (*see* **Note 5**).
2. Combine the following components in a sterile, RNase/DNase-free microcentrifuge tube:
    a. 2 µL of 10X StrataScript reaction buffer
    b. 1 µL of 20X dNTP mix, with aminoallyl dUTP (*see* **Note 6**)
    c. 1.5 µL of 0.1 *M* DTT
    d. 0.5 µL of RNase Block (40 U/µL)
    e. 1.0 µL StrataScript RT

3. Add this mixture to the RNA/Primer mixture from **step 1** and incubate at 48°C for 25 min.
4. Add another 1 µL of StrataScript RT and incubate at 48°C for an additional 35 min.
5. Add 10 µL of 1 *M* NaOH and incubate at 70°C for 10 min to hydrolyze RNA.
6. Cool to room temperature slowly; do not cool on ice.
7. Spin tube briefly to collect contents.
8. Add 10 µL of 1 *M* HCl to neutralize the solution (*see* **Note 7**).

3.3.2. cDNA Purification

The cDNA must be purified to remove unincorporated nucleotides, buffer components, and hydrolyzed RNA. Stratagene recommends an ethanol precipitation to purify the cDNA (*see* **Note 8**).

1. Add 4 µL of 3 *M* Na-acetate, pH 4.5, to the reaction in **Subheading 3.3.1., step 8**.
2. Add 1 µL of 20 mg/mL glycogen (*see* **Note 9**).
3. Add 100 µL of ice-cold 95% ethanol.
4. Store at –20°C for 1 h (probe can be stored overnight if desired) and centrifuge in a tabletop centrifuge at maximum speed for 30 min.
5. Remove ethanol by aspiration and wash with 70% ethanol.
6. Centrifuge 15 min at maximum speed, aspirate, and dry (*see* **Note 10**).

3.3.3. NHS-Ester-Containing Dye Coupling Reaction

In our lab, we use the mono-reactive Cy3 and Cy5 packs from Amersham Biosciences that contain derivatized CyDyes with only one reactive group on each dye molecule for accurate labeling of amine groups. Alexa fluor 555 and 647 from Molecular Probes are also suitable for this protocol.

1. Resuspend the cDNA pellet from **Subheading 3.3.2, step 6** in 5 µL of 2X coupling buffer (part of the FairPlay kit). Gently heat at 37°C for 15 min to aid in the resuspension process (*see* **Note 11**). Vortex gently to ensure the pellet is completely solubilized. Add 5 µL of fluorescent dye (e.g., Cy3 to the control and Cy5 to the experiment) to each cDNA (*see* **Note 12**).
2. Mix by gently pipetting up and down.
3. Incubate for 30 min at room temperature in the dark.

3.3.4 Dye-Coupled cDNA Purification

1. Add 90 µL of 1X TE Buffer to the labeled cDNA (*see* **Subheading 3.3.3., step 3**).
2. Combine 200 µL of DNA binding solution and 200 µL of 70% ethanol for each sample. Mix well by vortexing.
3. Add 200 µL of the mixture to the labeled cDNA and mix by vortexing.
4. Transfer the mixture to a microspin cup that is seated in a 2-mL receptacle tube (included in the kit). Exercise caution to avoid damaging the fiber matrix with the pipet tip. Snap the cap of the 2-mL receptacle tube onto the top of the microspin cup.

5. Spin the tube in a microcentrifuge at maximum speed for 30 s. The labeled cDNA is retained in the fiber matrix of the microspin cup.
6. Open the cap of the 2-mL receptacle tube, remove and retain the microspin cup, and discard the flow-through containing the uncoupled dye.
7. Add another 200 µL of the DNA-binding solution and ethanol mixture from **step 2** to the microspin cup. Snap the cap of the receptacle tube onto the top of the microspin cup.
8. Repeat **steps 5** and **6**.
9. Add 750 µL of 75% ethanol to the microspin cup. Snap the cap of the receptacle tube onto the top of the microspin cup.
10. Spin the tube in a microcentrifuge at maximum speed for 30 s.
11. Open the cap of the 2-mL receptacle tube, remove and retain the microspin cup, and discard the wash buffer.
12. Repeat **steps 9–11**.
13. Place the microspin cup back in the 2-mL receptacle tube and snap the cap of the receptacle tube onto the microspin cup.
14. Spin the tube in a microcentrifuge at maximum speed for another 30 s to completely dry the membrane. On removal from the centrifuge, make sure that all of the wash buffer is removed from the microspin cup.
15. Transfer the microspin cup to a fresh 1.5-mL microcentrifuge tube and discard the 2-mL receptacle tube.
16. Add 50 µL of 10 m*M* Tris-HCl, pH 8.5, directly onto the top of the fiber matrix at the bottom of the microspin cup.
17. Incubate the tube at room temperature for 5 min (*see* **Note 13**).
18. Snap the cap of the 1.5-mL microcentrifuge tube onto the microspin cup and spin the tube in a microcentrifuge at maximum speed for 30 s.
19. Repeat **steps 16–18** two times.
20. Open the lid of the microcentrifuge tube and recover the flow-through containing the purified labeled cDNA.
21. Pipet 100 µL of labeled cDNA into a sterile Uvette and determine the ssDNA concentration with Tris-HCl, pH 8.5, as reference (*see* **Note 14**).
22. Mix the labeled control cDNA and the experiment cDNA.

3.4. Prehybridization

Depending on how many arrays are to be prehybridized, two different volumes of prehybridization solution are recommended (**Table 3**).

The washing should be done array-wise and the centrifugation should be performed immediately after the last washing step (*see* **Note 15**).

1. Preheat the hybridization solution to 42°C in a Hellendahl staining trough.
2. Incubate the arrays for 45 min at 42°C.
3. Rinse each array for 15 s with water.
4. Dip the array in isopropanol.
5. Dry the array by centrifugation at 235*g* for 2 min.

Table 3
Prehybridization Buffer

Ingredients	A	B	Final concentration
Water	20.7 mL	92 mL	
20xSSC	22.5 mL	100 mL	5X
Formamide	45.0 mL	200 mL	50%
10% sodium dodecyl sulfate	900 µL	4 mL	0.1%
Bovine serum albumin (10 mg/mL)	900 µL	4 mL	0.1 mg/mL

A, 90 mL hybridization solution for five arrays; B, 400 mL hybridization solution for up to 25 arrays.

3.5. Hybridization

1. Precipitate the labeled cDNA from **Subheading 3.3.4, step 22** by adding 0.1 volumes of 3 *M* Na-acetate, pH 4.5, and 2.5 volumes 100% ethanol.
2. Store at –20°C for 1 h (probe can be stored overnight if desired) and centrifuge in a tabletop centrifuge at maximum speed for 30 min.
3. Remove ethanol by aspiration and wash pellet with 70% ethanol.
4. Centrifuge 15 min at maximum speed, aspirate the ethanol, and dry the pellet.
5. Pipet 10 µL of 3X SSC into the two holes of the Corning hybridization chamber (*see* **Note 16**).
6. Dissolve the precipitated labeled cDNAs in 65 µL of hybridization buffer, 1 µL of fish sperm DNA, and 1 µL of Oligo dA.
7. Incubate the solution 10 min at 80°C.
8. Centrifuge the solution to collect all vapor and pipet the 65 µL of target solution onto the end of the microarray.
9. Place the coverslip onto the microarray by first letting it touch the side of the array where the target solution is placed and then slowly lowering it down until it covers the array area. The target solution should cover the entire array area and there should not be any air bubbles present (*see* **Note 17**).
10. Place the slide in the hybridization chamber, close the chamber, and submerge it in the water bath over night at 37°C (*see* **Note 18**).

3.6. Washing

After hybridization, the microarray is washed to remove unbound target. During washing, the transitions from the baths should be performed swiftly so that the microarray does not dry before processing is finished (*see* **Note 19**).

1. Remove the microarray from the hybridization chamber and plunge it into 2X SSC, 0.1% SDS until the coverslip glides off. Remove the coverslip.
2. Shake the microarray in fresh 2X SSC, 0.1% SDS for 5 min.

3. Shake the microarray in 0.1X SSC, 0.1% SDS for 5 min.
4. Shake the microarray in 0.1X SSC for 5 s.
5. Repeat **step 4** four times.
6. Shake the microarray in 0.01X SSC for 5 s.
7. Dry arrays by centrifugation at 235*g* for 5 min and proceed to scanning (*see* **Note 20**).

3.7. Scanning and Quantification

The fluorescent-labeled cDNA targets that are bound to the spotted probes are detected by a confocal laser scanner (ScanArray 4000XL in our case). The microarray is scanned for Cy3 and Cy5 successively (*see* **Notes 21** and **22**). The fluorescent dyes are excited by laser light of pertinent wavelength and emission is detected by a photomultiplier (*see* **Note 23**). To obtain images well suited for signal quantification, the brightness must be adjusted by setting the laser power. Signals should be as bright as possible, but spots must not be saturated (indicated by white coloring). It might be necessary to scan at two different laser-power settings: one setting at which most spots give bright signals, but a few—such as some of the highly expressed genes—are saturated, and another setting at which no saturation is seen, but most spots give weak signals. The following protocol only works with the ScanArray 4000XL. Refer to the ScanArrayExpress manual for more detailed instructions on creating scan protocols or quantitation protocols or for answers to general questions.

1. Start ScanArray Express 3.0 and switch on lasers of required wavelengths by clicking on the grey buttons.
2. Press "Scan," choose Scan type and select a protocol by clicking on the bars. Each user should have his or her own protocol group, as it contains an image-autosave protocol which will save all images into the user's folder.
3. Press "Start." Next, the spot and background intensities of the scanned images are quantified by SanArray Express 3.0 (*see* **Note 24**). Before quantifying, select the control image via "File": "Set Control Image."
4. Press "Quantitate" and select a quantitation protocol.
5. Because the positions of the subarrays may vary from slide to slide, check the alignment of the subarrays by pressing "Adjust Template and Register Images." If necessary, move or rotate all or individual subarrays by dragging with the mouse; press "OK" when ready.
6. Press "Start."
7. The resulting quantification file must be checked for mis-spots. Individual spots can be moved and resized by dragging with the mouse, and their status (Good, Bad, Absent, Not Found, Found) may be changed by right-clicking on the spot. The distribution plot may help to spot irregularities when Footprint, Diameter, Median background, or Background Standard deviation are plotted.
8. Save the comma separated value (CSV) file by selecting the spreadsheet tab and choosing "File": "Save As."

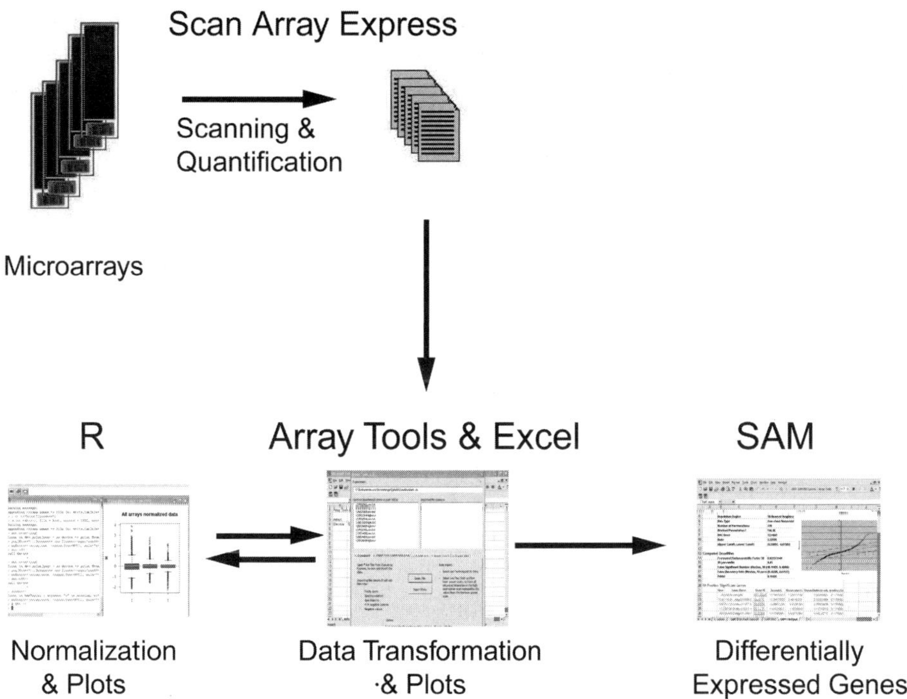

Fig. 2. Hybridized microarrays are scanned and quantified using Scan Array Express. This data are then exported to Microsoft Excel®, where additional analysis can be done and plots are easily created. Normalization and automatic plot generation are performed in R, differentially expressed genes are determined with significance analysis of microarrays (SAM).

3.8. Data Analysis

High-density microarrays require automated data analysis because the number of datasets is too large for manual processing. Commercial and academic software is available for the performance of numerous tasks. We use Scan Array Express 3.0 for scanning and quantification, R and various BioConductor packages for data normalization, and SAM to identify differentially expressed genes (*see* **Fig. 2**). To permit combination of the commercial solution with the free BioConductor packages, SAM, and other software that is being developed by academia, an add-in for Microsoft Excel® was written in VisualBasic. After signal quantification, the data are filtered for negative controls, saturated signals, and spiked controls of the SpotReport validation system. Normalization is performed through locally weighted polynomial regression (Lowess) fitting, provided by BioConductor for the statistical programming environment R. Normalization is necessary, because the labeling efficiency is not equal for Cy3 and Cy5 and also varies from reaction to reaction. Lowess fitting could possibly reduce this dye bias; however, it must be taken into account that the bias is intensity-dependent. The Excel add-in SAM uses a permutational approach to identify differentially expressed genes and to estimate the false discovery rate (FDR). SAM calculates a significance value for every gene by comparing the true ratio of experiment vs control with the "noise" ratios that are calculated from permutations of all microarrays in the experiment. In SAM, the accepted FDR can be chosen. The output of SAM is a list of differentially expressed genes that complies with the set FDR and a plot of all genes in which the observed difference in expression is plotted against the difference expected from the random noise.

Significant genes can be further analyzed by different bioinformatics tools. "Caryoscope" gives a graphical overview of gene locations on chromosomes (http://dahlia.stanford.edu:8080/caryoscope/index.html), "GOAT" analyzes GO enrichment for gene lists (http://dictygenome.bcm.tmc.edu/software/GOAT/), and "Compare" compares different ArrayTools experiments for accordance in significant genes (http://www.uni-koeln.de/med-fak/biochemie/transcriptomics/).

Start Excel and import ArrayTools to Excel: press "Tools": "AddIns": "Browse": Array Tools. Repeat this procedure with SAM.

1. Click on the "Array Tools" button.
2. Create a new Experiment file for the data from the microarrays of your experiment (*see* **Note 25**).
3. Open the ScanArray Express CSV files. The suffix must be changed to *.txt first.
4. ArrayTools works with one or two files per microarray. You now have the option of importing individual data files or selecting pairs of data files for high- and low-power laser scans, in which case the saturated values of the high-power laser scan will be replaced by nonsaturated values form the low-power laser scan.
5. Export the data to R. Copy the GAL file and all three of the other file types that are required into the same directory. The paths of the files are listed in the R Commands.R file and are used by R:
   - *.spot files containing the M (ratios) and A (intensities) values (these files are generated by ArrayTools).
   - A GAL file for R (this file is identical to the GAL file for ScanArrayExpress except that it must not contain a header).
   - An R Commands.R that contains a commands script for R (generated by ArrayTools).
   - An Arrays.txt file that lists the microarrays of the experiment (generated by ArrayTools).

6. Start R and copy the contents of the R Commands file into the R Console. R normalizes the data, writes it to Rout barcode.txt files, and creates several control plots *(19–22)*.

7. Close R and return to Array Tools. Select the microarrays of your experiment and import their corresponding Rout files.

8. Export your data to SAM.

9. Make sure the regional settings of your computer are set to English before starting SAM. Start SAM by clicking the SAM button and choose "Response Type: One class Response," "Number of Permutations: 1000" *(23)* (*see* **Note 26**).

10. With the SAM Plot Controller, you have to set a delta value according to the number of falsely reported genes you are prepared to accept, list Significant Genes, and list Delta Table (*see* **Note 27**).

4. Notes

1. GenePix Array List files describe the size and position of blocks, the layout of feature indicators in them, and the names and identifiers of the printed substances associated with each feature-indicator. They can be created in Excel by saving an Excel spreadsheet as Text (Tab delimited). GAL files consist of two sections: the header and data records. The header contains all the structural and positional information about the blocks and the data records contain all the name and identifier information for each spot. GenePix Pro assigns block numbers such that the top leftmost block on the image is block #1, and the block numbers increase from left to right and then from top to bottom. The Gal file for R must not contain a header.

2. Do not take too many cells for the RNA isolation because this might block the column that is used in later steps of the protocol. Never use more than 2×10^7 cells for the MiniPrep (1×10^8 cells for MidiPrep). The expected yield for the MiniPrep is 100 µg and for the MidiPrep 1 mg of RNA.

3. The lysis step is very important for further RNA isolation. Extend the lysis step to 10 min if there is a large pellet left or to ensure that all cells are lysed.

4. Centrifuging down the nuclei should result in a small, greyish pellet.

5. Incubating at 70°C destroys the secondary structure of the RNA. Oligo d(T) primers anneal to the RNA when cooling down on ice.

6. Aminoallyl dUTP is incorporated into cDNA by RT. The resulting aminoallyl-containing DNA is subsequently labeled with amine-reactive fluorescent dyes that bind to the aminoallyl group.

7. After generation of the aminoallyl-labeled cDNA, the reaction mixture is treated with NaOH to hydrolyze the RNA. HCl is used to neutralize NaOH. Work carefully to ensure identical volumes of NaOH and HCl.

8. Incomplete removal of the Tris-HCl and ethanol will result in lower aminoallyl-dye-coupling efficiency. Care must be taken to ensure that the pellet is completely dry, indicating complete removal of the ethanol, before proceeding to the dye-coupling reaction.

9. The glycogen is used as a co-precipitant to ensure complete precipitation of the cDNA.

10. A vacuum dryer can be used to speed up the process, but do not overdry. If the pellet is overdried, it will be difficult to get the cDNA back into solution. Loss of cDNA for further steps will be the result.

11. A visible precipitate may be seen in the 2X coupling buffer. Incubate the buffer at room temperature or 37°C to resolubilize the precipitate before use.

12. Cy3 and Cy5 are supplied lyophilized, and must be resuspended in 50 µL DMSO before use. Use the high-purity DMSO provided in the FairPlay kit. DMSO is hygroscopic and will absorb moisture from the air. Water absorbed from the air will react with the NHS ester portion of the dye and significantly reduce or eliminate dye:cDNA-coupling efficiency. To reduce absorption, allow the dye to reach room temperature before opening and store the DMSO at room temperature. Do not leave either the dye or the DMSO uncapped when not in use. During storage, tightly cap the resuspended dye and store at –20°C in the dark. The unused dye can be stored for up to 2 mo.

13. In the presence of a chaotropic salt (introduced by the DNA-binding solution, included in the kit), the dye-coupled cDNA binds to the silica-based fiber matrix seated inside the microspin cup. Washing steps are employed to remove buffer salts and uncoupled fluorescent dye from the bound cDNA. Finally, the cDNA is eluted from the matrix using a low-ionic strength solution.

14. The result should be handled with care so as to reduce the likelihood of inaccuracy. For further analysis, an aliquot of the cDNA may be electrophoresed in an agarose gel. To determine whether the fluorescent dye has been coupled, the cDNA is visualized with a laser scanner. To determine whether the RNA template strand has been completely removed, the cDNA is stained with ethidium bromide following electrophoresis and visualized with an ultraviolet transilluminator. *See* FairPlay Labeling Kit Protocol for details.

15. Prehybridization is used to block reactive groups on the surface of the slide that can bind labeled target DNA nonspecifically.

16. This is important to maintain proper humidity inside the chamber. Drying out will increase background on the slide.

17. This is a critical step. Make sure that, after pipetting the target solution to the array, the solution does not run down from the slide, because then dirt may pour in or too much solution may pour out. Also, avoid bubbles by any means. Practice several times before performing this step for the first time.

18. Keep the chamber with the slide in a perfectly horizontal position so that the solution does not run to any side of the slide, because this could lead to a signal gradient on the processed slide.

19. Drying out will lead to dramatically increased background because additional color will bind to the slide and cannot be washed away.

20. It is best to proceed to scanning immediately. If necessary, the microarray slides can be stored under vacuum in the dark for several weeks; however, signal intensity will decrease.

21. The possible resolution depends on the scanner. Currently, most scanners provide resolutions down to 10 or 5 µm/pixel. As a rule of thumb, a minimum of 10 pixels per spot is recommendable. That means a resolution of 10 µm/pixel is enough for arrays with spot sizes larger than 100 µm.

22. Cy3 and especially Cy5 will start to degrade when exposed to high humidity, high temperature, and elevated levels of ozone. Be careful under these conditions: delay the experiment, or process the arrays in the early morning and as quickly as possible. To avoid vanishing dyes, DyeSaver from Genisphere may be helpful (http://www.genisphere.com/).

23. The photomultiplier power should always be set at 70 to 80%, otherwise the intensity of the background increases more than the intensity of the spots.

24. If the background is too high after washing and scanning, wash again according to the protocol in **ref. *24***.

25. ArrayTools works with a minimum of two slides. Increase the number of slides to ensure significant results (six slides are recommendable). For microarray design issues like dye swap, biological and technical variation, and other topics, *see* **ref. *25***.

26. The higher the number of permutations, the more significant the result. One thousand permutations are a good choice. More permutations might lead to a system crash. For more information about the permutations and the "Response Type: One class response," see the SAM manual.

27. The software adds three more worksheets to the workbook. There is one hidden sheet called SAM Plot data that should be left alone. The sheet named SAM Plot contains the plot that the user can interact with. The sheet named SAM Output is used for writing any output. Initially, a slider pops up along with a plot that allows one to change the Δ parameter and examine the effect on the false-positive rate. We always choose the number of genes with the lowest false-positive rate, in order to be on the safe side for further analysis. Positive significant genes are labeled in red on the SAM plot, negative significant genes are green. When you have settled on a value for Δ, click on the "List Significant Genes" button for a list of significant genes. The "List Delta Table" button lists the number of significant genes and the false-positive rate for a number of values of Δ. Please note that all output tables are sent to the worksheet named SAM Output, erasing values previously present in the worksheet. While the slider is present, all interaction with the workbook is only possible via the slider.

Acknowledgments

We would like to thank Patrick Farbrother for his contributions to the data analysis workflow and Adrian Schreyer for preparing **Fig. 1**. This work was supported by the Deutsche Forschungsgemeinschaft and by Köln Fortune.

References

1. Schulze, A. and Downward, J. (2001) Navigating gene expression using microarrays—a technology review. *Nat. Cell Biol.* **3,** E190–E195.

 2. Schena, M., Shalon, D., Davis, R. W., and Brown, P. O. (1995) Quantitative monitoring of gene expression patterns with a complementary DNA microarray. *Science* **270,** 467–470.
 3. Schena, M., Shalon, D., Heller, R., Chai, A., Brown, P. O., and Davis, R. W. (1996) Parallel human genome analysis: microarray-based expression monitoring of 1000 genes. *Proc. Natl. Acad. Sci. USA* **93,** 10,614–10,619.
 4. Shalon, D., Smith, S. J., and Brown, P. O. (1996) A DNA microarray system for analyzing complex DNA samples using two-color fluorescent probe hybridization. *Genome Res.* **6,** 639–645.
 5. Lipshutz, R. J., Morris, D., Chee, M., et al. (1995) Using oligonucleotide probe arrays to access genetic diversity. *Biotechniques* **19,** 442–447.
 6. Lipshutz, R. J., Fodor, S. P., Gingeras, T. R., and Lockhart, D. J. (1999) High density synthetic oligonucleotide arrays. *Nat. Genet.* **21,** 20–24.
 7. Duggan, D. J., Bittner, M., Chen, Y., Meltzer, P., and Trent, J. M. (1999) Expression profiling using cDNA microarrays. *Nat. Genet.* **21,** 10–14.
 8. Kwast, K. E., Lai, L. C., Menda, N., James, D. T., 3rd, Aref, S., and Burke, P. V. (2002) Genomic analyses of anaerobically induced genes in *Saccharomyces cerevisiae*: functional roles of Rox1 and other factors in mediating the anoxic response. *J. Bacteriol.* **184,** 250–265.
 9. DeRisi, J., Penland, L., Brown, P. O., et al. (1996) Use of a cDNA microarray to analyse gene expression patterns in human cancer. *Nat. Genet.* **14,** 457–460.
10. Heller, R. A., Schena, M., Chai, A., et al. (1997) Discovery and analysis of inflammatory disease-related genes using cDNA microarrays. *Proc. Natl. Acad. Sci. USA* **94,** 2150–2155.
11. Eisen, M. B., Spellman, P. T., Brown, P. O., and Botstein, D. (1998) Cluster analysis and display of genome-wide expression patterns. *Proc. Natl. Acad. Sci. USA* **95,** 14,863–14,868.
12. Van Driessche, N., Shaw, C., Katoh, M., et al. (2002) A transcriptional profile of multicellular development in *Dictyostelium discoideum*. *Development* **129,** 1543–1552.
13. Dyrskjot, L. (2003) Classification of bladder cancer by microarray expression profiling: towards a general clinical use of microarrays in cancer diagnostics. *Expert. Rev. Mol. Diagn.* **3,** 635–647.
14. Kibler, K., Nguyen, T. L., Svetz, J., et al. (2003) A novel developmental mechanism in *Dictyostelium* revealed in a screen for communication mutants. *Dev. Biol.* **259,** 193–208.
15. Girardot, F., Monnier, V., and Tricoire, H. (2004) Genome wide analysis of common and specific stress responses in adult *Drosophila melanogaster*. *BMC Genomics* **5,** 74.
16. Spellman, P. T., Sherlock, G., Zhang, M. Q., et al. (1998) Comprehensive identification of cell cycle-regulated genes of the yeast *Saccharomyces cerevisiae* by microarray hybridization. *Mol. Biol. Cell* **9,** 3273–3297.
17. Urushihara, H., Morio, T., Saito, T., et al. (2004) Analyses of cDNAs from growth and slug stages of *Dictyostelium discoideum*. *Nucleic Acids Res.* **32,** 1647–1653.

18. Eichinger, L., Pachebat, J. A., Glockner, G., et. al. (2005) The genome of the social amoeba *Dictyostelium discoideum. Nature* **435**, 43–57.

19. Dudoit, S., Yang, Y. H., and Bolstad, B. (2002) Using R for the analysis of DNA microarray data. *R News* **2**, 24–32.

20. Dudoit, S. and Yang, Y. H. (2003) Bioconductor R packages for exploratory analysis and normalization of cDNA microarray data, in *The Analysis of Gene Expression Data: Methods and Software* (Zeger, S. L., ed.). Springer, New York: pp. 455–468.

21. Bland, J. M. and Altman, D. G. (1986) Statistical methods for assessing agreement between two methods of clinical measurement. *Lancet* **1**, 307–310.

22. Smyth, G. K., Yang, Y. H., and Speed, T. (eds.) (2002) Statistical issues in cDNA Microarray Data Analysis, in *Functional Genomics* (Brownstein, M. J. and Khodursky, A., eds.). Humana Press, Totowa: pp. 111–136.

23. Tusher, V. G., Tibshirani, R., and Chu, G. (2001) Significance analysis of microarrays applied to the ionizing radiation response. *Proc. Natl. Acad. Sci. USA* **98**, 5116–5121.

24. Martinez, M. J., Aragon, A. D., Rodriguez, A. L., et al. (2003) Identification and removal of contaminating fluorescence from commercial and in-house printed DNA microarrays. *Nucleic Acids Res.* **31**, E18.

25. Yang, Y. H. and Speed, T. (2002) Design issues for cDNA microarray experiments. *Nat. Rev. Genet.* **3**, 579–588.

6

Proteomic Analysis of *Dictyostelium discoideum*

Udo Roth, Stefan Müller, and Franz-Georg Hanisch

Summary

The social amoeba *Dictyostelium discoideum* is already known as a model organism for a variety of cellular and molecular studies. Now that the genome sequencing project has been completed and different tools with which to overexpress or knock out genes are available, this species has also moved into the spotlight of functional genomics studies. Consequently, this genomic sequence information can now be exploited to realize *D. discoideum* proteomics projects. Here, we present validated protocols adapted for analysis of the *D. discoideum* proteome. The workflow described in this chapter comprises two-dimensional polyacrylamide gel electrophoresis for protein separation and peptide mass fingerprint (matrix-assisted laser desorption/ionization time-of-flight mass spectrometry) for protein identification.

Key Words: 2D-PAGE; in gel digestion; peptide mass fingerprint; proteomics; database; protein identification; MALDI-TOF mass spectrometry.

1. Introduction

Proteomics is among the several "-omics" under investigation nowadays, and is defined as the analysis of the entirety of all proteins present in a given biological system at a certain time point or state. In contrast with the genome, the proteome of an organism is highly dynamic, because the expression of many genes is highly dependent on the cellular or environmental context. Furthermore, the number of different protein species in a cell resulting from mRNA splicing or posttranslational modifications (PTM) exceeds the number of genes significantly. For this reason, and as a result of the fact that the mRNA level of a transcribed gene does not necessarily reflect its actual protein level, proteomics can deliver new insights into cellular regulation mechanisms or protein functions. In this chapter, we present a proteomics workflow adapted for analysis of the *Dictyostelium discoideum* proteome.

From: *Methods in Molecular Biology, vol. 346:* Dictyostelium discoideum *Protocols*
Edited by: L. Eichinger and F. Rivero © Humana Press Inc., Totowa, NJ

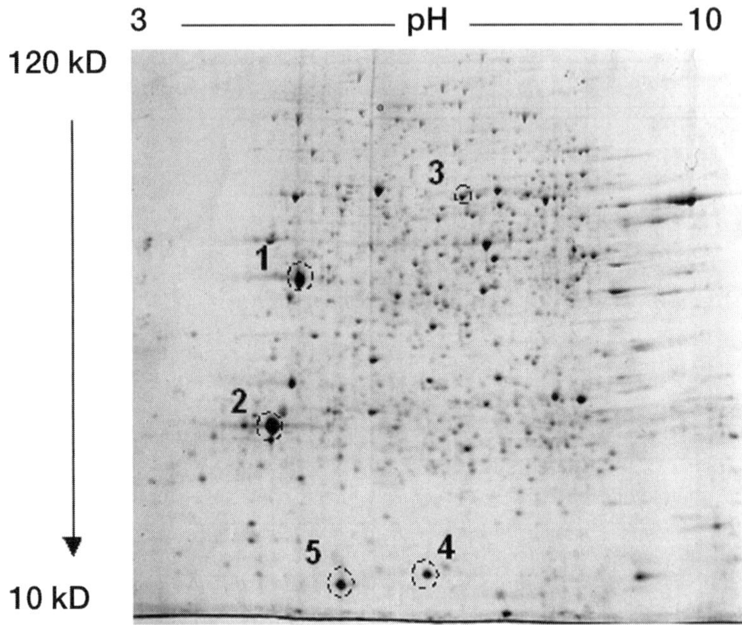

Fig. 1 Two-dimensional polyacrylamide gel electrophoresis of a *Dictyostelium discoideum* cell lysate. Two hundred fifty micrograms of protein was loaded on an immobilized pH gradient strip (24 cm, pH 3.0–10.0 nonlinear) via the rehydration method as described in the text. After electrophoresis, the gel was fixed and stained overnight according to the colloidal Coomassie (CBB G-250) protocol. The indicated spots were picked and identified as the following proteins: (**1**) act12; (**2**) cadA; (**3**) mitochondrial processing peptidase α-subunit; (**4**) SOD; (**5**) cofilin.

Despite the promising alternative of complementary technologies (e.g., multidimensional liquid chromatography coupled to mass spectrometry [MS], stable isotope labeling, protein or antibody arrays) that have been developed recently, two-dimensional polyacrylamide gel electrophoresis (2D-PAGE), which can be routinely applied for quantitative expression profiling of protein mixtures (*see* **Fig. 1**), is still the method of choice for proteomic analyses. The fast and sensitive identification of proteins from 2D spots is accomplished by matrix-assisted laser/desorption ionization (MALDI)-time-of-flight (TOF) MS and subsequent database searching (*see* **Fig. 2**).

2D-PAGE technology separates proteins according to their charge and molecular mass, thus allowing the separation and quantification of, typically, several thousand proteins in complex sample mixtures from which to create a protein map of a given sample. Furthermore, it can provide information about

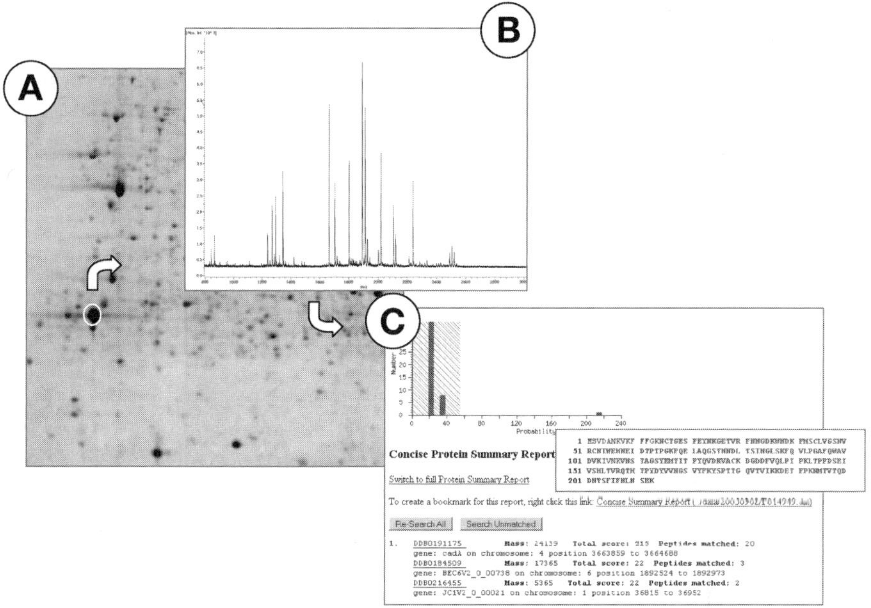

Fig. 2. General protein identification workflow. (**A**) Excision and trypsin digestion. (**B**) Extraction and matrix-assisted laser desorption/ionization time-of-flight mass spectrometry of tryptic peptides. (**C**) Processing of raw spectra, generation of peaklist and database search.

the occurrence of protein isoforms, which may result from PTM regulating different activity states of a protein. In parallel, 2D-PAGE on a micro-preparative level in combination with MS (peptide mass fingerprint, peptide fragment fingerprint) serves as a means for protein identification by permitting the analysis of single spots, which represent highly purified proteins in the gel.

Typically a 2D-PAGE experiment is comprised of the following steps: sample preparation, isoelectric focussing (IEF), sodium dodecyl sufate (SDS)-PAGE, gel staining, and image analysis.

Sample preparation is one, if not the most critical, point in the 2D-PAGE workflow. Ideally, a sample contains only proteins, but salts or other charged molecules, or contaminants such as nucleic acids, polysaccharides, and lipids, may be present and may interfere with protein gel electrophoresis. Although a sample preparation protocol suitable for all sample types would be highly desirable, so far a "one for all" method has not been developed yet. Generally, the proteins must be solubilized, fully denatured, and reduced in a 2D-compatible cell lysis buffer (*see* **Note 1**). The proteins are then separated in the first dimension by denaturing IEF. With this kind of equilibrium gel electrophoresis, proteins are separated in a pH gradient according to their isoelectric point. Originally, the pH gradient was established using mobile carrier ampholytes, but extensive effort had to be put into this system in order to obtain good resolution and reproducibility of the pH-gradients *(1,2)*. On the other hand, this technique still suffered from greater deviations within batch-to-batch or lab-to-lab comparisons of 2D protein maps. These limitations, in addition to the problematic focusing of basic proteins using carrier ampholytes, have been largely overcome by the introduction and commercialization of immobilized pH gradient (IPG) strips *(3)*. Additionally, a greater variety of pH gradients, covering wide (pH 3.0–11.0), medium (e.g., pH 4.0–7.0), and narrow (one pH unit) ranges, as well as different strip sizes (7–24 cm), are commercially available now.

In the second dimension, the proteins are separated according to their molecular weight by classical SDS-PAGE, for which a variety of chamber systems developed by several suppliers can be employed. Typically, proteomic experiments are carried out by running multiple gel replicates of a sample at the same time. For this reason, vertical chamber systems, which, for instance, in the case of the Ettan Dalt system (GE Healthcare, former Amersham Biosciences), allow one to run up to12 gels in parallel, have been developed. The best resolution in the second dimension is achieved by gradient gels (*see* **Note 2**).

The last step of 2D-PAGE is comprised of gel staining and computer-aided image analysis. To visualize the protein spots separated in the gel matrix, different stains with their different characteristic features (in terms of sensitivity, dynamic range, hardware required, MS compatibility and reproducibility,) can be employed (*see* **Note 3**).

For in-depth studies such as investigations of differentially regulated proteins, it is recommended that the gels are scanned and the resulting images are analyzed with the help of 2D image analysis software. Currently, several 2D image analysis software packages are available. These programs contain algorithms for automatic spot detection and quantification as well as spot matching across a number of gels analyzed in parallel (*see* **Note 4**).

The aim of this chapter is to give an overview of the typical proteomics workflow and to enable other scientists working with *D. discoideum* to produce high-quality 2D gels. It is also meant to provide helpful hints for choosing an appropriate 2D-PAGE setup and for avoiding pitfalls. The protocols are modified according to the 2D manual by GE Healthcare, which can be downloaded from the GE homepage. 2D-PAGE is based on commercially available IPG strips and on the use of a standard sample preparation procedure for lysates of *D. discoideum* cells in their vegetative, single-celled amoeboid phase. Moreover, the electrophoresis conditions described have been established for analytical gels using an overview gradient (pH 3.0–10.0, nonlinear) and a homogeneous 12.5% polyacrylamide gel for the second dimension. For preparative protein loads of up to 1 mg protein or narrow pH gradients, the focusing times must be increased significantly.

2. Materials

The hardware employed consists of IPGphor, Manifold, IPG strip holders, Ettan Dalt six (all GE Healthcare), and an Umax Powerlook III flatbed scanner. Reagents of best quality and deionized water ($dH_2O > 18$ MΩ) should be used for all buffers. IPG strips and electrophoresis machines may be purchased from GE Healthcare or Bio-Rad.

2.1. Sample Preparation and Determination of Protein Content

1. Lysis buffer: 7 *M* urea, 2 *M* thiourea, 4% CHAPS, 10 m*M* Tris base, 2% Pharmalytes or Immobilines (pH 3.0–10.0, respectively), and complete protease inhibitor cocktail (Roche). Immediately before use, add dithiothreitol (DTT) to a final concentration of 50 m*M*. The buffer should be aliquoted and stored at –20°C.
2. Determination of protein content: After completion of the sample preparation, measure the protein content with the help of the 2D-Quant kit (GE Healthcare) using ovalbumin or bovine serum albumin as standard protein (*see* **Note 5**).

2.2. Isoelectric Focusing

1. Rehydration solution: 8 *M* urea, 2% CHAPS, 0.002% bromophenol blue. Store the buffer in aliqots at –20°C and, immediately before use, add DTT (20 m*M*) and the appropriate immobilines (0.5%, v/v) depending on the pH gradient used (here pH 3.0–10.0, nonlinear).
2. Depending on the IEF method, prepare strip holders (rehydration method) or the Manifold equipment assembly.

2.3. SDS-PAGE

1. Acrylamide stock solution (30% T, 2.6% C): 30% acrylamide, 0.8% *N,N'*-methylene-bisacrylamide. Store at 4°C and take care to avoid contact with the unpolymerized solution, because acrylamide is a neurotoxin.
2. Gel buffer: 1.5 *M* Tris-HCl, pH 8.8.
3. SDS solution (10%).
4. 10% Ammonium persulfate solution (APS).
5. TEMED.
6. SDS-PAGE running buffer: 25 m*M* Tris base, 192 m*M* glycine, 0.1% SDS. No further pH adjustment is required.
7. Equilibration buffer: IPG strip equilibration buffer: 50 m*M* Tris-HCl (pH 8.8), 6 *M* urea, 30% (v/v) glycerol, 2% SDS, 0.01% bromophenol blue. Aliquot the buffer (10 mL) and store at –20°C. Thaw aliquots before use and add 1% DTT to make up equilibration solution A and 4% iodoacetamide for equilibration solution B.

8. Agarose sealing solution: 0.5% low-melting-point agarose, 0.002% bromophenol blue in SDS-PAGE running buffer. Add the solid compounds and dissolve the agarose by heating the buffer carefully by constant stirring with a heating stirrer. Do not allow the solution to boil over.

2.4. Gel Staining With Colloidal Coomassie

1. Fixing solution: 40% ethanol, 10% acetic acid.
2. Incubation solution: 17% ammonium sulfate, 20% methanol, 2% phosphoric acid.
3. Staining: Solid Coomassie Brilliant Blue G-250.
4. Washing solution: 20% methanol.
5. Storage solution: 5% acetic acid.

2.5. Spot Excision

1. Light box.
2. 1.5-mm punching tool (Techne AG , Burkhardtsdorf, Germany).
3. Optional: Automatic spot-picking device.

2.6. In-Gel Digestion

1. Clean 0.5-mL Eppendorf tubes.
2. Centrifugal evaporator.
3. Optional: thermomixer for 48×0.5 mL tubes.
4. Reagents: Water, acetonitrile, ammonium bicarbonate, iodine acetamide, trypsin, trifluoroacetic acid.

2.7. Mass Spectrometry

1. MALDI-TOF mass spectrometer including acquisition and spectra processing software.
2. Targets with hydrophobic coating and hydrophilic patches for sample application (600 µm AnchorChip™ targets, Bruker Daltonic, Bremen, Germany).
3. 2,5-dihydroxybenzoic acid (DHB), 1 mg/mL in 0.1% trifluoroacetic acid–acetonitrile 2:1 (prepare fresh).
4. Peptide mix covering the mass range from 1000 to 3500 Da. Prepare 1 pmol/µL stock in 0.1% trifluoroacetic acid.

2.8. Database Searching

1. Web server and client computer with network access.
2. Peptide mass fingerprint search engine (e.g., MASCOT, Matrix Science) installed on local Web server.
3. Dicty primary protein sequence database, downloadable from, http://dictybase.org/db/cgi-bin/dictyBase/download/blast_databases.pl or latest releases of NCBInr, downloadable from ftp://ftp.ncbi.nih.gov/blast/db/FASTA/nr.gz.

3. Methods

3.1. Sample Preparation

It is important to work as quickly as possible during all stages of the sample preparation. Keep the sample on ice during the preparation procedure.

1. Transfer *D. discoideum* cells (AX2) grown in nutrient medium into 50-mL plastic tubes, carefully spin the cells down (5 min at 180g), and resuspend the cells in 200 mL of Soerensen phosphate buffer.
2. Centrifuge again, repeat the washing step, and aliquot the cells in aliquots of 1×10^7 cells.
3. Pellet the cells as described previously, remove excess Soerensen medium, and add 100 µL of lysis buffer.
4. Pipet the cell suspension vigorously up and down to ensure complete cell lysis and shearing of DNA, which may interfere with subsequent IEF (*see* **Note 6**).
5. After cell lysis, incubate the lysate for 10 min on ice and subsequently spin down at 20,000g at 4°C in a benchtop centrifuge for 10 min to remove cell debris. Transfer the supernatant in a new 1.5-mL Eppendorf tube and determine protein content, which is usually approx 600 µg protein per 1×10^7cells.

3.2. Isoelectric Focusing (First Dimension)

3.2.1. Rehydration Method

1. Pipet a volume containing 250 µg protein in a new 1.5-mL tube and dilute it with rehydration buffer to a final volume of 450 µL (*see* **Note 7**).
2. Spread the mixture evenly in a strip holder and cover it with a 24-cm IPG-strip (pH 3.0–10.0, nonlinear; gel-side down).
3. Put 1 mL of cover solution (mineral oil) on each strip and close the strip holder with the lid.
4. Subsequently, the strip holders are placed on the IPGphor and the following focusing program is applied:

   > 12 h at 50 V (Rehydration)
   > 1 h at 200 V
   > 1 h at 500 V
   > 1 h at 1000 V
   > 1.5 h gradient at 8000 V
   > Final focusing step: 8000 V for 64000 Vh

3.2.2. Cup-Loading Method

1. Pipet 450 µL of rehydration buffer *without the sample* into an empty strip holder and place an uncovered IPG strip (gel-side down) on the liquid.
2. Add 1 mL of cover fluid and let the strip rehydrate overnight. Put the "Manifold" tray on the IPGphor and transfer the strip (gel-side up) into an empty slot of the tray.

3. Assemble the electrodes and attach the sample cup close to the anode.
4. Add 1 μL of a diluted bromophenol blue solution to the protein aliquot to track the correct loading of the sample and carefully pipet the sample under the cover fluid into the sample cup.
5. Apply the following focusing program:

   1 h at 120 V
   1 h at 300 V
   6 h gradient 1000 V
   3 h gradient 8000 V
   Final focusing step: 8000 V for 32,000 Vh

6. After the run is completed, check the position of the bromophenol blue band (a yellow-colored band should be visible at the anode), remove the strips from the strip holders or the "Manifold" tray, and rinse them briefly in dH_2O to remove the mineral oil from the strip surface (*see* **Note 8**).
7. Strips can now be equilibrated directly for the second dimension or stored at −80°C for a couple of days or even weeks without loss of quality.

3.3. SDS-PAGE Gel Casting

1. Cast an appropriate number of large-format acrylamide gels (12.5%) and let them polymerize over-night.
2. The following is a recipe that will allow pouring of six gels in parallel:

   Acrylamide (30 : 0.8%) (*see* **Note 9**): 175 mL
   dH_2O: 134 mL
   Tris-HCl (1.5 *M*; pH 8.8): 105 mL
   SDS (10 %) : 4.2 mL
   TEMED : 139 μL

3. Mix the solution by stirring, but avoid introduction of air bubbles into the solution, which can affect polymerization.
4. Start the polymerization reaction by adding 2.1 mL of APS solution (10%).
5. Pour the gels immediately and overlay the acrylamide solution with isopropanol or water-saturated butanol to ensure a flat gel surface.
6. Cover the gel cassettes with moist paper towels and a layer of plastic wrap to prevent evaporation.
7. The next day, harvest the gel cassettes, rinse them with dH_2O, and fill the remaining space between the gel surface and the top of the cassettes with SDS running buffer.

3.4. IPG Strip Equilibration and SDS-PAGE (Second Dimension)

Before loading the focused proteins on the second dimension (SDS gel), the strips must be equilibrated in an appropriate SDS buffer. Simultaneously, the equilibration step can be exploited to provide a further reduction and the alkylation step to inactivate the redox-active sulfhydryl groups.

1. To equilibrate the strips, put the strips with the gel side facing up into a rehydration tray and add 3 mL of buffer A to each strip.
2. Incubate the strips on a shaker (40 rpm) for 15 min.
3. Replace buffer A with buffer B and repeat the incubation for another 15 min.
4. After the equilibration is finished, drain excess buffer solution from the strips by placing them for a few seconds with their edges on a piece of filter paper, and subsequently apply the strips to the surface of the SDS-PAGE with the help of a spatula. Take care that the strips are in tight contact with the gel surface and make sure that no air bubbles are trapped between the strip and the gel surface.
5. Seal the strips with 1 mL of liquid low-melting-point agarose. After the agarose gel is set, insert the cassettes into the electrophoresis chamber.
6. Finally, pour the running buffer into the upper chamber and start the electrophoresis.
7. For SDS-PAGE, we recommend the following running conditions:

   Phase I: Start 3 W per gel for 45 min to ensure a proper and smooth transfer of the proteins from the strip into the SDS gel.
   Phase II: 10–12 W per gel until running front has reached the bottom of the gel.

8. After the run is complete, disconnect the power supply, remove the gel cassettes from the chamber, and incubate the gels in fixing solution on a shaker (32 rpm) overnight.

3.5. Gel Staining With Colloidal Coomassie

Staining with colloidal Coomassie is a robust, convenient technique possessing a reasonable dynamic range (*see* **Note 3**). Compared with the usual alcoholic Coomassie staining protocol, the colloidal variant is more sensitive because it produces an almost clear background, so no excessive destaining steps, which may lead to disappearance of faint protein spots, are necessary.

1. To stain the gels, wash them in dH_2O for 30 min and subsequently place them in 500 mL of incubation solution per gel.
2. After an equilibration period of 60 min, add 330 mg of solid Coomassie G-250 to the solution and stain the gels with gentle shaking overnight. Typically, the Coomassie particles do not dissolve completely, thus some of them will stick on the gel surface during the staining process.
3. After staining is completed, wash the gels in the washing solution for approx 1 min to remove the dye particles.
4. Transfer the gels into storage solution and scan them on a suitable flatbed scanner at 300 dpi using a calibrated scanning software that supports 16-bit greyscale images (*see* **Note 10**).

3.6. Spot Picking

Wear a lab coat and gloves, do not bend over the gel, and spare no effort to avoid contamination with keratins. It might be useful to work under an acrylic screen to protect the gel from dust. Keep in mind that the concentration of protein is at least as important as the absolute amount, and try to pick the core of the spots to keep the protein:gel ratio as high as possible. We found that a 1.5-mm punching tool is a good choice for most gel dimensions. Smaller tools (e.g., 1 mm) might be useful for mini 2D gels (7-cm strips in the first dimension).

1. Rinse gels with several changes of water for at least 2 h (*see* **Note 11**).
2. Prepare a printout of the gel image and mark spots of interest.
3. Place the gel on a clean glass plate and put the assembly on a transilluminator (*see* **Note 12**).
4. Punch out the spots and store them in 0.5-mL Eppendorf tubes.
5. Remove remaining liquid and keep the tubes at 4°C for up to 1 wk or at –20°C for prolonged storage.

3.7. In-Gel Digestion

The given protocol is designed for 1.5 mm × 1 mm spots. Increase volumes to process larger pieces of gel (e.g., 1D bands). Take care to avoid contamination with keratin (*see* **Subheading 3.6.**).

1. Spin down samples in a benchtop centrifuge and remove residual fluid.
2. Add 25 µL of water–acetonitrile (1:1) and place tubes in a thermomixer for 15 min at 30°C with shaking at 800 rpm.
3. Remove supernatant and dry spots in a centrifugal evaporator for 15–30 min.
4. Add 25 µL of 10 m*M* DTT in 25 m*M* ammonium bicarbonate and incubate for 45 min at 45°C with shaking at 800 rpm.
5. Add 25 µL of 50 m*M* iodine acetamide in 25 m*M* ammonium bicarbonate and incubate at room temperature (22°C) in the dark for 30 min.
6. Spin down the tubes and remove residual fluid. Add 25 µL of water–acetonitrile (1:1) and place tubes in thermomixer for 15 min at 30°C with shaking at 800 rpm. Repeat once.
7. Spin down and remove residual fluid. Add 25 µL acetonitrile and incubate for 5 min at room temperature. Add 25 µL of 10 m*M* ammonium bicarbonate and incubate for another 15 min.
8. Spin down, remove residual fluid, and dry spots in a vacuum centrifuge for 30 min.
9. Place tubes on ice and ad 10 µL of ice-cold 10 ng/mL trypsin in 10 m*M* ammonium bicarbonate. Incubate on ice for 30 min.
10. Replace excessive trypsin solution by 5 µL of 10 m*M* ammonium bicarbonate and incubate at 30°C for 4– 16 h in an oven.
11. Add 5–10 µL of 1% trifluoroacetic acid. Place tubes in a thermomixer and extract peptides for 30 min at 30°C with shaking at 800 rpm (*see* **Note 13**).
12. Prepare MALDI target immediately (*see* **Subheading 3.8.**) and store remaining sample at –20°C.

3.8. MALDI-TOF Mass Spectrometry

The most popular matrices for MALDI-TOF measurements of peptides and protein digests are α-cyano-4-hydroxycinnamic acid (HCCA) and 2,5-dihydroxybenzoic acid (DHB). HCCA is particularly favored in many labs and can be used in thin-layer or dried-droplet preparations on conventional stainless steel or AnchorChip targets. The sample preparation protocol given as follows describes the use of DHB with a 600-μm AnchorChip and a Reflex IV MALDI-TOF mass spectrometer as it is used in in the authors' laboratory. This preparation method has turned out to be robust and has given excellent and consistent results over several years (*see* **Note 14**).

1. Clean AnchorChip by sonication in 50% methanol for 15 min. Rinse with water and sonicate for another 15 min in water. Dry target in a stream of air and store in a clean box.
2. Deposit 2 μL of DHB solution on a 600-μm spot. Immediately add 1 μL of peptide extract, mix briefly, and let dry.
3. Dissolve peptide standard mixture in DHB solution at a concentration of 50 fmol/μL. Deposit 2 μL on calibrant positions and let dry.
4. Acquire peptide standard spectrum and calibrate the instrument.
5. Acquire sample spectra on the adjacent positions.
6. Process spectra and store peak lists. Check the spectra for trypsin autolysis peaks. Use them for an internal recalibration when the mass accuracy is not satisfactory (*see* **Note 15**).

3.9. Database Searching

The latest release of the NCBInr protein database contains 14,957 sequences for *D. discoideum*. An alternative database is the "primary protein" database, which is available from the dictyBase website and contains 14,228 sequences. The entries in both databases are identical to a large extent, and the majority are hypothetical protein sequences that were generated by gene prediction from genomic data. From the authors' point of view, it is advisable to use the NCBInr because it is professionally cross-referenced to other databases and allows the quick retrieval of additional information (*see* **Note 16**).

We use an in-house license of MASCOT 1.9 for peptide mass fingerprint searches (*see* **Note 17**). XML-formatted peak lists are generated by Flexanalysis 3.0, and batch searches are submitted by client programs such as MASCOT demon or Biotools 3.0. For standard searches in NCBInr or Dicty primary protein, fixed modifications are set to *carbamidomethyl*, optional modifications to *Methionine oxidation*, and missed cleavages to *1*. Mass tolerance is set to *150 ppm* with external calibration or *60 ppm* with internal calibration.

4. Notes

1. The sample preparation should be as simple and prepared as quickly as possible to increase reproducibility. Additionally, protein modifications during sample preparation must be minimized, because they might result in artificial spots on the gel.
2. Only commercially available IPG strips guarantee high reproducibility of the gradient; self-cast gradients may differ from batch to batch. Another way to increase the resolution of particular molecular weight ranges in the gel is to vary acrylamide concentrations or the use of different running buffer system (Tris/glycine *[4]*, Tris/Tricine *[5]*).
3. The "workhorse" among the dyes is still Coomassie Blue-based protein staining, because the staining protocols are robust and provide a good reproducibility and MS compatibility. The major drawback of this staining technique is the relatively low sensitivity, with a detection threshold of approx 50 ng per spot for the most sensitive variety (colloidal Coomassie *[6]*). More sensitive are different silver-staining protocols (100 pg/spot), but these typically possess a low dynamic range and lack reproducibility. Furthermore, the most sensitive silver staining protocols render the protein incompatible with MS analysis. Commercially available fluorescence stains, like Sypro Ruby (Molecular Probes), Deep Purple (GE Healthcare), or a stain based on a ruthenium complex developed by Rabilloud *(7)* are excellent with regard to the features mentioned previously; however, special, expensive scanning hardware (laser scanner, e.g., Typhoon [GE Healthcare] or FLA-5000 [Fuji]) is required to visualize the signals. Another possibility, introduced as the so-called differential gel electrophoresis (DIGE) technique, is based on the prelabeling of proteins with fluorescing cyanine dyes on lysine or cysteine residues, but because of the complexity of the methodology, it is not be discussed in this chapter.
4. Different commercial software programs, such as Proteomweaver (Definiens), Delta 2D (Decodon), Image Master 2D (Phoretix), or PD Quest (Bio-Rad), with slightly different spot-matching strategies are available. Although they all work quite reliably, time-consuming manual intervention is often required in order to correct erroneous results.
5. Alternatively, other detergent-compatible protein quantification assays may be used; however, ampholytes and reducing agents may interfere with these.
6. If higher cell densities are used, e.g., for higher protein loads of preparative gels, an additional sonication step (5×10 s) may be applied to increase fragmentation of DNA.
7. The protein load depends on the pH gradient, strip length, and staining technique used. Generally, longer and more narrow gradients require a higher protein load. The same holds true for stains with lower sensitivity, such as Coomassie stains. For example, for an analytical gel using a pH 3.0–10.0 gradient, a protein load of 250 µg should be applied for Coomassie staining, whereas 125 µg is sufficient if a fluorescence stain is applied.
8. A diffuse bromophenol blue band, as well as extensively swollen areas at both ends of the strips together with precipitates, point to excess salt in the sample,

which interferes with IEF. Most often, this effect can be observed immediately after starting the run using a current exceeding 15–20 µA. In these cases, the salt concentration is sometimes too high to reach the voltage of the final focusing step and to yield sharply focused spots over the whole pH range. This problem can be overcome by desalting the sample (precipitation, ultrafiltration) prior to IEF. However, it should be kept in mind that additional sample preparation steps may lead to artificial changes in the protein composition or protein losses of the original sample.

9. On a large gel format (24 × 20 cm), normal acrylamide without any kind of plastic backing tends to tear under mechanical strain. For this reason, we recommend the use of commercially available precast gels on plastic backings. On the other hand, these backings typically cause problems in the reading out of fluorescence signals, because the laser beam cannot pass the plastic without dispersion. Self-cast gels can be strengthened by using Rhinohide (Molecular Probes), which is an additive that makes the gels more rigid. Another possibility is the use of Duracryl (Proteomics Solutions) instead of acrylamide, which produces very stable slab gel matrices. Besides the high price, another major drawback of Duracryl is an electrophoresis effect resulting in vertical streaking of high-molecular-weight proteins (so called "noses") and pronounced swelling during the preparation process for automatic spot picking.

10. Although proteins are fixed in the gels, the intensity of the staining tends to bleach within a few days. To avoid loss of information in terms of faint protein spots, we recommend scanning the gels after a short equilibration time in the storage solution. The scanner surface should be tightened to avoid dripping of gel liquid into the apparatus. Furthermore, a transmissive scanning system is required in order to create reasonable gel images. 2D image analysis software programs prefer 16-bit over 8-bit scans because of the much higher number of gray values (65536 instead of 256), which allows a better quantification. In terms of resolution, 300 dpi are sufficient.

11. Note that the colloidal Coomassie stain will fade when the gels are stored in pure water for a long time.

12. An ultraviolet transilluminator can be used to excise spots from fluorescently stained gels. Wear appropriate skin and eye protection! Work quickly because the stain fades when exposed to high-energy ultraviolet light. Furthermore, the gel will warm up and dry rapidly. In practice, manual spot picking from fluorescently stained gels is limited to a few spots. For the excision of large numbers of spots from complex 2D patterns, it is recommended that an automatic spot-picking device be used.

13. Up to 50% acetonitrile in the extraction buffer is compatible with subsequent dried-droplet preparation using DHB or HCCA and may faciliate the extraction of hydrophobic peptides that will not dissolve efficiently in water alone. However, if the remainder of the extract is used for liquid chromatography/MS analysis or cleanup by pipet tip chromatography, the acetonitrile must be removed by evaporation. Evaporation is always a critical step and might lead to a loss of peptides when the sample is dried completely.

14. In terms of accuracy and resolution, a DHB preparation is inferior to thin-layer or dried-droplet preparations using HCCA. However, we have compared both matrices for peptide mass fingerprinting and obtained a better success rate in terms of positive protein identifications with DHB. The reason for this is that more protein-related peaks appear in the spectra acquired from DHB preparations. Another advantage of the DHB preparation is that it is much more convenient than HCCA thin-layer or dried-droplet preparations on AnchorChips, which require removal of the sample or additional wash steps. Furthermore, we found that a sample cleanup by pipet tip chromatography is not necessary and does not usually improve the results.

15. It pays to keep a record of signals that appear frequently and are not related to the sample. Use them to create background lists and subtract them from the sample peak list. Keep in mind that every peak that does not fit to the protein hit decreases the score in peptide mass fingerprint searches.

16. A copy of NCBInr is automatically added when MASCOT is installed the first time. However, this version is probably not up-to-date and must be replaced by the latest version, which can be downloaded from ftp://ftp.ncbi.nih.gov/blast/db/FASTA/nr.gz. A script for automatic downloads (db_update.pl) is available in MASCOT. The *Dictyostelium* primary protein database must be installed manually. Download and decompress the *"dicty_primary_protein"* FASTA protein sequence database. Create a new folder for the *Dictyostelium* data base in the MASCOT sequence directory. Create three subfolders named *"new," "current,"* and *"old."* Copy the database file into *"new,"* rename the file extension to *".fasta,"* and move the file to *"current."* Open the *"database maintenance"* page from the MASCOT main page, choose *"new database,"* and enter the full path to the fasta file and a name for the database. Deactivate *"local ref file"* and set *"taxonomy source"* to *"none."* Appropriate parse rules to extract the accession and description string from the fasta file must be added. ">\(.*\)/*Protein*" works for the accession string and ">.*/*Protein*/ \(.*\)" is suitable to extract the description.

17. Meanwhile, there is a good choice of software tools that can be used for peptide mass fingerprint searches. Free versions of MASCOT, ProFound, MsFit and Aldente are accessible on public web servers:

    MASCOT: http://www.matrixscience.com/search_form_select.html
    ProFound: http://prowl.rockefeller.edu/profound_bin/WebProFound.exe
    MsFit: http://prospector.ucsf.edu/ucsfhtml4.0/msfit.htm
    Aldente: http://au.expasy.org/tools/aldente/

    Note that these free versions do not offer full functionality. Batch searches, the use of local databases, editing of potential modifications, or the interaction with the MS processing software require the purchase and installation of an in-house license on a local web server.

References

1. O'Farrell, P. H. (1975) High resolution two-dimensional electrophoresis of proteins. *J. Biol. Chem.* **250,** 4007–4021.

2. Klose, J. (1975) Protein mapping by combined isoelectric focusing and electrophoresis of mouse tissues. *Humangenetik* **26,** 231–243.
3. Görg, A., Weiss, W., and Dunn, M. J. (2004) Current two-dimensional electrophoresis technology for proteomics. *Proteomics* **4,** 3665–3685.
4. Laemmli, U. K. (1970) Cleavage of structural proteins during the assembly of the head of bacteriophage T4. *Nature* **227,** 680–685.
5. Schägger, H. and von Jagow, G. (1987) Tricine-sodium dodecyl sulfate-polyacrylamide gel electrophoresis for the separation of proteins in the range from 1 to 100 kDa. *Anal. Biochem.* **166,** 368–379.
6. Neuhoff, V., Arold, N., Taube, D., and Ehrhardt, W. (1985) Improved staining of proteins in polyacrylamide gels including isoelectric focusing gels with clear background at nanogram sensitivity using Coomassie Brilliant Blue G-250 and R-250. *Electrophoresis* **9,** 255–262.
7. Rabilloud, T., Strub, J. M., Luche, S., van Dorsselaer, A., and Lunardi, J.(2001) A comparison between Sypro Ruby and ruthenium II tris (bathophenanthroline disulfonate) as fluorescent stains for protein detection in gels. *Proteomics* **1,** 699–704.

II

BASIC METHODS AND MOLECULAR GENETIC TECHNIQUES

7

Cultivation, Spore Production, and Mating

Hideko Urushihara

Summary

Dictyostelium discoideum proliferates as solitary amoebae, constitutes multicellular structures called fruiting bodies, and mates to form macrocysts depending on environmental conditions. All of these processes can be easily induced in the laboratory. The amoebae are normally cultured with food bacteria, but the strains with mutations in *axe* loci can proliferate in nutrient media without bacteria. The strains can be stored either as spores or amoebae. Synchronous development of fruiting bodies is initiated by depleting the culture media or food bacteria. Synchronous development of macrocysts is achieved by mixing the cells of heterothallic strains separately cultured in darkness to induce the sexual maturation.

Key Words: *Dictyostelium discoideum*; 2-member culture; axenic culture; frutingbody formation; macrocyst formation; strain storage.

1. Introduction

Dictyostelium discoideum is a micro-organism in the soil known for its unique life cycles (*1*). It normally feeds on bacteria as solitary amoebae. On starvation, however, about a hundred thousand amoebae gather by chemotaxis to cAMP to form a multicellular slug. The slug migrates toward light to reach the surface, and finally culminates as a fruiting body composed of a spore mass and a stalk to lift it up. This process of fruiting body formation contains many important aspects of cellular and developmental biology such as signal transduction, cell locomotion and interaction, cell differentiation, and pattern formation. These issues will be dealt in detail in the later chapters of this book. *D. discoideum* also undergoes sexual development: Under dark and submerged conditions, the amoeboid cells become sexually mature and fuse with appropriate mating-type cells and become zygotes. The zygotes then secret cAMP to gather the surrounding cells, engulf the cells as nutrients, and develop into

From: *Methods in Molecular Biology, vol. 346:* Dictyostelium discoideum *Protocols*
Edited by: L. Eichinger and F. Rivero © Humana Press Inc., Totowa, NJ

dormant structures called macrocysts *(2–4)*. This process is especially interesting in terms of the mechanisms of cell recognition and membrane fusion.

One of the great advantages of *D. discoideum* as a model organism is its easiness of handling: No special equipments or trainings are necessary. All of the previously mentioned processes of fruiting body and macrocyst formation can be easily induced in the laboratory. In addition to the two-member culture, with the food bacteria reflecting the field conditions, cultivation without bacteria in axenic media, which is suitable for defined biochemical or molecular biological experiments, is possible for some strains. The purpose of this chapter is to describe the basic laboratory protocols for cultivation and maintenance, storage, and sexual and asexual development of *Dictyostelium*.

2. Materials

All solutions, culture wares, and instruments that touch the *Dictyostelium* cells should be sterilized before use unless sterility is absolutely unnecessary, as in the case of sampling for sodium dodecyl sulfate (SDS)-polyacrylamide gel electrophoresis (PAGE). In order to reflect the conditions of soil, ionic strength of the buffers or salines is low (10–20 mM). Many natural isolates, laboratory strains, and mutants, including gene knockouts, are available from the *Dictyostelium* Stock Center (http://dictybase.org/StockCenter/StockCenter.html) (*see* Chapter 4).

2.1. Culture and Maintenance of the Strains

1. SM-medium agar *(5)*: 10.0 g glucose, 10.0 g proteose peptone, 10.0 g yeast—extract, 1.9 g KH_2PO_4, 1.0 g K_2HPO_4, 0.5 g $MgSO_4$, 15.0 g agar, deionized water (dH_2O) to 1 L. Autoclave and pour into culture dish until about one-half full. Store the agar plates in a cold room, sealed in a container to prevent drying.
2. A-medium agar (modified SM agar): 5.0 g glucose, 10.0 g proteose peptone, 0.5 g yeast extract, 0.9 g KH_2PO_4, 0.3 g Na_2HPO_4, 15.0 g agar, dH_2O to 1 L. Autoclave and pour to dish as in item 1 (*see* **Note 1**).
   SM- and A-media can be prepared without agar for liquid culture of bacteria.
3. Food bacteria: *Klebsiella aerogenes* and *Echerichia coli* (strain *B/r*) are the most commonly used strains. They can be maintained either in normal bacterial culture media such as Luria broth (LB) or in SM- or A-medium (described previously) with or without agar. Most of the *Dictyostelium* strains show preferences for food bacteria, but they may gradually adapt to unfavorable ones.
4. HL5 (a liquid medium for axenic culture) *(6)*: 14.0 g glucose, 7.0 g yeast extract, 14.0 g proteose peptone, 0.5 g KH_2PO_4, 0.5g Na_2HPO_4, dH_2O to 1 L. Autoclave and add streptomycin sulfate to 50 μg/mL after cooling (*see* **Note 2**).

2.2. Storage of Strains

1. Spore storage solution: 10% (W/V) nonfat dried milk.
2. Amoeba storage solution: 20% dimethylsulfoxide (DMSO) (cell-culture grade) in HL5.

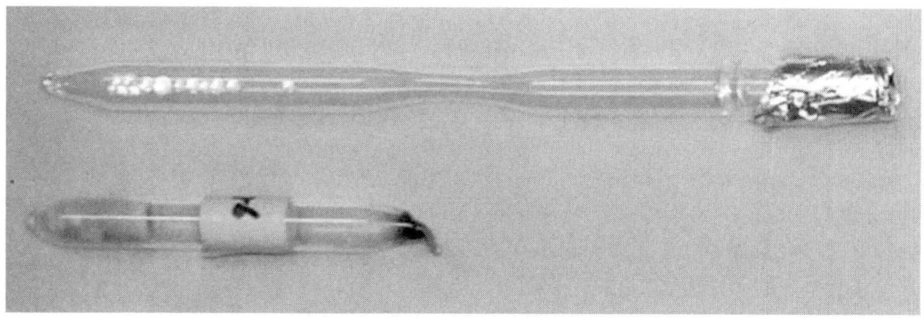

Fig. 1. A spore-storage ampoule (up) and a stock of freeze-dried spores (down).

3. Silica gel: white fine granules. We use 28–200 mesh from Wako Pure Chemical Industries (Tokyo, Japan).
4. Glass ampoules: these are commercially available, but we make our own for fun. Cut and flame-seal the tip of a Pasteur pipet and elongate the middle part to create a narrow portion. Insert five to six glass beads (ϕ = 2–3 mm) to prevent boiling under vacuum and to enhance spore mixing. Cover the top with aluminum foil and sterilize in an oven (*see* **Fig. 1**).

2.3. Asexual Development

1. Bonner's salt salution (BSS) *(7)*: 0.6 g NaCl, 0.75 g KCl, 0.3 g $CaCl_2$, dH_2O to 1 L.
2. KK2 buffer: 10 mM KH_2PO_4 and 10 mM K_2HPO_4, mix 1:1, pH 6.4.
3. Plain agar: 15.0 g agar, dH_2O to 1 L. Autoclave and pour into culture dish as in **Subheading 2.1., step 1**.
4. Membrane filter (0.45 µm) or filter paper: boil for 30 min or soak in ethanol overnight. Rinse several times with tap water before use. Filter properties such as black or white, with or without grids are optional, depending on experimental purposes. Normally, products from any company work well (*see* **Note 3**).

2.4. Sexual Development

1. There are heterothallic (*mat A1*, *mat A2*, and *mat A3*) and homothallic strains in *D. discoideum* *(8,9)*. The representative strains for each are listed in Table 1.
2. LP agar: 10.0 g lactose, 10.0 g proteose peptone, 15.0 g agar, dH_2O to 1 L.
3. Bactomilk: centrifuge a fresh overnight culture of *K. aerogenes* at 3000*g* for 10 min. Discard supernatant, resuspend the pellet in a small amount of BSS, and then add BSS to 15% of the original culture volume. Store at 4°C.

3. Methods

3.1. Culture and Maintenance of the Strains

Because *Dictyostelium* favors lower temperatures, temperature control of the incubator or the room itself is important. Care should be taken not to expose

Table 1
Mating Types and Representative Strains in *Dictyostelium discoideum*

Mating type	Strain
Heterothallic	
mat A1 (*mat A*)	NC4 (and derivatives: AX2, AX3, etc.), WS472, WS583
mat A2 (*mat a*)	V12 (and derivatives: V12M2, HM1, etc.), WS7, WS5656
mat A3 (*bisexual*)	WS2162, WS112B
Homothallic	AC4, ZA3A

Mating types in the parentheses are synonyms. The *mat A3* strains were originally described as bisexual because they formed macrocysts with a strain of either mating type (*mat A* or *mat a*) but not by themselves *(9)*. However, because *mat A* and *mat a* strains are also "bisexual" by this criterion, it seems more appropriate to employ the third mating type rather than to introduce the concept of bisexuality. Homothallic strains can form macrocysts in a clonal population.

the amoebae to temperatures higher than 25°C. The temperature for cultivation and incubation is 22°C throughout, unless otherwise described. Sterile handling is required, as is usual for many other microbes. Two standard culture protocols, two-member culture on agar surface with food bacteria and shaking culture in an axenic medium, are described here. The former is much easier for maintenance because minor contamination of bacteria or fungi does not spoil the entire culture. Moreover, the cells do not die from starvation, but gather and develop to fruiting bodies containing dormant spores, except for the non-spore-forming mutants. The axenic culture is suitable for analytical studies, but it is possible only for the special strains with mutations at the *axe* loci *(10)*. Strains AX2 *(11)* and AX3 *(12)* and their derivatives belong to this category.

3.1.1. Culture With Bacteria on Agar Surface

1. Prepare a fresh culture of bacteria. This can be achieved by inoculating a single colony of bacterium in the SM- or A-medium and culturing overnight at 30–37°C.
2. On a clean bench, pick up one to five spore balls, located at the top of fruiting bodies, using a platinum transfer loop that has been flame-sterilized and dipped into sterile water and which therefore carries a liquid film. Suspend the spore balls in a small volume (e.g., 1 mL) of bacterial culture.
3. Vortex to disperse the spores.
4. Inoculate an aliquot of this suspension on an SM agar plate (0.1–0.2 mL to a 9-cm plate) and spread using a sterile spreader.
5. Put the plate into an incubator. Sealing or inverting the plate is optional.
6. The amoebae germinated from the spores eat up the bacteria in 40–48 h, after which cell aggregation and morphogenesis start. Fruiting bodies are formed after additional 24 h or so (*see* **Note 4**). Spores remain viable for 2–4 wk under the

low-humidity conditions. Restart the culture within this period. Water absorption reduces the recovery of viable spores (*see* **Note 5**).

3.1.2. Culture in Axenic Media

1. On a clean bench, pick up several spore balls as described in **Subheading 3.1.1.**, **step 2** and suspend them in a small volume (2–3 mL) of HL5. Do not to touch the agar surface where the bacteria may remain.
2. Incubate for 2–3 d or until the culture medium becomes turbid.
3. Expand the culture in a new Erlenmeyer flask and incubate on a gyratory shaker (120–140 rpm) or a reciprocal shaker (120 strokes/min). The volume of culture medium should be one-tenth to one-quarter of the flask nominal volume.
4. Monitoring the cell density is recommendable for maintaining a good cell condition. A Fuchs-Rosenthal type of hemocytometer is convenient for this purpose. In our hands, KAX3 cells divide approximately once in every 8 h at densities between 1×10^5 cells/mL and 1×10^7 cells/mL.
5. Dilute the culture to 1/100 in fresh HL5 for the routine maintenance. Continuation of subculturing over 1 mo is not recommendable. Subculturing over 3 mo should be avoided.

3.1.3. Cloning of Cells

1. Plaque cloning: Prepare a suspension of spores or amoebae in BSS or bacterial culture at low densities (e.g., 500 cells/mL) by serial dilution. Spread approx 100 cells with bacteria on the nutrient agar and place in an incubator. Plaques appearing in the bacterial lawn on the third or fourth day of culture represent the clonal populations (*see* **Note 6**).
2. Limited dilution: Dilute the amoebae of an axenic strain in HL5 to 5 cells/mL and dispense 0.1 mL of this suspension to each well of a 96-well titer plate. Culture for a week or until colonies are visible in about half of the wells. Use the cells in single-colony wells as clonal populations.

3.2. Storage of Strains

Because the growth-phase cells of *D. discoideum* are haploid and proliferate with a short doubling time, genetic variation quickly accumulates, resulting in heterogeneity of the population. Strain maintenance as fruiting bodies is recommendable again, because many mutations can be effectively eliminated through the process of spore formation. To maintain mutant strains without spore-forming abilities, culture the cells at lower temperatures so that cell proliferation in a given period is kept minimal. In either case, a long-term continuation of the culture should be avoided.

3.2.1. Storage of Spores in Silica Gel

1. Collect the spore balls from 1- to 2-d-old fruiting bodies in a tube containing the cold spore storage solution (1 mL for a 9-cm plate).

2. Vortex to disperse the spores and cool on ice.
3. Dispense 0.1 mL of the suspension into a screw-cap tube containing 2–3 g sterilized fine silica gel; cap the bottle and seal with parafilm.
4. Vortex for even distribution of spores and store in a desiccator. Normally, this stock can be revived within a year or two.
5. When reconstituted, shake off a portion of silica granule onto a nutrient agar plate and culture with bacteria.

3.2.2. Storage of Freeze-Dried Spores

1. Prepare a spore suspension in the spore-stock solution as described under **Subheading 3.2.1.**
2. Dispense 0.2-mL aliquots into glass ampoules prepared as described under **Subheading 2.2.**, label, and freeze at –80°C.
3. Dehydrate the contents using a lyophilizer and seal the ampoule by burning the narrow portion while maintaining the vacuum conditions.
4. Store the ampoules in a cold and dry place. These freeze-dried spores can last practically forever.
5. For reconstituting, open the ampoule by cutting the top portion, add a small amount of bacterial culture, and spread on the nutrient agar plate.

3.2.3. Storage of Amoebae

1. Transfer the axenic culture of amoebae at the logarithmic phase in a centrifuge tube. For strains without the spore-forming ability, recover the cells from the agar surface using a rubber policeman or equivalent.
2. Spin down the cells at 2000g for 3 min, and remove the supernatant.
3. Add cold HL5 medium to give the approximate density of 1×10^7 cells/mL. Resuspend the cell pellet and keep the tube on ice.
4. Add an equal volume of the amoeba storage solution, mix well, and cool on ice.
5. Dispense 0.5- to 1-mL aliquots of the cell suspension to cryotubes.
6. Put the tubes in a foam container and store at –80°C. This obviates the quick-freezing of cells, which may cause damage to the cells on thawing and thereby contribute to greater viability.
7. To restart the culture, thaw the tube in water and transfer the contents to nutrient agar plates or directly to HL5 (for axenic strains).

3.3. Fruiting Body Formation (Synchronous Asexual Development)

Cultivation of cells on nutrient agar plates with bacteria eventually leads to fruiting body formation. To analyze the intermediate steps, however, induction of synchronous development is necessary. This can be achieved as follows.

1. Prepare growth-phase cells either from an axenic culture or from the two-member culture. In either case, the media or food bacteria should be removed completely by repeated centrifuge-washing in BSS.
2. Adjust the cell density to 1×10^7 cells/mL and keep on ice until use (*see* **Note 7**).

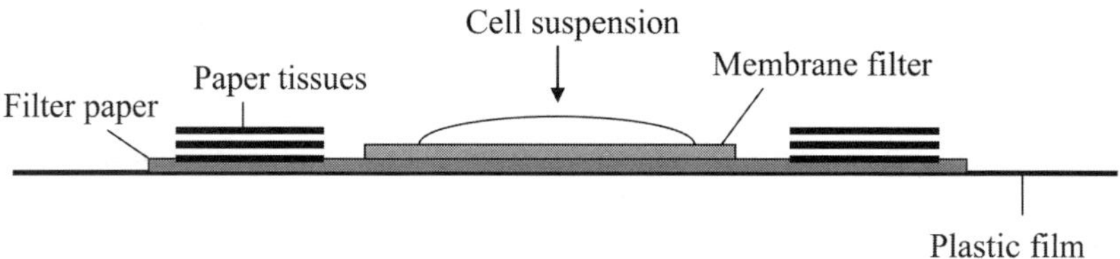

Fig. 2. Loading cell suspension onto the membrane filter (side view). Excess liquid on the membrane filter goes down through the filter paper and is absorbed up by paper tissues.

3. Put the pretreated nitrocellulose filter (or filter paper) on the larger filter paper, which should be wet with water.
4. Evenly drop 0. 3 mL of cell suspension onto the filter and absorb the excess water by firmly attaching the dry paper tissues to the open space of the filter paper around the nitrocellulose filter (*see* **Fig. 2**).
5. Transfer the filter cell-side up to either a wet filter paper pad or to the surface of plain agar.
6. Put the whole plates in a sealed container together with wet paper tissues to maintain the high humidity, and incubate.
7. Cell aggregation, slug migration, culmination, and fruiting body formation are observed at roughly 8, 12, 16, and 24 h, respectively for KAX3. This time course varies considerably with strains and culture history. In general, cells cultured in an axenic medium develop more rapidly than those cultured with bacteria.

3.4. Macrocyst Formation (Sexual Development)

Macrocyst formation is a sexual process of *Dictyostelium* (*13*) known to occur under dark and submerged conditions (*3*). It involves the steps of sexual maturation of the amoebae, cell fusion followed by nuclear fusion to form zygotes, cell aggregation around the zygotes, and phagocytosis of those cells by the zygotes. Because cell aggregation around the zygotes is mediated by the chemotaxis to cAMP (*4*), which is known to be induced by starvation, rich nutrients tend to inhibit macrocyst formation. Basically, macrocyst formation is possible whenever the amoebae of appropriate mating type strains (*see* **Table 1**) co-exist in a culture, although its extent varies with strains and environmental conditions. Mutually compatible heterothallic strains, KAX3 (*mat A1*) and V12 (*mat A2*), are used in the protocols that follow. The simple protocols for mated cultures are described first, followed by more defined protocols for synchronous sexual development.

3.4.1. Standard Mated Culture

1. Collect the spores of KAX3 and V12 in BSS as described under **Subheading 3.1.1.**

2. Mix them at an equal ratio and inoculate on an LP agar plate with 0.1 mL of bacterial culture.
3. Cover the surface with a small amount of water (0.5 mL for a 9-cm dish).
4. Seal the plate with parafilm to avoid drying, wrap it with aluminum foil to shut out the light, and incubate for 4 d or longer.

3.4.2. Variations of Mated Culture

1. Spread 0.1–0.2 mL of the bacterial culture on a nutrient agar plate. Transfer or spot the amoebae of KAX3 and V12 to the plate separately and incubate, preventing a high concentration of light. Macrocysts are observed in the region where plaques of the two strains merged. This method is useful for determining the mating types of unknown strains.
2. Prepare cell suspensions of KAX3 and V12 at 1×10^5 cells/mL each in $0.1 \times$ diluted bactomilk. Mix 1:1 and dispense aliquots in wells of a multiwell titer plate (24-well or 96-well). The recommendable volume is 0.5 mL and 0.1 mL for a 24-well and a 96-well plate, respectively. Wrap the plate with aluminum foil and incubate for 3–4 d.

3.4.3. Preparation of Maturation-Inducing Conditioned Medium **(14)**

1. Collect the amoebae of both mating type strains from the agar surface as described in **Subheading 3.4.3., step 2**.
2. In a 200-mL Erlenmeyer flask, mix 20 mL of bactomilk, 6×10^6 cells of KAX3 and V12, 1 mL of conditioned medium (CM) (if available), and BSS to 40 mL.
3. Wrap the flask with aluminum foil to shut out the light and incubate on either a gyratory shaker (120 rpm) or a reciprocal shaker (120 strokes/min) for 24 h.
4. Check the formation of large giant cells in the culture.
5. Transfer the culture to a centrifuge tube and spin down the cells at 4°C ($3000g$ for 10 min).
6. Collect the supernatant and freeze at –20°C for 16 h or longer.
7. Thaw the supernatant and centrifuge again to remove any remaining cell debris.
8. Collect the supernatant and store at –20°C (*see* **Note 8**).

3.4.4. Synchronous Cell Fusion

1. Prepare the growth-phase cells of KAX3 and V12 from the two-member culture (*see* **Note 9**).
2. Scrape the amoebae from the agar surface using a platinum transfer loop, suspend in BSS, and determine the cell density.
3. In a 50-mL Erlenmeyer flask, inoculate 3×10^6 cells of KAX3 in 5 mL of BSS and add 5 mL of bactomilk (*see* **Note 10**).
4. Wrap the flask with aluminum foil to shut out the light and culture on either a gyratory shaker (120 rpm) or a reciprocal shaker (120 strokes/min) for 15 h.
5. Culture the V12 cells in the same way except for the addition of 0.5 mL of the CM (*see* **Subheading 3.4.2.**) to enhance the fusion competence. Reduce the volume of BSS to 4.5 mL.

6. At the end of culture, cool the flasks on ice at least for 1 h before exposure to light. Then, transfer the culture to centrifuge tubes and spin down the cells (2000g for 10 min).
7. Re-suspend the cell pellet with cold BSS and centrifuge (1500g for 3 min) again. Repeat the centrifuge-wash until the bacteria are completely removed (*see* **Note 11**).
8. Adjust the cell density to 5×10^6 cells/mL for each strain.
9. Mix 0.1–0.5 mL of each cell suspension 1:1 in a small test tube siliconized in advance to prevent cell attachment on the surface, and incubate on a gyratory shaker (120 rpm).
10. At the end of incubation, add EDTA to 5 mM to stop cell fusion, which requires Ca^{2+}.
11. For numerical treatment, determine the numbers of total particles in a given volume using a hemocytometer. The progress of cell fusion is expressed as a decrease of N_t/N_0 ratio, where N_0 and N_t are particle numbers at times 0 and t, respectively. If possible, monitoring the decrease of unfused cells is more direct. Cell fusion starts immediately and reaches a plateau within 30 min *(15)*. Those cells giving a high (>60%) fusion rate are called fusion-competent cells.

3.4.5. Synchronous Development of Zygotes (16)

1. Mix equal numbers of fusion-competent cells of KAX3 and V12 in a glass-bottom dish (IWAKI, Japan) at a density of 2×10^5 cells/cm^2, cover the bottom surface with a small amount of BSS, and put in an incubator.
2. Extensive cell fusion is observed during the initial 1 h of incubation, but gradually the cell size decreases as a result of cytokinesis. Zygotes with large nuclei are detected at 8 h, and cell aggregation around them is observed at 12–16 h. Early macrocysts appear after 24 h (*see* **Fig. 3**).

4. Notes

1. For convenience, we make a 20X strength buffer for A-medium with KH_2PO_4 and Na_2HPO_4 alone. A-medium buffer (20X): 18.0 g KH_2PO_4, 6.0 g Na_2HPO_4, dH2O to 1 L. Use 50 mL for 1 L. If the 1.5% agar is not hard enough, increase the concentration to 2.0%.
2. Axenic strains can be maintained at a much-reduced growth rate in a dish without shaking. The medium depth should be kept shallow (e.g., 5 mL in a 9-cm dish). The FM medium *(17)* is a completely defined medium for axenic culture and is especially useful for radiolabeling the cells. The recipe can be obtained through the dictyBase (http://dictybase.org/techniques/fm-medium.html).
3. We normally use black cellulose acetate filters with grid from ADVANTEC (Tokyo, Japan). Black filters help observation of development, but if the structures are to be stained, a white filter is appropriate. If long slugs are desired (e.g., for cutting experiments), the coarse filter paper is better.
4. The culture length for fruiting body formation varies with strains. In addition, the thickness of nutrient agar affects bacterial growth and thereby the number of amoebae collectable and the timing of starvation as well.

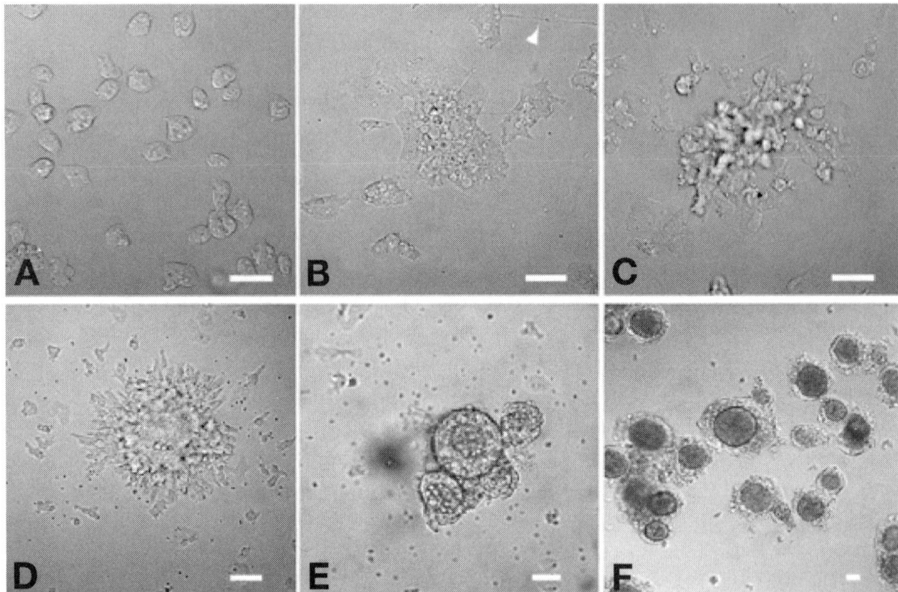

Fig. 3. Synchronous sexual development. (**A**) Sexually mature amoebae at 0 h. (**B**) Cell fusion and cytokinesis (arrowhead) observed at 1 h of incubation. (**C**) Cell aggregation at 12 h. (**D**) Formation of large cell aggregates at 16 h. (**E**) Early macrocysts at 24 h. (**F**) Mature macrocysts at 48 h. (Reproduced from **ref. *16***, with permission from Blackwell.) The bar represents 20 µm.

5. One may ocasionally encounter trouble in the two-member culture. If the agar surface is covered by a thick lawn of bacteria even after 3 d, there are no live amoebae. It is likely that the stock fruiting bodies used did not contain viable spores. In other cases, the culture is transparent, with occasional tiny fruiting bodies. This is most probably due to poor bacteria growth. The bacterial culture should be renewed. Homothallic strains form macrocysts in a clonal population and it is difficult to obtain good fruiting bodies. Relatively dry plates and strong light help the asexual development.

6. Counting and diluting the cell suspensions does not always prove accurate. Therefore, it is safe to prepare extra plates with 10× and 0.1× numbers of cells.

7. If developmental timing is very important, do not to raise the temperature of cell suspension while handling. Experimental manipulation should be quick.

8. On preparation of the CM as described, the cells can be re-used. Resuspend the precipitated cells at **Subheading 3.4.2.**, **step 5** with 40 mL of BSS, transfer to a 200-mL Erlenmeyer flask, and incubate for additional 3 h. Collect the supernatant in the same way as in **Subheading 3.4.2.**, **steps 5–8**. The two ĊM preparations are equally effective because the relevant components are secreted by the fused cells.

9. If this culture is started from cells cultured in HL5, they do not become fusion-competent. The reason for this is currently unknown.
10. Fusion-competency of cells decreases by starvation. Therefore, to obtain highly fusion-competent cells, it is safe to keep cell densities at the end of culture below 5×10^6 cells/mL.
11. The first centrifugation is at a higher speed and for a longer time. This is because the presence of excess bacteria interferes with the recovery of amoebae. Cells quickly lose fusion-competence in the light if the temperature is raised.

References

1. Kessin, R. H. (2001) Dictyostelium: *Evolutional Cell Biology, and the Development of Multicellularity.* Cambridge University Press, New York.
2. Blaskovics, J. C. and Raper, K. B. (1957) Encystment stages of *Dictyostelium. Biol. Bull. Woods Hole Mass.* **113,** 58–88.
3. Erdos, G. W., Raper, K. B., and Vogen, L. K. (1976) Effects of light and temperature on macrocyst formation in paired mating types of *Dictyostelium discoideum. J. Bacteriol.* **128,** 495–497.
4. O'Day, D. H. and Durston, A. J. (1979) Evidence for chemotaxis during sexual development in *Dictyostelium discoideum. Can. J. Microbiol.* **25,** 542–544.
5. Sussman, M. (1966) Biochemical and genetic methods in the study of cellular slime mold development, in *Methods in Cell Physiology* (Prescott, D., ed.) Academic, New York: pp. 397–409.
6. Cocucci, S. M. and Sussman, M. (1970) RNA in cytoplasmic and nuclear fractions of cellular slime mold amebas. *J. Cell Biol.* **45,** 399–407.
7. Bonner, J. T. (1947) Evidence for the formation of the aggregates by chemotaxis in the development of the slime mold *Dictyostelium discoideum. J. Exp. Zool.* **106,** 1–26.
8. Erdos, G. W., Raper, K. B., and Vogen, L. K. (1973) Mating types and macrocyst formation in *Dictyostelium discoideum. Proc. Natl. Acad. Sci. USA* **70,** 1828–1830.
9. Robson, G. E. and Williams, K. L. (1980) The mating system of the cellular slime mould *Dictyostelium discoideum. Curr. Genet.* **1,** 229–232.
10. North, M. J. and Williams, K. L. (1978) Relationship between the axenic phenotype and sensitivity to ω-aminocarboxilic acids in *Dictyostelium discoideu m. J. Gen. Microbiol.* **107,** 223–230.
11. Watts, D. J. and Ashworth, J. M. (1970) Growth of myxamoebae of the cellular slime mould *Dictyostelium discoideum* in axenic culture. *Biochem. J.* **119,** 171–174.
12. Loomis, W. F. (1971) Sensitivity of *Dictyostelium discoideum* to nucleic acid analogues. *Exp. Cell Res.* **64,** 484–486.
13. MacInnes, M. A. and Francis, D. (1974) Meiosis in *Dictyostelium mucoroides. Nature* **251,** 321–324.
14. Saga, Y. and Yanagisawa, K. (1983) Macrocyst development in *Dictyostelium discoideum.* III. Cell-fusion inducing factor secreted by giant cells. *J. Cell Sci.* **62,** 237–248.

15. Saga, Y., Okada, H., and Yanagisawa, K. (1983) Macrocyst development in *Dictyostelium discoideum*. II. Mating type-specific cell fusion and acquisition of fusion-competence. *J. Cell Sci.* **60,** 157–168.
16. Ishida, K., Hata, T., and Urushihara, H. (2005) Gamete fusion and cytokinesis preceding zygote establishment in the sexual process of *Dictyostelium discoideum*. *Develop. Growth Differ.* **47,** 25–36.
17. Franke, J. and Kessin, R. (1977) A defined minimal medium for axenic strains of *Dictyostelium discoideum. Proc. Natl. Acad. Sci. USA* **74,** 2157–2161.

8

Parasexual Genetics Using Axenic Cells

Jason King and Robert Insall

Summary

Normally, vegetative *Dictyostelium* grow as haploid cells. Occasionally, two haploid cells fuse together during normal growth, forming a diploid cell containing both parental sets of chromosomes within a single nucleus. The diploid state is reasonably stable, and the growth, development, and general behavior of diploids are similar to their haploid parents. However, during normal growth of diploids, cells may spontaneously lose one copy of each chromosome at random and revert back to a haploid state containing a selection of chromosomes from both parents. This diploid cycle therefore allows nonsexual recombination between two different mutant strains.

Diploid cells have multiple practical uses. They allow the generation of double and multiple knockouts, and are particularly useful for strains that are sick or difficult to generate using molecular genetics. They provide a means of manipulating genes that are lethal when disrupted in haploids. In diploids, it is possible to isolate heterozygous knockouts with no phenotype and then introduce a further mutant allele. These cells can then be segregated to yield haploid progeny with an effective gene replacement. Similarly, diploids made from different parent strains offer a means of examining the effects of different genetic backgrounds and overriding strain-specific phenotypes. A number of other uses are possible, making parasexual genetics potentially even more versatile.

Key Words: Parasexual cycle; diploid; genetic recombination; linkage mapping.

1. Introduction

The isolation of diploid strains and subsequent re-segregation of recombinant haploid cells can be used for a number of experimental purposes (*1*). Historically, these techniques were widely used for the mapping of mutant alleles to chromosomal linkage groups. Although the publication of a fully sequenced genome combined with techniques for targeted gene mutation make this largely obsolete, the recombination of mutant strains and the advantages of diploidy

From: *Methods in Molecular Biology, vol. 346:* Dictyostelium discoideum *Protocols*
Edited by: L. Eichinger and F. Rivero © Humana Press Inc., Totowa, NJ

offer a powerful complement to the standard genetic techniques available in *Dictyostelium*.

Because of the many potential uses of parasexual genetics, different combinations of parental strains and selections are required, depending on the experimental objective. In this chapter, we describe several such schemes, but careful planning of parasexual experiments with forethought is essential, with particular attention to the experimental uses of strains generated.

The parasexual cycle has been well studied and is fairly well understood. Diploids occur naturally in any population of growing cells, at low levels—typically one in 10^5 to 10^6 cells in any population is diploid. Parasexual genetics as a means of recombination is therefore dependent on the availability of suitable markers to select and maintain the diploid cells from out of the haploid background. This is achieved by combining two selections, with each parent containing a single selectable marker. Applying both selections simultaneously kills both haploid parents, but allows fused diploid cells containing both markers to survive.

1.1. Uses of Parasexual Genetics

Nonsexual recombination and isolation of diploid strains has numerous uses. First, strains containing mutations on different chromosomes can easily be fused to isolate diploid cells containing both mutant alleles. If the mutations are recessive, both will be complemented and the resulting diploid will have no mutant phenotype. However, this strain can then be segregated to produce haploid cells with both mutant alleles and therefore allows the isolation of strains with multiple mutations from pre-existing strains (*see* **Fig. 1**).

Diploid strains also provide a useful tool for the manipulation of genes essential for cell survival. As diploid cells contain two copies of each gene, it is possible to make a heterozygous knockout. As such strains also contain a wild-type allele, the mutation will again be complemented and therefore it is possible to disrupt essential genes with no apparent phenotype. The advantage of this is that it is then possible to introduce a further copy of the gene (i.e., tagged, with a specific mutation, green fluorescent protein [GFP]-fusion, and so on) and then segregate the diploid selecting for cells with the disrupted chromosomal allele with the extra, mutant copy as a gene replacement. Therefore, diploid strains allow the straightfoward manipulation of essential genes. In addition, the isolation of a heterozygous knockout also provides a test for the lethality of mutations as if this were the case: on segregation, only haploid progeny containing the wild-type allele will be produced.

In the same vein, the ability to produce a heterozygous knockout with no phenotype is also useful for the isolation of mutants with severe phenotypes and slow growth. Such mutants can be difficult to isolate by conventional meth-

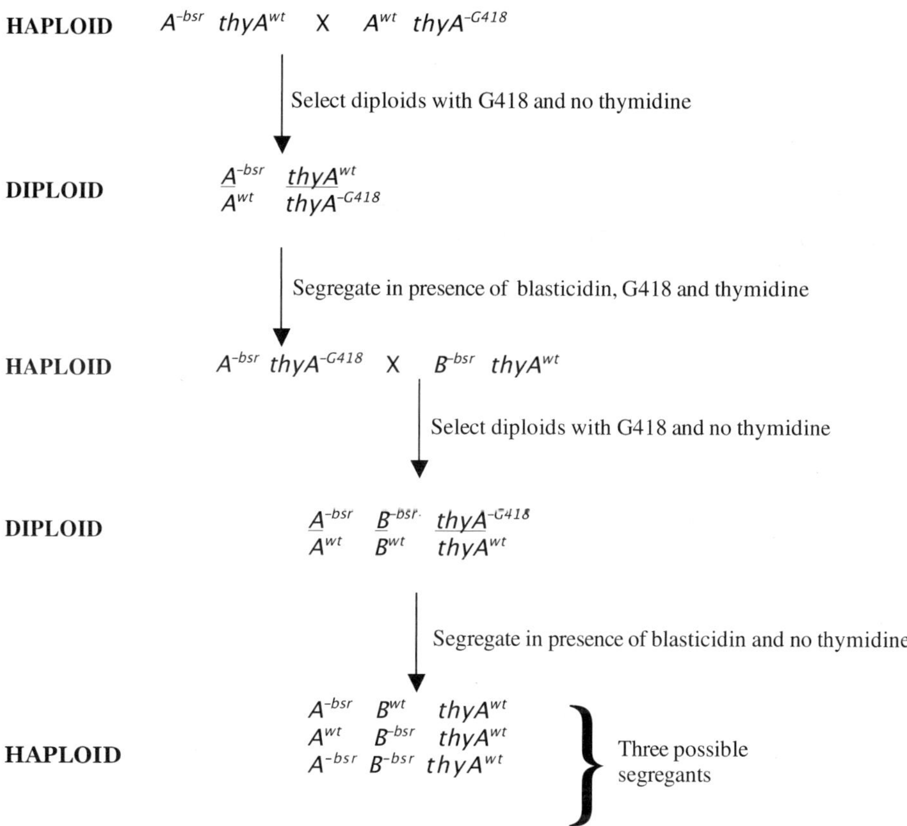

Fig. 1. A typical scheme for the production of multiple knockout strains. **A** and **B** can be any two genes previously disrupted using blasticidin, or any selectable marker except G418, so long as they are on different chromosomes.

ods, but as the phenotype will be hidden until segregation in a diploid cell, this is not a problem. Heterozygous knockouts can then be segregated while maintaining selection for the mutant allele, thus selecting for appropriate haploid cells. In addition, as the haploid phenotype is uncovered at segregation, the cells have less time in which to adapt to a mutation, and therefore phenotypes which are rapidly lost may be stronger.

1.2. Selection of Diploids From Haploid Parents

Historically, several different combinations of selections have been used. In nonaxenic strains, two parents with temperature-sensitive mutations in different, complementary genes were widely used *(2,3)*. Unfortunately, these selections have not proved suitable for handling axenic strains. We have found that axenic strains in general, and axenic diploid strains in particular, grow poorly and are unstable at the elevated temperatures needed for selection. We have therefore developed a set of markers that work well under axenic conditions. In practice, any pair of dominant-acting selectable markers (for example, antibiotic resistance genes and auxotrophic markers) can be used to isolate diploids. However, it should be noted that if the selectable markers are located at different genetic loci, then eventually haploid progeny will arise that are resistant to both markers. This is particularly severe when the markers are on different chromosomes, as doubly resistant haploids will emerge constantly. Therefore, for the maintenance of a true diploid population, it is necessary to use two different selectable markers that are found at the same genetic locus.

This can be achieved in different ways. For example, the Ax3-derived strain DH1 and its relative JH8 each carry a deletion in the *pyr56* gene on chromosome 3 *(4)*, and therefore require exogenous uracil (20 μg/mL) for axenic growth. Another strain, JH10, requires added thymidine (100 μg/mL) to grow as a result of reinsertion of *pyr56* into the *thyA* gene, also on chromosome 3 *(5)*. Thus, neither parent is able to grow in unsupplemented FM minimal medium *(6)*, whereas a diploid formed by fusion between both cells will contain a single normal copy of *thyA* and *pyr56* and are able to grow (*see* **Fig. 2**). As both selectable markers reside at the same locus, haploid progeny containing both genes are unable to segregate and therefore the diploid cells can be cultured indefinitely *(7)*, unless there is a chromosomal dislocation or other exceptionally rare event.

A second way of generating a pair of marked strains has been achieved by disruption of *thyA* with a drug selectable marker such as G418, blasticidin, or hygromycin *(8)*. In strain IR110, for example, *thyA* is inactivated by disrupting it with the G418 resistance gene. Therefore, this strain is both a thymidine auxotroph and resistant to G418. Fusions between these and any type of unmarked cells will therefore be able to grow in unsupplemented medium in the presence of G418, whereas both parents will die. Again, as both selectable markers are found at the same genetic locus, any haploid cells formed by segregation would be killed by one of the selections *(7)*.

1.3. Segregation of Haploids From Diploid Parents

When the selection is not maintained, diploid cells will, over time, segregate to give haploids with all possible combinations of parental chromosomes. Therefore, if the diploid strain contains a selectable marker that is recessive (that is to say, does not allow diploid heterozygotes with one wild-type and one marked chromosome to survive selection), it is possible to select for haploid progeny against the diploid background. Two such markers, potentially at least, are *cycA* and *pyr56*. *cycA* is a point mutant that confers resistance to cyclo-

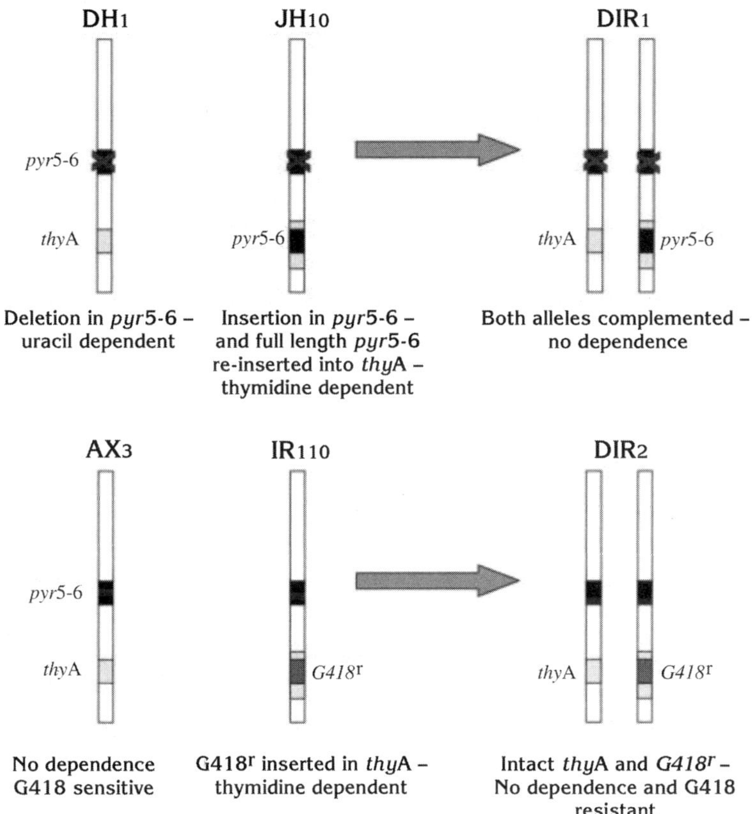

Fig. 2. Schematic for the selection of stable diploid strains. A representation of chromosome 3 is shown, indicating the various selections. Note that in both cases both selectable markers are in the same loci on the same chromosome.

heximide *(9)*, but cells containing one wild-type and one *cycA* allele are as sensitive to cycloheximide as the wild type. *Pyr56* deletions, as described earlier, make cells requiring uracil for growth. Mutant cells are, however, resistant to the metabolic poison 3-fluoroorotic acid (FOA) *(10,11)*. Therefore, in the presence of FOA and uracil, diploids containing one marked copy of the gene will be killed, leaving only haploid cells containing the marked allele.

If selections for haploids are not available, most diploid strains are very stable, and spontaneous segregation occurs at such a low rate that it is impractical to merely screen a large number of subclones. The segregation process is therefore accelerated by treating diploids with microtubule inhibitors such as benlate (benomyl) and thiabendazole *(12,13)*. Although several such inhibitors are effective, many have a low solubility in aqueous solutions, and we have therefore found thiabendazole to be the easiest to work with, especially in axenic cultures. Growth of diploids in low-thiabendazole concentrations presumably puts the mitotic spindle under enough stress to cause occasional loss of chromosomes, leading to an aneuploid cell; as aneuploids grow exceptionally poorly *(14)*, they resolve fairly rapidly into true haploids. Thiabendazole and benlate treatment work well in both bacterial plates and axenic culture. Bacterial plates are more efficient, but segregation in liquid allows isolation of haploids in the presence of drug selections, and therefore facilitates selection of a desired combination of marked chromosomes from the random pool.

2. Materials

2.1. Apparatus

In addition to standard *Dictyostelium* culture apparatus, no additional equipment is required for the selection and segregation of diploid cells; however, because of the unstable nature of diploids, it is highly advisable to either image mitotic spreads or analyze by flow cytometry all resultant strains to confirm ploidy.

For cytological imaging, a standard fluorescence microscope capable of at least 200× magnification and a filter set giving excitation at 350 nm and detection at 488 nm is required.

2.2. Chemicals

1. FM minimal medium *(6)*, available preformulated from Formedium Ltd., Norwich, England.
2. KK2 buffer: 16.5 mM KH_2PO_4, 3.8 mM, pH 6.2.
3. SM agar: 10 g/L proteose peptone, 1 g/L yeast extract, 10 g/L glucose, 1.9 g/L KH_2PO_4, 1.3 g/L $K_2HPO_4 \cdot 3H_2O$, 0.49 g/L $MgSO_4$, 17 g/L agar.
4. HL5 medium: 14 g/L proteose peptone, 7 g/L yeast extract, 13.5 g/L glucose, 0.5 g/L Na_2HPO_4, 0.5 g/L KH_2PO_4, pH 6.4.
5. Drugs for selection: Uracil, thymidine, blasticidin S, hygromycin sulphate or G418. All stocks dissolved in standard KK2 buffer; store at –20°C.
6. Thiabendazole: 2 mg/mL stock in dimethylsulfoxide (DMSO). Store at 4°C.
7. Chemicals for fixation: Ethanol, glacial acetic acid.
8. Mounting medium (Vectashield, Vector Laboratories Inc., or similar).
9. 4',6-Diamidino-2-phenylindole dihydrochloride (DAPI): 1 mg/mL in water (light sensitive). Store at –20°C.
10. Nocodazole: 10 mg/mL in DMSO, store at –20°C.
11. Propidium iodide: 50 µg/mL in PBS (light sensitive). Store at –20°C.
12. Phosphate-buffered saline (PBS): 0.14 M NaCl, 2.7 mM KCl, 1.5 mM KH_2PO_4, 8.1 mM Na_2HPO_4.
13. RNAse.

3. Methods

3.1. Fusion of Haploid Cells and Selection of Diploids

1. Mix 5×10^6 cells of each parental strain in 10 mL of HL5 containing any nutritional additions required (e.g., uracil or thymidine). Do not include any antibiotics for selections (e.g., G418, hygromycin or blasticidin) and grow in shaking flasks overnight. For positive and negative controls, also shake two flasks containing 5×10^6 cells of each parent strain alone in separate flasks. Therefore, five flasks are required for an appropriate selection.
2. After 16–24 h of shaking, transfer cells to Petri dishes and leave for 1 h to adhere.
3. When all cells have adhered, aspirate off medium and replace with appropriate selective medium, i.e., FM for uracil selection (*see* **Note 1**), and HL5 with antibiotic for others (working concentrations: 20 µg/mL uracil, 100 µg/mL thymidine, 10 µg/mL blasticidin S, 30 µg/mL hygromycin sulfate [*see* **Note 2**], 20 µg/mL G418). The same selective conditions should be used for the negative control plates as for the actual diploid plate. To check for cell viability, the two positive control plates should contain any selections and nutritional additions required for growth of the parental strain.
4. Incubate at 22°C for 10–14 d feeding with fresh selective medium every 2–3 d. Colonies should appear after approx 1 wk. Typically, 100–200 colonies will form in each dish. Within the limitations of the genetic background discussed previously, these can then be cultured indefinitely as normal, with approximately wild-type growth rates, and subcloning to isolate single colonies is not required (*see* **Notes 3** and **4**).

3.2. Segregation of Haploid Cells From Diploids: Segregation on Bacterial Lawns

1. Pour standard SM agar plates with the addition of 2 µg/mL thiabendazole (from a 2 mg/mL stock in DMSO). This should be added to cooled agar, just prior to pouring.
2. Spread a range of 20–200 diploid cells/plate together with a suspension of *Klebsiella aerogenes*. The amount of suspension should be enough to leave the plate just moist, but with no excess liquid.
3. After 24 h, invert the plates and leave in a moist environment for 1–2 wk. Growth is much slower than usual and plating efficiency is reduced to approx 30%.
4. After 2 wk, the colonies should have some faster-growing sectors as a number of different haploid recombinants are formed.
5. With a sterile toothpick or inoculation loop, pick a number of colonies from their very edges into axenic culture and screen for growth (or lack of) in appropriate selective conditions (most conveniently done in multiwell tissue-culture plates). Each colony derives from a diploid, so different sectors will contain different segregants. Normally, >90% of clones isolated will be haploid.
6. After a primary screen, all correctly identified clones must be re-cloned on fresh SM agar (without thiabendazole) to ensure that they do not contain a mixed population of segregants.

7. To ensure that the cells isolated are indeed haploid, they should be further verified by cytological analysis (*see* **Subheadings 3.4.**, **3.5.**, and **Note 5**).

3.3. Segregation of Haploid Cells From Diploids: Segregation in Axenic Medium

1. Seed 3×10^6 cells into a 10-cm Petri dish in 10 mL of HL5.
2. Add thiabendazole to a final concentration of 5 µg/mL and any selections or nutritional additions required by the desired haploid cells and, where possible, cause undesirable ones to perish.
3. Incubate at 22°C for 3 d. During this period, growth is almost arrested and many cells will die.
4. Wash the cells twice with fresh medium and leave to recover from the treatment for 3–4 d, maintaining all desired selections.
5. Plate the cells out clonally, either on SM agar plates with *Klebsiella aerogenes* or in 96-well plates. Plating efficiency is variable but normally approx 50%.
6. Screen a number of clones for growth (or lack of) in appropriate conditions to identify the desired segregants.
7. After identification, positive clones should be re-cloned to ensure complete removal of mixed cultures due to unresolved aneuploids (*see* **Subheadings 3.4.**, **3.5.**, and **Note 5**).

3.4. Cytological Staining for Verification of Ploidy

1. Seed 5×10^6 cells in a 5-cm dish containing acid-washed coverslips.
2. Incubate for 2 h in 5 mL of HL5 medium (including nutritional additions if required) to allow cells to adhere.
3. Aspirate off medium and replace with 5 mL of HL5 containing 33 μM nocodazole (*see* **Note 6**). Nocodazole stock is 10 mg/mL in DMSO; predissolve the stock dropwise in medium while swirling before use. One microliter stock per milliliter medium gives the correct concentration.
4. Leave for 2 h.
5. Place coverslips in prechilled distilled water for 10 min.
6. Aspirate off water and fix cells with freshly made, ice-cold solution of 3:1 v/v ethanol/glacial acetic acid for 1 h.
7. Aspirate off and re-fix in ethanol/acetic acid as before for 10 min.
8. Remove most of the fix and air-dry coverslips.
9. Mount coverslips cell-side down on glass slides in 3 µL of DAPI/mounting medium. Vectorshield (Vector laboratories) works well, but any mountant should suffice. Mix 10 µL of stock with 1mL mountant for working concentration.
10. Observe under a fluorescence microscope (*see* **Notes 7** and **8**).

3.5. Fluorescence-Activated Cell Sorting Analysis of Ploidy

1. Seed 3×10^6 cells in 10-cm dishes (in normal growth medium) and leave overnight.
2. Harvest and pellet the cells by centrifugation at 200*g* for 2 min.

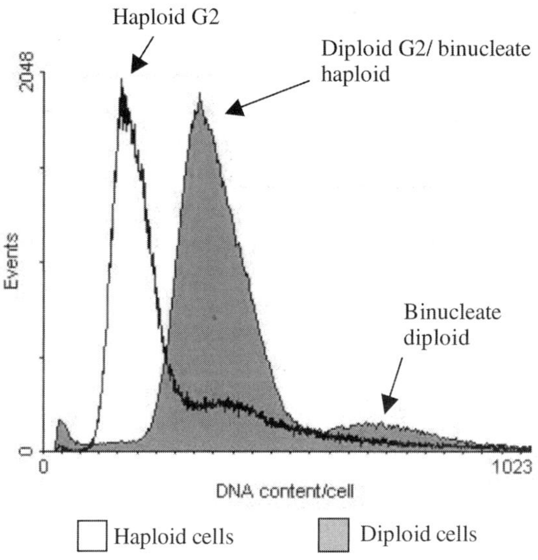

Fig. 3. A typical fluorescence-activated cell sorting graph for haploid and diploid cells. In each case, the largest peak represents cells in G2 with a smaller peak at roughly double the DNA content for binucleate cells.

3. Fix by resuspension in 10 mL of 100% methanol prechilled to −15°C and vortex for 30 s.
4. Wash cells twice in 10 mL of PBS, aspirating the supernatant carefully, as the pellet is soft.
5. Pellet the cells by centrifugation as described previously and aspirate off PBS.
6. Resuspend pellet in 100 µL of 200 µg/mL RNAse and transfer to a 1.5-mL microfuge tube.
7. Leave for 20 min at room temperature.
8. Add 400 µL of 50 µg/mL propidium iodide and mix by gently pipetting up and down.
9. Keep samples on ice prior to fluorescence-activated cell sorting (FACS) analysis.
10. Sample by FACS machine using an excitation wavelength of 488 nm. A count of 10,000 cells is sufficient for accurate analysis. Diploid cultures will give a peak for G2 cells of approximately twice the fluorescence of haploid cells (*see* **Note 9** and **Fig. 3**).

4. Notes

1. Normal, unsupplemented HL5 axenic medium contains traces of uracil and, although *pyr56*-null cells will not flourish, they can persist in cultures for an

extended time. Therefore, for efficient selection against these cells, growth in FM minimal medium is required.

2. The concentration of hygromycin sulfate required for efficient selection can be variable between strains maintained by different laboratories. Therefore, it may be necessary to titrate the amount used to ensure that all nonresistant cells will die.

3. Stable diploid cells can easily be transformed by standard electroporation protocols. However, because they are slightly larger than haploid cells, we have found that transformation efficiency is reduced. This can easily be improved by increasing the voltage applied by approx 20%.

4. Normally, unless the cells have been segregated, kept in culture for an extended time, or transformed, it is generally not worthwhile to re-clone diploids after selection, as all cells should have the same genetic background. In addition, even after cloning, the spontaneous nature of the parasexual cycle means that there will always be a small population of segregants.

5. Occasionally, during segregation an aneuploid cell will form and, instead of losing further chromosomes to become haploid, a duplication of the single orphaned chromosome will occur. This leads to a diploid cell with the apparent phenotype of a haploid. Therefore, it is advisable to examine any resultant strain by cytological staining or FACS analysis to verify ploidy.

6. Cytological analysis of *Dictyostelium* is hampered by the low mitotic index of growing cells. Normally, <1% of cells stained will have condensed chromosomes suitable for counting. Therefore, a pre-incubation with nocodazole is used, blocking the cells at mitosis and raising the mitotic index to approx 20%.

7. *Dictyostelium* chromosomes are small, and generally, magnification of 200× is required for clear imaging. In addition, deconvolution of a Z-series can be helpful for clarification if necessary.

8. Although the *Dictyostelium* genome is split among 6 chromosomes, imaging of mitotic spreads clearly shows 7 bodies in haploid cells and 14 in diploids *(15,16)*. Also, cells with any number from 5 to 20 are often observed at low frequency. This discrepancy has been attributed to aggregation of the ribosomal DNA into a single chromosome-sized body which disintegrates during fluorescent *in situ* hybridization (FISH) *(17)*.

9. The *Dictyostelium* cell cycle is unusual in that G1 is very short and therefore the majority of the cells in any population will be in G2. Therefore, when looking at the DNA content distribution of a population of cells, the largest peak will be that of cells in G2 *(18)*. Often, a further peak at approximately twice the fluorescence per cell is observed. This is due to binucleate cells, and therefore the overall distribution seen should not be confused with that normally seen for mammalian and other cell types in which most cells are in G1.

References

1. Loomis, W. F. (1987) Genetic tools for *Dictyostelium discoideum. Methods Cell Biol.* **28,** 31–65.

2. Sussman, R. R. and Sussman, M. (1963) Ploidal inheritance in the slime mould *Dictyostelium discoideum*: haploidization and genetic segregation of diploid strains. *J. Gen. Microbiol.* **30**, 349–355.

3. Katz, E. R. and Sussman, M. (1972) Parasexual recombination in *Dictyostelium discoideum*: selection of stable diploid heterozygotes and stable haploid segregants (clones-temperature sensitive-ploidy-fruiting bodies-spore-slime mold). *Proc. Natl. Acad. Sci. USA* **69**, 495–498.

4. Faure, M., Camonis, J. H., and Jacquet, M. (1989) Molecular characterization of a *Dictyostelium discoideum* gene encoding a multifunctional enzyme of the pyrimidine pathway. *Eur. J. Biochem.* **179**, 345–358.

5. Hadwiger, J. A. and Firtel, R. A. (1992) Analysis of G alpha 4, a G-protein subunit required for multicellular development in *Dictyostelium*. *Genes Dev.* **6**, 38–49.

6. Franke, J., and Kessin, R. (1977) A defined minimal medium for axenic strains of *Dictyostelium discoideum*. *Proc. Natl. Acad. Sci. USA* **74**, 2157–2161.

7. King, J. and Insall, R. H. (2003) Parasexual genetics of *Dictyostelium* gene disruptions: identification of a ras pathway using diploids. *BMC Genet.* **4**, 12.

8. Egelhoff, T. T., Brown, S. S., Manstein, D. J., and Spudich, J. A. (1989) Hygromycin resistance as a selectable marker in *Dictyostelium discoideum*. *Mol. Cell Biol.* **9**, 1965–1968.

9. Katz, E. R. and Sussman, M. (1972) Parasexual recombination in *Dictyostelium discoideum*: selection of stable diploid heterozygotes and stable haploid segregants. *Proc. Natl. Acad. Sci. USA* **69**, 495–498.

10. Kalpaxis, D., Zundorf, I., Werner, H., et al. (1991) Positive selection for *Dictyostelium discoideum* mutants lacking UMP synthase activity based on resistance to 5-fluoroorotic acid. *Mol. Gen. Genet.* **225**, 492–500.

11. Loomis, W. F., Jr. (1971) Sensitivity of *Dictyostelium discoideum* to nucleic acid analogues. *Exp. Cell. Res.* **64**, 484–486.

12. Williams, K. L. and Barrand, P. (1978) Parasexual genetics in the cellular slime mould *Dictyostelium discoideum*: haploidisation of diploid strains using ben late. *FEMS Microbiol. Lett.* **4**, 155–159.

13. Welker, D. L. and Williams, K. L. (1980) Mitotic arrest and chromosome doubling using thiabendazole, cambendazole, nocodazole, and ben late in the slime mould *Dictyostelium discoideum*. *J. Gen. Microbiol.* **116**, 397–407.

14. Welker, D. L. and Williams, K. L. (1987) Recessive lethal mutations and the maintenance of duplication-bearing strains of *Dictyostelium discoideum*. *Genetics* **115**, 101–106.

15. Brody, T. and Williams, K. L. (1974) Cytological analysis of the parasexual cycle in *Dictyostelium discoideum*. *J. Gen. Microbiol.* **82**, 371–383.

16. Zada, H. (1977) Analysis of karyotype and ploidy of *Dictyostelium discoideum* using colchicine induced metaphase arrest. *J. Gen. Microbiol.* **99**, 201–208.

17. Sucgang, R., Chen, G., Liu, W., et al. (2003) Sequence and structure of the extra-chromosomal palindrome encoding the ribosomal RNA genes in *Dictyostelium*. *Nucleic Acids Res.* **31**, 2361–2368.

18. Weijer, C. J., Duschl, G., and David, C. N. (1984) A revision of the *Dictyostelium discoideum* cell cycle. *J. Cell Sci.* **70**, 111–131.

9

Slug Phototaxis, Thermotaxis, and Spontaneous Turning Behavior

Paul R. Fisher and Sarah J. Annesley

Summary

Dictyostelium slug phototaxis and thermotaxis are readily assayed phenotypes that reflect with great sensitivity and specificity the interactions of environmental stimuli with morphogenetic signaling systems controlling the collective movement of slug cells. Methods are described for conducting and recording phototaxis, thermotaxis, and spontaneous turning experiments, although it is pointed out that spontaneous turning rates are not easily measured in most strains of molecular biological interest. Both phototaxis and thermotaxis can be assayed qualitatively for rapid screening of prospective mutants and quantitatively for detailed phenotypic analysis. Both types of assay are simple to conduct, but require care to avoid the potentially misleading effects of other factors such as cell density and extraneous thermal and chemical gradients that might influence slug behavior. The quantitative analysis and statistical testing of conclusions are carried out using directional statistics, because traditional statistical methods for linear data are inappropriate and potentially misleading when applied to directional data. The appropriate statistical methods are given for measuring maximum likelihood estimates and confidence intervals for the average direction (μ) and the accuracy of orientation (κ) for unidirectional orientation, as well as the two preferred directions ($\pm\alpha$) and accuracy of orientation in bidirectional phototaxis. In addition, tests for uniformity ($\kappa = 0$), for equality of κ in the two-sample and multisample cases, and for bidirectional phototaxis vs unidirectional phototaxis are described. These methods are readily implemented in the R environment for statistical computing and the functions required to do so are described and provided.

Key Words: *Dictyostelium*; phototaxis; thermotaxis; spontaneous turning; directional statistics; von Mises distribution.

From: *Methods in Molecular Biology, vol. 346:* Dictyostelium discoideum *Protocols*
Edited by: L. Eichinger and F. Rivero © Humana Press Inc., Totowa, NJ

1. Introduction

The multicellular stage of the *Dictyostelium* life cycle—the slug—is a small, motile organism that is formed by the aggregation of starving amoebae and exhibits highly sensitive and exquisitely accurate orientation behaviors in response to physical (temperature gradients and light) and chemical stimuli (reviewed in **ref. *1***). In the soil context, the slugs' phototactic and thermotactic behaviors would take them to the surface, where overhead light and a concomitant drop in humidity induce fruiting body formation (culmination). This makes excellent biological sense. The surface is doubtless the best place from which to disperse spores to new environments that are (hopefully) well furnished with abundant bacterial food supplies, yet slugs risk dessication and death should they tarry too long in the harsh environment there before culminating. To navigate the subterranean passages of the soil microenvironment from their place of origin to the surface, *Dictyostelium* slugs use a photosensory apparatus whose sensitivity approaches that of the human retina (responding to energy fluxes as little as 1 μW/cm^2 *(2)*, and a thermosensory system that is unrivalled in sensitivity to shallow temperature gradients (responding to as little as 0.04°C/cm) *(3)*.

Although not as intensively studied or as well understood as chemotaxis in the single-cell stages of the life cycle, slug phototaxis and thermotaxis have been the subject of experimental investigation for many decades *(1)*, beginning with the seminal studies by Raper in 1940 *(4)*, Bonner et al. in 1950 *(5)*, and Francis in 1964 *(6)*. Slugs detect the light direction by means of a lens effect—light refracted at the roughly cylindrical surface of the slug is focused onto its distal side, so that when a slug turns towards a lateral light source, it is turning away from its most intensely illuminated side! A putative receptor for slug phototaxis was isolated many years ago, based on similarities between its absorption and photooxidation action spectra and the phototaxis action spectrum *(7–9)*. It was identified as a haem protein, but its postulated role in photoreception has not been verified by molecular genetic means. The nature of the thermoreceptor is unknown, as is the exact mechanism of motility of the slug itself. However, slug movement is probably the collective result of movement through the extracellular matrix of the many individual amoebae within it. The slug amoebae communicate with one another by means of extracellular chemical signals that organize and control their motility. Light (or temperature gradients) modulate this communal, intercellular signalling system to elicit slug turning during phototaxis (or thermotaxis). The key features of the signal transduction pathways controlling slug turning behavior emerged from a combination of behavioral, pharmacological, and genetic approaches *(1)*. They are illustrated in **Fig. 1** and may be summarized as follows:

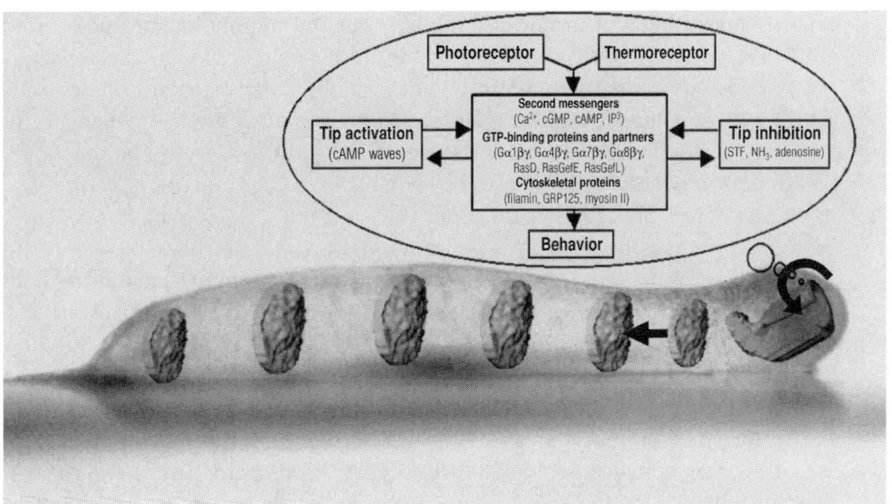

Fig. 1. Signaling pathways controlling slug phototaxis and thermotaxis. A lateral view of a slug migrating on a water agar surface is shown. The "thought bubbles" emanating from the tip indicate that the tip controls slug behavior via the indicated pathways. Signals from photoreceptor and thermoreceptor converge early and thence control the concentrations of the intracellular second messengers cyclic 3', 5'-adenosine monophosphate (cAMP), cyclic 3', 5'-guanosine monophosphate (cGMP), Ca^{2+}, and possibly inositol triphosphate (IP_3). The evidence for IP_3 involvement is the pharmacological effect of Li^+, whose target could also (or instead) be glycogen synthase kinase 3 (GSK3). Heterotrimeric G proteins and the small guanosine triphosphate-binding protein RasD are involved in transducing the signals. These in turn modulate the tip activation and inhibition signals that determine the position of the slug tip. Transient lateral imbalances between tip activation and inhibition induced by this means by light and temperature gradients cause temporary lateral shifts in tip position and thence slug turning because the slug "follows its nose." Depending on whether tip activation or inhibition dominates the response, the slug turns either towards or away from the light source, or up or down the temperature gradient. Sign reversals in slug turning responses result from switches in the balance between control by tip activation and inhibition. This explains direction-dependent sign reversals in phototaxis that cause bidirectional phototaxis (*see* text) and temperature-dependent sign reversals in thermotaxis that cause movement toward the warmth or the cold depending on the temperature. Tip activation signals are believed to be carried by three-dimensional spiral scroll waves of extracellular cAMP analogous to the two-dimensional cAMP waves that mediate aggregation. Candidate tip inhibitors are Slug Turning Factor (STF), ammonia, and adenosine. (Modified from **Fig. 3** of **ref. *1*.**)

- *Dictyostelium* slugs sense light and temperature gradients separately.
- Photosensory and thermosensory transduction pathways converge early, upstream of most signal transduction components, which accordingly function to regulate both phototaxis and thermotaxis.

- Sign reversals in slug turning responses occur in both phototaxis and thermotaxis. In thermotaxis, they result in temperature-dependent switches between positive and negative thermotaxis. In phototaxis, slugs that happen to be traveling directly toward the light source subsequently turn away from it, whereas slugs traveling at large angles to the light source will turn toward it. The outcome of this direction-dependent switch in behavior is that slugs aim not directly toward the light source, but at points on either side of it (bidirectional phototaxis).
- Slug behavior is controlled by the slug tip.
- Slug turning is mediated by transient, lateral shifts in slug tip position as determined by a temporary, local imbalance between the coupled processes of tip autoactivation and autoinhibition. Light intensity and temperature gradients across the slug tip cause turning responses by altering the balance between tip activation and inhibition, for example by stimulating differential production of tip inhibitor. This modulation of the tip activation/inhibition system occurs because, after converging, signals from the photoreceptor and thermoreceptor interact with the same central components as the tip activation and inhibition signals. Sign reversals can be economically explained because the turning behavior of the slug (toward or away from the light or the warmth) will depend on whether tip activation or inhibition dominates the response to light or temperature gradients.
- Tip activation signals are carried by extracellular cyclic 3', 5'-adenosine monophosphate (cAMP) waves, now known to be scroll-shaped, whereas tip inhibition signals may be borne by one or more of the secreted molecules NH_3, adenosine, and a small unidentified substance, Slug Turning Factor (STF). The inhibition signals may act primarily by influencing the pacemaker frequency associated with the cAMP waves. *Dictyostelium* has four cAMP receptors—G protein-coupled receptors, at least two of which are known from mutant phenotypes to be essential for normal slug phototaxis and/or thermotaxis.
- The central signal transduction components include second messengers (cAMP, inositol polyphosphates, cyclic 3', 5'-guanosine monophosphate [cGMP], and Ca^{2+}) along with associated proteins (including protein kinase A [PKA]), as well as signaling proteins (including heterotrimeric and small guanosine triphosphate [GTP]-binding proteins) and cytoskeletal proteins (including filamin). The grouping of these various components into a single central box in **Fig. 1** reflects the fact that it has not yet been determined exactly how they interact with one another. For some, the evidence that they play a role is pharmacological (inositol polyphosphates and Ca^{2+}), but for most, there is genetic evidence (i.e., based on mutant phenotypes) that they are essential for normal slug behavior.

Two things should be clear from the foregoing and from **Fig. 1**: that we have only the broadest of outlines of the signaling pathways controlling slug orientation behavior, and that many of their important components are likely to be identical or similar to those involved in amoeboid chemotaxis and in multicellular morphogenesis. In fact, because slug orientation behavior is the outcome of regulating the slug tip position, the analyses of phototaxis and thermotaxis are highly sensitive assays of the interaction between environmental signals and morphogenetic processes in the slug. Phototaxis and thermotaxis measurements can make a valuable contribution to future functional studies of the molecules encoded in the *Dictyostelium* genome. This article is aimed at assisting in this endeavor by describing the methods involved in making such measurements properly.

2. Materials

1. Sterile saline: 0.585 g/L NaCl, 0.75 g/L KCl, 0.4 g/L $CaCl_2 \cdot 2H_2O$.
2. SM agar: 1 g/L $MgSO_4 \cdot 7H_2O$, 2.2 g/L KH_2PO_4, 1g/L K_2HPO_4, 10 g/L Oxoid agar, 10 g/L Oxoid Bacteriological Peptone, 1 g/L Oxoid Yeast Extract, 10 g/L glucose. Autoclave as two separate 500-mL solutions, one containing the glucose and one containing the other ingredients, then mix them aseptically after autoclaving, before pouring the plates. Separate autoclaving prevents caramelization of the glucose.
3. Water agar: 1.5% (w/v) agar in H_2O.
4. Charcoal agar: water agar supplemented with 0.5% (w/v) activated charcoal, pH adjusted to 6.5 with HCl/NaOH. The charcoal in the agar absorbs stray light, resulting in improved phototaxis *(10,11)*.
5. Coomassie Blue stain: 0.6% (w/v) Coomassie Brilliant Blue R (Sigma-Aldrich Inc.) in ethanol/acetic acid/water (5:1:4, v:v:v), used to stain slugs and slug trails *(10)*.
6. Spatula-style wooden toothpicks (*see* **Fig. 2**).
7. Heat bar: an aluminum slab (114 cm × 83 cm × 2 cm) insulated below and on all sides with 4 cm polystyrene foam (*see* **Fig. 3**). The similarly insulated lid is placed onto the bar—this produces a closed chamber that is just deep enough to accommodate the plates. The dimensions of the apparatus are such that it can hold eight rows of 11 plates, each row at a different temperature ranging in arbitrary units from T = 1 to T = 8. The temperature gradient is established by water pumped through copper pipes (external diameter 12 mm) that run through each of the 114-cm long ends of the aluminum bar. The pipes are connected via insulated hosing to two Hetofrig pumping waterbaths (Heto, Birkerød, Denmark), one held (under standard conditions) at 10°C and the other at 30°C. The whole apparatus is held either within a temperature-controlled room or a room in which temperatures do not fluctuate significantly and are within a few degrees of 21°C. Separate calibration experiments showed that under these standard conditions, the temperature gradient at the surface of the agar plates is 0.2°C/cm and the temperatures at the centers of the plates range from 14°C (T = 1) to 28°C (T = 8) *(12)*.
8. Matte black polyvinylchloride (PVC) boxes (98 mm external diameter, 22.6 mm external height, 2 mm thick PVC) were manufactured locally either with (for phototaxis) or without (for thermotaxis and spontaneous turning) a 4-mm-diameter hole drilled through the side (*see* **Fig. 2**). Running 2 mm inside its perimeter, the base of each box has a 3 mm wide × 3 mm high lip which holds the

Fig. 2. Experimental setup for phototaxis. Top panel: An inoculated charcoal agar plate in a black polyvinylchloride box ready to be closed up for phototaxis. The plate is incubated in a lighted constant temperature room and light entering the hole in the side acts as the light source. Bottom left panel: Close-up of the edge of a growing *Dictyostelium* colony on a *Klebsiella aerogenes* lawn. About 5 mm of growth (arrowed) were scraped from the growing edge of the colony for inoculation onto water or charcoal agar. This illustrates the amount of amoebae to be inoculated—a larger inoculum may produce misleading results due to cell density effects. Bottom right panel: The blunt end of a spatula-style toothpick for scraping amoebae for inoculation.

base of a Petri dish in position when it is placed inside the box (*see* **Fig. 2**). The matte black color of the box is designed to absorb stray light, which can interefere with phototaxis.

9. Summagraphics MM1201 digitizing tablet (Summagraphics Corporation, Fairfield, CT) connected to a SUN workstation (running the SunOS 5.7 operating system). Any equivalent device allowing the manual digitizing of x,y coordinate positions and their storage in plain text files would serve the purpose.

10. Clear PVC discs (84.5 mm diameter $\times$ 0.2 mm thickness) were manufactured locally as a large, single-batch special order.

11. R, a free software statistical and graphical computing environment: R is freely available from the R Foundation for Statistical Computing, as are the associated

Fig. 3. Heat bar for thermotaxis experiments. The top panel shows the heat bar in operation with the lid on. The bottom panel shows the surface of the bar with two columns of plates laid along the surface pending incubation. The apparatus consists of a 2-cm thick aluminum heat bar insulated above, below, and on all sides with a 4-cm thick layer of polystyrene foam. Various water agar plates prepared as described in **Subheading 3.1.4., steps 1** and **2** were placed at intervals inside individual black polyvinylchloride boxes (without a hole in the side) along the surface in pairs at eight positions corresponding to eight different temperatures. The temperature at the agar surface at the centre of the plate ranges along the length of the heat bar from 14°C to 28°C, with the temperature varying by 2°C per 9-cm plate *(12)*. The temperatures are regulated by two pumping water baths located under the heat bar. One of the pumps is set at 10°C and the other is set at 30°C. These two pumps are each connected to a set of pipes which pump the water throughout the heat bar producing a heat gradient. The heat gradient was verified through the use of thermocouples placed on the surface at the center of each plate.

statistics packages that provide some of the functions used in the R scripts provided here. R and associated packages can be downloaded from the Comprehensive R Archive Network (CRAN) via the R Home Page at http://www.r-project.org/ (*see* **Subheading 3.2.3.**). Precompiled versions are available for PC, Macintosh, and Linux platforms, whereas the source code is available for porting to other platforms. The directional statistics analysis described here requires the *circular*, *CircStats*, and *Bhat* packages as well as the R base package. These packages are also available through CRAN (*see* **Subheading 3.2.3.** for installation instructions).

12. *DirStats*, a free software package written by one of us (P. R. F.) for analysis of *Dictyostelium* phototaxis and thermotaxis in the R environment: *DirStats* is a small extension for R that contains the functions described in this article. It and the necessary auxiliary functions and statistical tables can be downloaded from http://www.latrobe.edu.au/mcbg/DirStats.R (*see* **Subheading 3.2.3.**).

3. Methods

3.1. Experimental Setup

3.1.1. Qualitative Tests

Qualitative tests are initially used to establish the general phototactic and thermotactic nature of a particular strain of *D. discoideum*. The number of amoebae used in these experiments is not calculated, yet a small amount must be used in order to avoid the effects of high cell density, which is known to impair phototaxis and thermotaxis *(10)*. For a more detailed analysis and collection of statistical data, a quantitative test should be performed.

1. Grow *D. discoideum* cells on 30-mL SM agar plates (*see* **Note 1**) plated with cells from a fresh *Klebsiella aerogenes* lawn (less than 1 wk old) as a food source at 21°C ± 1.0°C. The *D. discoideum* colonies arise as plaques on the *K. aerogenes* lawn with the amoebae present at the edges of the colonies. Aggregation and differentiation occurs in the center of the colonies, as the food source (*K. aerogenes*) has been depleted in these areas.

2. Scrape a small quantity of amoebae growing from the edges of *D. discoideum* colonies using a sterile, flat-edged toothpick (*see* **Fig. 2**) and inoculate it onto a water agar (thermotaxis) or charcoal agar (phototaxis) plate. Charcoal agar is used for phototaxis experiments (except when testing the effects of pharmacological agents that might be adsorbed by the charcoal), as the charcoal absorbs any stray light, which may interfere with slug phototaxis *(11)*. It is important not to inoculate too many amoebae onto the plate, as phototaxis and thermotaxis are impaired at high cell densities because of the accumulation of high concentrations of STF *(10)*. An approximate guide to the amount of cell growth that should be plated can be observed in **Fig. 2**. An inoculum this size provides just enough amoebae to form several (<10) slugs.

3. After inoculation, place the charcoal agar plates for phototaxis into individual, covered, black PVC boxes with a 4-mm-diameter hole drilled in one side (*see* **Fig. 2**). The plates are incubated in the same orientation at 21°C ± 1.0°C for 48 h facing a lateral light source (ambient room light in a constant temperature room is appropriate) (*see* **Note 2**). Place the water agar plates for thermotaxis into covered PVC boxes with no hole (and thus no light source) and incubate them on an aluminum heat bar for 70 h (*see* **Note 3**). The starving amoebae aggregate to form slugs, which migrate over the agar surface, leaving behind a trail of collapsed slime sheath.

3.1.2. Quantitative Tests: General Protocol

Quantitative thermotaxis and phototaxis tests use known densities of *D. discoideum* amoebae inoculated onto water or charcoal agar plates. This allows proper account to be taken of cell density-dependent effects on slug behavior and ensures that differences caused by mutational defects or other treatments, such as the addition of pharmacological agents, are not due simply to different cell densities.

1. Make a suspension in sterile saline of approx 2×10^6 *D. discoideum* amoebae/mL using cells scraped from a growth plate as in qualitative tests, along with (in a separate tube) a milky suspension of bacteria (appearance similar to fat-reduced milk) harvested with a sterile loop from a fresh *Klebsiella aerogenes* lawn (less than 1 wk old).
2. Inoculate each plate with 0.1 mL of each of the bacterial and the amoebal suspensions (*see* **Note 4**). The inoculum is spread evenly over the entire surface of the plate and allowed to dry in the laminar flow hood, and finally the plates are incubated for 48 h at 21°C or until the lawns are clearing (*see* **Note 5**).
3. Harvest the amoebae from the clearing mass plates using a sterile glass spreader with 5 mL sterile saline to gently scrape the amoebae from the plate surface, and suspend them in the saline. Wash the cells free of bacteria by repeated centrifugation at the lowest speed in a bench-top centrifuge (~600g for 3 min). After the final wash, retain the amoebal pellet as the high-density, neat suspension. Prepare a 1:1000 dilution (typically 10 μL in 10 mL) in sterile saline for cell counting later. The neat amoebal suspension should contain approx 1×10^9 amoebae per mL.
4. Prepare dilutions of appropriate densities (discussed below) in sterile saline and inoculate 20 μL per plate onto water agar (thermotaxis) or water agar supplemented with 0.5% charcoal (phototaxis) in a 1-cm² area in the middle of each 30-mL plate (*see* **Note 6**). It is important to inoculate the amoebae in the center of the plate, as gas diffusion into and out of the Petri dish leads to the development of radial gradients of ammonia and other gases that may influence slug behavior *(1)*. Ammonia itself is a slug repellent that accumulates in the air space between slug tips and physical barriers such as the Petri dish wall, with the result that

slugs will tend to steer away from the edges of the plate. The 1-cm^2 area is measured (*see* **Note 7**) and the amoebae are spread evenly over this area using a pipet tip held at a low angle to the surface to spread the inoculum evenly.

5. Allow the agar plates to dry thoroughly in a laminar flow hood, as amoebae in a wet inoculate will not undergo multicellular development.

3.1.3. Phototaxis

During phototaxis, *D. discoideum* slugs sense light direction by a lens effect which was first shown by Francis in 1964 *(6)*. Light passing through the convex surface of the slug into its body is refracted and focused onto the slug's distal side, producing a lateral intensity gradient across the slug. The light focused onto the slug's distal side induces the production there of STF (a low-molecular repellent whose chemical identity is unknown) so that the light intensity gradient is translated into an extracellular signal gradient within the slug's tissues *(10)*. The slug is able to measure these gradients and turn toward the light source, but at high cell densities, the accumulation in the agar of high background STF concentrations interferes with this. The resulting dependence of phototactic (and thermotactic) accuracy on cell density means that quantitative phototaxis experiments must be conducted at known cell densities.

1. To conduct quantitative phototaxis experiments, prepare a dense amoebal suspension which contains approx 1×10^9 amoebae/mL as described under **Subheading 3.1.2.**
2. Make a separate (1:1000) dilution for counting purposes, but leave the counting until later to minimize the time the amoebae are exposed to high densities and the potential development of anoxic conditions.
3. Prepare aliquots of a series of standard dilutions (usually 100%, 80%, 60%, 40%, 20%, 10%, and 5%) and use them immediately for inoculation onto each charcoal agar plate (20 µL onto 1 cm^2; *see* **Subheading 3.1.2.**, **steps 4** and **5**). Duplicate plates for the 100%, 80%, 60%, and 40% dilutions are prepared. The lower dilutions (20%, 10%, and 5%) contain fewer amoebae and therefore produce fewer slugs, so 4–6 plates should be made from each of these dilutions. This ensures that enough slugs are produced to facilitate statistically accurate measurements of the slug behavior.
4. After the phototaxis plates have been inoculated, determine the cell counts that were used by using a hemocytometer to count cells in the separate counting dilution that was made earlier. This means that exact densities are not controlled beforehand but are measured after the plates have been set up. However, the standard nature of the procedure with the use of the packed cell pellet as the dense suspension ensures that the densities are in approximately the required range.
5. Place the charcoal agar plates into individual covered black PVC boxes with a drilled 4-mm-diameter hole on one side and incubate them in the same orienta-

tion at 21°C ± 1.0°C for 48 h facing a lateral light source (fluorescent room light). The starving amoebae aggregate to form slugs, which migrate over the agar surface, leaving behind a trail of collapsed slime sheath.

6. To achieve a permanent record of the trails, use clear PVC discs to blot the plates—slugs, fruiting bodies, and slime trails adhere to the plastic discs. Stain the discs for 5 min with Coomassie Blue and then rinse them gently in running water (*see* **Note 8**). Use the stained discs to digitize the data, which are analyzed statistically as described in the later sections.

3.1.4. Thermotaxis

D. discoideum slugs exhibit negative and positive thermotaxis, both of which promote migration of the slug to the soil surface for optimal spore dispersal. During the day, the surface temperature of the soil is higher than that of the subsurface regions, and slugs would move upwards to the surface as a result of positive thermotaxis. At night, the temperature of the soil surface falls faster than that of the subsurface layers so that as temperatures decline, the temperature gradients become inverted. *Dictyostelium* slugs accommodate to this by switching to negative thermotaxis at lower temperature—a behavioral adaptation that presumably means they will still migrate to the surface under these conditions. At temperatures close to the growth temperature (usually 21°C in laboratory conditions), slugs move toward the heat as a result of positive thermotaxis. At higher and lower temperatures, the accuracy of orientation is decreased until a critical point is reached, at which the slugs switch to negative thermotaxis. Thus quantitative analysis of slug thermotaxis requires measurements to be conducted over a range of temperatures. Like phototaxis, thermotaxis is impaired at high cell densities so that it is important in quantitative experiments to conduct the experiment at known densities. For most purposes, it is not necessary to use a range of cell densities, as the information derived from such an experiment would almost certainly duplicate the information obtained from quantitative phototaxis analysis (*see* **Subheading 3.1.3.**).

1. Prepare an amoebal suspension as described under **Subheading 3.1.2.** and dilute it to 20% of the neat density—approx 2×10^8 amoebae per milliliter (as determined by counting of separate parallel dilution, as in phototaxis experiments).
2. Inoculate a 15-µL aliquot onto a 1-cm² area in the middle of each water agar plate to achieve a target inoculum density of approx 3×10^6 cells/cm². Thermotaxis becomes significantly impaired only at cell densities well above this target density, so that in most circumstances, comparisons will be valid despite slight differences in inoculum density in different suspensions. Should more exact correspondence of inocula be necessary, the counts can be conducted after dilution but immediately before plating so that the size of the inoculum aliquot can be adjusted accordingly. Use at least duplicate plates at each temperature to assay thermotaxis for each strain or treatment.

3. Allow the plates to dry in a laminar flow hood and place them into individual covered PVC boxes (with no hole, so no light source) on the heat bar.
4. Incubate the plates in the heat bar for 70 h. The longer incubation time for thermotaxis compared with that for phototaxis is to allow for slower development and slug migration at low temperatures. When the plates are removed from the heat bar, take care not to disturb the condensation that forms on the lids of plates, particularly at the higher temperatures (T = 7 and T = 8).
5. Transfer the slugs to PVC discs as described in **Subheading 3.1.3.**, **step 6**.

3.1.5. Spontaneous Turning and Average Speed

In darkness, in the absence of external thermal or chemical gradients, *Dictyostelium* slugs forming at low cell densities in an inoculum area at the center of a plate will migrate in uniformly random directions. However, although the originally chosen direction is uniformly random, the organism tends to continue migrating in that direction thereafter. This phenomenon, called "persistence," arises because slug movement is polarized by the morphology and mechanism of movement of the organism. The morphological polarity (two recognizably different ends) coincides with a behavioral polarity—the slugs have a front and a rear, a tip and a tail, and they move only in the direction in which the tip is pointing. Why, then, in the absence of any external stimuli, do slugs change direction at all? Spontaneous slug turning experiments are designed to address this issue, which has been raised and studied in only a single paper *(13)*. One possible explanation for spontaneous turning by slugs was that it represented turning responses to fluctuations in the thermal and chemical gradients surrounding the tip. This was shown not to be the case. Spontaneous turning signals do not arise externally, but internally (not unlike "machine noise" in an electrical measuring device). A mutant severely impaired in slug phototaxis and thermotaxis showed no change in its spontaneous turning rates, as would have been expected if these were responses to external gradients. On the other hand, fluoride ions, which block signal transduction pathways in both phototaxis and thermotaxis, also blocked spontaneous turning. These results showed that spontaneous turning signals arise in the internal signaling pathways downstream of the mutant defect (and so are unaffected by it), but upstream of the site of action of fluoride ions. In other words, spontaneous turning arises from signaling events that occur spontaneously in a defined part of the signal transduction pathways controlling slug behavior. The most likely part of a signaling pathway at which such spontaneous signals might arise is at positive feedback sites, where autocatalytic processes can amplify small stochastic fluctuations in a signal to the point at which they produce measurable responses downstream. To assay spontaneous turning in slugs requires measurement of the straightness of trails (change in the variance of directions per millimeter) and the average speed of migration (millimeter per hour). This allows a measure of the rate of change in directional variance in radians^2 per hour—the spontaneous turning rate, the biological equivalent of a rotational diffusion constant *(13)*.

1. Prepare a suspension of washed amoebae at a density of 2×10^8 cells/mL as for thermotaxis experiments (*see* **Subheading 3.1.4.**, **step 1**), inoculate the center of each water agar plate with 15 µL of this suspension to achieve a target density of 3×10^6 cells/cm^2, and place each plate into an individual black PVC box (without a lateral light source—no hole) as for thermotaxis.
2. Incubate the plates at 21°C overnight until the slugs have formed (in darkness at low densities in the absence of external directional stimuli) and a majority have migrated beyond the edges of the original inoculum area.
3. Mark the positions of each individual slug with a slash in the agar next to its rear extremity and then incubate for a further 10–12 h, at which time the slugs and their trails are blotted onto PVC discs (*see* **Subheading 3.1.3.**, **step 6**). Before removing the PVC disc from the plate, mark it with an indelible, fine-point felt pen on its upper surface at the site of each slash in the agar indicating the earlier slug positions.
4. After staining and drying the discs, digitize the individual slug trails along their full lengths from the marked position to the end of the trail (*see* **Subheading 3.2.1.**). Exclude those slugs that have culminated or begun to culminate and use only those trails where the slug was still migrating at the end of the incubation period.
5. From the digitized trails, measure the total distance traveled and change in directions over successive 1-mm intervals along the length of the slug trail from the time it was marked to the end of the additional incubation period. From the change in direction between successive 1-mm trail segments, calculate the variance of the Wrapped Normal distribution $\sigma^2 = -2\log\overline{C}$ where $\overline{C}$ is the mean cosine of the change in direction. Calculate the spontaneous turning rate $T = \sigma^2 v$ where v is the mean speed (total distance/time of incubation after trail marking).

The analysis of spontaneous turning is technically difficult in strains that culminate too readily after disturbance (*see* **Note 9**).

3.2. Data Collection and Analysis

3.2.1. Digitizing

To analyze slug behavior statistically requires the movement of many individual slugs to be measured or digitized. One approach would be time-lapse cinematography that is nonintrusive, i.e., unwanted light stimuli or temperature gradients to which the slugs might respond must be excluded. Although feasible in principle, this is likely to be difficult and expensive in practice and it has not been done. However, because slugs leave trails that can be blotted

Fig. 4. Summagraphics Digitizing Tablet Model 120M for manually digitizing slug trails that have been blotted onto clear polyvinylchloride discs, stained with Coomassie Blue, rinsed, and dried. The tablet is connected to a SUN Workstation where the collected data is stored for subsequent analysis. The tablet is set up to send a single pair of coordinates each time the button is pressed, as long as successive coordinate pairs are separated by a minimum distance of 0.1 mm. The start and end points of individual slug trails are digitized and stored in a plain text file, one slug per line in the format "n x1 y1 x2 y2," where n is the slug (line) number, x1 y1 are the xy coordinates of the start point, and x2 y2 are the coordinates of the end point.

onto PVC discs and stained, the simpler, cheaper alternative is to manually digitize slug trails. We use a Summagraphics MM1201 digitizing tablet connected to a SUN workstation (*see* **Fig. 4**), but any equivalent device would be suitable.

1. Place the stained, dried discs with the slug trails blot onto the digitizing tablet so that the direction toward the light (in phototaxis) or warmth (in thermotaxis) is 0°. The *DirStats* package of R functions we have written for analysis of the data (*see* **Subheading 3.2.3.**) includes automatic tests of whether this is so and chooses the appropriate statistical methods accordingly.
   Use the tablet to digitize just the start and end points of individual slug trails, rather than the entire trail left by each slug (an exception is the measurement of spontaneous turning rates—*see* **Subheading 3.1.5.**). Store the data in a plain text file in five columns, with the columns separated by white space (spaces or tabs).

Each line of the file thus contains the following data for each slug in the digitized sample—n x1 y1 x2 y2—where n is the line (or slug) number, x1,y1 are the *xy* coordinates of the start point measured in millimeters, and x2,y2 are the *xy* coordinates of the end point of the trail. The file thus contains as many lines as there are slugs in the sample and is laid out as a simple table of slug numbers and paired *xy* coordinates (*see* **Note 10**).

3.2.2. Statistical Analysis

The statistical techniques required for directional data differ from those commonly employed for linear statistical analysis. The reason is simple: Linear data can, in principle, extend infinitely in both directions on the number line; directional data cannot. Because of its periodic nature, the uncritical application of the usual statistical techniques such as calculation of sample means can, therefore, lead to misleading outcomes. To cite an extreme example, for a collection of amoebae traveling toward an attractant source located in the direction 0°, the mean direction will be 180°—precisely in the opposite direction. Measurements of standard deviations of directions can be similarly misleading and are prone to be altered dramatically, for example, by the simple choice of a different direction as zero.

For the foregoing reasons it is essential that statistical analysis of directions be based on appropriate statistical methods *(14)*. Such methods may be based on one or more of several statistical distributions for directional data—referred to as distributions on the circle because they deal with the distributions of data points on the perimeter of a circle rather than on a line. The most widely used and most broadly applicable of these is the von Mises or Circular Normal distribution, so-called because it shares many properties and analogous uses with the familiar Gaussian bell-shaped curve or Normal distribution on the line. In an analogous manner to the Normal distribution, the von Mises distribution can be characterized by two parameters, a mean direction (μ) and a dispersion parameter ($1/\kappa$). In the case of the von Mises distribution, it is usual and more convenient to deal with κ itself—the concentration parameter that is used in the study of slug phototaxis and thermotaxis as a measure of the accuracy of orientation. The remainder of this article is focused on the practical use of directional statistics to assay phototaxis, thermotaxis, and other orientation behaviors in *Dictyostelium*.

Over the last three decades, several books have been written that summarize aspects of directional statistics. One of the first of these was *Directional Statistics* by K.V. Mardia *(14)* and the most recent of them is *Topics in Circular Statistics* by S. R. Jammalamadaka and A. SenGupta *(15)*, who also wrote an accompanying software package called *CircStats* (*see* **Subheading 3.2.3.**). These, along with the primary statistical literature, formed the main source for the methods described here. These methods have been used to study *Dictyostelium* slug behavior since the late 1970s, and the first report using them appeared in 1981 *(10)*. In the following sections, we describe the main methods needed to find the "best" point estimates of the mean direction (μ) and the accuracy of orientation (κ), along with confidence intervals. In addition, we describe how to conduct tests for whether there is significant orientation in any particular direction, and whether accuracies of orientation are the same in two or more different samples.

Fisher and Williams *(16)* reported in 1981 that phototaxis is actually bidirectional (slugs actually aim at the angles $\pm\alpha$ to either side of the light source), as described under **Subheading 1.** However, this only becomes apparent when the two preferred directions are sufficiently well separated and the accuracy of orientation sufficiently high (*see* **Fig. 5**). We therefore include sections on testing the statistical significance of bidirectional phototaxis and on estimating the associate values for the preferred directions ($\pm\alpha$) and accuracy of orientation in those directions (κ), along with confidence intervals. For the purpose of making these methods more accessible, the software for performing these tasks, which was originally written in Fortran, has been rewritten as scripts for the statistics and graphics computing environment R *(17)*.

3.2.2.1. THE RAYLEIGH TEST FOR UNIFORMITY

Before estimates of the mean direction (μ) and accuracy of orientation (κ) can be meaningfully made, it is important to know that there is in fact a preferred direction. In most cases, this is obvious and does not require a statistical test. However, when the accuracy of orientation approaches zero, the distribution approaches the uniform distribution on the circle in which all directions are equally probable. Tests for uniformity have this as their null hypothesis (H_o)—that all directions are equally probable, so the true distribution is perfectly flat, there is no preferred direction, and the concept of a mean direction is inapplicable. There are a variety of tests for uniformity, and several of them are supplied as part of the *CircStats* and *circular* packages available with R. Here, we describe the most widely used of them, the Rayleigh test.

If all directions are equally probable (H_o), then the sum of the cosines of the directions (C) will equal zero, as will the sum of the sines of the directions (S) and, as a consequence, the value of R $[(S^2 + C^2)^{1/2}]$ will also be zero. R is referred to as the resultant and is one of the key sample statistics for directional data. The alternative hypothesis (H_1), tested in the Rayleigh test, is that there is some preferred direction. In this case, S or C, and consequently R, can take on values up to the size of the population or sample under consideration (n). To adjust for this scaling effect with different sample sizes, the average value of R in the selected sample is used ($\bar{R} = R/n$). $\bar{R}$ will range from 0 in the case of the

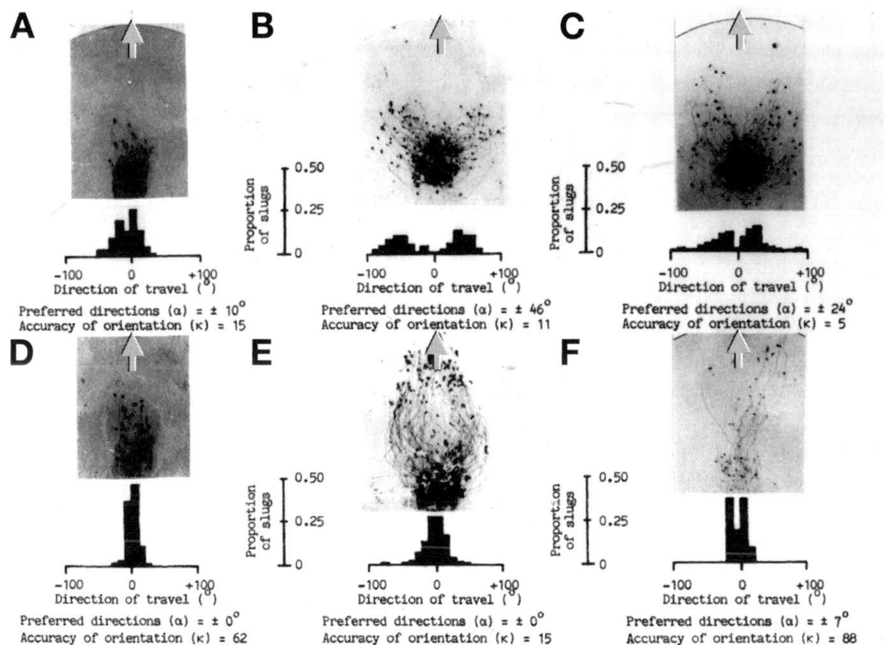

Fig. 5. Bidirectional phototaxis by phototaxis mutant HU410 (**A–C**) and its parent strain X22 (**D–F**). The figure illustrates the various circumstances (high density, slug turning factor [STF], lack of activated charcoal, mutations in phototaxis genes) that make the bidirectional nature of phototaxis more apparent by increasing the preferred angle of deviation ±α from the direction toward the light source. Slugs were formed from amoebae of strain X22 or HU410 and allowed to migrate phototactically, transferred to polyvinylchloride discs, stained, and analyzed using the unidirectional and bidirectional von Mises models as described under **Subheading 3.2.2.** The accuracy of orientation (κ) and the preferred angle of deviation (α) from the direction of the light source (0° marked by arrows) is shown. (**A**) Strain HU410, water agar, cell density = 6.4×10^5 cells/cm^2, 78 slugs analyzed from two plates. (**B**) Strain HU410, water agar, cell density = 9.6×10^6 cells/cm^2, 119 slugs analyzed from two plates. (**C**) Strain HU410, charcoal agar, cell density = 9.6×10^6 cells/cm^2, 162 slugs analyzed from two plates. The two halves of the distribution are skewed inwards by slugs readjusting their preferred directions as they move further from the origin. (**D**) Strain X22, water agar, cell density = 1.1×10^6 cells/cm^2, 94 slugs analyzed from five plates. (**E**) Strain X22, water agar, cell density = 1.7×10^7 cells/cm^2, 176 slugs analyzed from two plates. (**F**) Strain X22, charcoal agar, cell density = 5.3×10^5 cells/cm^2, 33 slugs analyzed from two plates, STF = 4300 Phototaxis Interference Units (PIU) per 30 mL plate (*10*). Crude STF exudate from strain X22 was filtered through an Amicon UM05 filter (excludes at molecular weight >500) and applied to the entire surface of each plate (*10*). When dry the plates were inoculated and incubated for phototaxis as usual. (Modified from **Fig. 1** of **ref. *16*.**)

uniform distribution to 1 in the case where all directions are identical. Thus the Rayleigh test is a measure of how likely the observed sample of angles would be under the null hypothesis that the sample is drawn from a uniformly distributed population; i.e., we test H_0: $\overline{R} = 0$ vs H_1: $\overline{R} > 0$.

Perform the Rayleigh test as follows (*see* **Note 11**):

1. If μ is known, calculate p value = $1 - p(Z) + f(Z)[(3Z - Z^3)/(16n) + (15Z + 305Z^3 - 125Z^5 + 9Z^7)/(4608n^2)]$
   where
   $\overline{C}_\mu = [\Sigma \cos(x - \mu)] / n$,
   $Z = \overline{C}_\mu (2n)^{1/2}$,
   n = total number of slugs, each x = direction for a different slug in the sample,
   p(Z) is the probability under the Normal distribution of values of x less than Z,
   and f(Z) is the value of the Normal distribution function for input x equal to Z (e.g., for n = 30 and $\overline{C}_\mu$ = 0.253: Z = 1.96, p(Z) = 0.975, f(Z) = 0.05847, and p value = 0.0248).
   Note that if μ is known to be $0°$, $\overline{C}_\mu$ collapses to $\overline{C}$, the mean cosine of the sample directions.
2. If μ is not known, calculate p value = Ye^{-Z}
   where

$$S = \sum \sin(x), \; C = \sum \cos(x), \; \overline{R} = \frac{1}{n}(S^2 + C^2)^{1/2}, \; Z = n\overline{R}^2,$$

   if (n < 50) Y = $[1 + (2Z - Z^2)/(4n) - (24Z - 132Z^2 + 76Z^3 - 9Z^4)/(288n^2)]$ otherwise Y = 1.
3. If p value is less than a selected critical value (say 0.05), reject the null hypothesis that there is no preference for any particular direction ($\kappa = 0$).

The Rayleigh test is included in the *CircStats* and *circular* packages as functions called *rayleigh.test* (*circular* package), *r.test*, and *v0.test* (*CircStats* package). The first of these (*rayleigh.test*) actually combines the functionality of *v0.test* and *r.test*, which are intended for the situations when μ is known or unknown, respectively. We have included the Rayleigh test in the *acorn* (for accuracy of orientation analysis based on the ordinary von Mises distribution for unidirectional data) function, which we provide for calculating directional statistics and plotting the observations for a single sample (*see* **Fig. 6**).

3.2.2.2. Estimates of the Mean Direction μ and the Concentration Parameter κ

The best, most probable unbiased estimate for μ is the sample mean direction $\overline{x}$. This is the angle whose tangent is $\overline{S}/\overline{C}$, i.e., x = $\begin{cases} \text{arcant } (\overline{S}/\overline{C}) \text{ if } \overline{S} > 0 \\ \text{arcant } (\overline{S}/\overline{C}) + 360 \text{ if } \overline{S} < 0 \end{cases}$

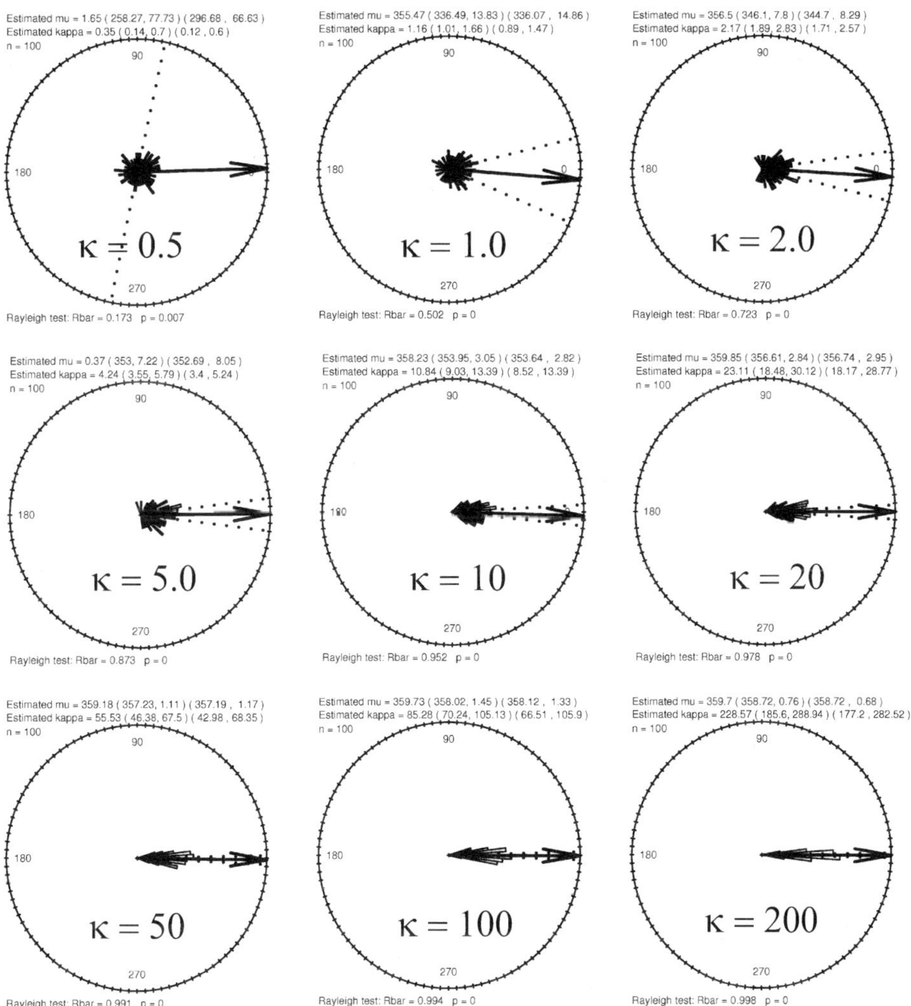

Fig. 6. Examples of the analysis of orientation when there is a single preferred direction (0°). Computer-generated samples containing 100 observations each were generated for the von Mises distribution with μ = 0° and κ as indicated in the figure. Each plot shows a rose diagram (circular histogram) of the directions, plus the mean direction (solid arrow) and its bootstrap confidence interval (dotted lines). The text above each rose diagram shows the estimates from the sample of μ and κ followed in parentheses first by the bootstrap confidence intervals and second by the traditional confidence intervals calculated as described in the text (*see* **Subheading 3.2.2.3.**). 99% confidence intervals for μ and 90% confidence intervals for κ are shown. The text below each rose diagram shows the Rayleigh test statistic and associated significance probability that the data came from a uniform distribution (i.e., κ = 0). The hypothesis of uniformity is rejected in every case shown.

The best, most probable unbiased estimate for κ is the value of $\hat{\kappa}$ that fulfills the equation

$$\overline{R} = \frac{\displaystyle\sum_{r=0}^{r=\infty} (r+1)^{-1}(r!)^{-2}\left(\frac{1}{2}\hat{\kappa}\right)^{2r+1}}{\displaystyle\sum_{r=0}^{r=\infty} (r!)^{-2}\left(\frac{1}{2}\hat{\kappa}\right)^{2r}}$$

Because the foregoing equation can only be solved numerically, calculate $\hat{\kappa}$ using the following approximations (accurate to two or three significant figures).

1. If μ is unknown, calculate
    a. for $\overline{R} < 0.53$, $\hat{\kappa} = 2\overline{R} + \overline{R}^3 + (5\overline{R}^5)/6$
    b. for $0.53 \leq \overline{R} < 0.85$, $\hat{\kappa} = 1.39\overline{R} + 0.43/(1-\overline{R}) - 0.4$
    c. for $0.85 \leq \overline{R}$, $\hat{\kappa} = 1/(\overline{R}^3 - 4\overline{R}^2 + 3\overline{R})$.

2. When μ is known, replace $\overline{R}$ in the above equations with $\overline{C} = \dfrac{\sum \cos(x - \mu)}{n}$.

We have included the calculation of these point estimates for μ and κ in the *acorn* function, which calls *mle.vonmises* from the *circular* package (*see* **Fig. 4**).

3.2.2.3. Confidence Limits for μ and κ

Two basic methods are available for calculating confidence limits for μ and κ, one based on the probability distribution functions for the sample statistics $\overline{C}, \overline{S}, \overline{R}$, and $\overline{x}$ and one based on the bootstrap principle. The former, traditional method has been used by Fisher and colleagues in slug phototaxis studies since the late 1970s and is, in general, more powerful (so that confidence limits are usually a little narrower) because it is based on the known properties of the distributions of the sample statistics. Unfortunately, the range of values included in the statistical tables of Mardia *(14)* restricts the choices of confidence levels and, for some combinations of small sample size, μ, and κ, the confidence limits cannot be determined at all by this means. The second method (bootstrapping) is more general and robust in that it does not depend on conformity to an underlying theoretical distribution, but instead uses the sample itself to provide empirical information about the distribution. This is done by randomly drawing (with replacement) new samples from the data in the original sample, calculating the sample statistics of interest. Because the resampling is random, some of the original observations will be missing and others will be repeated in the redrawn sample. By collecting very large numbers of such randomly redrawn samples, empirical distributions for the sample statistics are built up and can be used to calculate confidence limits. Bootstrapping became feasible with the advent of sufficient computing power to carry out, typically, at least 1000 such resampling experiments. The *CircStats* and *circular* packages in R include functions for calculating bootstrap confidence limits for μ and κ. In the *acorn* function, we have provided calculations and output of both the traditional and the bootstrap confidence intervals.

1. Confidence intervals for μ. Calculate confidence intervals for μ as follows:
   a. When κ is unkown (as is always the case in phototaxis and thermotaxis experiments), calculate a traditional 99% confidence interval for μ for samples containing 8 to 30 slugs from the curves in Appendix 2.7b of Mardia *(14)*.
   b. If n > 30 and $\kappa' = n\bar{R}$ $\hat{\kappa}$ < 10, calculate a traditional 99% confidence interval for μ as $\bar{x} \pm \delta$, where and $\delta = 180 - \delta'$ and δ' is obtained from Appendix 2.6 of Mardia *(14)* using the value of κ' and $\alpha = 0.01$.
   c. If n > 30 and $\kappa' = n\bar{R}$ $\hat{\kappa} \geq 10$, calculate a traditional 99% confidence interval

      for μ as $\bar{x} \pm \delta$, where $\delta = 2.576\sqrt{\kappa'-1/2}$.
   d. To determine bootstrap confidence intervals for μ at the desired level of significance, generate at least 1000 random samples redrawn with replacement from the original experimental sample of directions. From these randomly redrawn samples, calculate the desired cutoff values for which the desired proportion (e.g., 0.95 and 0.05) of estimates of μ are greater than the cutoff.

   We have digitized the curves in Appendix 2.7b of Mardia *(14)*, converted them to tables, and incorporated them into R. We have similarly incorporated values from Mardia's Appendix 2.6 into R. The *acorn* function uses the foregoing formulae and interpolates linearly within these tables as appropriate to generate traditional 99% confidence intervals for μ (*see* **Fig. 6**). In addition, *acorn* calculates 99% confidence intervals for μ by the bootstrap method using the function *mle.vonmises.bootstrap.ci* in the *circular* package (*see* **Fig. 6**). The desired confidence levels can be changed in *acorn* by the user for the bootstrap, but not the traditional method.

2. Confidence intervals for κ. Calculate lower (κ_L) and upper (κ_U) confidence limits for κ as follows:
   a. If μ is known and $\hat{\kappa} < 2$, find a traditional 90% confidence interval (κ_L, κ_U) for

      given n and $\bar{C} = \dfrac{\sum \cos(x - \mu)}{n}$ by interpolation in **Table 1** of Stephens *(18)* at

      $\alpha = 0.05$ and 0.95, respectively.
   b. If μ is known and $\hat{\kappa} \geq 2$, find a traditional 90% confidence interval for κ by calculating

      $$\kappa_L = \frac{1 + \sqrt{1 + 3a}}{4a} \text{ and } \kappa_U = \frac{1 + \sqrt{1 + 3b}}{4b}$$

      where $a = n(1 - \bar{C}_u)/\chi_n^2(0.95)$ and $b = n(1 - \bar{C}_u)/\chi_n^2(0.05)$

and χ_n^2 are the critical values of the χ^2 distribution with n degrees of freedom and $\alpha = 0.95$ or 0.05 as indicated.

  c. If μ is unknown and $\hat{\kappa} < 2$, find a traditional 90% confidence interval for κ (κ_L, κ_U) for given n and $\overline{R}$ by interpolation in **Table 2** of Stephens *(18)* at $\alpha = 0.05$ and 0.95, respectively.

  d. If μ is unknown and $\hat{\kappa} \geq 2$, find a traditional 90% confidence interval for κ by calculating

$$\kappa_L = \frac{1 + \sqrt{1 + 3a}}{4a} \quad \text{and} \quad \kappa_U = \frac{1 + \sqrt{1 + 3b}}{4b}$$

where $a = n(1 - \overline{R}) / \chi_{n-1}^2(0.95)$ and $b = n(1 - \overline{R}) / \chi_{n-1}^2(0.05)$

where χ_{n-1}^2 are the critical values of the χ^2 distribution with $n - 1$ degrees of freedom and $\alpha = 0.95$ or 0.05 as indicated.

  e. To determine bootstrap confidence intervals for μ at the desired level of significance, generate at least 1000 random samples redrawn with replacement from the original experimental sample of directions. From these randomly redrawn samples, calculate the desired cutoff values for which the desired proportion (e.g., 0.95 and 0.05) of estimates of κ are greater than the cutoff.

We have incorporated **Tables 1** and **2** of Stephens *(18)* into R and they and the foregoing formulae are used as appropriate by the function *acorn* to generate traditional 90% confidence limits for κ. In addition, *acorn* calculates 90% confidence intervals for κ by the bootstrap method using the function *mle.vonmises.bootstrap.ci* in the *circular* package (*see* **Fig. 6**). The desired confidence levels can be changed in *acorn* by the user for the bootstrap, but not the traditional method.

3.2.2.4. Tests for the Equality of κ

It is often necessary to test whether the accuracies of phototaxis or thermotaxis (κ) are the same in two or more different samples. The tests are based on approximations to the distribution of the sample statistic $R = n\overline{R}$.

In the two-sample case: The null hypothesis H_0 to be tested is that $\kappa_1 = \kappa_2 = \kappa$. The approximations lead to three different test statistics depending on the value of $\overline{R}$:

  1. For $\overline{R} < 0.45$, calculate $Q = \dfrac{\dfrac{2}{\sqrt{3}}\left\{\arcsin(1.224745\overline{R}_1) - \arcsin(1.224745\overline{R}_2)\right\}}{\sqrt{(n_1 - 4)^{-1} + (n_2 - 4)^{-1}}}$

The critical values and significance probabilities for Q are those of the Normal distribution.

  2. For $0.45 \leq \overline{R} \leq 0.7$, calculate $Q = \dfrac{g(\overline{R}_1) - g(\overline{R}_2)}{0.89325\sqrt{(n_1 - 3)^{-1} + (n_2 - 3)^{-1}}}$

where $g(\overline{R}) = \dfrac{\sinh^{-1}(\overline{R} - 1.0894)}{0.25789}$ and $\sinh^{-1}(x) = \log(x + \sqrt{x^2 + 1})$

The critical values and significance probabilities for Q are those of the Normal distribution.

3. For $\bar{R} > 0.7$, calculate $Q = \dfrac{(n_1 - R_1)(n_2 - 1)}{(n_2 - R_2)(n_1 - 1)}$.

The critical values and significance probabilities are those of the $F_{n_1-1,\,n_2-1}$ distribution with $n_1 - 1$ and $n_2 - 1$ degrees of freedom.

We provide an R function called *vmtests.on.files* (<u>v</u>on <u>M</u>ises distribution–based <u>tests on files</u> containing data sets for which Œ is hypothesized to be equal in the two-sample and multisample cases), which performs the two sample test using the appropriate statistic and then outputs the $\bar{R}$ for the combined samples, the test statistic Q, and the associated significance probability.

In the multisample case: The null hypothesis (H_o) to be tested is that for q samples of directional data, $\kappa_1 = \kappa_2 = \ldots = \kappa_i = \ldots \kappa_q = \kappa$. Approximations to the distribution of R lead to three different test statistics depending on the value of $\bar{R}$. For all three, the critical values and significance probabilities are those of χ^2_{q-1}, the chi square distribution with q -1 degrees of freedom.

1. For $\bar{R} < 0.45$, calculate $U = \sum \omega_i g_i^2 - (\sum \omega_i g_i)^2 / \sum \omega_i$

where $\omega_i = 4(n_i - 4)/3$, $g_i = \sin^{-1}(1.224745\bar{R}_i)$.

2. For $0.45 \le \bar{R} \le 0.7$, calculate $U = \sum \omega_i g_i^2 - (\sum \omega_i g_i)^2 / \sum \omega_i$

where $\omega_i = (n_i - 3)/0.7979$, $g_i = \sinh^{-1}\left\{(\bar{R}_i - 1.0894/0.25790)\right\}$,

$\sinh^{-1}\{X\} = \log\left(X + \sqrt{X^2 + 1}\right)$.

3. For $\bar{R} > 0.7$, calculate $U = \dfrac{v \log\left\{\left(n - \sum n_i \bar{R}_i\right)/v\right\} - \sum v_i \log\left\{\left(n_i - n_i \bar{R}_i\right)/v_i\right\}}{1 + d}$,

where $n = \sum n_i$, $v_i = n_i - 1$, $v = n - q$, $d = \dfrac{\sum v_i^{-1} - v^{-1}}{3(q - 1)}$.

The R function *vmtests.on.files* that we provide determines whether a two sample or a multisample test is required and, based on the value of $\bar{R}$, selects and calculates the appropriate test statistic before returning $\bar{R}$, the test statistic, and the associated significance probability.

3.2.2.5. TEST FOR BIDIRECTIONAL ORIENTATION

Phototaxis by *D. discoideum* slugs is actually bidirectional—the slugs actually aim in directions $\pm\alpha$ to either side of the direction of the light source *(16)*. In other words, phototaxis data are actually samples drawn from a mixture of two distributions with equal concentration parameters (κ) and preferred directions $+\alpha$ and $-\alpha$ (with the direction towards the light source at $0°$). In such a mixture, the combined distribution is actually unimodal (i.e., has a single peak)

when $\alpha \leq \cos^{-1}\left(-\dfrac{1}{2\kappa} + \sqrt{1 + \dfrac{1}{4\kappa^2}}\right)$. Thus, the bidirectional nature of phototaxis

only becomes apparent when α and κ are large enough to yield a clearly bimodal distribution of sample directions (*see* **Fig. 5**). This occurs in the presence of certain pharmacological agents, at high cell densities in many strains and in mutants *(16)*, but under other circumstances, the bidirectional von Mises distribution may not be a significantly better statistical description of the sample data than the best fitting unidirectional distribution. In fact, Bartels *(19)* showed that as α approaches $0°$ and as κ and the sample size get smaller, statistical estimates of α and κ become biased and confidence limits become misleading. For this reason, use of the bidirectional statistical model should be restricted to those situations when the null hypothesis that $\alpha = 0°$ (i.e., the ordinary unidirectional model) is rejected. The statistical test for testing bidirectional phototaxis against the null hypothesis that $\alpha = 0°$ belongs to the general class of tests known as the likelihood ratio (LR) tests. These tests essentially determine whether the observed sample data is significantly more likely under one hypothesis, H_o (in this case that $\alpha = 0$), than another, H_1 (in this case that $\alpha > 0$). **Figure 7** shows examples of the unidirectional and bidirectional analysis of random samples from bidirectional von Mises distributions with the indicated values for α and κ. The significance probabilities (p) indicate the probability that the data could have arisen under the unidirectional model on the left rather than the bidirectional model on the right.

1. To perform the LR test for the hypothesis of bidirectional vs unidirectional phototaxis, calculate the statistic $LR = 2(L_{\alpha=0} - L_{\alpha>0})$

$$\text{where } L(\alpha,\kappa) = -\sum \log \frac{\left(e^{\kappa\cos(x_i+\alpha)} + e^{\kappa\cos(x_i-\alpha)}\right)}{4\pi I_0(\kappa)}, \quad L_{\alpha=0} = L(0,\hat{\kappa}), \quad L_{\alpha>0} = L(\tilde{a},\tilde{\kappa}),$$

$I_0(\kappa)$ is the Modified Bessel function of κ of the first kind of order 0, $\hat{\kappa}$ is the maximum likelihood estimate for κ in the ordinary von Mises distribution (*see* **Subheading 3.2.2.2.**),

α and κ are the maximum likelihood estimates of α and κ in the bidirectional distribution (*see* **Subheading 3.2.2.6.**).

The R environment includes the required Bessel function *bessel1*. The critical values and significance probabilities for LR are those of χ_1^2, the chi-square distribution with a single degree of freedom. In the function *bimstat* (<u>bi</u>modal <u>stat</u>istics based on the von Mises distribution for bidirectional data), which we provide, the LR test for bidirectional orientation is performed and the LR statistic, as well as the corresponding significance probability, are output along with estimates of α and κ (*see* **Fig. 7**).

3.2.2.6. ESTIMATES OF THE PREFERRED DIRECTIONS ($\pm\alpha$)
AND ACCURACY OF ORIENTATION (κ) IN BIDIRECTIONAL PHOTOTAXIS

To estimate α and κ when the data are significantly bidirectional, the method of maximum likelihood is used. This means that the values of α and κ are found that maximize the likelihood function and thus minimize the negative loglikelihood function $L(\alpha,\kappa)$ defined in **Subheading 3.2.2.5., step 1**. These values ($\tilde{\alpha}$ and $\tilde{\kappa}$) represent the best, most probable estimates of α and κ and can only be found numerically.

1. To calculate the maximum likelihood estimates for α and κ, use the Newton method, a standard numerical procedure for finding the minima of an arbitrary function. This method requires the input of initial estimates of α and κ. Because the Newton method works most efficiently when the initial estimates are already close to the correct values, use another method, the Davidon-Fletcher-Powell optimization algorithm, to obtain estimates of α and κ for input to the Newton procedure.

Functions for both the Newton (*newton*) and the Davidon-Fletcher-Powell (*dfp*) methods are provided in the freely available *Bhat* package for R. The function *bimstat*, which we provide, uses initial values of $\alpha = 1$ and $\kappa = 1$ as input to *dfp*, whose output estimates are then provided as input to *newton*. The final output from *newton* includes $\tilde{\alpha}$, $\tilde{\kappa}$, and bootstrap 95% confidence intervals for both (*see* **Fig. 7**, right column).

3.2.3. Graphical Representation of Data and Results of Statistical Analysis

The implementation of the statistical and graphical analysis described here is based on the R Environment for Statistical Computing. To install the necessary software, the following steps should be followed:

1. Download and install R from CRAN via the R Home Page (http://www.r-project.org/).
2. Run R, and from the *Packages* menu select *Install package(s) from CRAN*, then select and install the *Bhat*, *CircStats*, and *circular* packages.
3. Download *DirStats* from http://www.latrobe.edu.au/mcbg/DirStats.R, saving it as a plain text file called *DirStats.R* in the R home folder.
4. To install *DirStats*, run R and type the command *source("DirStats.R")* and save R on exit. The functions described in the text will then be available permanently.

In *DirStats*, we provide three major functions for the graphical and statistical analysis of slug phototaxis and thermotaxis data: *acorn*, *bimstat*, and

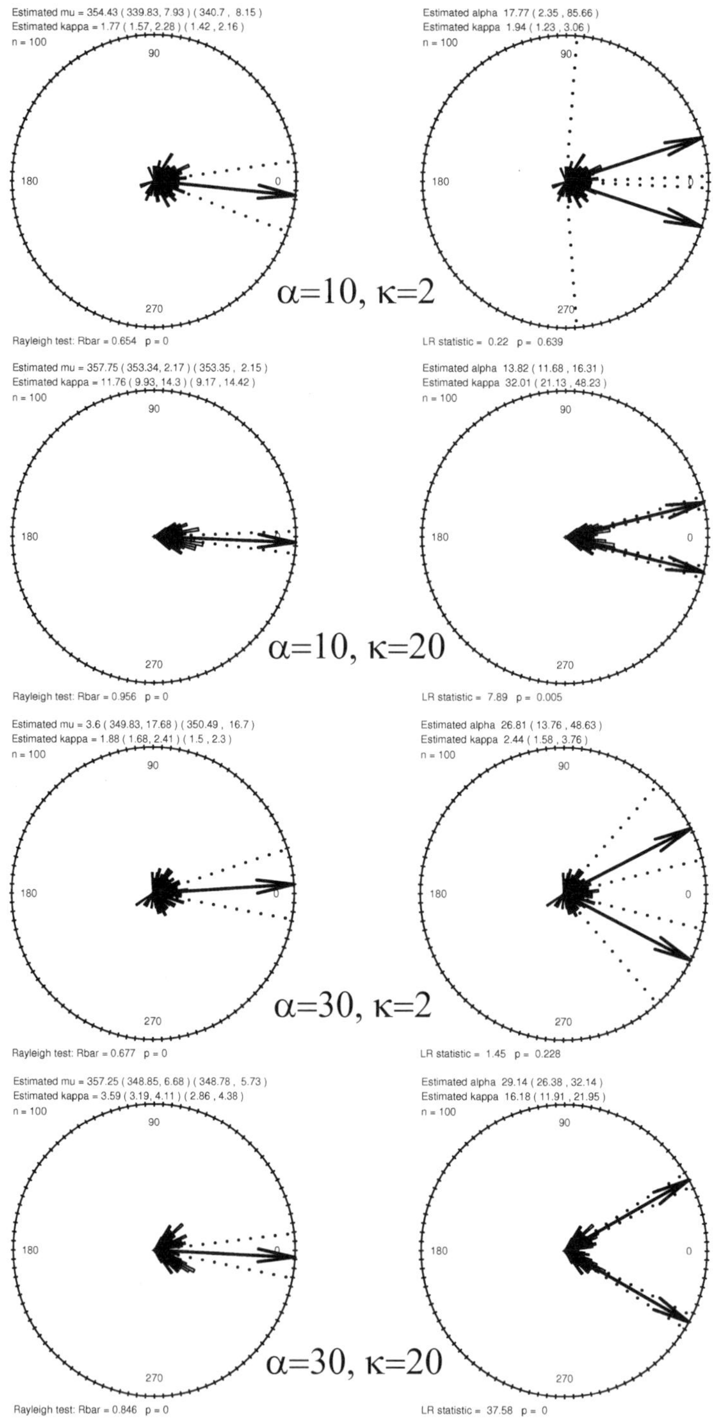
Estimated mu = 354.43 (339.83, 7.93) (340.7 , 8.15)
Estimated kappa = 1.77 (1.57, 2.28) (1.42 , 2.16)
n = 100
90
180
0
270
α=10, κ=2
Rayleigh test: Rbar = 0.654 p = 0

Estimated alpha 17.77 (2.35 , 85.66)
Estimated kappa 1.94 (1.23 , 3.06)
n = 100
90
180
0
270
LR statistic = 0.22   p = 0.639

Estimated mu = 357.75 (353.34, 2.17) (353.35 , 2.15)
Estimated kappa = 11.76 (9.93, 14.3) (9.17 , 14.42)
n = 100
90
180
0
270
α=10, κ=20
Rayleigh test: Rbar = 0.956 p = 0

Estimated alpha 13.82 (11.68 , 16.31)
Estimated kappa 32.01 (21.13 , 48.23)
n = 100
90
180
0
270
LR statistic = 7.89 p = 0.005

Estimated mu = 3.6 (349.83, 17.68) (350.49 , 16.7)
Estimated kappa = 1.88 (1.68, 2.41) (1.5 , 2.3)
n = 100
90
180
0
270
α=30, κ=2
Rayleigh test: Rbar = 0.677 p = 0

Estimated alpha 26.81 (13.76 , 48.63)
Estimated kappa 2.44 (1.58 , 3.76)
n = 100
90
180
0
270
LR statistic = 1.45 p = 0.228

Estimated mu = 357.25 (348.85, 6.68) (348.78 , 5.73)
Estimated kappa = 3.59 (3.19, 4.11) (2.86 , 4.38)
n = 100
90
180
0
270
α=30, κ=20
Rayleigh test: Rbar = 0.846 p = 0

Estimated alpha 29.14 (26.38 , 32.14)
Estimated kappa 16.18 (11.91 , 21.95)
n = 100
90
180
0
270
LR statistic = 37.58 p = 0

vmtests.on.files. Both *acorn* and *bimstat* produce a circular plot of the data along with estimates of the relevant parameters (μ and κ for *acorn*, α and κ for *bimstat*), along with confidence limits (as described in the preceding sections) and tests of their significance (against H_0 that $\kappa = 0$ in *acorn* using the Rayleigh test, against H_0 that $\alpha = 0$ in *bimstat* using the LR test). In each case, the data are in the form of one or more files containing a data set consisting of digitized start and endpoints for slug trails in the form "n x1 y1 x2 y2" as described under **Subheading 3.1.** We now describe briefly the use of these functions and provide examples in **Figs. 6–8**.

3.2.3.1. How to Use *acorn*

To analyze a data set using the ordinary unidirectional von Mises distribution in R type the command:

> *acorn(file,rose=T,stack=F,ml=F,shrink=1.2,kappa.alpha=0.1,*
> *mu.alpha=0.01,tracks=T, type = "n")*

using the appropriate values for each of the arguments to produce the desired outcome. In the preceding line, the arguments taken by the command *acorn* are included in parentheses, along with their default values. Except for the "file" argument, they all have predefined default values so that their inclusion in the

---

Fig. 7. *(opposite page)* Examples of the analysis of orientation when there are two preferred directions ($\pm\alpha$). Computer-generated samples containing 100 observations each were generated for the bidirectional von Mises distribution with $\alpha = 10°$ or $30°$ and $\kappa = 2$ or 20 as indicated in the figure. Each plot shows a rose diagram (circular histogram) of the directions, plus estimates of the preferred directions (solid arrows) and their bootstrap confidence intervals (dotted lines). The left-hand member of each pair shows the results of analysis of the sample based on the unidirectional von Mises model and the right-hand member shows the analysis based on the bidirectional model. The text above each rose diagram in the left column shows the estimates from the sample of μ and κ followed in parentheses first by the bootstrap confidence intervals and second by the traditional confidence intervals calculated as described in the text. Ninety-nine percent confidence intervals for μ and 90% confidence intervals for κ are shown for the unidirectional model. The text below each rose diagram in the left column shows the Rayleigh test statistic and associated significance probability that the data came from a uniform distribution (i.e., $\kappa = 0$). The hypothesis of uniformity is rejected in every case shown. The text above each rose diagram in the right column shows the maximum likelihood estimates for α and κ in the bidirectional model, along with their 95% confidence intervals. The likelihood ratio statistic and associated significance probability indicate the likelihood that the sample data could have come from the unidirectional rather than the bidirectional model. The unidirectional model is clearly rejected only in the two cases where κ was large.

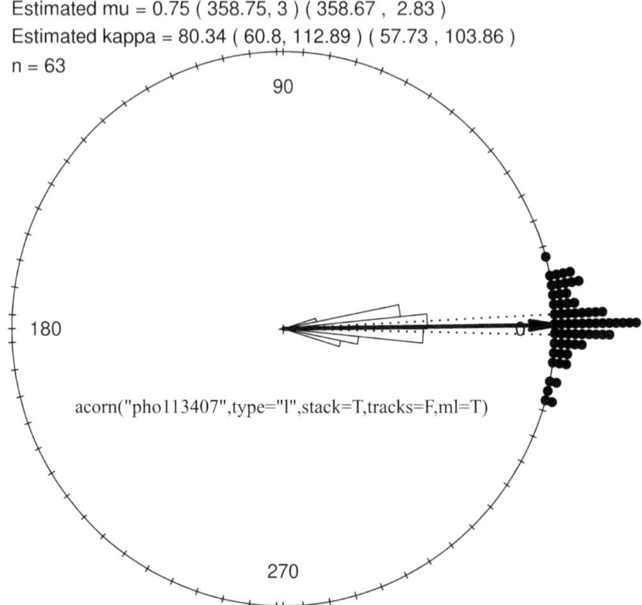

pho113407
Estimated mu = 0.75 (358.75, 3) (358.67 , 2.83)
Estimated kappa = 80.34 (60.8, 112.89) (57.73 , 103.86)
n = 63
90
180
270
acorn("pho113407",type="l",stack=T,tracks=F,ml=T)
Rayleigh test: Rbar = 0.994   p = 0

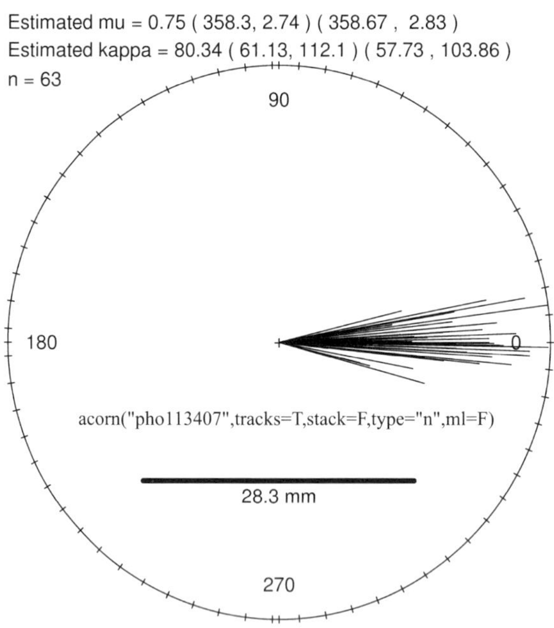

pho113407
Estimated mu = 0.75 (358.3, 2.74) (358.67 , 2.83)
Estimated kappa = 80.34 (61.13, 112.1) (57.73 , 103.86)
n = 63
90
180
270
acorn("pho113407",tracks=T,stack=F,type="n",ml=F)
28.3 mm
Rayleigh test: Rbar = 0.994   p = 0

command is optional—if a given argument is omitted, its default value is used. Thus, the simplest form of the *acorn* command is acorn(file="my_data_file_name"). Regardless of which arguments are provided to it, *acorn* returns a single series of numbers (a vector in R parlance) consisting in order of the sample size (n), the sample statistic ($\bar{R}$), estimates of the mean direction (μ), a bootstrap confidence interval for the mean direction (μ_L, μ_U), a traditional confidence interval for the mean direction (μ_L, μ_U), the accuracy of orientation (κ), a bootstrap confidence interval for the accuracy of orientation (κ_L, κ_U), a traditional confidence interval for the accuracy of orientation (κ_L, κ_U), and the significance probability in the Rayleigh test for H_0 that $\kappa = 0$. All of the optional arguments to *acorn* control the output plot as follows:

- If rose = F, no plot is produced.
- If stack = T, individual angles are plotted in the perimeter of the circle stacked in 2°-wide bins.
- If ml = T, the mean direction (solid arrow) and bootstrap confidence limits (dotted lines) are drawn.
- If tracks = T, a solid line is drawn to represent each individual observation, along with a scale bar.
- If type = "n," the plot is produced, but without the rose, whereas type = "l" plots the rose.
- kappa.alpha and mu.alpha are the level of confidence for maximum likelihood estimates.
- Shrink is a numerical shrinkage factor to reduce the size of the plot (sometimes needed to accommodate stacks).

Figure 8 illustrates the various options for the graphical output from *acorn* and *bimstat* using phototaxis data from slugs formed on charcoal agar at a density of 5×10^6 cells/cm^2.

3.2.3.2. How to Use *bimstat*

To analyze a data set using the bidirectional von Mises distribution in R, type the command:

> *bimstat(file,alpha=1,kappa=1,kappa1,rose=T,stack=F,ml=F,
> tracks=T,shrink=1.2,type="n")*

Fig. 8. *(opposite page)* Examples of the different graphical output options for the *acorn* and *bimstat* functions. The command that generated the output is shown as text inside each rose diagram. The file containing the digitized phototaxis results that were analyzed was called "pho113407." In both cases shown here, the *acorn* function was used, but *bimstat* for bidirectional analysis provides the same graphical output options (*see* **Subheadings 3.2.3.1.** and **3.2.3.2.**).

using the appropriate values for each of the arguments to produce the desired outcome. As with *acorn*, most arguments to *bimstat* are optional and take the indicated default values if they are omitted from the command. The exceptions are the data file name (in the form file= "my_data_file") and the numeric value of kappa1, which is the maximum likelihood estimate of κ returned by *acorn*. The value of kappa1 is required for the LR test of H_0 that $\alpha = 0$. The arguments alpha and kappa are initial values to be used as starting points for numerical determination of the maximum likelihood estimates of α and κ by the *dfp* and *newton* commands (*see* **Note 12**). The other arguments to *bimstat* control the plot output and have the same meaning as the corresponding arguments to *acorn* (*see* **Subheading 3.2.3.1.**).

3.2.3.3. How to Use vmtests.on.files

To test whether κ is equal for two or more data sets, each of which is contained in a separate file in the standard format (one line per slug in the format n x1 y1 x2 y2, as described in **Subheading 3.2.1.**, **step 2**) we provide the command *vmtests.on.files* (*see* **Note 13**). This command takes two arguments. The first is a collection of names of the files containing the data sets to be tested. The second, "save," takes the value T (i.e., true) by default and determines whether the results of running the *acorn* command on the set of files should be retained and saved in an R matrix called "acorn.results."

To use *vmtests.on.files*, type a command in R of the form:

> *vmtests.on.files(files=c("my_first_file","my_second_file",*
> *"my_third_file"), save=F)*

The output of *vmtests.on.files* is three numbers: $\bar{R}$ for the combined sample, the test statistic Q (in the two sample case), or U (in the multisample case) and the associated significance probability. If the significance probability is below a chosen critical value (say 0.05), reject the null hypothesis that κ is equal for all samples.

4. Notes

1. The plates are thick so that they support growth of a luxurient bacterial lawn and therefore a correspondingly high density of amoebae in the growing edge of the *Dictyostelium* colony.
2. It is worthwhile, when establishing phototaxis assays in a new location, to carry out some control experiments using containers for plates that do not provide any entry for light. If the slugs migrate in an oriented manner on these plates, then there are temperature or other gradients present that can control the behavior and influence the outcome of phototaxis experiments. In such cases, it is useful to try several locations in the room or incubator until one is found that lacks significant assymetric stimuli.

3. It is worthwhile to check the plates after several hours for the formation of a water droplet at the inoculum site. If this happens, the plate should be allowed to dry in the laminar flow cabinet and the incubation then continued. Droplet formation sometimes happens, especially on freshly prepared plates, and if not corrected would prevent aggregation.

4. To achieve the most even spread possible, the amoebae are inoculated after the bacteria and the inocula are spread over the plate surface in batches of no more than 10 plates.

5. The optimum stage for harvesting cells is when clearing is well advanced in the central regions of the plates, but there are still no signs of aggregation. However, plates can still be used if some aggregates have begun to form in limited areas of the plate.

6. The use of volumes of agar lower than this produces thinner plates, with the result that higher concentrations of metabolites accumulate in the agar, and phototaxis and thermotaxis begin to be impaired at lower cell densities.

7. For water agar plates, this is easily achieved by placing under the plate a template consisting of a Petri dish lid that has been marked with a 1-cm^2 area. For charcoal agar, it must be done by visual comparison with a template placed alongside the plate being inoculated.

8. By placing the discs face-down onto the surface of the Coomassie Blue staining solution, rather than immersing them, the intensity of the background staining of the plastic is halved.

9. The experiment requires disturbance of the slugs (to mark their early positions). This makes it extremely difficult to do these experiments in the strains of most interest in the "molecular era," the commonly used axenic strains and most of their derivatives. These are so sensitive to disturbance that almost every slug will cease migration and fruit shortly or immediately after its trail has been marked. For this reason, we do not describe the analysis of spontaneous turning in any further detail here. The development of suitably nonintrusive time-lapse cinematography methods for slugs would overcome the problem.

10. Manual digitizing of slug trails in this way is simple in principle, but complicated in practice (especially at high densities) by the criss-crossing of slug trails and the fact that, in practice, it is generally difficult, if not impossible, to follow a given trail all the way to its original start point in the 1-cm^2 area of the inoculum site. The digitizing thus requires a certain amount of practice and skill. The following points may be helpful.

    a. The datum of interest is the overall direction of travel of the slug, and its trail from where it first becomes visible (usually, as it emerges from the inoculum site) is as good a representation of its behavior as any other. The approximate point at which the trail leaves the inoculum square is therefore useful as a start point, if the true start point cannot be identified.

    b. The operator should not succumb to the temptation to arbitrarily designate the center of the inoculum area as the start point for all slugs. The reason can be illustrated by a simple, if extreme example. Imagine that slugs start migrating from sites randomly distributed over the 1-cm^2 inoculum site, that every slug migrates perfectly toward the light, and that every slug, having crossed the front edge of the inoculum area, migrates only 1 mm further before stopping. If the center were used as the "start point" for every slug, the measured directions to their end points would range from just under 45° leftward of the correct direction to just under 45° rightward of it—an arc of nearly 90°, which is clearly not representative of the true behavior. On the other hand, if each trail's start point is measured at the point at which it becomes visible—let us say, at the leading edge of the inoculum site—the measured directions will be a much better representation of the true behavior.

c. It is not critical that every slug trail be measured or that every individual measurement be absolutely accurate. Especially at high cell densities, there are so many slugs and trails that this may be humanly impossible. What is important is that the distribution of directions of the slug trail measurements that are made is a fair and accurate representation of the actual behavior. Therefore, it is important to avoid the temptation to measure only or mostly the slugs at the extremes of the distribution whose trails are easy to follow visually. If most slugs traveled within ±5° of the direction toward the light, most of the measurements should also fall in this range even if they resulted from less accurate digitizing of the individual start points because of crowded, criss-crossing trails.

d. Outliers: The occasional slugs whose behavior is clearly unrepresentative of the overall population can exert an undue influence on the statistical measurements for purely mathematical reasons. For example, an occasional slug formed at the back of a high-density inoculation site might first exit the inoculation area at the back only to cease migration and fruit before turning around toward the light and traveling in a direction more representative of the whole population. The choice by such a slug exits from the rear was probably influenced by repellants in the crowded inoculum site; once away from them, the slug would turn toward the light and behave in a more representative manner. However, if the slug culminates before it can do this, its original choice of starting direction will be "fixed." Such a slug will be easily observed and its trail easy to digitize, tempting the operator to include it in a sample of measurements that misses many of its fellows, whose behavior may be more representative but harder to pick out in the crowd. Such a slug may have been only 1 in 200 on the plate, but could end up as 1 in 50 measurements. For this reason, as well as because of their unduly large mathematical impact on the statistical analysis, outliers may exert a strongly misleading influence on the result. If they do, they should be excluded without compunction.

e. The graphical presentations of results produced by the software described under **Subheadings 3.2.3.1.** and **3.2.3.2.** will allow new operators to check visually that what they digitized is indeed a fair and accurate representation of the data. It is worthwhile for new operators to practice on a set of results on stained discs, to determine whether their results are both reproducible and a good representation of the actual behavior of the slugs. This will allow them to gain a feel for the process and gain confidence in the results.

11. To a biologist, the formulae used to calculate *p* values and sample statistics may seem strangely arbitrary. However, this is not so. They have been derived by statisticians from the mathematical properties of the appropriate distributions and then, in cases where the solutions are mathematically intractable, valid polynomial or other approximations are derived and thoroughly tested before being reported in the primary statistical literature.

12. Sometimes for particular combinations of alpha, kappa, and a given data set, *bimstat* will fail as a result of arithmetic singularities or similar problems. This is an inherent problem in numerical optimization algorithms such as the Newton (*newton*) and the Davidon-Fletcher-Powell (*dfp*) methods. It can almost always be overcome by providing different initial values for alpha and kappa, but the user might sometimes need to try several different combinations of values before the function succeeds. A simple call to *bimstat* that does this might look as follows: *bimstat("my_data_file," alpha=1, kappa=2, kappa1 = 5.234)*.

13. The function *vmtests.on.files* is actually a front end for the function *vmtests* which take as its input a matrix whose first two columns contain sample sizes and mean resultants ($\bar{R}$) for the data sets to be tested. A matrix containing these values is generated in *vmtests.on.files* by running *acorn* on each file and storing the result in the matrix *acorn.results*.

References

1. Fisher, P. R. (2001) Genetic analysis of phototaxis in *Dictyostelium*, in *Photomovement. ESP Comprehensive Series in Photosciences Vol 1*. (Häder, D.-P. and Lebert, M., eds.). Chapter 19. Elsevier Science Ltd., Amsterdam: pp. 519–559.
2. Poff, K. L. and Häder, D.-P. (1984) An action spectrum for phototaxis by pseudoplasmodia of *Dictyostelium discoideum. Photochem. Photobiol.* **39**, 433–436.
3. Poff, K. L. and Skokut, M. (1977) Thermotaxis by pseudoplasmodia of *Dictyostelium discoideum. Proc. Natl. Acad. Sci. USA* **74**, 2007–2010.
4. Raper, K. B. (1940) Pseudoplasmodium formation and organization in *Dictyostelium discoideum. J. Elisha Mitchell Sci. Soc.* **56**, 241–282.
5. Bonner, J. T., Clarke, W. W., Neely, C. L., and Slifkin, M. K. (1950) The orientation to light and the extremely sensitive orientation to temperature gradients in the slime mold *Dictyostelium discoideum. J. Cell. Compar. Physiol.* **36**, 149–158.
6. Francis, D. W. (1964) Some studies on phototaxis of *Dictyostelium. J. Cell. Comp. Physiol.* **64**, 131–138.
7. Poff, K. L., Butler, W. L., and Loomis, W. F. (1973) Light-induced absorbance changes associated with phototaxis in *Dictyostelium. Proc. Natl. Acad. Sci. USA* **70**, 813–816.
8. Poff, K. L., Loomis, W. F., and Butler, W. L. (1974) Isolation and purification of the photoreceptor pigment associated with phototaxis in *Dictyostelium discoideum. J. Biol. Chem.* **249**, 2164–2168.

9. Poff, K. L. and Butler, W. L. (1974) Spectral characteristics of the photoreceptor pigment of phototaxis in *Dictyostelium discoideum. Photochem. Photobiol.* **20,** 241–244.

10. Fisher, P. R., Smith, E., and Williams, K. L. (1981) An extracellular chemical signal controlling phototactic behavior by *D. discoideum* slugs. *Cell* **23,** 799–807.

11. Fisher, P. R. and Williams, K. L. (1981) Activated charcoal and orientation behaviour by *Dictyostelium discoideum* slugs. *J. Gen. Microbiol.* **126,** 519–523.

12. Fisher, P. R. and Williams, K. L. (1982) Thermotactic behaviour of *Dictyostelium discoideum* slug phototaxis mutants. *J. Gen. Microbiol.* **128,** 965–971.

13. Fisher, P. R., Grant, W. N., Dohrmann, U., and Williams, K. L. (1983) Spontaneous turning behaviour by *Dictyostelium discoideum* slugs. *J. Cell Sci.* **62,** 161–170.

14. Mardia, K. V. (1972) *Statistics of Directional Data.* Academic Press, London & New York.

15. Jammalamadaka, S. R. and SenGupta, A. (2001) *Topics in Circular Statistics.* World Scientific Publishing Co. Ptg. Ltd., Singapore.

16. Fisher, P. R. and Williams, K. L. (1981) Bidirectional phototaxis by *Dictyostelium discoideum* slugs. *FEMS Microbiol. Lett.* **12,** 87–89.

17. The R Foundation for Statistical Computing. (2005) *The R Project for Statistical Computing.* http://www.r-project.org/index.html.

18. Stephens, M. A. (1969) Tests for the von Mises distribution. *Biometrika* **56,** 149–160.

19. Bartels, R. (1984) Estimation in a bidirectional mixture of von Mises distributions. *Biometrics* **40,** 777–784.

10

Purification Techniques of Subcellular Compartments for Analytical and Preparative Purposes

Laurence Aubry and Gérard Klein

Summary

In the following protocols, broken cells are the starting material of all downstream purifications of functional organelles or intact subcellular membranes. The choice of the breakage method has direct and deep repercussions on the quality of subsequent steps. Breaking vegetative amoebae by shear stress with a steel ball cell cracker preserves the integrity of subcellular organelles and in particular that of lysosomes, the rupture of which is very deleterious to further purifications.

In this chapter, we propose purification schemes for plasma membrane, nuclei, mitochondria, and endocytic compartments. Plasma membranes are purified without any cell coating by partition between aqueous polymer phases. Nuclei and mitochondria are purified by differential centrifugations in adequate buffer conditions. Endosomes are magnetically isolated after feeding the cells with colloidal iron dextran and phagosomes by flotation on a sucrose gradient after feeding amoebae with latex beads.

As analytical approaches, we propose procedures to label the plasma membrane and the endo-lysosomal compartments by biotinylation and to separate early and late compartments on a Percoll gradient.

Key Words: Plasma membrane; nuclei; mitochondria; endocytic compartment; gradient; purification, *Dictyostelium*.

1. Introduction

Despite the increasing use of imaging techniques, cell fractionation remains a useful tool to determine, in addition to the localization of molecules, the functional capacity and biochemical properties of individual compartments or organelles.

Classical fractionation techniques rely on density gradient centrifugation and differential velocity centrifugation of the membranes or organelles after homogenization of the cell. Partitioning in aqueous polymer two-phase systems

From: *Methods in Molecular Biology, vol. 346:* Dictyostelium discoideum *Protocols*
Edited by: L. Eichinger and F. Rivero © Humana Press Inc., Totowa, NJ

is an alternative for the fractionation of biological material such as plasma membranes. Purification of endosomal compartments (pinosomes and phagosomes) benefits from the possibility to specifically load these organelles with a marker that can be directly used for purification (iron-dextran, latex beads).

A clue for the purification of the most intact compartments, as determined by their protein profile, is the use of a cell breakage method that maximally preserves the integrity of organelles. Inclusion in the buffers of protease inhibitor cocktails that lower the potency of potentially aggressive lysosomal enzymes released by damaged lysosomes is a recommended precaution.

2. Materials

Except when mentioned otherwise, all steps should be performed at 4°C and solutions prechilled in an ice bucket.

2.1. Cell Breakage

Balch and Rothman have introduced a new device to break cells: the cell cracker *(1)*. The cell suspension is forced through a bore containing a steel ball via attached syringes. This ball-bearing homogenizer has the advantage of permitting rapid breakage of large volumes of cell suspension without damaging subcellular compartments. We have achieved efficient cell breakage of *Dictyostelium* amoebae with this device *(2)*.

1. Buffer 1: 20 mM 2-morpholinoethanesulfonic acid (MES)-NaOH, pH 6.5.
2. Buffer 2: 0.25 M sucrose, 1 mM ethylenediamine tetraacetic acid (EDTA), 20 mM MES-NaOH, pH 6.5. We classically use this buffer to break cells, but it can be modified depending on the subsequent purification steps and the organelles to be purified.
3. Cell counter (Beckman-Coulter).
4. Cell cracker (*see* **Fig. 1**). Instructions for its fabrication are given in **ref. 1**. It is also commercially available from the EMBL machine shop in Heidelberg.

2.2. Purification of Plasma Membrane

A number of methods for the isolation of *Dictyostelium discoideum* plasma membranes have been described (for review, *see* **ref. 3**). We prefer a method that uses an aqueous two-phase dextran-polyethylene glycol system *(4)*.

1. General buffer (Sörensen buffer): 14.6 mM KH$_2$PO$_4$, 2.0 mM Na$_2$HPO$_4$, pH 6.1.
2. Protease inhibitor cocktail (*see* **Note 1**):
   a. Phenylmethanesulfonyl fluoride (PMSF): stock solution at 100 min dry ethanol. Active against trypsin, chymotrypsin. Working dilution: 1/100 (*see* **Note 2**).
   b. Aprotinin: stock solution at 2 mg/mL in 10 mM HEPES, pH 8.0. Active against kallikrein, trypsin, chymotrypsin, plasmin. Working dilution: 1/1000.

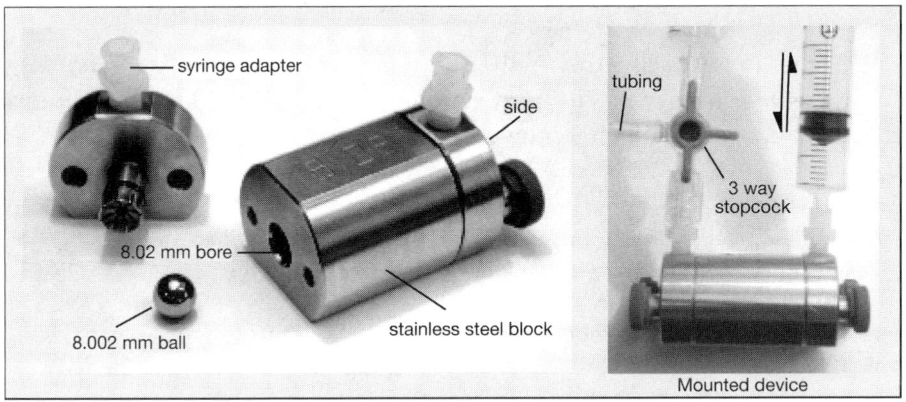

Fig. 1. Cell breakage. The cell suspension is forced through a ball-containing cylinder (18-μm gap) by a push-pull movement on the connected syringes, leading to cell disruption by shear stress. This device maximally preserves the integrity of subcellular organelles.

   c. Leupeptin: stock solution at 2 mg/mL in distilled water. Active against plasmin, trypsin, papain, cathepsin B. Working dilution: 1/1000.

   d. Pepstatin A: stock solution at 1 mg/mL in ethanol. Active against pepsin, cathepsin D. Working dilution: 1/1000.

Stock solutions can be stored at –20°C for months.

3. 20% (w/w) Dextran 500 (M_w 500,000, Amersham Biosciences). Dissolve progressively 50 g of dextran by fractionwise addition of dextran to 200 mL of distilled water.

4. 30% (w/w) polyethylene glycol 40,000. Dissolve 75 g of polyethylene glycol (PEG) in 175 mL of distilled water.

5. 0.22 *M* sodium phosphate, pH 6.5. Prepare a 0.22 *M* solution of NaH_2PO_4 by dissolving 30.36 g of sodium dihydrogen phosphate monohydrate in a final volume of 1 L distilled water. The pH of such a solution should be close to 4. Prepare a 0.22 *M* solution of Na_2HPO_4 by dissolving 31.24 g of anhydrous disodium hydrogen phosphate in a final volume of 1 L distilled water. The pH of such a solution should be close to 9.0. Mix about 700 mL of the Na_2HPO_4 solution and 1000 mL of the NaH_2PO_4 solution to obtain a 0.22 *M* sodium phosphate buffer, pH 6.5. Volumes are indicative and should be adjusted to reach pH 6.5 for the final solution.

6. In a separating funnel, mix 200 g of the 20% dextran solution, 103 g of the 30% PEG solution, 333 mL of the 0.22 *M* sodium phosphate buffer, pH 6.5, and 179 mL of distilled water. Allow the two phases to separate overnight in the cold. The upper and lower phases are then recovered independently and stored at 4°C.

2.3. Purification of Nuclei

The method described by Nellen is our favorite to purify nuclei from *Dictyostelium* strains *(5)*. The purification is further polished by an additional centrifugation step on a discontinuous sucrose gradient *(6)*.

2.3.1. General Solutions

1. Spermidine trihydrochloride (FW 254.6) (Sigma). Spermidine is a polycation that condenses and stabilizes chromatin by interaction with the polyanionic DNA.
2. NP40: nonionic detergent
3. PMSF solution: *see* **Subheading 2.2.**, **item 2**.
4. Buffer 3: prepare 150 mL of buffer containing 5% (w/v) sucrose, 20 mM KCl, 40 mM MgCl$_2$, 50 mM 4-(2-hydroxyethyl)piperazine-1-ethane sulfonic acid (HEPES)-NaOH, pH 7.5, 0.1% β-mercaptoethanol (14 mM), 0.038 mg/mL spermidine (0.15 mM), 0.2 mM PMSF.
5. Buffer 4: mix 90 mL of buffer 3 and 10 mL of Percoll (Amersham Biosciences) (*see* **Note 3**).
6. Sörensen buffer (*see* **Subheading 2.2.**, **item 1**).

2.3.2. Sucrose Step Gradient

Prepare the following solutions:

1. 50% Sucrose (w/w) in 5 mM Tris-HCl, pH 7.5: 5 g sucrose + 50 μL 1 M Tris-HCl, pH 7.5 + 5 mL distilled water.
2. 65% Sucrose (w/w) in 5 mM Tris-HCl: 6.5 g sucrose + 50 μL 1 M Tris-HCl, pH 7.5 + 3.5 mL distilled water.
3. 80% Sucrose (w/w) in 5 mM Tris-HCl: 8 g sucrose + 50 μL 1 M Tris-HCl, pH 7.5 + 2 mL distilled water (*see* **Note 4**).

Chill the solutions before preparation of the sucrose step gradients.

2.4. Purification of Mitochondria

Most purification techniques of mitochondria rely on differential centrifugations steps. Titration of adenine nucleotide translocator sites in mitochondria purified according to the procedure below gives values similar to data in yeast and rat liver mitochondria *(7)*.

Buffer 5: prepare 100 mL of 0.25 M sucrose, 10 mM Tris-HCl, pH 7.5 and protease inhibitors (*see* **Note 2** and **Subheading 2.2.**, **item 2**).

2.5. Purification of Endocytic Compartments

2.5.1. Endo-lysosomes: Magnetic Purification

Magnetic fractionation of *Dictyostelium* compartments of the endo-lysosomal pathway has been introduced by the group of Steck *(8)* and used in our group to characterize the proteome of endocytic compartments *(9)*. The protocol can be adapted in pulse-chase variations to specifically load endosomes, lysosomes or postlysosomal compartment as assessed by their specific pH *(10,11)*.

2.5.1.1. GENERAL SOLUTIONS AND MATERIAL

1. Buffer 6: 5 m*M* glycine-KOH pH 8.5, 100 m*M* sucrose.
2. Magnetic column: a plastic column (50 mL) filled with stainless steel wool (scour pad).
3. Magnet: a permanent magnet (10 × 5 × 40 mm) of 0.2 Tesla custom-made by UGIMAG (Saint Pierre-d'Allevard, France).

2.5.1.2. PREPARATION OF COLLOIDAL IRON PARTICLES (MAGNETITE-DEXTRAN)

1. Mix 5 mL of 1.2 *M* FeCl$_2$ with 5 mL of 1.8 *M* FeCl$_3$.
2. Add dropwise 5 mL NH$_4$OH (30%) while stirring at room temperature.
3. Gather the precipitate along the beaker with a permanent magnet. Wash the precipitate once with 60 mL of 5% (v/v) NH$_4$OH and twice with distilled water. Smell of ammoniac should be undetectable.
4. Resuspend the precipitate in 40 mL of HCl (0.3 *N*) and stir for 30 min.
5. Add 2 g of solid dextran T-40 (Amersham Biosciences) and keep stirring for at least 30 min (*see* **Note 5**).
6. Dialyze against distilled water in the cold room for 2 days with regular changes of water (*see* **Note 6**).

2.5.1.3. IRON CONTENT ASSAY **(12)**

1. In a glass test tube, add 50 µL of concentrated nitric acid and 50 µL of concentrated sulfuric acid (*see* **Note 7**) to the iron-dextran sample. This step mineralizes the sample.
2. Heat the tube for 5 min with a Bunsen burner or, more safely, in a test tube heater at 250°C. Whitish fumes appear immediately, followed by brown nitrous fumes (*see* **Note 8**). A color-free viscous liquid remains in the tube.
3. Let the tube cool down and add 0.75 mL of saturated sodium acetate, 1.25 mL of distilled water, 0.1 mL of 5% (w/v) sodium ascorbate in 3 *M* NaOH, and 0.4 mL of 0.1% (w/v) aqueous bathophenanthroline. Ascorbate will reduce ferric iron into ferrous iron that reacts with bathophenantroline to give a colored complex.
4. Read the optical density at 535 nm. The molar extinction coefficient is 0.01 *M*$^{-1}$cm^{-1}.

2.5.2. Phagosomes: Latex Beads and Flotation

Phagosome purification exploits their floating properties in sucrose gradients after loading the cells with latex beads *(13,14)*.

2.5.2.1. GENERAL SOLUTION

1. Sorbitol.
2. Sörensen buffer: *see* **Subheading 2.2.1.**

3. Buffer 7: 20 m*M* HEPES-KOH, pH 7.2, 0.25 *M* sucrose, protease inhibitor cocktail (*see* **Subheading 2.2.2.**).
4. Buffer 8: 20 m*M* HEPES-KOH, pH 7.2, 20 m*M* KCl, 2.5 m*M* MgCl$_2$, 1 m*M* dithiothreitol (DTT), 20 m*M* NaCl.
5. Buffer 9: 25 m*M* HEPES-KOH, pH 7.2, 1.5 m*M* Mg-Acetate, 1 m*M* NaHCO$_3$, 1 m*M* CaCl$_2$, 25 m*M* KCl, 1 m*M* ATP, 1 m*M* DTT, 100 m*M* sucrose and protease inhibitors (*see* **Subheading 2.2.2.**).
6. Latex beads: 0.8 μm diameter (Sigma).

2.5.2.2. Sucrose Step Gradient

Prepare the following solutions:

1. 10% Sucrose (w/w) in 3 m*M* imidazole-HCl: 1 g sucrose + 60 μL 0.5 *M* imidazole-HCl, pH 7.4 + 9 mL distilled water.
2. 25% Sucrose (w/w) in 3 m*M* imidazole-HCl: 2.5 g sucrose + 60 μL 0.5 *M* imidazole-HCl, pH 7.4 + 7.5 mL distilled water.
3. 35% Sucrose (w/w) in 3 m*M* imidazole-HCl: 3.5 g sucrose + 60 μL 0.5 *M* imidazole-HCl, pH 7.4 + 6.5 mL distilled water.
4. 62% Sucrose (w/w) in 3 m*M* imidazole-HCl: 6.2 g sucrose + 60 μL 0.5 *M* imidazole-HCl, pH 7.4 + 3.8 mL distilled water (*see* **Note 3**).

Chill the solutions before preparation of the gradients.

2.6. Plasma Membrane and Endocytic Compartment Labeling

The plasma membrane and the connected endocytic compartments can be labeled with a variety of probes. Biotinylation using sulfoNHS-biotin presents practical advantages over iodination, as no radioactivity is on hand and residues susceptible to be biotinylated are much more abundant. A derivate of sulfoNHS-biotin that carries a reducible disulfide bond is available and allows selective labeling of internal endocytic compartments after labeling, endocytic turnover, and stripping of the labeled proteins still resident at the plasma membrane *(15)*.

1. Sulfo-NHS-SS-biotin (Pierce).
2. Sörensen buffer (*see* **Subheading 2.2.1.**) adjusted to pH 7.2 or pH 8.0.
3. Buffer 10: 50 m*M* glutathione, 75 m*M* NaCl, 75 m*M* NaOH, 1% bovine serum albumin (BSA).

2.7. Endosome/Lysosome Separation (Percoll Gradient)

Fractionation of the different compartments of the endocytosis pathway is complementary to imaging using compartment-directed antibodies. Percoll gradients separate light endosomes from heavy lysosomes and postlysosomal compartment *(16)*, and the fluid-phase transit through these successive compartments can be followed with a fluorescent fluid-phase marker such as fluorescein isothiocyanate (FITC)-dextran.

1. FITC-dextran (Sigma FD-70S, M_w 70,000): prepare a fresh stock at 10 mg/mL in growth medium and filter through a 0.22-μm Millipore filter to remove insoluble particles.
2. Buffer 2: *see* **Subheading 2.1.**
3. Percoll auto-forming gradient: Prepare 21 mL of a 24% Percoll (Amersham-Biosciences) solution in Buffer 2.
4. Buffer 11: 100 mM Na_2HPO_4, pH 9.0, 0.25% Triton X-100.

3. Methods

3.1. Cell Breakage

1. Mount the cell cracker as shown in **Fig. 1** (*see* **Note 9**). Adjust a three-way stopcock, two 5-mL syringes and Teflon inlet/outlet tubing. Store on ice to cool down the device.
2. Count cells in a cell counter (*see* **Note 10**). Optimally, cells should be harvested from a log-phase culture ($2-8 \times 10^6$ cells/mL). Collect about 1×10^9 cells.
3. Tare an empty centrifuge tube.
4. Wash the cells in ice-cold buffer 1 by centrifuging 5 min at 1500g at 4°C.
5. Determine the weight of the wet cell pellet (*see* **Note 11**).
6. Add a volume of buffer 2 equivalent to the weight of wet cells. The cell titer after resuspension of the cells is close to 5×10^8 cells/mL. Determine the titer on a diluted aliquot in a cell counter.
7. Aspirate the cell suspension into the first syringe. Connect the two syringes and push/pull 10 times to break the cells. Eject the broken cell suspension.
8. A new counting gives an estimate of the cell-breaking yield (equals counting × diluting factor after breakage over counting × diluting factor before breakage). If the yield is around 50% or less, start breakage procedure again at **step 7**.
9. Dilute the suspension by addition of one to four volumes of buffer 2. Centrifuge 5 min at 1500g to eliminate unbroken cells and nuclei. The supernatant represents a postnuclear fraction to be used in further purification steps.

3.2. Purification of Plasma Membrane

1. Wash $2-4 \times 10^9$ cells in Sörensen buffer containing 1 mM PMSF. Resuspend at a cell concentration of 5×10^8 cells/mL in Sörensen buffer containing protease inhibitor cocktail.
2. Break the cells as described in **Subheadings 2. 1.** and **3.1.** Dilute the suspension, eliminate unbroken cells by a 5 min centrifugation at 1500g.
3. Centrifuge the postnuclear supernatant at 15,000g for 15 min.
4. Resuspend the pellet in 10 mL of ice-cold upper phase (*see* **Subheading 2.2.3.** preparation of the two-phase system). Homogenize with a Potter homogenizer. Add 10 mL of ice-cold lower phase (*see* **Subheading 2.2.3.**, preparation of the two-phase system). Mix with a pipet and centrifuge for 5 min at 10,000g at 4°C.
5. Carefully collect the material at the interphase and resuspend in 10 mL of upper phase. Add 10 mL of lower phase, mix gently and centrifuge for 5 min at 10,000g.

6. Repeat **step 5** once.
7. Collect carefully the interphase, taking care to remove a minimal volume of upper and lower phases (less than 2.5 mL). Add 10 mL of Sörensen buffer plus PMSF (*see* **Note 12**). Resuspend the membranes and centrifuge for 10 min at 15,000*g*.
8. The pellet, representing plasma membranes, can be resuspended in any buffer compatible with the following experiment (*see* **Note 13**).

3.3. Purification of Nuclei

1. Collect 5×10^9 cells from an exponential culture by centrifugation at 1500*g* for 5 min.
2. Wash the cells twice with 500 mL ice-cold Sörensen buffer and resuspend the cells in 50 mL of buffer 4.
3. Add 2.5 mL of 20% NP-40 and spin for 5 min at 3000*g*.
4. Eliminate the supernatant and resuspend the pellet in 20 mL of buffer 4. Centrifuge for 5 min at 150*g*.
5. Centrifuge the supernatant for 5 min at 3000*g*.
6. Eliminate the supernatant and resuspend the pellet in 20 mL of buffer 3. Centrifuge for 5 min at 3000*g*.
7. Resuspend the pellet in 1 mL of buffer 3.
8. Prepare a sucrose step-gradient by successively layering 1.5 mL of 80% sucrose, 3 mL of 65% sucrose, 3 mL of 50% sucrose, and 1 mL of the nuclei preparation.
9. Spin for 1 h at 60,000*g* in a swinging-bucket rotor (type SW41, Beckman).
10. Recover the material at the 65–80% sucrose interface. Purified nuclei can be diluted in buffer 3, spun 5 min at 3000*g* and resuspended in an appropriate buffer.

3.4. Purification of Mitochondria

1. Wash vegetative amoebae in ice-cold buffer 5 and break them with the cell cracker (*see* **Subheadings 2.1.** and **3.1.**)
2. Dilute the homogenate by addition of 5 volumes of buffer 5. Centrifuge the suspension for 5 min at 1200*g* to remove intact cells and nuclei.
3. Centrifuge the postnuclear supernatant from **step 2** for 15 min at 8000*g*.
4. Resuspend the pellet in buffer 5 (in the same volume as **step 2**) and centrifuge the suspension for 5 min at 1200*g* to remove remaining intact cells and nuclei.
5. Repeat **step 3** to pellet mitochondria. Resuspend the pellet in a minimal volume of buffer 5 and measure the protein content of the suspension (*see* **Note 14**).

3.5. Purification of Endocytic Compartments

The purification protocols described below take advantage of a preloading of cells with specific probes. Depending on the needs, the endocytic pathway can either be fully loaded by a 2-h charge or subjected to a pulse-chase consisting of a short pulse period (5–10 min) in the presence of the probe followed by a chase period of variable length in fresh medium.

3.5.1. Endo-Lysosomes: Magnetic Purification

1. Harvest 1×10^9 cells from a log phase culture by centrifugation (5 min, 1500g) and resuspend at 1×10^7/mL in axenic medium containing colloidal iron particles (1 mg iron/mL) (*see* **Note 15**). Keep at 21°C on a rotary shaker (120 rotations/min) for at least 120 min to label the entire endocytic pathway (*see* **Note 16**).
2. Stop endocytosis by addition of 100 mL of ice-cold buffer 6 supplemented with 0.5% BSA (*see* **Note 17**).
3. Harvest the cells (1000g, 5 min) and wash three times with 100 mL of the same buffer.
4. Resuspend the pellet in 5 mL of buffer 6 containing 2 mM EDTA and protease inhibitor cocktail. Break the cells in a cell cracker (*see* **Subheadings 2.1.** and **3.1.**).
5. Increase the volume to 10 mL, centrifuge 5 min at 1500g to remove unbroken cells and nuclei. Circulate the postnuclear supernatant through the magnetic column placed in the magnetic field (*see* **Fig. 2**) with a peristaltic pump at low rate (1 mL/min) for 30 min.
6. Wash the column with 100 mL of ice-cold buffer 6 containing 2 mM EDTA and remove the column from the magnetic field.
7. Elute the retained fraction with 40 mL of buffer 6 containing 2 mM EDTA and pellet the purified compartments by centrifugation (15,000g, 40 min).

3.5.2. Phagosomes: Latex Beads and Flotation

1. Harvest 1×10^9 cells in log phase of growth by centrifugation (5 min, 1500g). Resuspend in 5 mL of ice-cold Sörensen buffer containing 120 mM sorbitol, pH 8 and add 2×10^{11} latex beads to allow adsorption of the beads onto the plasma membrane (i.e., 200 beads/cell).
2. Transfer the suspension to 95 mL of culture medium at 21°C (cell density is now 1×10^7 cells/mL) and keep shaking (120 rpm) for 5 to 120 min to allow progression of ingested beads along the phagocytic pathway.
3. Stop phagocytosis by addition of 300 mL of ice-cold Sörensen buffer/sorbitol and centrifuge 5 min at 150 g to pellet the cells. Most external beads will remain in the supernatant. Repeat washing twice (*see* **Note 18**).
4. If a chase period is required to reach a specific compartment, resuspend rapidly the cell pellet in 100 mL of culture medium at 21°C and keep shaking for the duration of the chase period. Then, wash extensively as described in **step 3**.
5. Resuspend the cell pellet in buffer 7 and break the cells using a cell cracker (*see* **Subheadings 2.1.** and **3.1.**).
6. The postnuclear supernatant is treated with 10 mM ATP-Mg for 15 min on ice to remove most actin/myosin and associated proteins.
7. Adjust the supernatant to 40% sucrose by addition of a 62% sucrose solution.
8. Load the 40% sucrose supernatant (10 mL) on top of 1-mL cushion of 62% sucrose. Add sequentially 2 mL of 35% sucrose, 2 mL of 25% sucrose and 2 mL of 10% sucrose solutions. Centrifuge the gradient in a swinging bucket rotor (type SW28.1, Beckman) for 3 h at 100,000g.

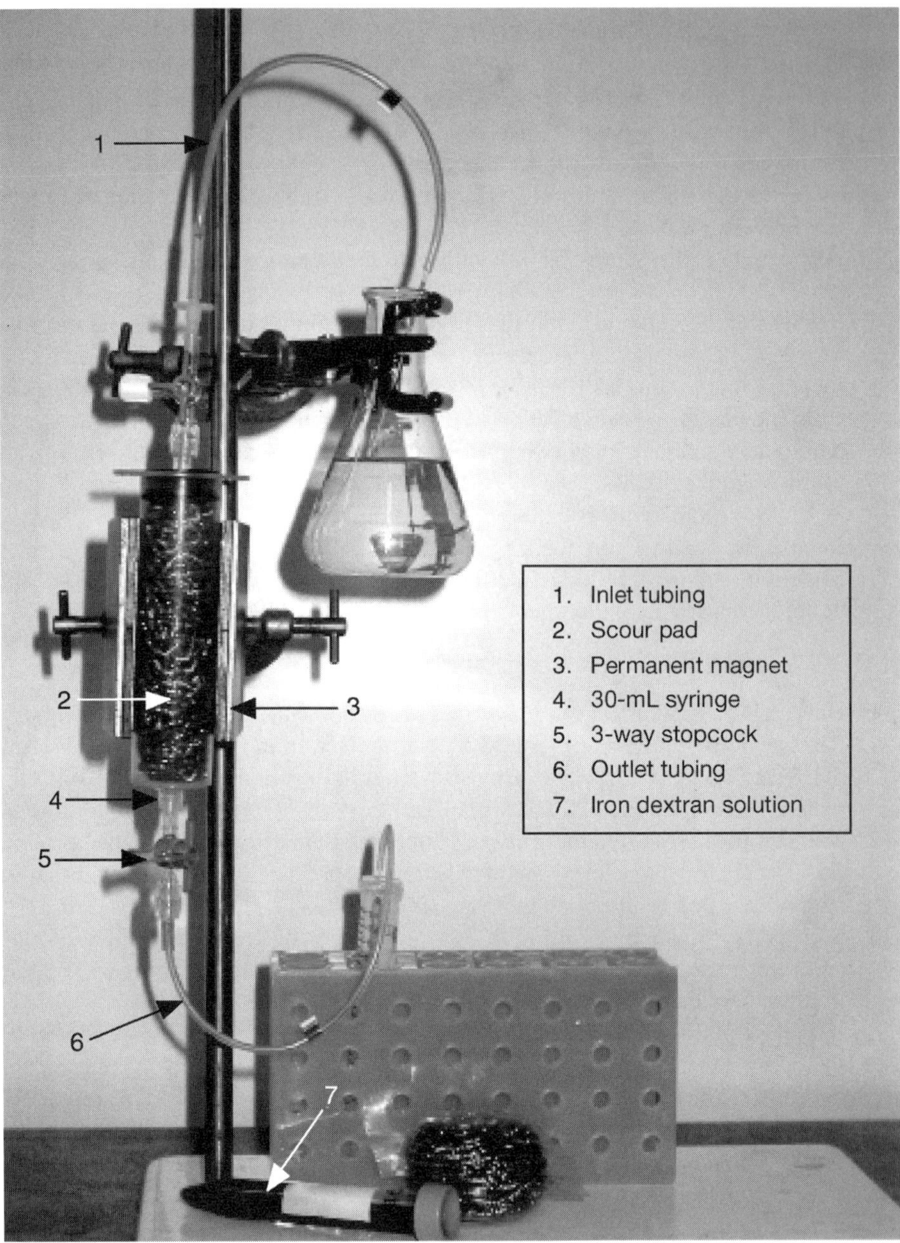

Fig. 2. Magnetic purification of endocytic compartments. Cells preloaded with magnetite-dextran are broken using a cell cracker (*see* **Fig. 1**). The postnuclear supernatant is passed through a column containing a stainless steel scour pad maintained in the air gap of a permanent magnet. Endocytic compartments retained on the column are eluted after extensive washes by removing the magnet.

9. Collect the 10–25% sucrose interface, dilute in 30 mL of buffer 8, and centrifuge for 1 h at 100,000*g*.
10. The membranous pellet can be resuspended in buffer 9 and stored at –80°C for further analysis.

3.6. Plasma Membrane and Endo-Lysosome Labeling by Biotinylation

1. Cells harvested from a log phase culture are washed in ice-cold Sörensen buffer, pH 8.0 and resuspended at a density of 10^7 cells/mL in Sörensen buffer containing 0.25 mg/mL of sulfo-NHS-SS-biotin (*see* **Note 19**).
2. Label proteins of the plasma membrane by a 30 min incubation at 4°C.
3. Wash twice in ice-cold Sörensen buffer, pH 7.2 containing 40 m*M* NH_4Cl to quench any remaining biotinylation reagent.
4. If endocytic compartment labeling is required, resuspend the cells in prewarmed (21°C) culture medium (*see* **Note 20**) and let endocytosis proceed for adequate times that allow biotinylated proteins to reach the appropriate compartment *(10,19)*.
5. Stripping of plasma membrane labeling by reduction of the disulfide bond of sulfo-NHS-SS-biotin can be performed at this stage by incubation of the cells in ice-cold buffer 10 for 30 min.
6. Wash twice in Sörensen buffer, pH 7.2 before use in the subsequent steps.
7. Biotinylated proteins can be visualized by Western blot using avidin-horseradish peroxidase (HRP).

3.7. Endosome/Lysosome Separation (Percoll Gradient)

1. Harvest 10^9 cells and resuspend in 90 mL of culture medium.
2. Let cells recover for 10 min on a rotary shaker before addition of FITC-dextran (1 mg/mL final concentration).
3. Let cells internalize the fluid-phase marker for 2 h at 21°C to allow complete labeling of the endocytic compartments.
4. Stop endocytosis by a cold shock by addition of 100 mL of ice-cold buffer 1 containing 0.5% BSA (*see* **Note 17**) and wash twice in the same buffer by resuspension/centrifugation (1500*g*, 5 min).
5. Resuspend the pellet in buffer 2 and proceed to cell breakage as described under **Subheadings 2.1.** and **3.1.**
6. Load 4 mL of the postnuclear supernatant onto the Percoll gradient (*see* **Subheading 2.7.**).
7. Centrifuge the gradient in a TFT70 Kontron rotor at 33,000*g* for 1 h (*see* **Note 21**).
8. Elute the gradient by 1.5-mL fractions from the bottom of the tube by insertion of a 100-µL capillary tube connected to a collecting pump (*see* **Note 22**).
9. Remove 200 µL of each fraction and lyse membranes in buffer 11 (*see* **Note 23**). Measure the fluorescence of the samples at an excitation wavelength of 470 nm and an emission wavelength of 520 nm. Subtract fluorescence values of samples from a blank gradient with nonlabeled cells (*see* **Note 24**). Endocytic compartments mainly follow a bimodal distribution with endosomes in the light fraction and lysosomes/postlysosomes in the heavy fraction (*see* **Note 25**).

3.8. Assays of Contaminating Organelles

Purification of a given compartment is estimated from the purification factor and yield based on the recovery of a compartment-specific activity. Its purity is estimated by assaying the activities of contaminating organelles. Classical protocols assay acid phosphatase for lysosomes, 4-chloro-7-nitrobenzofurazan (NBD)-sensitive ATPase (assayed at pH 7.0) for the contractile vacuole, oligomycin- or azide-sensitive ATPase for mitochondria, vanadate-sensitive ATPase for plasma membrane, aspartate alanine aminotransferase for cytosol, RNA for rough endoplasmic reticulum, DNA for nuclei. References for the different assays can be found in earlier work *(17,18)*.

4. Notes

1. Alternatively, ready-to-use inhibitor cocktails such as Complete® (Roche) are commercially available.
2. PMSF readily hydrolyses in water within 1 h at 4°C. Prepare solutions containing PMSF just prior to use.
3. Low viscosity of dense Percoll solutions enables the use of low-speed centrifugations for the purification of nuclei.
4. Sucrose solutions above 60% (w/w) are made by gently heating (no caramel!) on a magnetic stirrer.
5. Colloidal iron in solution is not attracted by the magnet.
6. Colloidal iron can be stored at 4°C at a concentration of 10 mg iron/mL in distilled water for at least 2 mo.
7. Use sulfuric acid with an iron concentration below 0.1 ppm.
8. Handling a boiling mixture of nitric and sulfuric acids is potentially dangerous. Hold the tube with wooden tongs. Work under a hood, and wear protective gloves and a face shield.
9. In our hands, use of an 8.02-mm cylinder with an 8.002-mm ball gives optimal results.
10. Isoton® used for counting red blood cells should be diluted five times to count *Dictyostelium* cells. Cells can also be counted in a hemocytometer.
11. 10^9 cells equals approx 1 g wet cells which equals approx 70 mg protein.
12. At this concentration, dextran and polyethylene glycol no longer form a two-phase system.
13. Bound actin and myosin can be removed by treating membranes for 15 min with 10 m*M* ATP-Mg on ice and centrifuging again.
14. It is essential to resuspend mitochondria at the highest possible protein concentration, especially if mitochondria are to be frozen and conserved. A protein concentration of mitochondrial protein of 40 mg/mL or above is adequate.
15. The iron probe is used at an optimal concentration of 1 mg iron/mL. At a lower concentration, vesicles are not completely magnetically retained and higher concentrations result in an excessive fragility of the compartments.

16. Colloidal iron does not affect *Dictyostelium* endocytic activity. FITC-dextran can be added as a second fluid-phase probe at 1 mg/mL to estimate the overall yield of vesicle purification (*see* **Subheading 2.7.**).
17. BSA is included in the buffer compositions to diminish the hydrophobic interaction of the probe with the plasma membrane.
18. Sorbitol (120 mM) increases buffer density and reduces sedimentation of noningested beads.
19. The sulfo-NHS-SS-biotin probe used to label the plasma membrane proteins reacts with unprotonated amine groups. Be sure to use alkaline environment (pH 8.0) and buffers devoid of reactive amines. The probe is also highly sensitive to reductive conditions. Use of DTT and β-mercaptoethanol is prohibited.
20. To prevent lysosomal degradation of labeled proteins, the lysosomotropic weak base NH$_4$Cl should be added to the medium during endocytosis, as it alkalinizes lysosomes to a pH at which proteases are less active.
21. The shape of the auto-formed Percoll gradient strictly depends on the rotor geometry (angle and radius), time and speed of centrifugation. Note that changes in any of these parameters will generate a different gradient.
22. Collection of fractions should be performed at a minimal speed and with a small-diameter tubing to avoid remixing of the fractions during elution.
23. FITC fluorescence is dependent on pH and is maximal above pH 8.0.
24. Because Percoll contribution to the overall fluorescence is nonnegligible, especially for the bottom fractions, it is advised to subtract the fluorescence value of a blank gradient for each fraction.
25. This protocol can be adapted to label specific subcompartments of the endocytic pathway on a pulse-chase basis *(10,19)*.

Acknowledgments

This work was supported in part by the Commissariat à l'Energie Atomique, the Centre National de la Recherche Scientifique, the Université Joseph Fourier Grenoble, and the Ministère pour la Recherche et la Technologie (ACI Biologie du Développement et Physiologie Intégrative). The authors would like to thank Michel Satre for constructive discussions and support.

References

1. Balch, W. E. and Rothman, J. E. (1985) Characterization of protein transport between successive compartments of the Golgi apparatus: asymmetric properties of donor and acceptor activities in a cell-free system. *Arch. Biochem. Biophys.* **240,** 413–425.
2. Laurent, O., Bruckert, F., Adessi, C., and Satre, M. (1998) In vitro reconstituted *Dictyostelium discoideum* early endosome fusion is regulated by Rab7 but proceeds in the absence of ATP-Mg^{2+} from the bulk solution. *J. Biol. Chem.* **273,** 793–739.

3. Goodloe-Holland, C. M. and Luna, E. J. (1987) Purification and characterization of *Dictyostelium discoideum* plasma membranes. *Methods Cell Biol.* **28,** 103–128.

4. Brunette, D. M. and Till, J. E. (1971) A rapid method for the isolation of L-cell surface membranes using an aqueous two-phase polymer system. *J. Memb. Biol.* **5,** 215–224.

5. Nellen, W., Datta, S., Reymond, C., et al. (1987) Molecular biology in *Dictyostelium*: tools and applications. *Methods Cell Biol.* **28,** 67–100.

6. Charlesworth, M. C. and Parish, R. W. (1975) The isolation of nuclei and basic nucleoproteins from the cellular slime mold *Dictyostelium discoideum*. *Eur. J. Biochem.* **54,** 307–316.

7. Bof, M., Brandolin, G., Satre, M., and Klein, G. (1999) The mitochondrial adenine nucleotide translocator from *Dictyostelium discoideum*. Functional characterization and DNA sequencing. *Eur. J. Biochem.* **259,** 795–800.

8. Rodriguez-Paris, J. M., Nolta, K. V., and Steck, T. L. (1993) Characterization of lysosomes isolated from *Dictyostelium discoideum* by magnetic fractionation. *J. Biol. Chem.* **268,** 9110–9116.

9. Adessi, C., Chapel, A., Vincon, M., Rabilloud, T., Klein, G., Satre, M., and Garin, J. (1995) Identification of major proteins associated with *Dictyostelium discoideum* endocytic vesicles. *J. Cell Sci.* **108,** 3331–3337.

10. Aubry, L., Klein, G., Martiel, J. L., and Satre, M. (1993) Kinetics of endosomal pH evolution in *Dictyostelium discoideum* amoebae. Study by fluorescence spectroscopy. *J. Cell Sci.* **105,** 861–866.

11. Brenot, F., Aubry, L., Martin, J. B., Satre, M., and Klein, G. (1992) Kinetics of endosomal acidification in *Dictyostelium discoideum* amoebae. ^{31}P-NMR evidence for a very acidic early endosomal compartment. *Biochimie* **74,** 883–895.

12. Beinert, H. (1978) Micromethods for the quantitative determination of iron and copper in biological material. *Meth. Enzymol.* **54,** 435–445.

13. Desjardins, M., Huber, L. A., Parton, R. G., and Griffiths, G. (1994) Biogenesis of phagolysosomes proceeds through a sequential series of interactions with the endocytic apparatus. *J. Cell Biol.* **124,** 677–688.

14. Gotthardt, D., Warnatz, H. J., Henschel, O., Bruckert, F., Schleicher, M., and Soldati, T. (2002) High-resolution dissection of phagosome maturation reveals distinct membrane trafficking phases. *Mol. Biol. Cell* **13,** 3508–3820.

15. Chia, C. P. and Luna, E. J. (1989) Phagocytosis in *Dictyostelium discoideum* is inhibited by antibodies directed primarily against common carbohydrate epitopes of a major cell-surface plasma membrane glycoprotein. *Exp. Cell Res.* **181,** 11–26.

16. Aubry, L., Mattei, S., Blot, B., Sadoul, R., Satre, M., and Klein, G. (2002) Biochemical characterization of two analogues of the apoptosis-linked gene 2 protein in *Dictyostelium* discoideum and interaction with a physiological partner in mammals, murine Alix. *J. Biol. Chem.* **277,** 21,947–21,954.

17. Bof, M., Brenot, F., Gonzalez, C., Klein, G., Martin, J. B., and Satre, M. (1992) *Dictyostelium discoideum* mutants resistant to the toxic action of methylene diphosphonate are defective in endocytosis. *J. Cell Sci.* **101,** 139–144.

18. Steck, T. L. and Lavasa, M. (1994) A general method for plasma membrane isolation by colloidal gold density shift. *Anal. Biochem.* **223,** 47–50.

19. Aubry, L., Klein, G., Martiel, J.-L., and Satre, M. (1997) Fluid-phase endocytosis in the amoebae of the cellular slime mold *Dictyostelium discoideum*: mathematical modelling of kinetics and pH evolution. *J. Theor. Biol.* **184,** 89–98.

11

Generation of Multiple Knockout Mutants Using the Cre-*loxP* System

Alan R. Kimmel and Jan Faix

Summary

Dictyostelium discoideum is an exceptionally powerful system by which to study various aspects of modern cell and developmental biology. The completion of the genome sequencing project together with a high-efficiency of targeted gene disruption have enabled researchers to characterize many specific gene functions. However, as a result of many gene products with overlapping functions, there is great need to produce mutants carrying mutations in multiple genes. We have, therefore, developed a robust system and describe a protocol for the generation of multiple gene mutations in *Dictyostelium* by recycling the Blasticidin S selectable marker after transient expression of Cre recombinase.

Key Words: Cre recombinase; *Dictyostelium*; *loxP*; homologous recombination; gene replacement; multiple knockouts.

1. Introduction

The small genome of *Dictyostelium* makes it particularly advantageous for molecular manipulation. Targeted mutation rates using homologous recombination are often at a relatively high frequency (>20%), and methods exist that permit genome-wide insertional (restriction enzyme-mediated integration [REMI]) mutagenic screens and the recovery of targeted loci *(1)*. However, because the number of selectable markers in *Dictyostelium* is restricted and the ability to perform effective genetic crosses between strains is limited, it was difficult to create multiple mutations within a single cell. This had made it difficult, if not impossible, to study epistatic relationships among the approx 12,500 genes of the transcriptome *(2)* or to evaluate potential redundancies between various pathways.

To overcome these genetic limitations, we developed a very efficient system for the creation of multiple gene mutations within an individual *Dictyostelium*

From: *Methods in Molecular Biology, vol. 346:* Dictyostelium discoideum *Protocols*
Edited by: L. Eichinger and F. Rivero © Humana Press Inc., Totowa, NJ

cell by recycling a single selectable marker, Blasticidin S-resistance *(3)*, using the Cre-*loxP* system *(4,5)*. We created a universal gene targeting vector backbone with a *Bsr* cassette flanked (floxed) by *loxP* recombination sites; translational stop codons in all six reading frames were placed outside one *loxP* site *(6)*. Transient expression of the Cre recombinase removes the Blasticidin-resistance expression cassette from the disrupted gene, but leaves translational stop codons in all the reading frames. This intramolecular recombination event creates a nonsense mutation within the targeted gene. Because the resulting cells are Blasticidin-sensitive, they can be utilized for additional rounds of gene disruption or for REMI mutagenic screening. Further, many transformation vectors exist that direct temporal- or spatial-specific expression of mutated or tagged proteins (e.g., green fluorescent protein [GFP], red fluorescent protein [RFP]) in *Dictyostelium*, and *Bsr*-deleted cells remain G418-sensitive and can still be engineered for the regulated expression of specific protein variants, for global cDNA screening, for complementation expression, or for gene interference by RNA interference (RNAi) or antisense methodologies *(1,7–10)*.

Although we have not observed any abnormal phenotypes upon expression of Cre in *Dictyostelium*, it is prudent to compare the behavior of parental and Cre-recombined cells to ensure that *Bsr* deletion (or Cre-transformation) has not created a dominant (or secondary) phenotype. For example, we have frequently detected normal levels of mRNA expression after the floxed-*Bsr* is removed via Cre recombination. Although an in-frame nonsense codon has been incorporated into the gene, there is the potential for production of a truncated protein that may create a phenotype that is distinct from a loss-of-function mutation. Unless one is purposefully interested in the function of proteins with specific carboxyl-terminal truncations, floxed-*Bsr* insertions should generally be designed near the 5' end of the gene. The regulated expression of Cre using cell-specific promoters or the tetracycline-responsive system Cre-*loxP* may permit conditional gene disruption *(10)*. Thus, it may be possible to study the effects of gene loss at specific developmental stages or in specific cell-types or the function of essential genes. The Bsr-floxed cassette has other advantages. Once *loxP* sites have been inserted into a gene of interest, they are efficient recombination targets for the creation of gene knock-ins to express mutant or tagged protein variants or for novel promoter/reporter fusions *(4,5,11)*.

Finally, it should be noted that recombination between *loxP* sites has been used effectively in mammalian systems to induce gene expression by fusing a promoter and gene target that had been separated by a floxed-inactivating sequence, or conversely, to repress expression by deleting an element that is essential for transcription.

2. Materials

2.1. Generation of Targeting Vectors

1. *Escherichia coli* cells (e.g., DH5α).
2. Luria-Bertani (LB) broth: 10.0 g bacteriological tryptone (Difco), 5.0 g yeast extract, 5.0 g NaCl. Bring to 1 L with deionized water and adjust pH to 7.4 using 1 N NaOH.
3. Tetrahymena fractionation buffer (TFB) I: 100 mM RbCl, 50 mM MnCl$_2$, 10 mM CaCl$_2$, 30 mM potassium-acetate, 15% glycerol; adjust pH to 5.8 with saturated acetic acid. Do not autoclave, sterile-filter. Store at –20°C.
4. TFB II: 10 mM MOPS, 10 mM RbCl, 75 mM CaCl$_2$, 15% glycerol; adjust pH to 7.0 with 1 N NaOH, sterile-filter. Store at –20°C.
5. Plasmid vector DNA (pLPBLP), available from the Dicty Stock Center (*see* http:// dictybase.org/StockCenter/StockCenter.html).
6. Ethidium bromide (EtBr) solution, 10 mg/mL.
7. Agarose.
8. Polymerase chain reaction (PCR) primers.
9. Taq polymerase, 5 U/μL (Roche).
10. PCR buffer (10X): 500 mM KCl, 100 mM Tris-HCl, pH 8.3.
11. dNTPs (10X), 2 mM each.
12. TE buffer: 10 mM Tris-HCl, pH 8.0, 1 mM ethylenediamine tetraacetic acid (EDTA).
13. DNA loading buffer: 40.0 g sucrose, 0.5 g sodium dodecyl sulfate (SDS), 0.25 g bromophenol blue. Bring to 100 mL with TE-buffer.
14. Restriction nucleases.
15. Gel extraction kit (Qiagen).
16. PCR purification kit (Qiagen).
17. Alkaline phosphatase (Roche).
18. T4 DNA ligase, 1 U/ μL.
19. Ligation buffer (10X): 660 mM Tris-HCl, pH 7.5, 100 mM MgCl$_2$, 10 mM spermidine, 10 mM ATP, 20 mM dithiothreitol (DTT), 1.5 mg/mL bovine serum albumin (BSA). Commercially available 10X ligase buffers also work well.
20. Tissue culture medium (TCM) solution: 0.1 M CaCl$_2$, 0.1 M MgCl$_2$.
21. Ampicillin stock solution (1000X): 0.5 g ampicillin; bring to 10 mL with deionized water, sterile-filter. Store at 4°C.
22. LB-ampicillin-Plates: 15.0 g Bacto-agar (Difco); bring to 1 L with LB broth, autoclave, cool in a water bath to 50°C, add 1 mL ampicillin stock solution.
23. Qiaprep Spin Miniprep Kit (Qiagen)
24. Plasmid DNA Maxi Kit (Qiagen).

2.2. Transformation of Dictyostelium *Cells*

1. *Dictyostelium discoideum* cells (e.g., Ax2 or Ax3).
2. Ax medium: 14.3 g bacteriological peptone (L-34, Oxoid), 7.15 g yeast extract, 18.0 g maltose, 0.616 g Na$_2$HPO$_4$ · 2 H$_2$O, 0.486 g KH$_2$PO$_4$; bring to 1

L with deionized water and adjust pH to 6.7 with 1 N NaOH. Autoclave. Store at 4°C.

3. HL5 medium: 17.8 g bacteriological peptone (L-85, Oxoid), 7.2 g yeast extract, 0.54 g Na_2HPO_4, 0.4 g KH_2PO_4, 130 µL B12/folic acid mix; bring to 1 L with deionized water and adjust pH to 6.5. Autoclave and add 20 mL of 50% glucose. Store at 4°C.

4. B_{12}/folic acid mix: 5 mg B_{12} (cyanocobalamine), 200 mg folic acid; add 95 mL deionized water, adjust pH to 6.5 with 5 N NaOH and fill up the solution to 100 mL. Filter-sterilize and store at –20°C protected from light.

5. 17 mM Na-K-phosphate buffer, pH 6.0: 0.356 g Na_2HPO_4, 1.99 g KH_2PO_4. Bring to 1 L with deionized water, autoclave.

6. Electroporation buffer: 10 mM potassium phosphate, pH 6.1, 50 mM glucose, sterile-filtered.

7. Electroporation cuvets: 4-mm gap.

8. Electroporator XCell (Biorad).

9. Healing solution: 100 mM $CaCl_2$, 100 mM $MgCl_2$.

10. Blasticidin S stock solution (1000X): 10 mg/mL in deionized water, sterile filtered. Store at 4°C or freeze in aliquots.

11. *Klebsiella aerogenes*.

12. SM agar plates: 15.0 g Bacto-agar (Difco), 10.0 g bacteriological peptone (L-34, Oxoid), 1.0 g yeast extract, 10.0 g glucose, 1.0 g $MgSO_4 \times 7\ H_2O$, 2.2 g KH_2PO_4, 1.0 g K_2HPO_4, bring to 1 L and autoclave. Store at 4°C.

13. Sterile tooth picks.

14. Ampicillin/streptomycin stock solution (100X): 5 mg/mL of ampicillin, 4 mg/mL of streptomycin sulfate, sterile-filtered. Store at 4°C or freeze at –20°C. Commercially available 100X pencillin/streptomycin stock solutions also work well.

2.3. Validation of Knockout Mutants

1. High Pure PCR Template Kit (Roche).
2. Dimethylsulfoxide (DMSO).

2.4. Removal of the Floxed Bsr-Cassette

1. Plasmid vector DNA (pDEX-NLS-Cre), available from the Dicty Stock Center.
2. Phosphate agar plates: 15.0 g Bacto-agar (Difco); bring to 1 L with 17 mM Na-K-phosphate buffer, autoclave.

3. Methods

3.1. Generation of Targeting Vectors

3.1.1. Preparation of Chemically Competent E. coli Cells

1. Inoculate *E. coli* strain DH5α (or any other suitable *E. coli* host used in your laboratory) in 100 mL of LB broth and allow to grow at 37°C overnight.
2. Inoculate 5 mL of the overnight culture into 200 mL of fresh LB broth in a 1-L Erlenmeyer flask.

3. Grow the cells at 37°C at 220 rpm to an OD_{600} of approx 0.3–0.6. The best results are obtained when the cells are harvested at early log phase.

4. Chill the cells in an ice/water bath for approx 15 min. For all subsequent steps, keep the cells as close to 0°C as possible and chill all containers and centrifuges before adding cells. Transfer the cells to four sterile 50-mL centrifugation tubes and centrifuge at 4000g for 15 min at 4°C.

5. Carefully pour off and discard supernatant and resuspend the pellets in 50 mL of ice cold TFB I buffer and keep cells on ice for 1 h.

6. Centrifuge the cells at 4000g for 15 min at 4°C.

7. Carefully pour off and discard supernatant and resuspend the pellet in 8 mL of ice-cold TFB II buffer and keep cells on ice for 1 h.

8. Prepare 200-µL aliquots of the cells in prechilled 1.5-mL microfuge tubes and immediately shock freeze in liquid nitrogen.

9. Store competent cells at –80°C.

10. It is also possible to purchase *E. coli* cells that are competent for transformation from various companies.

3.1.2. Construction of Targeting Vector

1. Synthesize two primer pairs for the amplification of a 5' and 3' fragment of your gene of interest (*see* **Note 1**) that will generate PCR products of approx 500 bp for each fragment. At their 5' ends, the primers should carry recognition sites for restriction endonucleases to facilitate cloning in gene disruption vector pLPBLP (*see* **Fig. 1**). The unique cloning sites of vector pLPBLP are: *Kpn*I-*Sal*I-*Acc*I-*Cla*I-*Hin*dIII-*Sma*I-floxed *Bsr* cassette-*Pst*I-*Sma*I-*Bam*H-*Spe*I-*Not*I.

2. Amplify the 5' and 3' fragments from genomic DNA, plasmid DNA, or a cDNA library by PCR. Set up a 100-µL reaction consisting of: 0.1 µg of genomic template or plasmid DNA, 0.1 µM of each primer, 0.2 mM of each dNTP, and 10 µL of 10X PCR buffer. Bring to 99 µL with deionized water, add 1 µL of *Taq* DNA polymerase (*see* **Note 2**), and mix.

3. Confirm size and quantity of PCR products by analytical agarose gel electrophoresis and staining with EtBr.

4. Digest the PCR products with appropriate restriction enzymes overnight at 37°C in a final volume of 200 µL (*see* **Notes 3** and **4**).

5. Add 30 µL of DNA loading buffer and run the entire sample on preparative agarose gel for each fragment.

6. Cut out the band of interest with a razor blade and purify the appropriate DNA fragments using gel extraction kit.

7. Confirm quantity of purified PCR products by analytical agarose gel electrophoresis and staining with EtBr. Store the fragments at –20°C for later use.

8. Digest 10 µg of vector pLPBLP with two restriction enzymes compatible with the first fragment that shall be inserted for 2 h at 37°C in a final volume of 200 µL. Place the mix on ice and confirm completion of the digestion by running a small aliquot of the reaction in an agarose gel.

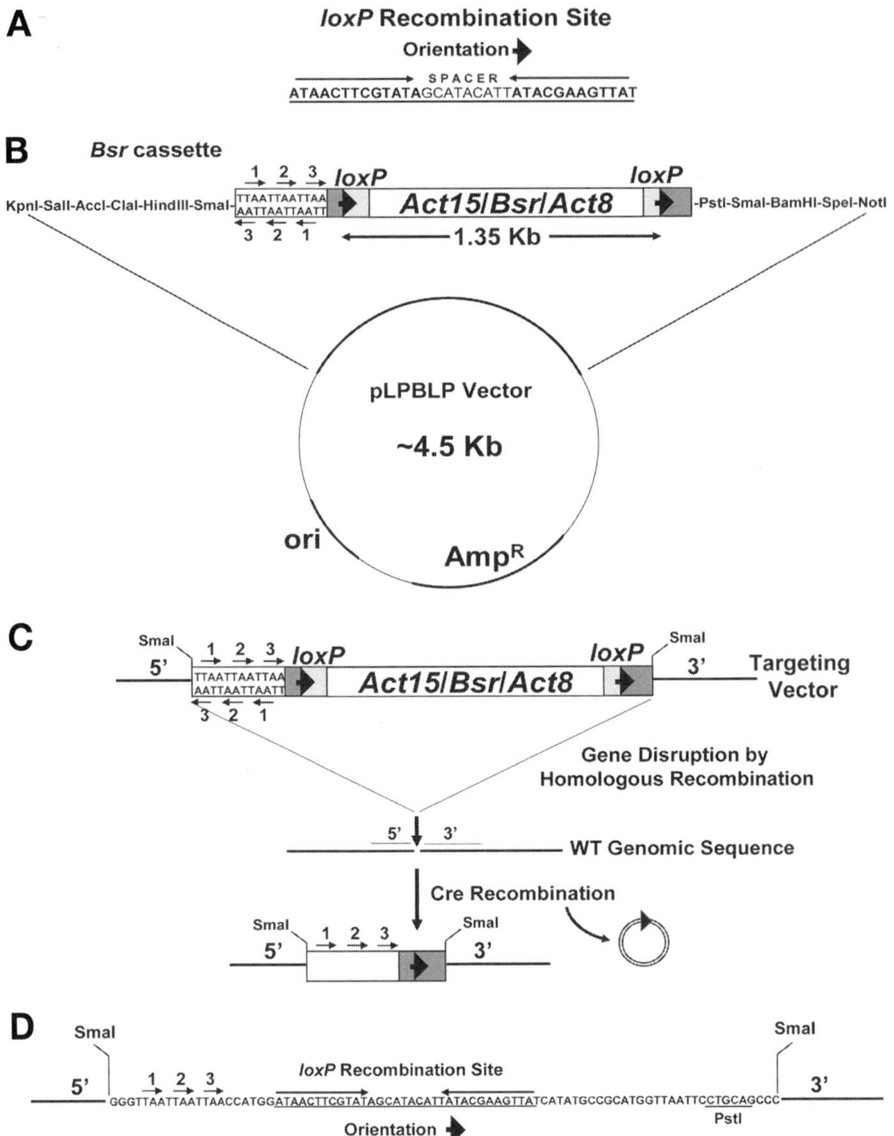

Fig. 1. A strategy for Cre-*loxP* recycling of the Bsr selectable marker. (**A**) The *loxP* recombination site includes a 13 bp inverted repeat separated by a spacer sequence. (**B**) The Floxed-Bsr cassette was constructed with *loxP* sites in the same orientation flanking both sides of the Blasticidin-resistance (Bsr) expression cassette. The *Bsr* gene is flanked by the *Act15* promoter and the *Act8* terminator. An oligonucleotide cassette was also added upstream with translational stop codons in all six reading frames. Restriction enzyme sites outside of the floxed-*Bsr* cassette permit the cloning

9. Stop the reaction by addition of fivefold volume (800 µL) of PB (included in PCR purification kit) and purify the DNA fragments using PCR purification kit.
10. Dephosphorylate cleaved vector pLPBLP with alkaline phosphatase for 1 h at 37°C in a final volume of 200 µL.
11. Stop the reaction by addition of 800 µL PB buffer and purify the DNA fragments using PCR purification kit.
12. Ligate 50–100 ng of cleaved and dephosphorylated vector pLPBLP with approximately fivefold molar excess of the first PCR fragment with T4 DNA ligase in 1X ligation buffer and a final volume of 20 µL overnight at 15°C.
13. Add 30 µL of TCM solution to the ligation mix and chill on ice.
14. Place frozen, chemically competent *E. coli* cells on ice and allow the cells to thaw slowly.
15. Add 200 µL of the cells to ligation mix and incubate on ice for 1 h.
16. Heat-shock cells for 1 min at 42°C.
17. Chill cells for 5 min on ice.
18. Add 800 µL of LB broth and incubate for 1 h at 37°C.
19. Plate cells on LB agar plates containing 50 µg ampicillin/mL and incubate at 37°C overnight.
20. Inoculate colonies in LB broth containing 50 µg/mL of ampicillin and grow the cells to the stationary phase over night at 37°C.
21. Prepare plasmid DNA with Qiaprep Spin Miniprep Kit (Qiagen) or other standard procedure and isolate derivatives of plasmid pLPBLP containing the first fragment.
22. Digest this plasmid with the second combination of restriction enzymes and perform treatment with alkaline phosphatase.
23. Ligate second fragment into this vector and repeat **steps 13–22** in order to isolate gene disruption plasmid pLPBLP containing both fragments flanking the *Bsr* cassette.
24. Inoculate cells containing the final gene disruption plasmid in 300 mL of LB broth containing 50 µg ampicillin/mL and incubate at 37°C over night.
25. Prepare plasmid DNA with Maxiprep Kit (Qiagen).
26. Digest approx 100 µg (this amount of plasmid DNA will allow performance of three *Dictyostelium* transformations) of plasmid DNA with a combination of

---

Fig. 1 *(continued)* of 5' and 3' gene sequences for targeted disruption. The pLPBLP gene targeting vector (~4.5 kb) is indicated with the relative positions of the Bsr cassette, a bacterial origin of replication (ori), and the ampicillin resistance gene (AmpR). **(C)** The floxed-Bsr cassette of pLPBLP is inserted within a gene sequence using the SmaI sites as indicated. Wild-type cells are transformed for gene disruption by homologous recombination and selected for resistance to Blasticidin S. Transient expression of Cre promotes recombination and deletion of sequences between the two *loxP* sites, leaving a 73-bp sequence that includes the translational stop cassette and a single *loxP* site. **(D)** The predicted sequence within the disrupted gene. The single *loxP* site, the stop codons, and various restriction sites are indicated.

restriction enzymes for 2 h at 37°C in a final volume of 250 μL, which liberates two DNA fragments: one fragment of approx 2.6 kbp containing the 5'-fragment-floxed-*Bsr* cassette-3'-fragment and the rest of the vector backbone.

27. Confirm completion of the digestion by running a small aliquot of the reaction in an agarose gel (*see* **Note 5**).
28. Stop the reaction by addition of 800 μL of PB buffer and purify the DNA fragments using PCR purification kit (Qiagen). It is not necessary to isolate and separate the *Bsr* cassette flanked by the 5' and 3' fragments from the vector backbone.
29. Elute the DNA fragments with 60 μL of 10 m*M* Tris-HCl, pH 8.0 (*see* **Note 6**).
30. Run 1 μL of the sample on an analytical 0.8% agarose to validate quality and quantity of the DNA fragments.
31. Store eluted DNA at –20°C for later use.

3.2. Transformation of Dictyostelium *Cells by Electroporation*

3.2.1. Preparation of Electrocompetent Dictyostelium *Cells*

1. Inoculate *Dictyostelium* cells at a concentration of $5–7 \times 10^5$ cells/mL into 200 mL of fresh axenic medium (Ax or HL5) in a 1000-mL flask. The cells may be washed off a plastic Petri dish or transferred from liquid media. The doubling time of *Dictyostelium* in shaken suspension is approx 8–10 h at 21°C.
2. Incubate the culture at 21°C for about 24 h, shaking at 150 rpm, and harvest the cells at a density of not more than 5×10^6 cells/mL.
3. Transfer 100 mL of the cells into two sterile, disposable, 50-mL centrifugation tubes and incubate on ice for 15 min.
4. Pellet the cells by centrifugation at 500*g* for 2 min at 4°C.
5. Carefully pour off and discard the supernatant and place the centrifugation tubes with the cell pellets on ice.
6. Pool the pellets and resuspend in 50 mL of ice-cold 17 m*M* Na-K-phosphate buffer, pH 6.0.
7. Pellet the cells by centrifugation at 500*g* for 2 min at 4°C, again pour off and discard the supernatant, and resuspend the pellet in 50 mL of ice-cold electroporation buffer.
8. Repeat **step 7** but resuspend the cells at a concentration of 1×10^7 cells/mL pellet in ice cold electroporation buffer. Keep the cells on ice and use as soon as possible for electroporation.

3.2.2. Electroporation of Dictyostelium *cells*

1. Pipet the DNA samples (approx 35 μg) to be electroporated into sterile 4-mm electroporation cuvets and place them on ice.
2. Add 700 μL of the competent cells to each DNA sample, mix gently, and incubate on ice.
3. Electroporate the cells using Biorad Xcell gene pulser preset protocol 3 for *Dictyostelium* (these conditions are: square wave, V = 1.0 kV, 10 μF, 1.0 ms pulse length, two pulses, 5-s pulse interval). The time constant should be approx

1 ms. The voltage should be approx 1.0 kV. This setup routinely yields hundreds of transformants.

4. Remove the cuvet from the chamber and plate the cells on the bottom of a Petri dish. Place the plate on a laboratory shaker and gently shake at approx 40 rpm for 15 min at room temperature.

5. Adjust the suspension to 2 m*M* CaCl$_2$ and 2 m*M* MgCl$_2$ with healing solution and continue shaking for another 15 min at room temperature.

6. Add 12 mL of axenic growth medium and allow the cells to recover overnight at 21°C.

7. Add Blasticidin S at a final concentration of 10 µg/mL and incubate at 21°C.

8. Select the transformants with Blasticidin S for approx 10–14 d at 21°C until colonies with a diameter of 1 mm are clearly visible.

9. Isolate clonal *Dictyostelium* cell lines by spreader dilution. For this, prepare a dense suspension of *K. aerogenes* that were grown on SM agar plates overnight at 37°C with a sterile-filtered 10 m*M* Tris-HCl, pH 8.0 solution.

10. Predry approximately five SM agar plates per transformation in the laminar flow bench for approx 30 min at room temperature and add 250 µL of the *K. aerogenes* suspension to the center of the agar plate.

11. Wash the transformants from the plastic surface by gently pipetting up and down and place one to three drops in a 1.5-mL microfuge tube filled with 1 mL of medium. Briefly vortex the cells to disrupt cell aggregates.

12. Add different amounts (approx 1, 2.5, 5, 10, and 20 µL) of this suspension into the drop of *K. aerogenes* and immediately spread the drop over the entire SM plate using a sterile Drygalski spatula.

13. Allow the plates to dry and subsequently incubate at 21°C.

14. After approx 3 d, individual plaque forming colonies will appear in the bacterial lawn.

15. After the colonies have reached a diameter of approx 1 mm, pick individual clonal cell lines with sterile toothpicks and transfer them into sterile 24-well plates containing axenic medium supplemented with 10 µg/mL of Blasticidin S and ampicillin/streptomycin solution to prevent bacterial growth.

3.3. Validation of Knockout Mutants

Two PCRs are performed to rapidly validate targeted gene disruption. One PCR design uses a primer from within the *Bsr* cassette and another primer outside of the targeting fragment. Another PCR examines the presence of the wild-type or disrupted gene using primers that flank the floxed-*Bsr* insertion sites (*see* **Fig. 2A,B**).

1. Inoculate 9-cm Petri dishes containing 12 mL of axenic medium supplemented with 10 µg/mL of Blasticidin S with individual clones from the 24-well plates and allow the cells to colonize the entire plate.

2. Prepare genomic DNA using the High Pure PCR Template Kit as described by the manufacturer, following the protocol for cultured cells.

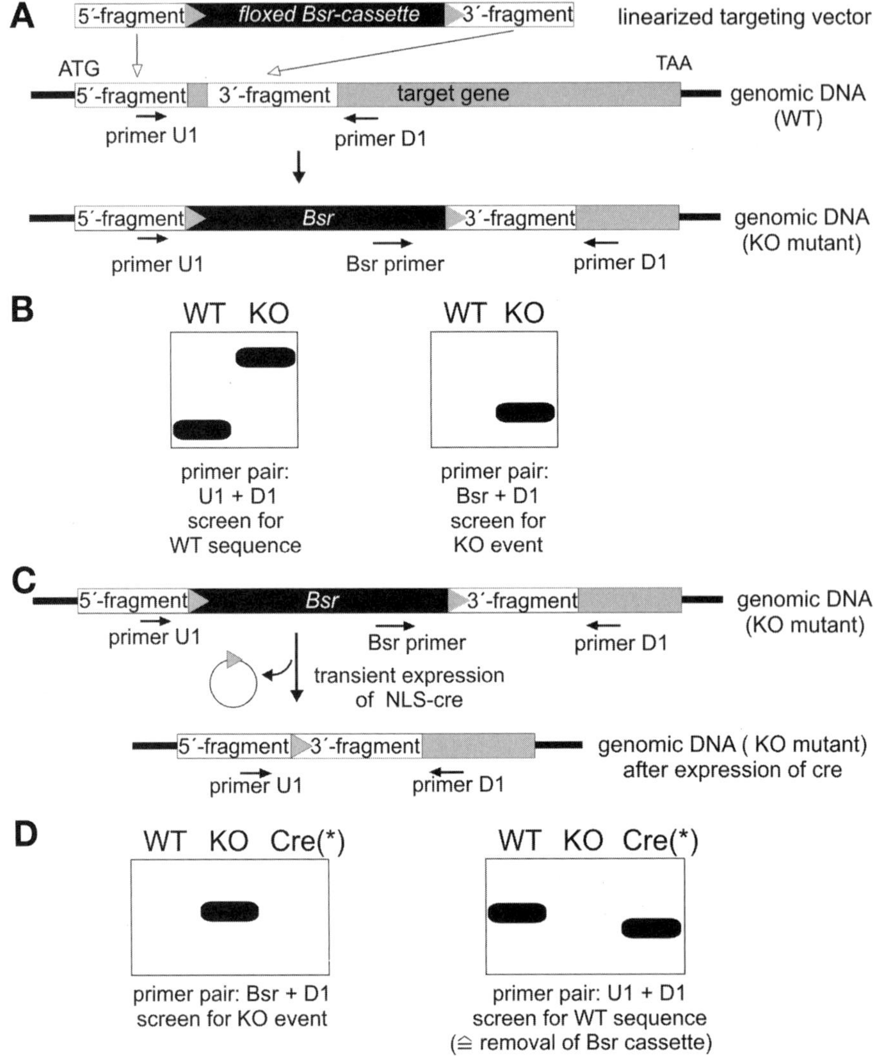

Fig. 2. Validation of targeted gene disruption and subsequent Cre-mediated recombination. (**A**) Generation of a null cell line from target gene. 5' and 3' specific sequences of the target gene are cloned into pLPBLP. The linear targeting vector is then used to disrupt the gene by homologous recombination. Depending on the spacing between the 5' and 3' fragments, targeted integration may cause a small deletion in the gene. (**B**) Schematic validation of targeted gene disruption by polymerase chain reaction (PCR). Both a wild-type and a Blasticidin-resistant knockout mutant are examined by PCR amplification employing the two different sets of primers indicated. Left panel: The primer combination of U1 and D1 (*see* **A**) identifies wild-type and homologously recombined sequences of the target gene. Upon homologous recombination, the target gene PCR product is approx 1.3 kbp larger than that of the wild-type. Right panel: the

3. Set up a 100-µL reaction consisting of: 0.1 µg of genomic template DNA, 0.1 µ*M* of each primer, 0.2 m*M* of each dNTP, and 10 µL of 10X PCR buffer. Bring to 99 µL with deionized water, add 1 µL of *Taq* DNA polymerase, and mix.

4. For a final product size of approx 1 kbp, perform the reaction using the following profile: 94°C for 30 s, 48°C for 60 s, 70°C for 90 s for 30 cycles. Allow a longer extension time in the last cycle.

5. For a final product size of approx 2.5 kbp, perform the reaction using the following profile: 94°C for 30 s, 48°C for 60 s, 68°C for 160 s for 30 cycles. Allow a longer extension time in the last cycle.

6. Validate targeted gene disruption by examination of aliquots of the two PCR reactions in 0.8% analytical agarose gels and EtBr staining.

7. Store the knockout mutants by either slowly freezing the cells in axenic medium containing 7% DMSO in liquid nitrogen or by first preparing and subsequently shock freezing spores in liquid nitrogen. Store the frozen cells in liquid nitrogen. The spores can be stored at –80°C.

3.4. Removal of the Floxed Bsr Cassette

1. Bsr knockout cells are electroporated with 35 µg of pDEX-NLS-cre as described in **Subheading 3.2.2.**, **steps 1–6**.

2. After the 24-h recovery period, G418 is added to a final concentration of 10– 20 µg/mL.

3. Selection is continued for 3–10 d (*see* **Note 7**).

4. After appearing of colonies, the cells are spreader diluted for clonal selection on SM agar plates containing *K. aerogenes* as described in **Subheading 3.2.2.**, **steps 9–14**.

5. After the colonies have reached a diameter of approx 1 mm, pick individual clonal cell lines with sterile toothpicks and transfer them in replica to two different plates. The first is a standard SM agar plate using *K. aerogenes* as the nutrient source. The second is a nonnutrient 17 m*M* Na-K-phosphate agar plate, pH 6.0, layered with 500 µL of a concentrated *K. aerogenes* suspension and 120 µL of Blasticidin S at 10 mg/mL. The final concentration of Blasticidin S in the agar plate is 40 µg/mL.

6. All *Dictyostelium* cells will grow on SM *K. aerogenes* plates, but only *Dictyostelium* that has retained the *Bsr* cassette will grow in the presence of Blasticidin S. A wild-type parental control should be always used for growth control.

Fig. 2 *(continued)* primer combination of Bsr and D1 (*see* **A**) specifically identifies only the homologous recombination event and, hence, is seen only in the knockout mutant. (**C**) Strategy for deletion of *Bsr* by transient expression of NLS-cre. Deletion of the floxed-*Bsr* cassette leaves a sequence of approx 70 bp (*see* **Fig. 1**). (**D**) PCR analysis of null cells following transient expression of NLS-cre (*). Most of the clonal cell lines that are sensitive to both Blasticidin and G418 for growth also lack the floxed *Bsr* cassette.

7. Inoculate cells not growing in Blasticidin S from corresponding SM *K. aerogenes* plates into sterile 24-well plates containing either axenic medium, axenic medium with Blasticidin S (10 µg/mL), or axenic medium with G418 (10 µg/mL) and incubate at 21°C for 24–48 h. The axenic media are supplemented with ampicillin/streptomycin solution to prevent bacterial growth.

8. Cells again not growing in Blasticidin S and G418 containing media are of potential interest. Usually, >95% of cells selected by these growth criteria will show appropriate Cre recombination. Inoculate these cells on 9-cm Petri dishes containing 12 mL of axenic medium and allow the cells to colonize the entire plate.

9. Isolate genomic DNA from the Blasticidin- and G418-sensitive cell lines and validate Cre-mediated recombination by PCR as described in **Subheading 3.3.**, **steps 1–6** using the same two primer pairs used to validate the knockout event (*see* **Fig. 2C,D**).

10. The cells devoid of the *Bsr* cassette can now be used for the next round of *Bsr*-mediated targeted gene disruption.

4. Notes

1. The orientation of the 5' and 3' fragments relative to the orientation of the *Bsr* cassette is not important for targeted gene disruption; however, both the 5' and 3' fragments must be inserted in the same orientation in targeting vector pLPBLP.

2. There are numerous thermostable polymerases currently available. This protocol has been written using *Taq* polymerase; however, several other enzymes can be used as well. The "proof-reading" activities of thermostable polymerases such as *Pfu* (Stratagene), *Vent* and *Deep Vent* (New England Biolabs) or *KOD* (Novagene) are not required.

3. Depending on the restriction nuclease used for cloning, the primers and the length of the primer overhang, it might be very important to digest the amplified DNA fragments overnight. This procedure significantly increases the number of appropriately digested primer ends and hence the number of *E. coli* transformants.

4. Digestions of DNA fragments at 37°C overnight should contain between 0.1 and 1 mM EDTA to avoid unspecific degradation of DNA by unspecific nucleases.

5. The restriction digest of the final gene targeting vector must be complete, otherwise the undigested gene targeting vector might integrate by a single cross-over event into the *Dictyostelium* genome. This event may considerably complicate the interpretation of the PCR data, and in most cases does not lead to disruption of the target gene.

6. It is recommended that EDTA-containing buffers not be used for this step. Millimolar concentrations of EDTA considerably inhibit the transformation efficiency of *Dictyostelium* cells. Thus, TE buffers should be avoided. Use instead 10 mM Tris-HCl, pH 8.0 or deionized water for final elution of the DNA prior to transformation of *Dictystelium* cells.

7. The duration of G418 selection is dependent on a number of parameters: the concentration of G418 used in the selection, the medium, the temperature, and probably also the strain or cell line used. The most important part of this step is the transient expression of Cre recombinase to remove background (cells not

transfected with pDEX-NLS-cre), and this must be optimized for each lab. A selection that is too short will yield knockout cells still harboring the *Bsr* cassette, whereas a selection that is too long will yield knockout cells lacking the *Bsr* cassette; however, these cells will have plasmid pDEX-NLS-cre stably integrated into their genome.

Acknowledgments

We are deeply grateful to our colleagues Lisa Kreppel, Gad Shaulsky, and Michael Schleicher for making it possible for us to develop this technology. This research was supported in part by the Intramural Research Program of the National Institutes of Health, the National Institute of Diabetes and Digestive and Kidney Diseases.

References

1. Kuspa, A. and Loomis, W. F. (1992) Tagging developmental genes in *Dictyostelium* by restriction enzyme-mediated integration of plasmid DNA. *Proc. Natl. Acad. Sci. USA* **89,** 8803–8807.
2. Eichinger, L., Pachebat, J. A., Glöckner, G., et al. (2005) The genome of the social amoeba *Dictyostelium discoideum. Nature* **435,** 43–57.
3. Sutoh, K. (1993) A transformation vector for *Dictyostelium discoideum* with a new selectable marker bsr. *Plasmid* **30,** 150–154.
4. Sauer, B. (2002) Cre/lox: one more step in the taming of the genome. *Endocrine* **19,** 221–228.
5. Branda, C. S. and Dymecki, S. M. (2004) Talking about a revolution: the impact of site-specific recombinases on genetic analyses in mice. *Dev. Cell* **6,** 7–28.
6. Faix, J., Kreppel, L., Shaulsky, G., Schleicher, M., and Kimmel, A. R. (2004) A rapid and efficient method to generate multiple gene disruptions in *Dictyostelium discoideum* using a single selectable marker and the Cre-loxP system. *Nucleic Acids Res.* **32,** e143.
7. Martens, H., Novotny, J., Oberstrass, J., Steck, T. L., Postlethwait, P., and Nellen, W. (2002) RNAi in *Dictyostelium*: the role of RNA-directed RNA polymerases and double-stranded RNase. *Mol. Biol. Cell* **13,** 445–453.
8. Robinson, D. N. and Spudich, J. A. (2000) Dynacortin, a genetic link between equatorial contractility and global shape control discovered by library complementation of a *Dictyostelium discoideum* cytokinesis mutant. *J Cell Biol.* **150,** 823–838.
9. Spann, T. P., Brock, D. A., Lindsey, D. F., Wood, S. A., and Gomer, R. H. (1996) Mutagenesis and gene identification in *Dictyostelium* by shotgun antisense. *Proc. Natl. Acad. Sci. USA* **93,** 5003–5007.
10. Blaauw, M., Linskens, M. H., and van Haastert, P. J. (2000) Efficient control of gene expression by a tetracycline-dependent transactivator in single *Dictyostelium discoideum* cells. *Gene* **252,** 71–82.
11. Hoess, R. H., Wierzbicki, A., and Abremski, K. (1986) The role of the loxP spacer region in P1 site-specific recombination. *Nucleic Acids Res.* **14,** 2287–2300.

Restriction Enzyme-Mediated Integration (REMI) Mutagenesis

Adam Kuspa

Summary

A method for the integration of linear DNA into the *Dictyostelium* genome is described. Restriction enzyme-mediated integration, or REMI, involves the transformation of cells with a mixture of plasmid DNA, linearized with a restriction enzyme, along with a restriction enzyme that is capable of generating compatible cohesive ends in the genome. The enzyme stimulates integration of the DNA into cognate restriction sites in the chromosomes, usually as a single-copy insertion event and with little collateral damage to the genome. REMI has proven useful for genetic screens and for placing genetic and molecular markers at particular points in the genome. Over the past 15 yr, REMI has been used to identify hundreds of interesting genes based on their mutant phenotypes.

Key Words: Transformation; genetic screen; insertional mutagenesis.

1. Introduction

Site-specific integration of linear DNA can be achieved by using restriction enzymes to stimulate DNA integration into cognate sites in the genome, as first demonstrated in *Saccharomyces cerevisiae (1)*. Restriction enzyme-mediated integration (REMI) has been developed for use in *Dictyostelium* as a powerful method for gene identification through insertional mutagenesis *(2)*. Because many different restriction enzymes can be used, REMI can be used to tag essentially any portion of the genome. The REMI protocol involves introducing a restriction enzyme along with the linear transforming DNA into the cells by electroporation. As long as the restriction enzyme used creates cohesive ends that are compatible with the ends of the transforming DNA, it will dramatically increase the frequency of integration into genomic restriction sites. Following the selection for stable transformants, the vast majority of clones can be shown

From: *Methods in Molecular Biology, vol. 346:* Dictyostelium discoideum *Protocols*
Edited by: L. Eichinger and F. Rivero © Humana Press Inc., Totowa, NJ

to have a single vector integrated into a single genomic restriction site that corresponds to the enzyme used *(3)*. The addition of restriction enzymes to the transformation experiment not only determines the sites of integration, but also stimulates integration 20- to 60-fold above what is observed without enzymes. The mechanism most likely involves infrequent chromosome breaks introduced into the chromosome by the restriction enzyme, followed by incorporation of the exogenous plasmid DNA during chromosome repair. Cleavage at sites where the plasmid does not integrate must be efficiently repaired because there is little evidence for increased rates of chromosomal rearrangements in REMI transformants *(3)*.

For a restriction enzyme to mediate integration, its recognition site must correspond to the site used to linearize the transforming plasmid. For instance, *Dpn*II (which generates GATC ends) will stimulate integration of DNA linearized with *Bam*HI (which generates GATC ends), but not DNA linearized with *Eco*RI (which generates AATT ends). The majority of the integration sites of *Dpn*II REMI transformants are *Dpn*II sites even when the plasmid is opened at a *Bam*HI site. Thus, the REMI enzyme determines the site of integration. To increase the range of chromosomal sites available to insertion, it is convenient to linearize plasmid DNAs with restriction enzymes that recognize six-basepair sites and use as REMI enzymes those that recognize the middle four basepairs of those sites. The six-basepair enzymes are optimal for plasmid linearization, because their sites are often unique in plasmid-sized DNAs, whereas the four-basepair sites are approximately 20-fold more frequent in the genome, and this provides better target site distribution. The most useful pairs of enzymes for REMI are shown in **Table 1**.

Once interesting REMI mutants are identified, the genomic sequences surrounding the insertion site are easily cloned so that the affected genes can be characterized. Cloning of the tagged genes opens up the possibility of molecular approaches to characterizing the physiological role of the gene product. It allows the formal genetic proof that the phenotype of the original mutant strain is the direct consequence of the insertional event by using the cloned fragment to recapitulate the insertion through homologous recombination. The cloned fragment can also be used to place the mutation in genetic backgrounds that may be useful for cell biological analyses or for recognition of specific cell types. The plasmid and flanking sequences can be isolated from each strain by restriction digestion of genomic DNA followed by selection of ligated, circularized plasmid in *Escherichia coli*.

Although the basic REMI procedure has remained the same since its introduction in 1992, several modifications have improved its utility. Dominant selectable markers, such as the blasticidin and hygromycin resistance cassettes, allow efficient selection of *Dictyostelium* transformants when integrated

Table 1
Useful Enzyme Combinations for Restriction Enzyme-Mediated Integration (REMI)

Restriction enzyme used to linearize DNA	Recognition site[a]	REMI restriction enzyme	Recognition site[a]
*Bam*HI	G/GATCC	*Dpn*II[b]	/GATC
*Bgl*II	A/GATCT	*Dpn*II[b]	/GATC
*Bcl*I	T/GATCA	*Dpn*II[b]	/GATC
*Eco*RI	G/AATTC	*Tsp*509I	/AATT
*Sph*I	GCATG/C	*Nla*III	CATG/

[a]The top strand of the DNA is shown in the 5' to 3' orientation and a slash indicates the position that the enzyme cuts.
[b]*Sau*3AI also will work in REMI.

at a single copy and work very well in REMI transformation *(4,5)*. REMI can also be used to make useful translational fusions into coding regions by using REMI plasmids engineered to have reporter genes such as *lacZ* or GFP near the insertion site *(6,7)*. It is also possible to make multiple mutations with the same marker by using resistance cassettes that are flanked by Cre recombination sites such that the marker can be excised from the genome after initial selection *(8)*.

2. Materials

2.1. Cell Culture (9)

1. HL-5 medium (Used for routine axenic growth of *Dictyostelium* cells.): 10 g/L proteose peptone number 2 (Becton Dickinson) (*see* **Note 1**), 5g/L yeast extract, 10 g/L glucose, 0.35 g/L Na_2HPO_4, 0.35 g/L KH_2PO_4, pH 6.4.
   Adjust pH with H_3PO_4. Sterilize by autoclaving for 40 min and remove from the autoclave immediately to minimize caramelization. Add antibiotics prior to use from the following stock solution. 100X Pen/Strep: 10,000 U/mL Penicillin G; 10 mg/mL Streptomycin sulfate.
2. SM agar medium (used for growth of *Dictyostelium* on bacterial lawns): 10 g/L Bacto-peptone (Becton Dickinson), 1 g/L yeast extract, 10g/L glucose, 1 g/L $MgSO_4$, 1.9 g/L K_2HPO_4, 0.6 g/L KH_2PO_4, 20 g/L Bacto-Agar (Becton Dickinson), pH 6.4.
   Adjust pH with H_3PO_4, sterilize by autoclaving and pour about 32–35 mL into each 100-mm Petri dish.

3. SM liquid medium (Used for growth of *Dictyostelium* food bacteria).
   Make just like SM agar, but leave out the agar. Innoculate a 50-mL culture in a 250-mL flask with *Klebsiella aerogenes* 2 d prior to use. Allow the culture to reach saturation at room temperature without shaking. The culture should be usable for up to 10 d. About 0.4 mL of this culture spread on a SM agar plate should form a thick lawn of bacteria within 48 h at room temperature.

2.2. Transformation and Selection

1. HF *electroporation* buffer *(10)*: 10 mM sodium phosphate, pH 6.1, 50 mM sucrose. The pH is adjusted to 6.1 by mixing monobasic sodium phosphate and dibasic sodium phosphate.
2. H50 electroporation buffer *(11)*: 20 mM HEPES, pH 7.0, 50 mM KCl, 10 mM NaCl, 1 mM Mg$_2$SO$_4$, 5 mM NaHCO$_3$, 1 mM NaH$_2$PO$_4$.
3. Blasticidin stock solution *(1000X)*: 4 mg/mL blasticidin S, filter sterilized and stored at 4°C.

2.3. Equipment

There are many types of electroporation devices, but two of the most commonly used ones are the Gene Pulser from BioRad Laboratories (Richmond, CA) and the ECM630 made by BTX, Genetronics Inc. (San Diego, CA). Each model differs in the controls for setting the voltage and capacitance, so these parameters should by optimized. The ranges of conditions that have proven useful for REMI are: a field strength of 2.5–10 kV/cm (e.g., 1 kV for a 0.4-cm wide cuvet), 20–100 Ω resistance (most models do not allow adjustment of resistance), and a charge capacitance of 3–50 μF. The desired time constant, or the halftime for the electric discharge, is 0.6 ms, but a range of 0.5–1.2 is adequate (*see* **Note 2**).

3. Methods

3.1. REMI Mutagenesis

3.1.1. REMI Mutagenesis With Blasticidin Selection

When the following steps are carried out, REMI transformants are obtained at a frequency of >10^{-5}. Numerous blasticidin-based *E. coli/Dictyostelium* shuttle plasmids are available that are based on the original blasticidin resistance cassettes *(4)*.

1. About 2 wk prior to the experiment, streak-out the *Dictyostelium* strain to be transformed on a bacterial lawn by spreading 0.4 mL of bacterial culture on an SM agar plate and using a sterile loop to streak *Dictyostelium* cells across. Colonies arising from single *Dictyostelium* cells should appear in 3–4 d. Start an axenic culture of *Dictyostelium* cells by inoculating cells from growing edge of a single colony into 2 mL of HL-5 and expand the culture to the amount needed for the number of transformations planned (usually 25–100 mL). The use of a clonal inoculum will help to ensure that the phenotypes observed in the transformants

are due to the REMI event and not due to genetic variation in the culture. The cells must be used for transformation only when they are in the mid-exponential phase of growth ($2–4 \times 10^6$ cells/mL) (*see* **Note 3**).

2. Prior to the day of REMI electroporation, prepare 200 µg of the REMI plasmid by digestion with the appropriate restriction enzyme (*see* **Table 1**), followed by phenol/chloroform purification and ethanol precipitation. Resuspend the linearized plasmid in TE buffer (10 m*M* Tris-HCl, pH 7.0, 1 m*M* ethylenediamine tetraacetic acid [EDTA]) at 1 µg/µL.

3. Chill the *Dictyostelium* culture from **step 1** by immersing the flask in ice for 15 min with occasional swirling. Spin the cells down in a 50-mL tissue culture tube by centrifuging at 1500–2500 *g* at 4°C for 4 min.

4. While the culture is chilling, cool the electroporation cuvets (0.4 cm gap-width) and large sterile glass test tubes (16 × 125 mm) on ice for at least 5 min. Gather the linearized plasmid and REMI restriction enzyme.

5. Decant the growth medium and leave the tube upside-down for a few seconds while carefully aspirating the remaining liquid from the walls of the tube, especially near the pellet and around the rim and edges of the tube. Knock the pellet loose and plunge the tube into ice. You should not spend more than a minute manipulating the tube before plunging the pellet on ice. Add sufficient ice-cold HF electroporation buffer to resuspend the cells to 10^7/mL.

6. Distribute 0.8-mL aliquots of cells to the cold glass tubes. Add 20 µg of plasmid and 40–200 U of the REMI restriction enzyme and mix by briefly swirling the tube (*see* **Note 4**). (For 0.2-cm gap-width cuvets, all amounts and electrical parameters should be halved.)

7. Put the DNA/cell mix in a cold cuvet and electroporate at 2.5 kV/cm by following the instructions provided by the manufacturer for the device being used. Expect a time-constant of 0.5 to 1.1 ms. Wait 5 s and then electroporate the suspension a second time using the same conditions.

8. Immediately after each electroporation, remove the cells from the cuvet back into the same glass tube that they came from and add 10 µL of 1 *M* MgCl$_2$ and 10 µL of 1 *M* CaCl$_2$. Let the mixture stand at room temperature for >10 min. The transformation efficiency decreases after the cells have been in HF buffer for more than 60 min. Approximately 12–20 aliquots from the same batch of cells can be processed in this time.

9. Distribute 0.2 mL of the cells into each of four standard Petri plates, each containing 10 mL of HL-5 medium. Control samples consisting of cells electroporated without DNA can be plated as well to test the drug selection. Incubate the cells at 22°C in a humid chamber.

10. Add selective drug after 16–20 h of incubation. For example, add 10 µL of blasticidin stock solution to each plate.

11. Colonies of REMI transformants will appear after leaving the plates undisturbed for 6–7 d. At this time, the colonies will be about 1–3 mm in diameter and contain about 10^4 cells. Replace the medium with fresh HL-5 plus drug, or harvest the cells at this time. After 6 d, the background of untransformed cells is negligible with blasticidin selection.

3.1.2. REMI Mutagenesis Using the H50 Method

An alternative to the standard transformation procedure has been described that works quite well with REMI *(11)*.

1. Harvest exponentially growing cells by centrifugation at 4°C as described above (*see* **Subheading 3.1.1.**, **step 3**). Wash twice in ice-cold H50 buffer and bring the cells to 5×10^7 per mL in H50 buffer.
2. Add 0.1 mL of cells to a small (Eppendorf) tube, on ice, and add 1–10 µg of plasmid DNA and 10–100 U of REMI restriction enzyme.
3. Transfer the mixture to a cold, 0.1-cm gap-width electroporation cuvet and electroporate at 0.85 kV/cm (25 µF if this parameter is adjustable) twice with a 5-s pause between pulses. Return the cuvet to the ice while processing the remaining tubes to let the cells recover for at least 5 min.
4. Remove the cells from the cuvet, allow them to recover, and distribute them to Petri dishes as described above (*see* **Subheading 3.1.1**, **steps 8** and **10**).

3.2. Harvesting REMI Transformants

1. After selection, collect the colonies by directing a stream of medium over the surface of each plate from a pipet. The cells from each plate may be kept separate or pooled.
2. Because the plating efficiency varies, the suspension of primary transformants should be diluted before plating. Make a series of four 10-fold dilutions into liquid SM medium by mixing 0.5 mL of the harvested cells into 4.5 mL of SM, mixing the suspension and repeating the process three times.
3. Spread 0.1, 0.2, and 0.4 mL of suspension from the three highest dilutions onto separate SM plates containing 0.4 mL of saturated bacterial culture (*see* **Subheading 2.1.**, **step 3**). Transformants appear as pinpoint plaques in about 3 d, and grow to form larger plaques over the next few days.
4. With a sterile loop, pick a 1–2-mm-diameter ball of cells from the edge of large plaques (3–6 mm in diameter) to tubes containing 2 mL HL-5 containing 1X Pen/ Strep and 1X blasticidin. Each transformant should grow to >10^6 cells/mL after 2 d of shaking at 22°C. Isolates that take longer than 3 d to grow to high titer in 2 mL of HL-5 are probably not true transformants.
5. Expand individual cultures to 25 mL and to a density of $1–4 \times 10^6$ cells/mL. Pellet the cells as described above, resuspended in HL-5 (without any antibiotics) to a total volume of 2 mL, add in 0.2 mL of fresh dimethylsulfoxide (DMSO), and freeze in two 1-mL aliquots at –80°C. The remainder of the culture can be expanded for further experimentation.

3.3. Molecular Cloning of the Site of Insertion and the Flanking DNA

The site of insertion is usually identified by direct cloning, using the conceptually simple method of plasmid rescue. The key to success in this procedure is in keeping the level of contaminating plasmid DNA low. Thus, from the time that the REMI transformant is isolated for DNA purification, it is recommended that no reusable glassware is used. Because molecular cloning is now performed with a variety of kits and buffers directly supplied by manufacturers, the procedure below only describes the important aspects of cloning REMI insertions.

1. Isolate genomic DNA from the REMI insertion strain, using any procedure that produces reasonably clean DNA.
2. Digest 1–2 µg of genomic DNA with a restriction enzyme that will not cut the REMI plasmid and purify the digestion products by phenol/chloroform purification and ethanol precipitation (*see* **Note 5**). Useful enzymes include *Bgl*II, *Eco*RI, *Cla*I, and *Spe*I.
3. Ligate about 1 µg of digested DNA in a 50-µL ligation reaction at 14°C for 12–16 h.
4. Transform *E. coli* by electroporation using 4 µL of the ligation reaction. SURE cells from Strategene (San Diego, California) support the growth of *Dictyostelium* DNA-containing plasmids quite well. This should yield 40–400 transformants.
5. Screen the cloned plasmids for the appropriate structure. Rescued plasmids are usually 2–8 kb larger than the REMI plasmid. Sequencing the DNA across the plasmid/genomic DNA borders should provide an unambiguous location of the insertion site.

4. Notes

1. Alternatively, Oxoid bacteriological peptone (Unipath; England) can be used. It is wise to buy a small amount of any particular peptone to make a test batch to test cell growth. Axenic *Dictyostelium* cells should have a doubling time of no more than 9 h in HL-5. Longer doubling times may indicate that the peptone is inadequate.
2. If the time constant is less than 0.5 ms,, the electrical discharge is occurring too rapidly and this indicates that the ionic strength of the electroporation mixture may be too high for the electrical parameters used.
3. In growing cells for transformation, the most reliable method is axenic culture in shaken suspension. Mid-exponential phase is the point in the growth curve at which cells display their shortest doubling time. For most strains growing in HL-5, this phase occurs between 2×10^6 and 4×10^6 cells/mL. Cells should not have been recently diluted to achieve this density; rather, they must grow up from an initial density of $<5 \times 10^5$ cells/mL (i.e., more than two cell doublings). An alternative method has been explored recently in our laboratory. Cells can be inoculated directly from colonies growing on bacterial lawns to 10 mL HL-5 in standard tissue culture dishes. It only takes about 1.5 d for the cells to reach confluence. To use these cells for electroporation directly, wash off all floating cells the day before the REMI transformation and replace the media with fresh HL-5. On the day of electroporation, harvest the cells, cool them on ice, wash them with electroporation buffer (H50 or HF buffer), and resuspend as described in the two protocols (*see* **Subheading 3.1.1.**, **steps 3** and **5; or Subheading 3.1.2.**, **step 1**).

4. In general, 10–200 U of enzyme work well in most cases. The REMI enzyme should be titrated when a new tube or a new enzyme supplier is used. This is especially true because different suppliers may define enzyme units differently. For many enzymes, 0–200 U produce an increasing number of transformants and additional enzyme above 200 U produces no additional transformants.

5. This includes the enzyme with which you linearized the plasmid. For a *Bam*HI cut plasmid that was transformed with *Dpn*II as the REMI enzyme, about 15% of the time the *Bam*HI site will be re-established. This is because about 15% of the time a G-C basepair will follow the genomic *Dpn*II site into which the plasmid integrates (GATCC), and this combined with the G residue from the cut *Bam*HI site on the REMI plasmid will recreate the *Bam*HI site at the point of insertion (*see* **Table 1**). Thus, in attempting to clone with *Bam*HI, some of the time the flanking genomic DNA will be separated from the shuttle plasmid on one side and this may make it difficult to sort out the genomic position of the insertion site.

Acknowledgments

I thank those members of the laboratory, past and present, who have contributed to the continuing refinement of the REMI procedure. I would particularly like to thank Guokai Chen, Chenyu Zhang, and Milligan Fossett for many helpful suggestions.

References

1. Schiestl, R. H. and Petes, T. D. (1991) Integration of DNA fragments by illegitimate recombination in *Saccharomyces cerevisiae*. *Proc. Natl. Acad. Sci. USA* **88,** 7585–7589.

2. Kuspa, A. and Loomis, W. F. (1992) Tagging developmental genes in *Dictyostelium* by restriction enzyme-mediated integration of plasmid DNA. *Proc. Natl. Acad. Sci. USA* **89,** 8803–8807.

3. Kuspa, A. and Loomis, W. F. (1994) REMI-RFLP mapping in the *Dictyostelium* genome. *Genetics* **138,** 665–674.

4. Adachi, H., Hasebe, T., Yoshinaga, K., Ohta, T., and Sutoh, K. (1994) Isolation of *Dictyostelium discoideum* cytokinesis mutants by restriction enzyme-mediated integration of the blasticidin S resistance marker. *Biochem. Biophys. Res. Commun.* **205,** 1808–1814.

5. Egelhoff, T. T., Brown, S. S., Manstein, D. J., and Spudich, J. A. (1989) Hygromycin resistance as a selectable marker in *Dictyostelium discoideum*. *Mol. Cell. Biol.* **9,** 1965–1968.

6. Kuspa, A. and Loomis, W. F. (1994) Transformation of *Dictyostelium*—Gene disruptions, insertional mutagenesis, and promoter traps. *Meth. Mol. Genet.* **3,** 3–21.

7. Chang, W. T., Gross, J. D., and Newell, P. C. (1995) Trapping developmental promoters in *Dictyostelium*. *Plasmid* **34,** 175–183.

8. Faix, J., Kreppel, L., Shaulsky, G., Schleicher, M., and Kimmel, A. R. (2004) A rapid and efficient method to generate multiple gene disruptions in *Dictyostelium discoideum* using a single selectable marker and the Cre-loxP system. *Nucleic Acids Res.* **32,** e143.

 9. Sussman, M. (1987) Cultivation and synchronous morphogenesis of *Dictyostelium* under controlled experimental conditions, in *Methods in Cell Biology* (Spudich, J. A., ed.), Vol. 28. Academic, Orlando, FL: pp. 9–29.
10. Howard, P. K., Ahern, K. G., and Firtel, R. A. (1988) Establishment of a transient expression system for *Dictyostelium discoideum. Nucl. Acids Res.* **16,** 2613–2623.
11. Pang, K. M., Lynes, M. A., and Knecht, D. A. (1999) Variables controlling the expression level of exogenous genes in *Dictyostelium. Plasmid* **41,** 187–197.

13

RNA Interference and Antisense-Mediated Gene Silencing in *Dictyostelium*

Markus Kuhlmann, Blagovesta Popova, and Wolfgang Nellen

Summary

Knockouts by homologous recombination are frequently used to investigate the function of genes in *Dictyostelium* and other organisms. Antisense-mediated gene silencing and RNA interference (RNAi) are convenient alternatives to reduce gene expression to different levels and to silence multigene families. We describe here the methods for efficient RNA interference in *Dictyostelium* and some useful mutant strains that enhance the success rate or may serve as convenient controls. We believe that it is helpful to also discuss failed attempts to optimize and expand the system because these are rarely discussed in the literature. In addition, a list of *in silico* and experimentally identified components in the RNAi and antisense pathway is presented.

Key Words: RNA interference; antisense; RdRP; Dicer; siRNA; *helF*.

1. Introduction

Since 1985 *(1)*, antisense-mediated gene silencing has been frequently used in *Dictyostelium*, especially to downregulate genes that are required for growth, to modulate gene expression, or to obtain conditional silencing. A drawback was that some genes were refractile to antisense and could not be silenced.

With the advent of RNA interference (RNAi) *(2)*, the method was immediately applied for *Dictyostelium* and proved successful *(3)*. The first applications did not show significant advantages over the established antisense procedure and were even slightly more laborious because in most cases, an additional cloning step was required. However, by examining silencing effects on several genes in comparison with antisense, it has turned out that RNAi is more reliable and has a significantly higher success rate. In addition, valuable tools have been developed to enhance the efficiency of silencing or to provide useful controls for silencing effects.

From: *Methods in Molecular Biology, vol. 346:* Dictyostelium discoideum *Protocols*
Edited by: L. Eichinger and F. Rivero © Humana Press Inc., Totowa, NJ

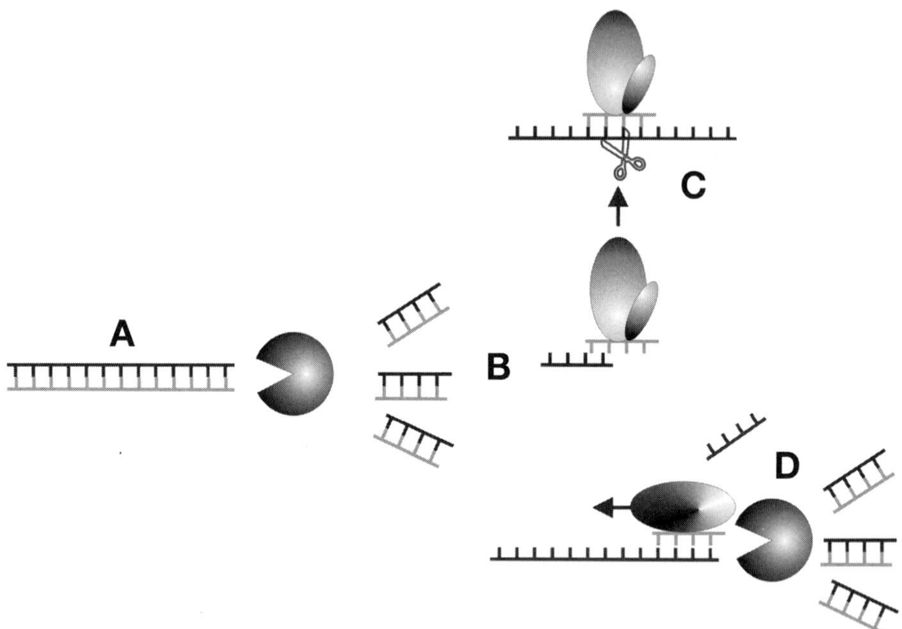

Fig. 1. The RNA interference mechanism. Double-stranded RNA (**A**) is diced to 21-nt long small interfering RNAs (siRNAs) (**B**) by Dicer. The fragments are unwound and one strand is transferred to the RNA-induced silencing complex (RISC) that targets a complementary mRNA. An argonaute protein within the RISC serves as slicer and cleaves the target opposite the center of the small RNA (**C**). Alternatively, the antisense strand of the siRNA may serve as a primer for an RNA-directed RNA polymerase (RdRP) on the mRNA and thus generate new double stranded RNA that is cleaved by Dicer (**D**).

The *Dictyostelium* genome contains most of the components that are known from other organisms to be involved or required for RNAi. The extensive (though still incomplete) knowledge on how the mechanism works (*see* **Fig. 1**) is in strong contrast to antisense-mediated silencing, in which the attempts to understand the pathway were very limited. Having the components at hand allows one to manipulate the system and to foresee further improvements as well as problems. For example, by a search in the genome database, we identified a homolog of the nuclease eri-1 that is known to degrade small interfering RNAs (siRNAs) in *Caenorhabditis elegans*. A knockout of this gene results in a significant enhancement of silencing efficiency *(4)*. The same effect most likely will be found in *Dictyostelium*, and could thus provide a further tool to improve the method. Knowing the rules that determine which strand of an siRNA is transferred to the putative RNA-induced silencing complex (RISC) *(5,6)* facilitates the design of short hairpin constructs (Larochelle, personal communication) and obviates the use of 21mers with an unfavorable 5' end of the antisense strand.

Table 1 gives a list of relevant genes that are either found by homology searches in DictyBase or have been experimentally confirmed to be involved in RNAi. Since RNAi, translational control by miRNAs and RNA-mediated chromatin remodeling as well as RNA-mediated DNA methylation share branches of their pathways, some of the listed genes may participate in multiple mechanisms. At this point, it cannot be unambiguously determined in which mechanism(s) some of the listed components may be involved.

In other organisms, RNAi protocols have led to a multitude of experimental shortcuts and interesting applications. Some of of these have been tried in *Dictyostelium*. We believe that unsuccessful attempts to apply these methods still provide valuable information. We therefore added in the notes a section on "failed experiments" (*see* **Notes 1–4**). We do not claim that these methods cannot be used at all in *Dictyostelium*, but just would like to point out that they are not as straightforward as may be assumed from reports in the literature on other organisms.

2. Materials

2.1. Preparation and Detection of siRNAs

1. Phosphate buffer (pH 6.7): 76.6 g/L KH_2PO_4, 99.2 g/L K_2HPO_4 (trihydrate). Store at 4°C.
2. Solution D: 4 M Guanidiniumisothiocyanate, 25 mM sodium-citrate (pH7), 0.5% [w/v]. N-laurylsarcosine, 100 mM β-mercapoethanol. Store at room temperature.
3. RNA-loading dye: 0.1% Bromophenolblue, 0.1% xylenexyanol in formamide, add 8 µL of a 10 mg/mL ethidium bromide stock solution per mL of dye. Store at 4°C.
4. Carbonate buffer: 80 mM $NaHCO_3$, 120 mM Na_2CO_3. Store at room temperature.
5. Hybridization buffer: 50% formamide, 50 mM Na_2HPO_4/NaH_2PO_4 buffer, pH 7.0, 7% sodium dodecyl sulfate (SDS), 0.3 M NaCl, 5X Denhardt's solution. Pass through sterile filter and store frozen or at 4°C.
6. 100X Denhardt's solution: 2% Ficoll 400, 2% polyvinylpyrrolidone, 2% bovine serum albumin (BSA) in water treated with diethylpyrocarbonate (DEPC). Store at –20°C.
7. DEPC-treated water: Add 0.1% diethylpyrocarbonate to distilled water, mix, and autoclave.
8. Polyacrylamide gels for small RNA separation: Denaturing polyacrylamide gels are set up with 19:1 acrylamide/bisacrylamide, 7 M urea in 0.5X Tris-acetate-ethylenediamine tetraacetic acid (EDTA) (TAE) buffer. Gel size should be in the range of 20 cm × 20 cm with a thickness of 1 mm or less.

Table 1
***Dictyostelium* Genes Involved in the RNA Interference (RNAi) Pathway**

Protein name/function	Gene name (accession no.)	Strains and Phenotypes	Reference to other systems
Dicer, RNaseIII-like enzyme, cleaves dsRNA to ~21mers, processes pre-miRNAs, **Drosha** processes pri-miRNAs	*drnA* DDB0220441	KO in AX2, no detectable phenotype	*dicer1, S. pombe*; *dcl-1, -2, N. crassa*; *dcr-1, C. elegans*; *dcl-1, -2, -3, -4, A. thaliana*; *dcr-1, dcr-2, D. melanogaster, Dicer, Homo sapiens*
	drnB DDB0231869	KO in AX2, no detectable phenotype	
RNA directed RNA polymerase, **RdRP**	*rrpA* DDB0191515	KO in AX2, negative for RNAi and antisense	*rrf-1, rrf-3, ego-1 C. elegans*; *SDE1/ SGS2, RDR2 A. thaliana*; *qde-1, N. crassa*
	rrpB DDB0220017	KO in AX2, negative for antisense RNA	
	rrpC DDB0216193	KO in AX4, impaired in development, negative for antisense RNA	
Argonaute, PPW-domain-protein, Different argonautes have slicer function in RISC and mediate translational inhibition in miRNA containing RNPs	*agnA* DDB0220136	C-term-GFP of isolated PPW domain	*ago1, S. pombe*; *qde-2, N. crassa*; *rde-1, ppw-1 C. elegans*; *AGO-1, A. thaliana*; *Ago-1, Ago-2, Aubergine, PIWI, D. melanogaster*; *ago-1, ago-2, ago-3, ago-4 Homo sapiens*
	agnB DDB0220437		
	agnC DDB0220438		

	agnD		
	DDB0218341		
	agnE		
	DDB0220439		
	agnF		
	DDB0220440		
Eri, RNaseIII-like enzyme, preferentially degrades siRNAs	*eriA* DDB0231961	C-term-GFP	*Eri-1*, *C. elegans*
DsRBD - RNA helicase	*helF* DDB0168963	KO in AX2, enhancement of RNAi efficiency, C-term. GFP, C-term. myc	closest relative: *drh-1/2*, *C. elegans*, KO impairs RNAi, others: *Armitage, Spindle E D. melanogaster; smg-2 mut-14 C. elegans, sde-3 A. thaliana*
R2D2, bridging protein between Dicer and RISC	not found in *Dictyostelium*		*RDE-4, C. elegans; HYL1 A. thaliana; R2D2, D. melanogaster*

dsRNA, double-stranded RNA; miRNAs, micro RNAs; *S. pombe*, *Saccharomyces pombe*; *N. crassa*, *Neurospora crassa*; *C. elegans*, *Caenorhabditis elegans*; *A. thaliana*, *Arabidopsis thaliana*; *D. melanogaster*, *Drosophila melanogaster*; KO, knockout; RISC, RNA-induced silencing complex; RNPs, ribonucleoproteins; GFP, green fluorescent protein; siRNAs, small interfering RNAs; dsRBD, double-stranded RNA binding domain.

9. SSC solution (20X stock): 3 *M* NaCl, 0.3 *M* sodium citrate, pH 7.0.
10. TAE buffer (50X stock): 2 *M* Tris, 1 *M* glacial acetic acid, 50 m*M* EDTA, pH 8.0.
11. TE buffer: 10 m*M* Tris-HCl, pH 7.4, 1 m*M* EDTA, pH 8.0.

3. Methods

3.1. Preparation of RNAi Constructs

Transformation vectors for RNAi-mediated gene silencing are constructed in the available vector systems using Geneticin resistance for selection in order to obtain high copy number. Vectors based on Blasticidin S resistance have been used but proved inefficient. Actin 6, actin 15, V18, and discoidin promoters have been successfully employed. Partially controlled silencing has been obtained with the folate-repressible discoidin I promoter *(7)*. Constructs are generated by standard cloning procedures. A fragment of the gene of interest ranging from 300 bp to 1000 bp or more is inserted downstream of the promoter in the vector in sense or antisense orientation. An in-frame fusion should be avoided (*see* **Note 5**). Two strategies for insertion of the corresponding antisense fragment may be used:

1. To create the hairpin loop, an unrelated spacer fragment of 300 bp to 900 bp is inserted downstream of the sense fragment. This could be any piece of DNA. e.g., part of a green fluorescent protein (GFP) or β-galactosidase gene (*see* **Fig. 2**). Avoid base compositions that would result in highly structured transcripts, repetitive sequences, and fragments that are already contained in the plasmid—these may result in recombination. Fragments of other *Dictyostelium* genes should also be avoided, because loop sequences may generate secondary siRNAs in some cases and thus result in undesired silencing of the gene from which the spacer sequence is derived. The same fragment as for the sense construct is then fused in antisense orientation downstream of the spacer (*see* **Note 6**).
2. Alternatively, a longer sense fragment is directly fused to a shorter antisense fragment with the single copy part of the sense fragment in the center of the inverted repeat. The nonrepeated part forms the loop and should be in the range of 300–800 bp. Intron sequences may be used for the loop, but should be avoided for the inverted repeat structure. In plants, the insertion of introns as spacers actually appears to enhance the silencing efficiency, but this has not yet been examined in *Dictyostelium*.

Constructs may also be done in the opposite orientation, i.e., antisense proximal, sense distal to the promoter. Obviously, this immediately avoids the problem of in-frame fusions (*see* above).

Larochelle and co-workers (personal communication) managed to clone inverted repeats without a spacer by the following procedure: polymerase chain reaction (PCR) products of the desired sequence were cut at one end at an endogenous or constructed restriction site, the other end contained the single

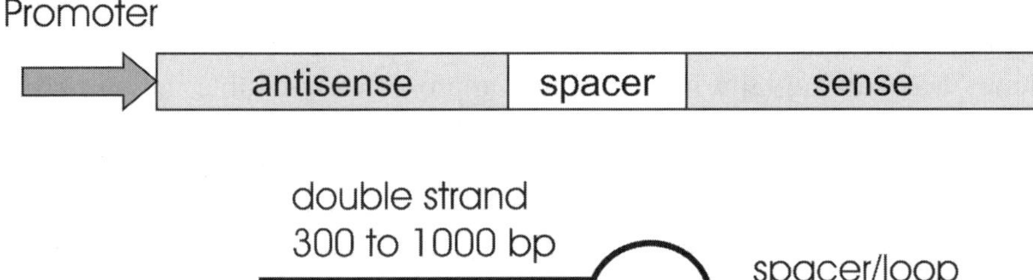

Fig. 2. Gene construct for transcription of hairpin RNA. Antisense, spacer, and sense fragments of a gene are fused as described in the text. Transcription from a strong promoter generates an RNA that folds back to form a hairpin. The double-stranded part of the hairpin is diced; the single-stranded loop sequence is probably degraded by endonucleases.

A-overhang usually generated by Taq polymerase. Ligation into a T-cloning vector thus results in a head-to-head fusion of two fragments. This product was then cloned into a *Dictyostelium* expression vector. It should be noted, however, that cloning inverted repeats appeared to be limited to sequences up to 200 bp per unit; longer fragments caused substantial problems when fused directly head-to-head.

Silencing was also obtained with constructs expressing short hairpin RNA (shRNA) containing an inverted repeat of 25 bp. Because these transcripts would only generate a single siRNA, it is important to design the base composition according to the rules of Khvorova *(6)* and Schwarz *(5)*. This means that the 5' end of the antisense RNA must form less stable base pairing than the 3' end. Gene sequences may be screen for the most promising shRNA construct at http://jura.wi.mit.edu/siRNAext/. At least for the *racE* gene investigated by Larochelle and colleagues, shRNAs provided the most homogeneous though not the highest silencing efficiency when single cells were examined.

RNAi vectors are transformed into *Dictyostelium* cells by standard procedures. No significant difference in efficiency was observed by either electroporation or calcium phosphate precipitation.

3.2. Downregulation of Silencing Resistant Genes in helF Knockout Mutants

Some genes (e.g., coronin) appear to be rather resistant to silencing by RNAi. This is reminiscent of, for instance, neuronal genes in *C. elegans* that are refractile to RNA interference. Using a strain disrupted in the *helF* gene, full or partial silencing of coronin could be obtained in approx 50% of the examined clones. For other genes that could be silenced in the wild type background, the efficiency was substantially improved and approximately double the number of knock-down clones could be obtained. This was true for the number of clonal isolates that showed reduced gene expression as well as for the average level of suppression (*see* **Note 7**).

HelF⁻ cells may be used as a standard strain for RNAi experiments. However, they are impaired in development: there is a delay at about 10 h after starvation, slugs are oversized, they overproduce stalk and the number of mature fruiting bodies is reduced (*see* **Note 7**). Nevertheless, the cells perform complete development and functional spores can be hartvested.

HelF⁻ cells were used in parallel to wild-type cells to investigate if RNAi is functional on developmental genes. The spore coat gene *sp96* was used as an example and successful silencing was observed in both strains (45% for wild-type and 90% for *helF⁻* cells).

3.3. Increased Silencing Efficiency by Cotransformation With Antisense Constructs

Substantial differences have been observed between silencing attempts by antisense RNA and by RNAi constructs. Usually, RNAi results in significantly more efficient silencing. Molecular analysis revealed that the mechanisms are similar but not identical, and that some components are used by one but not the other pathway. As an example, silencing by antisense RNA is not enhanced in *helF* knockout strains. Cotransformation of RNAi and antisense constructs increases the number of knock-down clones and the level of downregulation (*see* **Note 8**). Plasmids may simply be constructed in two steps. The first ligation of a gene (fragment) into an expression vector provides the antisense construct. In a second ligation, the sense orientation with a spacer (preferably a non-*Dictyostelium* sequence; *see* **Subheading 3.1.**) is added. The same or different promoters may be used and one or both plasmids may carry the resistance cassette. Integration of both plasmids into the genome should be confirmed by PCR. For the antisense construct, primers in the promoter and in the antisense sequence can be used whereas for the RNAi construct, a gene specific primer (either sense or antisense) and a primer in the spacer sequence are recommended. A unique non-*Dictyostelium* fragment as a spacer facilitates the confirmatory PCR.

3.4. Silencing of Essential Genes

Antisense, as well as RNAi, usually does not completely abolish expression of the target gene. In fact, in most experiments we find various degrees of silencing that range from (almost) complete to slight reductions on the RNA and the protein level. It is therefore assumed that essential genes may be down-regulated by siRNAs without causing a lethal phenotype. As a caveat, we will describe attempts to manipulate the *suvA* gene that encodes a putative histone methyl transferase (Essid and Nellen, unpublished). *SuvA* could not be knocked out by homologous recombination using several different constructs. Transformation with an RNAi construct resulted in several resistant colonies, but they never grew up to sufficient cell numbers to be analyzed. To examine if the vector construct by itself had some deleterious effect on the cells, we transformed it into *rrpA⁻* cells that are impaired in RNA interference. This resulted in the usual number of resistant colonies that would grow normally but, as expected, did not show any silencing. A second control was done with the *helF⁻* strain which shows enhanced silencing. In this case, not even a single primary transformant could be detected. We assume that any reduction of the *suvA* gene has dramatic consequences for the cell and does thus not allow for transformation with an RNAi construct. Nevertheless, it was unexpected that no transformants at all could be found in the wild-type background that were viable for a longer period of time. Usually, several RNAi transformants display no or incomplete silencing and should survive.

3.5. Silencing of Gene Families

As for antisense-mediated gene silencing, RNA interference affects gene families. This was clearly shown for discoidin (*discA* to *discD*) and thioredoxin (*trxA* to *trxC*). Theoretically, an approx 25-bp continuous identity between family members should be sufficient if the RNAi construct covers this region. It should, however, be considered that not all sequences are equally accessible, and that Dicer as well as RNA-directed RNA polymerases (RdRPs) that are involved in the pathway most likely have sequence preferences. More extensive continuous identities will therefore be required in most cases.

Specific silencing of single family members may be possible if they differ by a sufficiently long divergent sequence that could be targeted by the RNAi construct. However, it cannot be excluded that transitive silencing of other family members occurs: primary siRNAs annealing to the unique sequence in the messenger RNA (mRNA) may be extended by RdRPs in 5' direction along the mRNA and spread into sequences common to other family members. This would result in secondary siRNAs that could affect other genes. Because processivity of RdRPs appears to be low, a distance between the unique target sequence and common sequences of 100 bp should be sufficient to avoid transitive silencing.

3.6. Examination of Silencing Efficiency

Silencing efficiency may vary considerably from one target gene to the other and also when different silencing constructs are used for a single gene.

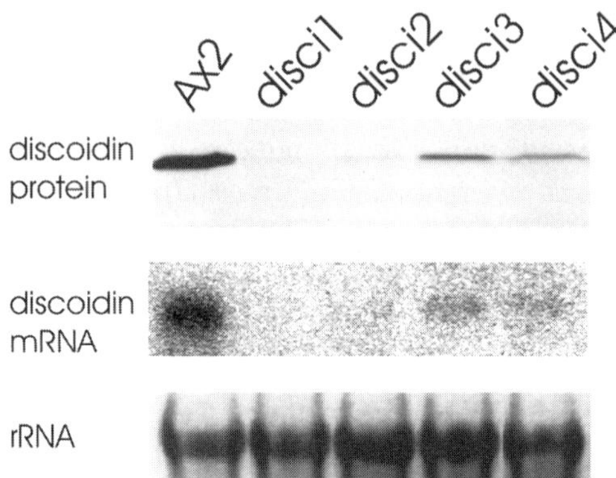

Fig. 3. Silencing of the discoidin gene family by RNA interference. Single clones were picked from Ax2 cells transformed with a hairpin gene construct directed against discoidin. Representative clones (disci 1 to disci 4) are shown in Western and Northern blots in comparison with the parent strain. Clones disci 3 and disci 4 show significantly reduced levels of discoidin mRNA and protein; these are considered to be partially silenced. Clone disci 2 shows very faint RNA and protein bands in long exposures. Setting an arbitrary threshold of 95% reduction, this clone still falls into the category "completely silenced." Clone disci 1 does not show any detectable signal on the RNA or the protein level.

Efficiency should be examined on two different levels: (a) the number of cell clones in a transformed population that displays any silencing and (b) the degree of silencing in different clones. For some targets, we obtained 5% of (partially) silenced clones, for others 70% or more. For some targets, many clones showed (almost) complete silencing, for others the majority of cells were only partially silenced.

If the target gene is expressed at sufficiently high levels and an antibody directed against the gene product is available, many clones can be easily screened for silencing by colony blots *(8)*. Usually, clones with different silencing levels will be readily detected by Western blotting of 10 to 20 different isolates. Northern blots may also serve to examine silencing efficiency. **Figure 3** provides an example for different silencing efficiencies of the discoidin gene. Since the hairpin transcripts are immediately diced, they are rarely detectable in blots and do not interfere with the quantitation of residual mRNAs (*see* **Note 9**). So far, we have never seen a discrepancy between the amount of mRNA and of protein.

The production of siRNAs may be examined to prove that silencing was due to RNAi; however, we find that detection is sometimes difficult.

3.7. Stability of Silencing

Once a silenced strain is identified, it should be immediately frozen for storage because, after prolongued growth, the silenced phenotype may revert. Frozen spores usually display the previous phenotype when thawed and grown up. Nevertheless, growing as well as freshly set up cultures should be re-examined from time to time to confirm that the gene of interest is still suppressed at the expected level.

3.8. Preparation of Small RNAs and Northern Blotting of siRNAs

siRNAs of 21 to 25 nt are indicative for gene silencing by RNA interference. It should be kept in mind that double-stranded RNA (dsRNA) in the form of hairpin transcripts can also lead to transcriptional gene silencing via RNA-mediated DNA methylation or RNA mediated chromatin remodelling. Although there is not yet experimental evidence for these mechanisms in *Dictyostelium*, they are not excluded but rather likely to be present (*see* Chapter 30). To provide evidence that a knock-down occurred on the level of RNA destabilization, demonstrating that siRNAs of the target gene can be detected may be required (*see* **Note 10**). Small RNAs may be detected in total RNA preparations, but because their abundance appears to be generally very low in *Dictyostelium*, enrichment for small RNAs is recommended. The following protocol describes the preparation of small RNAs from total cellular RNA and adaptation of the general Northern blot protocols for the detection of small RNAs.

3.8.1. Preparation of Small RNAs

Total cellular RNA is prepared by standard procedures (*9*).

1. 5×10^7 axenically grown cells are harvested by centrifugation at 540g for 10 min and 4°C; cells are washed once in phoshate-buffer.
2. Solubilize the pellet in 3 mL of solution D.
3. Add one-tenth volume of 3 M sodium acetate (pH 4.7).
4. Add one volume of phenol/chloroform/isoamylalkohol (25/24/1, Rotiphenol, Carl Roth GmbH, Karlsruhe, Germany) and separate phases by centifugation at 10,000g at 4°C for 30 min.
5. Precipitate total RNA with one volume of isopropanol from the aqueous phase for 30 min at –20°C and centrifuge at 10,000g for 10 min.
6. Wash precipitate with 70% ethanol and dry in air or in a speed vac.
7. Resuspend the pellet in 1 mL RNAse-free water pretreated with DEPC, and keep 50 µL as a control.
8. Add one volume of 10% polyethyleneglycol (PEG)-8000 (Serva, Heidelberg) in 1 M NaCl to the main fraction of the RNA, mix, and freeze at –20°C, centrifuge for 1 h at 4°C, 10,000g. High-molecular-weight RNAs are precipitated while small RNAs remain in the supernatant.

9. When the supernatant is transferred to a fresh tube, add one volume of isopropanol and precipitate small RNAs at –20°C overnight.
10. Collect small RNAs by centrifugation at 10,000g for 30 min at 4°C.
11. Wash pellet with 70% ethanol and dissolve in 100 µL formamide *(10)*.

For quality control, 5-µL aliquots of the complete control RNA and the small RNA preparation are mixed with RNA loading dye containing ethidium bromide and run on a 1.8% agarose gel containing 20 mM guanidiniumisothiocyanate (Carl Roth GmbH, Karlsruhe, Germany). Ribosomal RNAs should be reduced by at least 90% in the small RNA preparation and a strong enrichment of small RNAs in the range of tRNAs should be seen. RNA is quantitated by comparison with a dilution series of defined amounts of tRNA run on the same gel.

3.8.2. Polyacrylamide Gel Electrophoresis of Small RNAs

1. 20 µL of RNA-loading dye are added to 80 µL of RNA (50–100 µg), samples are heated to 65°C for 5 min and then placed on ice.
2. The samples are then loaded on a 15% polyacrylamide gel.
3. The gel is run in 0.5X TAE buffer at approx 400 V; gel plates should become hot. The run is terminated when the bromophenol blue dye has passed through at least three-quarters of the gel.

As a marker, a *Sau3a*-digested pGEM-Vector, radioactively labeled by a fill-in reaction with Klenow-fragment, may be used for convenience. One should, however, be aware that DNA fragments run differently from RNA fragments. Size determination may, therefore, be inaccurate, but is still sufficient for a crude estimate. Alternatively, size markers may be generated by in vitro transcription or obtained from commercial suppliers.

3.8.3. Blotting of Small RNAs

RNA is transferred from the gel to a membrane (*see* **Note 11**) by "tank-blot" in 0.5X TAE at 20 mA for 2 h or overnight. The buffer is constantly mixed by a stirring rod and the tank is cooled on ice.

After transfer, the RNA is cross-linked to the moist membrane by ultraviolet light with 150 mJ and immediately used for prehybridization. Alternatively, moist membranes may be wrapped in plastic wrap and stored at 4°C.

3.8.4. Probes and Hybridization

As probes for the detection of small RNAs, DNA oligonucleotides labeled by the kinase reaction with T4 polynucleotide kinase (MBI Fermentas, St. Leon-Rot, Germany) may be used according to the supplier's instructions. The forward reaction that requires dephosphorylation by shrip alkaline phosphatase (SAP) yields significantly higher incorporation. Buffers and instructions are supplied by the manufacturer. For unknown siRNAs, a mixture of different oligonucleotides is recommended because not all potential siRNAs are equally abundant. Alternatively, in vitro transcripts are generated by T7 or SP6 RNA polymerase from the DNA fragment of interest and continuously labeled with $[\alpha^{32}\text{-P}]$UTP.

1. Combine in a reaction tube: 5 µL (1 µg) of template DNA (PCR product with T7 promoter or appropriately linearized plasmid DNA), 1 µL of T7 or SP6 RNA polymerase (MBI Fermentas, St. Leon-Rot, Germany), 10 µL of transcription reaction buffer (supplied by the manufacturer), 10 µL of rNTP mix (5 mM each), 1 µL of RNasin (MBI Fermentas), 18 µL of DEPC-treated water.
2. Incubate for 40 min at 37°C.
3. Add 1 µL of RNase-free DNase and incubate for 15 min at 37°C.
4. Add 50 µL of phenol-saturated TE buffer.
5. Load sample on a sephadex G-50 spin column and centrifuge for 5 min at 580g and collect the flow-trough to remove unincorporated nucleotides.

In contrast to standard Northern blot probes, the in vitro-transcribed RNA is subjected to limited hydrolysis prior to hybridization:

1. For hydrolysis, 50 µL of labeled RNA are mixed with 750 µL of 200 mM carbonate buffer for 30 min for a 600-nt transcript. The time of carbonate treatment depends on the length of the transcript and can be calculated as follows: t = (Li – Lf)/ (k × Li × Lf) t: time of hydrolysis (min) Li: initial length of probe (kb) Lf: final length of probe (kb) k: rate constant = 0.11 kb × min^{-1} The average final length of the probe should be around 50 nt.
2. The reaction is stopped by neutralization with 50 µL of 3 M sodium acetate (pH 5.0).
3. Hydrolysis may be examined by running a small aliquot of the probe on a 15% polyacrylamide gel. A smear at the expected size range should be detected.
4. Before hybridization, the probe is denatured for 5 min at 95°C.
5. The membrane is prehybridized for 30 min (or longer) at 40°C in hybridization solution. The solution is exchanged and the radiolabeled probe is added. Hybridization is carried out at 40°C overnight.
6. After removal of the probe, the membrane is washed as follows: Rinse with 2X SSC/0.1% SDS, wash at 42°C for 5 min with 2X SSC/0.1% SDS, wash at 42°C for 5 min with 1X SSC/0.1% SDS, wash at 42°C for 5 min with 0.5X SSC/0.1% SDS. Monitor radioactivity in between washing steps and stop washing when the membrane appears "clean." Washes with 1X SSC and 0.5X SSC may not be required.
7. The membrane is exposed to an imager screen. Signals may be detectable on a phosphoimager after over night exposure but in some cases exposure times of up to one week are required (*see* **Note 12**).

4. Notes

1. Feeding RNaseIII deficient bacteria that express dsRNA was surprisingly successful in *C. elegans*. Similar attempts were made in *Dictyostelium* but did not result in any silencing.

2. Transitive silencing describes the generation of secondary siRNAs that could silence an endogenous gene. The method could be very useful when a preformed strain was established that expressed a hairpin RNA from a foreign gene fragment, e.g., β-galactosidase. A second transformation vector is then constructed that contains any gene fragment of interest fused 5' to the β-galactosidase fragment, the primary target gene. After introduction into cells that contain the hairpin construct, the gene fusion is transcribed from a strong promoter. SiRNAs derived from the hairpin and directed against the primary target could serve as primers for RdRPs and be elongated, eventually, into the adjacent gene fragment of interest. The resulting dsRNA would generate secondary siRNAs that would not only target the artificial gene fusion but also the corresponding endogenous gene. Transitive silencing is functional in *C. elegans* and plants *(12,13)*. Experiments in *Dictyostelium* suggest that transitivity can also be observed. However, the efficiency is low and the success rate is limited. In both plants and *C. elegans*, it has been shown that the processivity of RdRPs is apparently low and the amount of secondary siRNAs decreases rapidly with increasing distance from the fusion point. Although it is tempting to use RNAi approaches with this system, we do not recommend it. It should be noted that even in organisms where it has been reported to be functional, transitive silencing is not as frequently used as would be expected.

3. It may appear useful to transform a strain with an antisense construct and, if silencing is insufficient, to supertransform with a corresponding sense construct. In theory, both transcripts should form hybrids that will be targeted by Dicer. At least in our experience, this is either not the case or functions with very low efficiency. Cotransformation with sense and antisense vectors did not result in sufficient silencing of the target gene. This was probably due to a low frequency of hybrid formation, because sense and antisense transcripts could be readily detected in Northern blots.

4. It would be convenient to clone a sequence of interest in between two opposing promoters and generate sense and antisense transcripts from one and the same fragment. The advantage would be that hybridization should be facilitated, because both RNAs are synthesized in close vicinity. We have used opposing V18 and actin 6 promoters and did not observe any silencing of the target gene.

5. In-frame fusions of sense fragments in RNAi vectors may initiate translation and protect the fold-back construct from being diced. In one case, we observed that a simple disruption of the frame enhanced silencing significantly.

6. The loop in RNAi constructs is mainly used to prevent cloning problems with inverted repeat sequences, and not to facilitate folding of the hairpin. Do not try to construct the inverted repeat first and then insert the loop—this will most likely fail due to recombination in *E. coli*.

7. The molecular function of the *helF* gene product is not yet known. The pleiotropic effects in the gene disruption mutant may be due to defects in the processing of micro RNAs that appear to be present in *Dictyostelium* (Söderbom and Ambros, personal communication). That the disruption strain has an indirect effect on the expression of other genes cannot, therefore, be ruled out. Functional

analysis of RNAi showed that subthreshold levels of hairpin transcripts (as determined by run-on transcription) become effective in silencing in the *helF⁻* background. It thus appears that a failure of silencing is mainly due to low levels of hairpin expression and that these low levels become functional silencers in the absence of *helF*.

8. For some genes, we have observed that colony blots did not reflect silencing in axenic culture. Especially for discoidin, but also for coronin, colony blots were positive for the protein but in axenic culture, the cells were almost completely silenced. Interestingly, enhancement of silencing by cotransformation with an antisense construct or by RNA interference in the *helF* knockout strain was also reflected in colony blots in that many clones that lacked the protein could be identified.

9. RNAi transcripts are usually not detectable in Northern blots. Only in a few cases, weak bands corresponding to the size of the expected hairpin transcript have been observed. However, run-on experiments show that the constructs are strongly transcribed but apparently, they are immediately diced. Surprisingly, the primary foldback transcripts are also not found when the knock-down of a target gene was not successful. It therefore seems that dicing is not the limiting step for efficient knockdown experiments. Therefore, hairpins may be degraded by other nucleases that do not generate siRNAs or siRNAs are rapidly degraded by EriA related nucleases.

10. SiRNAs are produced in RNA interference experiments as well as in antisense RNA experiments. The two mechanisms cannot be qualitatively distinguished on this level. It appears, however, that more siRNAs are generated in antisense experiments.

    Detection of siRNAs requires that a target mRNA is present. Contrary to intuition, the majority of the siRNAs detected is not derived from hairpin transcripts generated by the RNAi construct that was introduced into the cells. When no target gene is expressed (e.g., when a GFP hairpin is introduced into a wild-type cell), no siRNAs are found. Consistent with this observation, the amount of detected siRNAs is independent of the transcription levels of the hairpin construct (as determined by run-on transcription). Detectable siRNAs are most likely diced products of dsRNA (secondary siRNAs) generated by RdRP on a target mRNA (secondary siRNAs).

11. Membrane brands and batches appear to be crucial for transfer and detection of small RNAs. There are positive reports on GeneScreen plus (NEN) but there are also membranes where no siRNAs can be detected at all. It is recommended that one test various membranes on a small RNA that is abundant (*11*) before deciding which one to use for the experiments.

12. SiRNAs for endogenous genes silenced by hairpin constructs are apparently present at very low levels: whereas U6 RNA can be detected after 3 to 4 h of exposure, only faint bands of, e.g., discoidin, are seen after 1 wk of exposure.

Acknowledgments

This work was supported by a grant from the Deutsche Forschungsgemeinschaft (Ne 285/8) and by the European Union (FOSRAK, 005120) to W.N.

References

1. Crowley, T. E., Nellen, W., Gomer, R. H., and Firtel, R. A. (1985) Phenocopy of discoidin I-minus mutants by antisense transformation in *Dictyostelium*. *Cell* **43**, 633–641.
2. Fire, A., Xu, S., Montgomery, M. K., Kostas, S. A., Driver, S. E., and Mello, C. C. (1998) Potent and specific genetic interference by double-stranded RNA in *Caenorhabditis elegans*. *Nature* **391**, 806–811.
3. Martens, H., Novotny, J., Oberstrass, J., Steck, T. L., Postlethwait, P., and Nellen, W. (2002) RNAi in *Dictyostelium*: the role of RNA-directed RNA polymerases and double-stranded RNase. *Mol. Biol. Cell* **13**, 445–453.
4. Kennedy, S., Wang, D., and Ruvkun, G. (2004) A conserved siRNA-degrading RNase negatively regulates RNA interference in *C. elegans*. *Nature* **427**, 645–649.
5. Schwarz, D. S., Hutvagner, G., Du, T., Xu, Z., Aronin, N., and Zamore, P. D. (2003) Asymmetry in the assembly of the RNAi enzyme complex. *Cell* **115**, 199–208.
6. Khvorova, A., Reynolds, A., and Jayasena, S. D. (2003) Functional siRNAs and miRNAs exhibit strand bias. *Cell* **115**, 209–216.
7. Blusch, J., Morandini, P., and Nellen, W. (1992) Transcriptional regulation by folate: inducible gene expression in *Dictyostelium* transformants during growth and early development. *Nucleic Acids Res.* **20**, 6235–6238.
8. Wallraff, E. and Gerisch, G. (1991) Screening for *Dictyostelium* mutants defective in cytoskeletal proteins by colony immunoblotting. *Methods Enzymol.* **196**, 334–348.
9. Maniak, M., Saur, U., and Nellen, W. (1989) A colony-blot technique for the detection of specific transcripts in eukaryotes. *Analyt. Biochem.* **176**, 78–81.
10. Hamilton, A. J. and Baulcombe, D. C. (1999) A species of small antisense RNA in posttranscriptional gene silencing in plants. *Science* **286**, 950–952.
11. Aspegren, A., Hinas, A., Larsson, P., Larsson, A., and Soderbom, F. (2004) Novel non-coding RNAs in *Dictyostelium* discoideum and their expression during development. *Nucleic Acids Res.* **32**, 4646–4656.
12. Alder, M. N., Dames, S., Gaudet, J., and Mango, S. E. (2003) Gene silencing in *Caenorhabditis elegans* by transitive RNA interference. *RNA* **9**, 25–32.
13. Van Houdt, H., Bleys, A., and Depicker, A. (2003) RNA target sequences promote spreading of RNA silencing. *Plant Physiol.* **131**, 245–253.

III

IMAGING AND LOCALIZATION METHODS

14

Application of Fluorescent Protein Tags as Reporters in Live-Cell Imaging Studies

Annette Müller-Taubenberger

Summary

The advent of the "GFP technology" together with advances in digital imaging and microscopic techniques has revolutionized our view of the living cell. Genetically encoded fluorescent proteins have become widely used as markers in living cells. The application of these fluorescent proteins as noninvasive tags revealed new aspects of protein dynamics and the biological processes they regulate. The modification of naturally occurring fluorescent proteins as well as the identification of new fluorescent proteins now provide researchers with a variety of useful fluorescent markers suitable for all kind of investigations in live-cell imaging studies.

This chapter provides an overview of the most widely used fluorescent proteins for the labeling of either individual proteins or compartments of *Dictyostelium discoideum* cells. Furthermore, aspects such as how to design a fusion protein as well as advantages and disadvantages of specific fluorescent proteins are discussed. Finally, a protocol from transformation to identification of positive clones is provided.

Key Words: *Dictyostelium*; expression vector; fluorescent protein; GFP; RFP; live-cell imaging; microscopy; photoactivation.

1. Introduction

The green fluorescent protein (GFP) has now been used for over a decade in order to visualize either subcellular compartments or to track specific molecules within cells *(1)*. The great success of this approach is that fluorescent protein indicators can be designed by choice and can be monitored in living cells. Thus, these probes allow one to follow the distribution and localization of specific proteins or to visualize responses of cells and even tissues or whole organisms to a variety of biological events or signals *(2–5)*.

From: *Methods in Molecular Biology, vol. 346:* Dictyostelium discoideum *Protocols*
Edited by: L. Eichinger and F. Rivero © Humana Press Inc., Totowa, NJ

Because the technique allows working with live cells, it changed the way we fundamentally look at cells. Before the introduction of fluorescent protein reporters, researchers were mostly restricted to fixed cell preparations that either could be stained with specific dyes or were labeled with antibodies coupled to fluorescent dyes. The use of fluorescent dyes to visualize proteins in living cells was more the exception than a general method. Various techniques that were employed successfully for mammalian cell lines, such as microinjection of dyes or antibodies into live cells, failed for *Dictyostelium*. Although progress in improving the technique of microinjection was reported *(6)*, *Dictyostelium* cells are difficult to inject, because these cells are relatively small (approx 10 µm) and adhere insufficiently. Other loading techniques had the disadvantage of low yields *(7)*.

Once the advantages of the GFP technology were recognized, its use became widespread. In parallel to this, the development of variants of GFP and the isolation of fluorescent proteins from other organisms proceeded *(8–10)*. Most fluorescent proteins were isolated from either *Hydrozoa* (hydromedusas like the jellyfish *Aequorea victoria*) or *Anthozoa* (coral animals). More recently, GFP-like proteins were also detected in other phyla such as *Arthropoda (9)* and other *Cnidaria*. Not all of the fluorescent proteins available commercially or described in the literature are suitable for live-cell imaging in *Dictyostelium*. This chapter describes the most commonly used fluorescent proteins for localization studies in *Dictyostelium*, and provides practical advice on how to design a fusion protein and how to analyze fluorescent protein-expressing cells. More specific technologies related to fluorescent proteins, such as fluorescence resonance energy transfer (FRET) or specific imaging techniques during cell movement or cell differentiation, are discussed in other chapters of this book.

2. Materials

2.1. Choice of the Fluorescent Protein

2.1.1. GFP and Variants

The term "fluorescent protein" describes those proteins that can become spontaneously fluorescent through the autocatalytic synthesis of a chromophore. The structure of GFP was determined as a β-can fold (an 11-stranded β-barrel) with an intrinsic chromophore *(1)*. The formation of the chromophore is an autocatalytic process *(11)*. The wavelengths of light absorbed by the chromophore depend on the local chemistry and, consequently, amino acid substitution in or near the chromophore can result in changes of the absorbed or emitted light.

GFP has been engineered to produce a palette of fluorescent proteins used for tagging approaches as well as for more specialist applications such as

Table 1
Fluorescent Proteins Used for Studies in *Dictyostelium*

Species	Fluorescent Protein	Excitation	Emission
Aequoria victoria			
green	wtGFP	395/475 nm	507 nm/503 nm
	S65T-GFP	475 nm	503 nm
	PA-GFP	399/475 nm	517 nm/504 nm
cyan	ECFP	434 nm	477 nm
yellow	EYFP	514 nm	527 nm
	DYFP	514 nm	527 nm
Heteractis crispa			
red	HcRed1	588 nm	618 nm
Discosoma striata			
red	mRFP1	584 nm	607 nm
	mRFPmars	585 nm	602 nm

wtGFP, wild-type green fluorescent protein; PA-GFP, photoactivatable green fluorescent protein; ECFP, enhanced cyan fluorescent ptoein; EYFP, enhanced yellow fluorescent protein; mRFP1, monomeric red fluorescent protein 1.

biosensors to measure Ca^{2+} concentration or pH. The latter uses will not be discussed here, but are mentioned to provide an example of the versatile possibilities offered by this protein. A broad range of fluorescent proteins with different emission spectra has been produced that spans almost the entire width of the visible spectrum (*see* **Table 1**).

The optimum temperature for maturation of wild-type (wt)GFP is 28°C. In order to optimize folding in mammalian cells, several mutations have been introduced into the wtGFP sequence, such as F64L or a valine residue after the start methionine, which is present in all of the "enhanced" (E) versions such as EGFP. However, for work with *Dictyostelium*, these alterations have little advantage over the S65T-GFP variant *(8)*, which provides a much brighter tag than the original wtGFP and is often the first choice when designing a fluorescent fusion protein.

Blue (BFP), cyan (CFP), and yellow (YFP) fluorescent proteins are all derivatives of GFP with emissions in the respective color range. BFP is not recommendable for the work with *Dictyostelium*, because cells round up within seconds as a result of the short wavelength excitation and therefore cannot be visualized over longer time periods. Likewise, CFP is only of restricted use, although it has been used successfully in some FRET studies in combination with YFP *(12)*.

YFP has similar characteristics as GFP. The commercially available EYFP variant was optimized for mammalian cells, but has been expressed in *Dictyostelium* cells fused to a number of different proteins. Citrine *(13)* and Venus *(14)* are YFP variants carrying additional amino acid changes that improved brightness and facilitated host protein folding. The DYFP variant (available from the author) derives from S65T-GFP and harbors fewer amino acid changes than EYFP (T65G, V68L, T203Y, and H231L).

2.1.2. Photoactivatable GFP

Another application is offered by the use of photoactivatable fluorescent proteins. In general, photoactivation is defined as the light-induced activation of an inert molecule to an active state, e.g., the ultraviolet light-induced release of a group from a "caged" compound. This method can also be employed within living cells *(15)*. Attempts to develop new fluorescent proteins also included photoactivatable fluorescent proteins, and recently, three different versions, photoactivatable GFP (PA-GFP) *(16)*, Kaede *(17)*, and kindling fluorescent protein 1(KFP) *(18)* have been reported.

The ability to "turn on" the fluorescence of a photoactivatable protein makes it an excellent tool for studying protein behavior within live cells. As fluorescence of the respective proteins appears only after photoactivation, results are not blurred by newly synthesized proteins. Furthermore, activation can be also restricted to a specific area to study the dynamic behavior of the fluorescent protein.

PA-GFP was developed by optimizing the photo-conversion properties of *Aequorea victoria* wtGFP *(16)*. wtGFP normally has a major absorption peak at about 400 nm and a minor peak at about 475 nm. The T203H mutation decreases the initial absorbance in the minor peak region and leads to an approx 100-fold increase of the 475-nm excitation peak after photoactivation. Prior to photoactivation, PA-GFP shows very little fluorescence with 488-nm excitation. After photoactivation, a stable fluorescence is detected. PA-GFP has been cloned into a *Dictyostelium* vector and has been used in combination with LimEΔcoil *(19)* to visualize filamentous actin structures after photoactivation (*see* **Fig. 1**).

2.1.3. Red Fluorescent Proteins

The color spectrum of fluorescent proteins has been also enlarged considerably by the introduction of red fluorescent proteins (RFPs). DsRed (BD Biosciences Clontech) was the first RFP introduced, and DsRed2 and DsRed-timer are variants of the original version. Compared with GFP, DsRed has several disadvantages. DsRed forms tetramers, and therefore is not very useful for studying protein dynamics. Furthermore, it matures very slowly through a

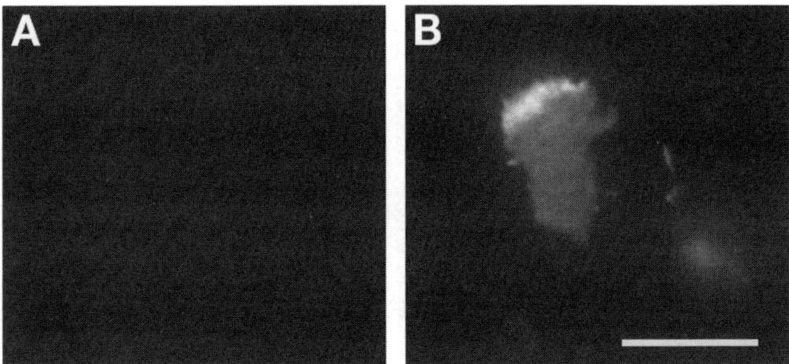

Fig. 1. Photoactivation of a photoactivatable (PA)-green fluorescent protein (GFP)-tagged protein *(16)* in *Dictyostelium* cells. *Dictyostelium* wild-type AX2 cells expressing PA-GFP-LimEΔcoil before (**A**) and after (**B**) photoactivation. PA-GFP-LimEΔcoil is used to visualize filamentous actin structures *(19)*. Fluorescence was recorded with an Olympus IX70 microscope equipped with a condensor for total internal reflection fluorescence (TIRF) microscopy (Visitron) modified for photoactivation, and a 60×/ 1.45 Plan-Apo oil objective. Photoactivation was achieved by irradiation for 1 s using a 100 W Hg^{2+}-lamp in combination with a D405/20 filter (500 DCXR; 475 LP; AHF Analysentechnik). The activated PA-GFP-LimEΔcoil was imaged using a 488-nm laser for excitation and a standard GFP filter set (excitation 488/10; DC 498; emission 505 LP; Chroma Technologies, Rockingham, VT). The bar represents 10 μm. The images were recorded in collaboration with Kurt I. Anderson at the Max-Planck-Institute for Molecular Cell Biology and Genetics, Dresden, Germany.

green intermediate. Our experience is that only the green intermediate forms in *Dictyostelium* cells, and to date there is no report on the use of this fluorescent protein in *Dictyostelium*.

HcRed1 is another RFP isolated from the *Hexacorallia* species *Heteractis crispa* (BD Biosciences Clontech) *(20)*. There are no reports so far on the use of this RFP in *Dictyostelium*, but work in our lab has shown that it can be used for instance in combination with a peroxisome targeting signal to selectively label peroxisomes (*see* **Fig. 2**). Furthermore, it has been tested to produce a fusion protein with the actin-binding domain of *Dictyostelium* filamin in order to visualize filamentous actin. But in contrast to the labeling of a cellular compartment, the use of HcRed1 for tracking protein dynamics is rather limited because of its low molar extinction coefficient, and hence its low brightness.

Another important improvement has been the development of monomeric red fluorescent proteins (mRFPs). By sequential mutagenesis of DsRed and exchange of 50 amino acid residues, an RFP—mRFP1—has been obtained that is a true monomer and matures within 1 h; however, it shows somewhat lower extinction coefficient, quantum yield, and photostability than DsRed

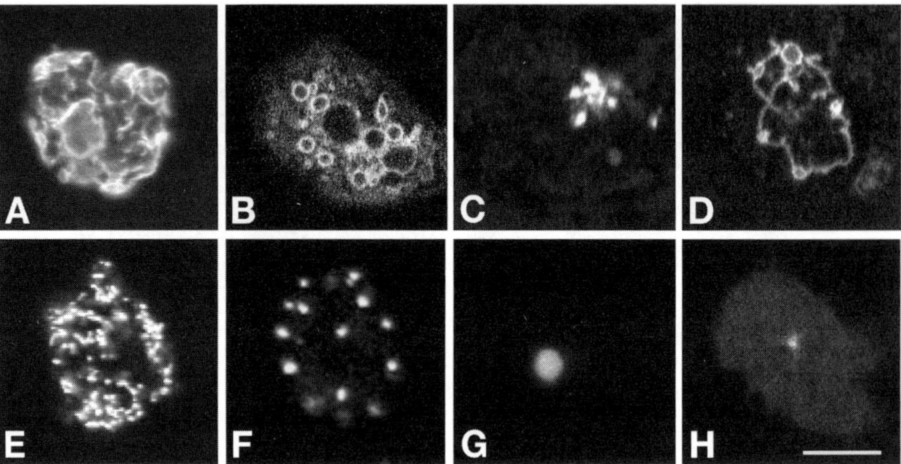

Fig. 2. Fluorescent fusion proteins localizing to specific organelles of *Dictyostelium* cells. Fluorescent fusion proteins were expressed in *Dictyostelium* and fluorescence was recorded using a confocal microscope (Zeiss LSM 410 or LSM 510 Meta). (**A**) A cell expressing calnexin-green fluorescent protein (GFP) *(29)* showing labeling of endoplasmic reticulum membranes including the nuclear envelope. (**B**) GFP(N)-golvesin is primarily associated with the membranes of endosomes *(31)*. (**C**) Golvesin(C)-GFP provides a highly specific marker of the Golgi apparatus, showing no association with endosomes *(31)*. (**D**) The contractile vacuole complex consisting of bladders and a network of connecting tubules is visualized using dajumin-GFP *(32)*. (**E**) Mitochondria labeled with a polypeptide probe carrying a targeting sequence at its N-terminus and a GFP-tag at its C-terminus *(33)*. (**F**) Peroxisomes were visualized by expression of a construct of HcRed1 tagged with a peroxisomal targeting sequence specific for peroxisomes (unpublished). (**G**) A nucleus of an interphase cell labeled with histone 2B-GFP *(34)*. (**H**) Centrosome visualized by expression of GFP-Spg1 (unpublished). The bar in **H** represents 5 µm, and magnifications in the other figures have been adapted accordingly except for **C**, where the bar corresponds to 3 µm.

(21). When expressed as a fusion protein with various cytoskeletal proteins in *Dictyostelium*, mRFP1 is rather dim. Its fluorescence is too weak to allow recording of image series with high spatial and temporal resolution, but it can be used for the labeling of compartments, such as for instance, the nucleus using mRFP1-histone-2B (unpublished results).

In order to design an improved mRFP for the use in *Dictyostelium*, seven amino acid changes that enhanced the brightness in the tetrameric DsRed1 variant RedStar *(22)* as well as amino acid changes introduced into the mRFP1 variant were combined to make a brighter RFP. In addition, the codon usage of

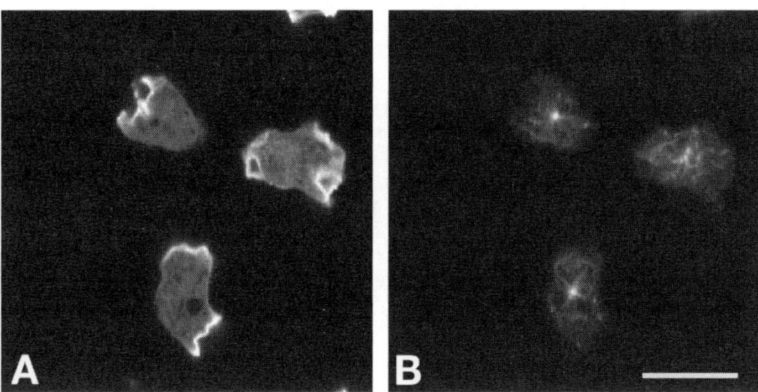

Fig. 3. Dual-color labeling of *Dictyostelium* cells. *Dictyostelium* cells are expressing mRFPmars-LimEΔcoil (**A**) to visualize filamentous actin structures, in combination with green fluorescent protein (GFP)-α-tubulin (**B**) to visualize microtubules. Fluorescence was recorded with a confocal microscope (Zeiss LSM 410) equipped with 488-nm argon and 543-nm helium–neon lasers and a 100×/1.3 Plan-Neofluar objective. BP510-525 and HQ607-682 filters were used to separate the emissions. Bar represents 10 μm.

the sequence was optimized according to the high A/T content of *Dictyostelium*. The gene encoding the new mRFP variant (mRFPmars) was completely synthesized from oligonucleotides *(23)*. The spectral properties of mRFPmars are similar to mRFP1, but compared with mRFP1, mRFPmars shows an enhanced brightness in vivo. mRFPmars proved to be suitable for monitoring the dynamics of cytoskeletal proteins in cell motility, mitosis, and endocytosis using dual-wavelength microscopy *(23,24)*.

2.2. Combinations of Fluorescent Proteins for Multi-Color Labeling

One major goal when using a combination of two or more fluorescent proteins for live-cell imaging is to achieve a good spectral resolution. The excitation maxima of GFP and YFP are about 20 nm apart, and in principle, newer microscopes can, at least in theory, separate these wavelengths. Practically, it is still difficult to distinguish between these adjacent emission wavelengths.

Cross-talk between filter sets causes major problems, especially when similar structures are labeled. Therefore, fluorescent proteins of the red or far-red spectra are of specific interest, as they can be clearly separated from GFP or YFP using suitable filter combinations *(23,24)*. An additional advantage when using RFPs is that *Dictyostelium* cells like many other cell types display reduced autofluorescence at longer wavelength excitations. An example for dual-wavelength microscopy of *Dictyostelium* cells expressing an mRFPmars fusion protein in combination with a GFP fusion protein is shown in **Fig. 3**.

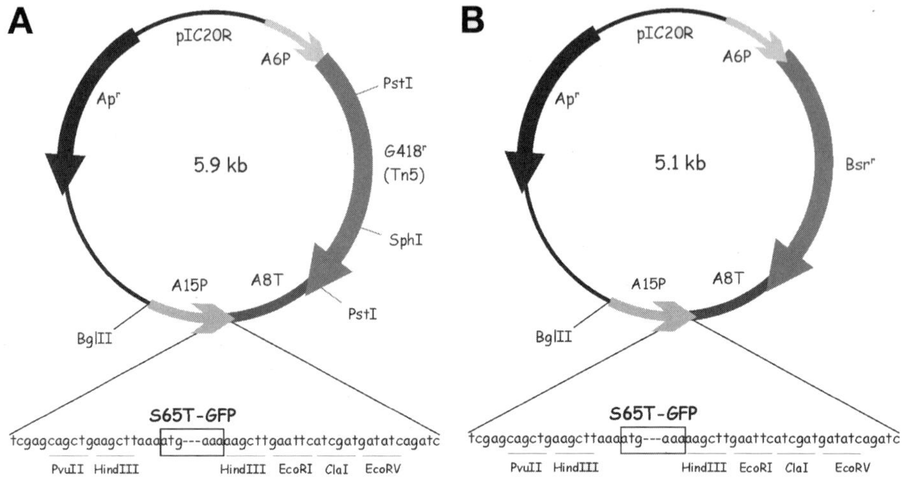

Fig. 4. Vectors suitable for expression of fluorescent fusion proteins in *Dictyostelium*. (**A**) Green fluorescent protein (GFP) expression vectors of the pDEX series *(26)* conferring resistance against G418 (neomycin), and (**B**) against blasticidin (bsr). Abbreviations: A15P, actin-15 promoter; A6P, actin-6 promoter; A8T, actin-8 terminator; Ap[r], ampicillin resistance. In the S65T-GFP sequence indicated, only the start codon, atg, and the last codon, aaa, are depicted.

2.3. Vectors for Expression of Fluorescent Fusion Proteins

For the expression of a fluorescent fusion protein, *Dictyostelium* cells must be transformed with a plasmid that allows expression of the fusion protein and carries resistance markers for selection in *Dictyostelium* as well as in *Escherichia coli*. Extrachromosomal vectors have been used successfully in different studies (e.g., *25*), and are preferable for the expression of larger fragments, but integrating vectors like the ones of the pDEX *(26)* (*see* **Fig. 4**) or the pDXA series *(27)* are recommended. The following features characterize them:

1. A *Dictyostelium expression cassette* that allows the joining of the gene of choice to a fluorescent protein-encoding sequence and expression of the fusion protein. The cassette consists of an actin-15 promoter, a multiple cloning site either at the N- or at the C-terminal end of the fluorescence protein-encoding gene allowing tagging of either end of the protein, and an actin-8 terminator.
2. A *resistance marker cassette* that allows positive selection of *Dictyostelium* cells that have taken up the introduced plasmid. Very often, it consists of an actin-6 promoter that directs a gene whose expression confers resistance against an antibiotic and an actin-8 terminator. Routinely, cassettes conferring resistance against blasticidin (bsr) or geneticin/neomycin (G418) are used for selection in *Dictyo-*

stelium, but hygromycin (hyg) provides an alternative when the two other markers are not usable.

3. A bacterial *selection marker cassette* that allows selection after transformation and amplification of the plasmid in *E. coli*.

The vectors shown in **Fig. 4** enable constitutive expression of fluorescent fusion proteins in *Dictyostelium*, and are randomly integrated into the genome and maintained by antibiotic selection. Vectors for work with *Dictyostelium* are available from the Stock Center at DictyBase (http://dictybase.org). Vectors containing fluorescent proteins are also commercially available (e.g., BD Biosciences Clontech; Stratagene; Qbiogene; Evrogen).

Plasmids should be purified using one of the commercially available isolation kits (e.g., Qiagen) or by CsCl-density gradient centrifugation. Linearization of the plasmid using a single cut restriction enzyme site is not mandatory, although some users prefer to transform linearized constructs to increase transformation efficiency.

2.4. Transformation of Dictyostelium Cells, and Preparation of Samples for Assaying Expression of Fluorescent Proteins

2.4.1. Dictyostelium Cell Culture and Transformation

Live-cell imaging techniques can be used for the analysis of fluorescent proteins in either wild-type or mutant *Dictyostelium* cells. For transformation of *Dictyostelium* with plasmids directing the expression of a fluorescent fusion protein, the classical Ca^{2+}-precipitation method is applicable *(28)* (or http://www.dictybase.org/techniques/), but for routine purposes, the electroporation method is recommended (*see* Chapter 11).

2.4.2. First Testing of Fluorescent Protein Expression

After the appearance of clones following transformation, cells must be tested for expression of fluorescent fusion proteins. This can be done by a quick test using cells from the primary transformation plates.

1. Cells from individual clones or from a mixture of cells (about 50–100 µL) are allowed to settle on a coverglass for about 10–15 min.
2. Cells are then washed twice with 17 m*M* Soerensen's phosphate buffer (PB), pH 6.0, by sucking up the liquid and replacing it immediately with buffer (*see* **Note 1**).
3. Cells can then be examined using an inverted fluorescence microscope within the next 30–60 min.

2.4.3. Chambers for Live-Cell Imaging

Once a clone expressing the fluorescent fusion protein of choice has been isolated, cells usually must be explored in more detail. To this end, self-made open chambers can be used that allow observation of cells over longer time periods (*see* **Note 2**). To assemble them, the following is needed:

1. Microscopic coverglasses, 50 × 50 mm, thickness 1.5 (Menzel GmbH & Co KG, 28116 Braunschweig, Germany), or alternatively standard coverglasses, e.g., 24 × 40 mm (Menzel). The microscopic coverglasses are cleaned using saran wipes. For specific purposes (e.g., total internal reflection fluorescence [TIRF] or reflection interference contrast microscopy), coverglasses should be cleaned by stirring them gently in 0.1 *N* HCl for several hours, washing them carefully in distilled water, and drying them with linen.

2. Silicone grease, heavy, high vacuum (e.g., from Merck, 64271 Darmstadt, Germany, cat. no. 1.07921.0100).

3. Plexiglas rings: these are usually self-made plexiglass rings with a diameter of about 2–5 cm, a wall thickness of 2.5 mm and a height of 3–4 mm. The plexiglass rings are sealed onto the coverglasses using silicone grease. Depending on the size of the chambers and the specific experiment, 0.5–2.0 mL of cells are allowed to settle and washed three times with PB as described above (*see* **Subheading 2.4.2.**).

Alternatively, tissue culture dishes with coverglass bottom are available either from MatTek corp., 35 mm, No. 1.5 (0.16–0.19 mm) coverslip (http://www.glassbottomdishes.com) or FluoroDish FD35, 35 mm, glass 23 mm, glass thickness 0.17 mm (http://www.wpiinc.com).

2.5. Microscopic Setup

2.5.1. Microscopes

For selection and analysis of cells transformed with vectors encoding fluorescent fusion proteins, a routine inverted microscope allowing fluorescence detection is mandatory. Upright microscope stages are only of limited use for live-cell investigations. Microscopes for live-cell investigations are, for example, the IM135 (Zeiss) or the IX51 (Olympus), or, if more options in respect to the microscope stage are required, the Axiovert 200 (Zeiss) or the IX71 (Olympus) are recommended. Other manufacturers offer comparable microscope stages.

Very often, live-cell imaging is combined with more specific techniques such as confocal microscopy, spinning disc microscopy, or TIRF microscopy *(19)*. Laser scanning microscopes (e.g., Zeiss LSM 510, or Leica TCS SP2) are providing a very good standard technology and will be suitable for most purposes, although for some investigations, more specialized instruments might be needed. The technology of microscopes is constantly improving. Therefore, recommendations are difficult, and depend also on the specific requirements of the user and the specimen to be analyzed.

2.5.2. Objective Lenses

The choice of an objective lens depends on the optical requirements of the instrumentation as well as on the specific requirements of a particular experiment. In general, for analysis of live cells in fluorescence microscopy, the recommendation is to choose an objective with a high numerical aperture (NA). The numerical aperture determines the resolution, and therefore directly influences the results of an experiment. Furthermore, the optical correction of an objective lens is very important, but cannot be discussed here in detail. As a rough guide, for microscopy of cells expressing one fluorescent protein, a 40× or 63× fluar objective with a NA of 1.3 will be sufficient for most experiments. For multi-color labeling experiments, a 40× or 63× plan-apochromat lens with a NA of 1.4 is more suitable.

2.5.3. Filters

In fluorescence microscopy, specific filter combinations must be employed to separate the light used for excitation from the emitted light. The choice of the appropriate filters depends on the excitation and emission properties of the fluorescent proteins analyzed in a given experiment. Confocal microscopes have, in general, more sensitive detectors than standard microscopes, allowing the filtering by quite narrow band-pass filters.

More noteworthy is the issue of filters in multicolor-tracking. A very important consideration is the possibility of filter cross-talk. Practically, this refers to the fact that different fluorescent proteins have overlapping emission spectra, which must be separated.

Techniques like *linear un-mixing* and specific software additions are available, but overlapping spectra still remain problematic, especially with co-localizing structures or when a high temporal resolution is required. Although the use of correct filter combinations may separate close emission spectra like those of GFP and YFP, quite often only the choice of fluorescent proteins that can be clearly separated will provide satisfactory results (e.g., GFP in combination with RFP). Well documented information about standard filter combinations for the most commonly used fluorescent proteins is provided by specialist microscopic filter retailers (e.g., Chroma Technology Corp.).

2.5.4. Cameras and Tracking Software

In recent years, improvements in optical and microscope performance have been accompanied by improvements in camera performance. The most widely utilized digital camera for fluorescence imaging of living cells is the charge-coupled device (CCD) camera. In practice, investigators choose between cooled CCD and intensified CCD (ICCD) cameras. Cooled CCD cameras can have very high quantum efficiency, but readout speeds are limited as a result of the readout noise of the CCD chip. Modern ICCD cameras can be used to sample full frames at video rate or faster, making them ideal detectors for single-molecule imaging in living cells. Recently, on-chip multiplication gain systems have become available (e.g., from Andor) that amplify the signals before read-out.

A large number of different cameras are commercially available, and it is impossible to give specific recommendations, because the brightness of the investigated specimen and the spatial and temporal resolution required influences the choice of a camera. In general, the spatial and temporal resolution of the detector and the camera must be compatible. High quantum efficiency and low readout noise are usually required. Quantum efficiency ranges from 0 to 80% depending on the wavelength of the photons entering the pixels. The quantum efficiency of CCD cameras is low in the blue wavelength range (350–450 nm) and high in the green and red wavelength range (450–650 nm). High read-out speeds generate more noise.

Specific software is often needed for the analysis of fluorescence images. Very often, software packages are provided with the camera or the microscope system (e.g., LSM 510 META). For general purposes and analyses, ImageJ is a freely available software package (http://rsb.info.nih.gov/ij/) that is constantly updated.

3. Methods

3.1. Designing a Fluorescent Fusion Protein

Fluorescent fusion proteins are being used to address a wide range of different questions in live cells, and for individual studies, it is an important requirement that a functional fusion protein is produced (*see* **Note 3**). The resultant fusion protein should be expressed at or close to the level of the endogenous protein without the tag; it should be full-length, and correctly localized (*see* **Note 4**). As GFP and other fluorescent proteins are about 27 kDa, it is important to consider potential effects of adding such a large tag in association with a protein under investigation.

There are two principal choices in adding a fluorescent tag: amino- or carboxyl-terminal addition (although internal insertion, or insertion after a leader sequence *[29]*, provide additional options). The choice will often depend on the location of critical regions of interaction or folding; therefore, knowledge of the domain structure of the investigated protein is helpful. If, for example, a critical interaction domain is located at the N-terminus, the logical consequence would be to add GFP to the C-terminus. In many cases, it might be beneficial to analyze amino- and carboxyl-terminal fusion constructs simultaneously and assess their correct localization and/or functions (*see* **Notes 3** and **4**).

An important prerequisite is that both the fluorescent protein and the protein under investigation are correctly folded by avoiding steric hindrance or folding interference. The fluorescent protein must fold correctly to be fluorescent, and the analyzed protein should fold correctly to maintain its endogenous functionality. In order to achieve this, a linker should be introduced between fluorescent protein and investigated protein. The linker should be sufficiently long and flexible, and can be optimized if necessary. Glycine confers flexibility and has the smallest side chain of all amino acids; the introduction of serine residues improves the solubility. Very often, also alanine residues are introduced. The length of the linker may vary between 6 and 18 amino acid residues. The introduction of a linker is less critical when *Aequorea* GFP is used, as its C-terminus is naturally quite floppy.

3.2. Advantages and Disadvantages of Fluorescent Proteins: Some Considerations

Recent advances in imaging tools and techniques as well as the collection of different fluorescent proteins have allowed researchers to monitor protein dynamics in living cells *(2–5)*. For live *Dictyostelium* cells, examples of labeling of different subcellular compartments by different fluorescent fusion constructs are shown in **Fig. 2**. The use of fluorescent fusion proteins in live-cell imaging studies continually provides us with new insights into the behavior of proteins, organelles, and cells. However, some points should be considered when using fluorescent proteins within live cells.

For many studies, it might be important to ensure that fluorescent fusion proteins are expressed at similar levels as the endogenous protein in order to avoid effects on the normal cellular function (*see* **Notes 4** and **5**). Expression of fusion proteins is often under control of the constitutive actin-15 promoter, and therefore, in individual cells, higher protein levels than directed by the endogenous promoter activity will be produced. Overexpression of any protein always risks disturbing the structural integrity of the cell. If the expressed protein is an enzyme, the additional protein levels may cause harmful elevated enzymatic activity, or the labeled protein may sequester upstream binding components or saturate downstream targets, causing altered cellular function and phenotypic change. To prevent an unbalanced situation, knock-in constructs should be used to express the fluorescent proteins under the endogenous promoter of the protein under investigation.

All of the anthozoan GFP-like proteins that have been characterized so far, including DsRed and HcRed1, form obligate oligomers. Although oligomerization does not strictly exclude the use of a fluorescent protein in reporting gene expression or marking a compartment, it may interfere with the function of the protein to which it is fused. Very often, the normal behavior and dynamic properties of the protein fused to the oligomerizing fluorescent protein are disturbed.

Another problem is the potential aggregation of fluorescent proteins, which impedes cellular functions and also leads to toxicity within cells. Different molecular mechanisms can cause fluorescent protein aggregation. Among them are excessive overexpression or improper folding of the fluorescent fusion protein. For mRFP1, clustering has been suggested to be caused by hydrophobic interactions between the fluorescent monomers.

The maturation of fluorescent proteins involves a multi-step folding process that consists of cyclization, dehydration, and oxidation, and in the case of DsRed and variants, of an additional autocatalytic step. DsRed has the disadvantage of residual green fluorescence. However, because of the improved maturation properties of the monomeric RFPs, RFP1 *(21)* and also the *Dictyostelium* variant mRFPmars *(23)* showed only minimal green fluorescence and therefore are perfectly suited for dual-color experiments.

In cases in which only individual cells or parts of cells are being examined, the use of photoactivatable probes *(16–18)* may be helpful. Before photoactivation, cells expressing photoactivatable proteins display little or no fluorescence in the spectral region that is used for detecting enhanced fluorescence. After photoactivation of a selected region, an increase in fluorescence is observed.

3.3. Construction of Vectors for Expression of Fluorescent Fusion Protein

3.3.1. Cloning of Protein-Encoding Sequences into Fluorescent Protein Expression Vectors

1. In most cases, protein-encoding fragments will be amplified by polymerase chain reaction (PCR) using either *Dictyostelium* cDNA or genomic DNA as template. This has the advantage that the cloning sites of choice can be added to the ends at will. Either a full-length sequence encoding the complete protein or only specific parts of it can be cloned. In the latter case, knowledge of the domain structure is very helpful to obtain a functional fusion protein that is folded properly.

2. The sequence of the protein of interest is cloned in frame either to the N- or C-terminus of a fluorescent protein encoded by the expression vector of choice (*see* **Subheading 2.3.**). Inclusion or exclusion of the stop-codon must be considered.

3. In order to ensure correct folding of the fusion protein very often a short linker sequence of 6 to 15 amino acid residues is inserted between the fluorescent protein and the protein of interest. The introduction of a flexible hydrophilic linker consisting of glycine, alanine, and serine or threonine residues can prevent steric hindrance (e.g., $[GGS]_{4x}$), but should be optimized if necessary. The linker amino acid residues should be included into the considerations when planning the oligonucleotides to amplify the sequence of the protein of interest.

4. Most *Dictyostelium* expression constructs contain the actin-15 or actin-6 promoter. Using an actin-15 promoter ensures expression of the fusion construct during growth and development, whereas the actin-6 promoter is much less active when the cells are grown on bacteria compared with axenically grown cells *(30)*. Other promoters that allow expression of the fusion construct during a specific time-point of development—such as, for instance, the contact-site A promoter *(26)*, which starts to direct expression during early aggregation—can be used if necessary.

5. Once the sequence encoding the protein of interest is cloned successfully into the appropriate fluorescent protein expression vector, it is advisable to confirm its correct sequence by sequence analysis to exclude that mutations have been introduced during the PCR amplification.

3.4. Expression of a Fluorescent Fusion Protein and Analysis of Transformants

1. Before the experiment is started, plasmid DNA allowing expression of the fluorescent fusion protein must be prepared. The plasmid DNA concentration should be between 0.5 and 3.0 µg/µL. The DNA should be dissolved either in electroporation buffer (50 mM sucrose, 10 mM potassium phosphate, pH 6.1), H$_2$O, or very low-salt buffer.

2. *Dictyostelium* cells should be grown either in shaking culture or on plate to a cell density not higher than 2–3 × 10^6 per mL.

3. On d 1, transform 2–3 × 10^7 *Dictyostelium* cells with 15–25 µg of plasmid DNA by electroporation or the calcium precipitate method.

4. On d 2, add the selective antibiotic.

5. On d 3–5 nontransformed cells will die. Change medium carefully if necessary.

6. On d 7–14, check by microscopic inspection for the appearance of clones. Once clones have been detected, cells from a clone can be either transferred with a pipet to a microtiter or Costar well and cultivated further. Alternatively, a mixture from the transformation plate is cloned by plating the cells on a bacterial lawn. Single clones are then transferred to Costar wells and cultivated in axenic medium with addition of the selective antibiotic and 1% Penicillin-Streptomycin (Gibco).

7. Individual clones are then tested by fluorescence microscopy for expression of the fluorescent fusion protein (*see* **Subheading 2.4.3.**).

4. Notes

1. Sometimes, the axenic medium is responsible for a high fluorescence background. For cases in which observation of cells under growth conditions is necessary, and the normal medium (e.g., HL5) provides problems as a result of the high fluorescent background, a medium with low fluorescence background has been described (*30* or http://www.dictybase.org/techniques/lowflo_medium.htm).

2. For imaging of more delicate structures, such as microtubules, the agar-overlay technique *(5)* can help to visualize specific structures.

3. Sometimes, it may be necessary to show that the fusion protein is functional. In these cases, specific assays can help to test the activity of a fluorescent fusion protein. For instance, an actin spin-down assay may prove that a fluorescent actin-binding protein is still able to interact with actin. Rescue of a mutant phenotype by expression of a fusion protein also indicates that the fusion protein is functional.

4. In some cases, the experiment may not lead to the expected result that is clear localization of the fusion protein. Sometimes, just the endocytic compartment is labeled, or the label appears diffusely in the cytoplasm (both localizations that may turn out to be the correct ones in some instances), or simply no label is detected at all. The reasons for such results can be numerous, but for troubleshooting a couple of tests should be performed to exclude that a fluorescent fusion protein for instance is degraded:

   a. Does the localization of the fluorescent fusion protein correspond to the localization of the endogenous protein? To answer this question, immuno-staining methods using protein-specific and fluorescent-protein specific antibodies in fixed cell preparations can be employed to compare the structures labeled.

   b. Western blotting of sodium dodecyl sulfate (SDS)-polyacrylamide gel electrophoresis (PAGE)-separated cell lysates using a fluorescent protein-specific antibody can determine whether the fusion protein is expressed, and whether it has the correct size. Polyclonal and monoclonal antibodies against GFP, HcRed, or DsRed are commercially available, e.g., from BD Biosciences Clontech or Molecular Probes.

   c. Alternatively, if a protein-specific antibody is available, the endogenous protein level can be compared with the level of the fluorescently tagged version of the protein.

   d. It is also advisable to sequence the construct and confirm that no changes have been introduced into the tagged gene.

5. Quite often, it is observed that only a certain percentage of cells of a clone express the fluorescent fusion protein. In general, expression levels of a fluorescent fusion protein may vary in different cells from the same clone.

Acknowledgments

I would like to thank Günther Gerisch (Max-Planck-Institut für Biochemie, Martinsried) for constant support and encouragement, and all members of his lab for collaboration. This work was supported by the Deutsche Forschungsgemeinschaft (SFB 413).

References

1. Tsien, R. Y. (1998) The green fluorescent protein. *Annu. Rev. Biochem.* **67,** 509–544.
2. Lippincott-Schwartz, J., Snapp, E., and Kenworthy, A. (2001) Studying protein dynamics in living cells. *Nature Rev. Mol. Cell Biol.* **2,** 444–456.
3. Lippincott-Schwartz, J. and Patterson, G. H. (2003) Development and use of fluorescent protein markers in living cells. *Science* **300,** 87–91.

4. Miyawaki, A., Sawano, A., and Kogure, T. (2003) Lighting up cells: labeling proteins with fluorophores. *Nat. Cell Biol.* **(Suppl),** S1–S7.
5. Gerisch, G. and Müller-Taubenberger, A. (2003) GFP-fusion proteins as fluorescent reporters to study organelle and cytoskeleton dynamics in chemotaxis and phagocytosis. *Meth. Enzymol.* **361,** 320–337.
6. Fukui, Y. (2000) Microinjection technique for *Dictyostelium.* http://pubmed. nwu.edu/~yoshifk/fukui.html.
7. Schlatterer, C., Knoll, G., and Malchow, D. (1992) Intracellular calcium during chemotaxis of *Dictyostelium discoideum*: a new fura-2 derivative avoids sequestration of the indicator and allows long-term calcium measurements. *Eur. J. Cell Biol.* **58,** 172–181.
8. Heim, R. and Tsien, R. Y. (1996) Engineering green fluorescent protein for improved brightness, longer wavelengths and fluorescence resonance energy transfer. *Curr. Biol.* **6,** 178–182.
9. Shagin, D. A., Barsova, E. V., Yanushevich, Y. G., et al. (2004) GFP-like proteins as ubiquitous metazoan superfamily: evolution of functional features and structural complexity. *Mol. Biol. Evol.* **21,** 841–850.
10. Zhang, J., Campbell, R. E., Ting, A. Y., and Tsien, R. Y. (2004) Creating new fluorescent probes for cell biology. *Nature Rev. Mol. Cell Biol.* **3,** 906–918.
11. Miyawaki, A., Nagai, T., and Mizuno, H. (2003) Mechanisms of protein fluorophore formation and engineering. *Curr. Opin. Chem. Biol.* **7,** 557–562.
12. Janetopoulus, C., Jin, T., and Devreotes, P. (2001) Receptor-mediated activation of heterotrimeric G-proteins in living cells. *Science* **291,** 2408–2411.
13. Griesbeck, O., Baird, G. S., Campbell, R. E., Zacharias, D. A., and Tsien, R. Y. (2001) Reducing the environmental sensitivity of yellow fluorescent protein. *J. Biol. Chem.* **276,** 29,188–29,194.
14. Nagai, T., Ibata, K., Park, E. S. Kubota, M., Mikoshiba, K., and Miyawaki, A. (2002) A variant of yellow fluorescent protein with fast and efficient maturation for cell-biological applications. *Nat. Biotechnol.* **20,** 87–90.
15. Politz, J. C. (1999) Use of caged fluorochromes to track macromolecular movement in living cells. *Trends Cell Biol.* **9,** 284–287.
16. Patterson, G. H. and Lippincott-Schwartz, J. (2002). A photoactivatable GFP for selective photolabeling of proteins and cells. *Science* **297,** 1873–1877.
17. Ando, R., Hama, H., Yamamoto-Hino, M., Mizuno, H., and Miyawaki, A. (2002) An optical marker based on the UV-induced green-to-red photoconversion of a fluorescent protein. *Proc. Natl. Acad. Sci. USA* **99,** 12,651–12,656.
18. Chudakov, D. M., Belousov, V. V., Zaraisky, A. G., et al. (2003) Kindling fluorescent proteins for precise *in vivo* photolabeling. *Nat. Biotechnol.* **21,** 191–194.
19. Bretschneider, T., Diez, S., Anderson, K., et al. (2004) Dynamic actin patterns and Arp2/3 assembly at the substrate-attached surface of motile cells. *Curr. Biol.* **14,** 1–10.
20. Gurskaya, N. G., Fradkov, A. F., Terskikh, A., et al. (2001) GFP-like chromoproteins as a source of far-red fluorescent proteins. *FEBS Lett.* **507,** 16–20.

21. Campbell, R. E., Tour, O., Palmer, A. E., et al. (2002) A monomeric red fluorescent protein. *Proc. Natl. Acad. Sci. USA* **99,** 7877–7882.

22. Knop, M., Barr, F., Riedel, C. G., Heckel, T., and Reichel, C. (2002) Improved version of the red fluorescent protein (drFP583/DsRed/RFP). *Biotechniques* **33,** 592–602.

23. Fischer, M., Haase, I., Simmeth, E., Gerisch, G., and Müller-Taubenberger, A. (2004) A brilliant monomeric red fluorescent protein to visualize cytoskeletal proteins in *Dictyostelium. FEBS Lett.* **577,** 227–232.

24. Diez, S., Gerisch, G., Anderson, K., Müller-Taubenberger, A., and Bretschneider, T. (2005) Subsecond reorganization of the actin network in cell motility and chemotaxis. *Proc. Natl. Acad. Sci. USA* **102,** 7601–7606.

25. Moores, S. L., Sabry, J. H., and Spudich, J. A. (1996) Myosin dynamics in live *Dictyostelium* cells. *Proc. Natl. Acad. Sci. USA* **93,** 443–446.

26. Faix, J., Gerisch, G., and Noegel, A. A. (1992) Overexpression of the csA cell adhesion molecule under its own cAMP-regulated promoter impairs morphogenesis in *Dictyostelium. J. Cell Sci.* **102,** 203–214.

27. Knetsch, M. L. W., Tsiavaliaris, G., Zimmermann, S., Rühl, U., and Manstein, D. (2002) Expression vectors for studying cytoskeletal proteins in *Dictyostelium discoideum. J. Musc. Res. Cell. Motil.* **23,** 605–611.

28. Nellen, W., Silan, C., and Firtel, R. A. (1984) DNA-mediated transformation in *Dictyostelium discoideum*: expression of an actin gene fusion. *Mol. Biol. Cell* **4,** 2890–2898.

29. Müller-Taubenberger, A., Lupas, A. N., Li, H., Ecke, M., Simmeth, E., and Gerisch, G. (2001) Calreticulin and calnexin in the endoplasmic reticulum are important for phagocytosis. *EMBO J.* **20,** 6772–6782.

30. Liu, T., Mirschberger, C., Chooback, L., et al. (2002) Altered expression of the 100 kDa subunit of the *Dictyostelium* vacuolar proton pump impairs enzyme assembly, endocytic function and cytosolic pH regulation. *J. Cell Sci.* **115,** 1907–1918.

31. Schneider, N., Schwartz, J.-M., Köhler, J., Becker, M., Schwarz, H., and Gerisch, G. (2000) Golvesin-GFP fusions as distinct markers for Golgi and post-Golgi vesicles in *Dictyostelium* cells. *Biol. Cell* **92,** 495–511.

32. Gabriel, D., Hacker, U., Köhler, J., et al. (1999) The contractile vacuole network of *Dictyostelium* as a distinct organelle: its dynamics visualized by a GFP marker protein. *J. Cell Sci.* **112,** 3995–4005.

33. Hüttig, A. (2004) DMIF1-ein Regulator der mitochondrialen F1F0-ATPase in *Dictyostelium discoideum*. Doctoral thesis. Ludwig-Maximilians-Universität München.

34. Gerisch, G., Faix, J., Köhler, J., and Müller-Taubenberger, A. (2004) Actin-binding proteins required for reliable chromosome segregation in mitosis. *Cell Motil. Cytoskel.* **57,** 18–25.

15

Investigating Gene Expression

In Situ *Hybridization and Reporter Genes*

Ricardo Escalante and Leandro Sastre

Summary

Coordinated cell type differentiation is essential for morphogenesis during *Dictyostelium* development. The specification of different cell types and the regulation of temporal and spatial patterns of expression of cell type-specific genes are important problems currently being addressed in many laboratories. Besides, determination of gene expression patterns provides significant information in the characterization of developmental mutants. Cell type-specific probes and well characterized promoters are available that allow the identification of most cell types during *Dictyostelium* development. Expression patterns can be studied by whole-mount *in situ* messenger RNA (mRNA) detection and by the use of reporter genes under the control of specific promoters. The most common *in situ* hybridization technique, based on nonradioactive ribo-probes that are hybridized to fixed whole-mounts prepared at different developmental stages, is described. Several reporter genes have been used to characterize gene expression patterns and to determine functional promoter elements. The *lacZ* gene, coding for the β-galactosidase enzyme, is the reporter most frequently used in *Dictyostelium* because both temporal- and spatial-patterns of expression can be easily determined. Generally used β-galactosidase detection methods are described.

Key Words: *In situ* hybridization; gene expression; expression pattern; β-galactosidase; reporter; promoter; X-Gal.

1. Introduction

Dictyostelium cell types arise soon after aggregation and rapidly organize themselves spatially in the developing structure (for a review, *see* **ref. *1***). *In situ* hybridization for mRNA detection and the use of cell-type-specific promoters fused to reporter genes provide spatial information at a cellular level for gene expression studies.

From: *Methods in Molecular Biology, vol. 346:* Dictyostelium discoideum *Protocols*
Edited by: L. Eichinger and F. Rivero © Humana Press Inc., Totowa, NJ

Whole-mount *in situ* hybridization has been used to determine the pattern of expression of newly identified genes as well as to study cell type differentiation and patterning in *Dictyostelium* mutants (*see* **refs.** *2* and *3* as representative examples). Here we describe the basic protocol for whole-mount *in situ* hybridization with Digoxigenin-RNA-labeled probes. A subsequent enzyme-catalyzed color reaction will produce an insoluble blue/purple precipitate that can be easily visualized. Although this is the most common method, variations on both labeling and detection have been described *(4)*.

Reporter genes are extensively used in *Dictyostelium* research. These genes are cloned in plasmid vectors under the transcriptional control of different gene promoters. Constructs are transformed in *Dictyostelium* cells, where the expression of the reporter gene can be easily detected. The expression pattern of the reporter gene indicates the activity of the gene promoter used in the construct. Therefore, reporter genes are used to characterize gene promoters and their functional regions (for example, *see* **refs.** *5* and *6*). A number of studies from several laboratories have characterized many useful *Dictyostelium* promoters, including constitutive and cell type-specific promoters. Reporter genes under the control of some of these promoters are used to identify different cell populations. For example, cells from a strain carrying a reporter gene construct can be identified in a mixture with cells from other strains and, therefore, the structures coming from each can be discerned *(7,8)*. Reporter gene constructs are also useful to identify cell types during development. For example, the differentiation of prestalk or prespore cells and their segregation in prestalk and prespore regions can be studied in a given mutant strain using reporter genes under the control of cell type-specific promoters *(9,10)*.

Several reporter genes have been used in *Dictyostelium*, including those coding for the enzymes chloramphenicol acetyl transferase, luciferase, and β-galactosidase. These enzymes were chosen as reporters because their activity can be easily determined in vitro using extracts of the transfected cells. The three reporter genes are of great utility in determining the transcriptional activity of gene promoters or fragments derived from them and have been used for their functional characterization. The study of promoter regions from genes regulated during development usually requires the establishment of spatial patterns of expression and, therefore, the determination of reporter gene activity in whole structures. Whereas the three reporter proteins mentioned previously can be detected in whole structures by immunohistochemistry using the appropriate antibodies *(11)*, β-galactosidase activity can be easily determined by histochemical methods. This advantage has made of the β-galactosidase-encoding gene, *lacZ*, the most popular reporter gene used in *Dictyostelium* research. β-galactosidase is very stable once expressed and accumulates in the cells, which also facilitates detection of its enzymatic activity. Stability of the protein also has some disadvantages when dynamic patterns of gene expression are to be characterized. It is possible that β-galactosidase is still present in the cells after the promoter under study has been silenced. Reporter genes coding for short-lived β-galactosidase *(12)* are used as reporters in the study of promoters whose activity is regulated within relatively short time periods *(13,14)* (*see* **Note 1**).

In this chapter, the methods used for the histochemical detection of β-galactosidase activity will be briefly described. β-galactosidase detection methods were originally developed by Dingermann et al. *(15)*, and have been modified by several authors thereafter. The methods detailed in this chapter were described by Wang et al. *(9)*. Alternative procedures for the fixation and permeabilization steps are discussed in the Notes section. Some plasmid vectors that contain the *lacZ* reporter gene are mentioned in **Notes 1** and **2**. Plasmid vectors that express *lacZ* under the control of some constitutive or cell type-specific promoters are described in **Note 3**. Methods for the detection of chloramphenicol acetyl transferase and luciferase activity have been described (*see*, for example, **refs. *6,11,16*,** and *17,18*, respectively). Fluorescent proteins are also increasingly used either as reporter genes or fused to other proteins under the control of their own promoters. Methods used for detection of fluorescent proteins are described by A. Müller-Taubenberger in Chapter 14 of this book.

2. Materials

2.1. Filter Development

1. HL-5 culture media: 14 g/L Tryptone (Difco Laboratories), 7 g/L yeast extract (Difco Laboratories), 0.35 g/L Na_2HPO_4, 1.2 g/L KH_2PO_4, pH 6.0 to 6.6. Autoclave, equilibrate to room temperature, and add sterilized glucose to 14 g/L from a stock solution of 224 g/L (25 mL of glucose stock solution for 400 mL of HL-5 media) and 2 mL of penicillin/streptomicin solution (Gibco) for 400 mL of media.
2. Absorbant pads and nitrocellulose filters from Millipore (cat. nos. AP1004700 and HAWP04700, respectively).
3. PDF buffer: 20 m*M* KCl, 9 m*M* K_2HPO_4, 13 m*M* KH_2PO_4, 1 m*M* $CaCl_2$, 1 m*M* $MgSO_4$, pH 6.4.

2.2. In Situ *Hybridization*

2.2.1. Digoxigenin-Labeled Riboprobes

1. DEPC-treated water. Add diethyl pyrocarbonate (DEPC) at 0.1% (v/v) to re-distilled water, and after 12 h of incubation at 37°C, autoclave the solu-

tion on a liquid cycle. DEPC may be a carcinogen and should be handled with care (*see* **Note 4**).
2. Dig-RNA Labeling Kit from Roche (cat. no. 1277073) (*see* **Note 5**).
3. SP6, T7, and T3 polymerases from Roche (including 10X transcription buffer).
4. Carbonate buffer (2X): 120 mM Na_2CO_3, 80 mM $NaHCO_3$, pH 10.2.

2.2.2. Fixing Dictyostelium *Structures*

1. Phosphate-buffered saline (PBS) 1X: 133 mM NaCl; 8 mM Na_2HPO_4, 2 mM KH_2PO_4, pH 7.4. A 10X stock can be prepared for storage.
2. Paraformaldehyde stock solution (Polysciences): prepare 20% (w/v) paraformaldehyde in water. To dissolve the paraformaldehyde, the solution must be heated to 60–65°C while stirring and a few drops of concentrated NaOH carefully added until it becomes dissolved. Filtrate the solution through a 0.2-μm syringe filter and make aliquots, which must be stored at –20°C.
3. 4% paraformaldehyde in PBS (from stock): thaw an aliquot of 20% paraformaldehyde stock solution and warm it up to 60–70°C until paraformaldehyde again gets into the solution. Add the corresponding amount of 10X PBS and water to obtain the desired concentration. Check that the final pH is kept around 7.5.
4. 4% (w/v) paraformadehyde in PBS (freshly prepared). This solution may be prepared directly at the indicated concentration. In this case, it must be freshly prepared for each experiment. The solution must be carefully heated until dissolved, and then cooled to room temperature for use.
5. Proteinase K stock solution: 20 mg/mL in H_2O. Keep at –20°C.

2.2.3. Hybridization

1. Prehybridization buffer: 4X SSC, 1X Denhardt's solution, 0.5 mg/mL sonicated denatured calf thymus DNA (Sigma), 0.25 mg/mL yeast RNA (Sigma), 60% formamide. 50X Denhardt's reagent: 1% w/v Ficoll Type 400 (Pharmacia), 1% w/v polyvinylpyrrolidone, 1% w/v bovine serum albumine (fraction V, Sigma).
2. Hybridization buffer: 4X SSC, 0.5 mg/mL denatured sonicated calf thymus DNA, 0.25 mg/mL yeast RNA, 60% formamide.
3. SSC 10X: 1500 mM NaCl, 150 mM sodium citrate, pH 7.0.

2.2.4. Anti-DIG-Alkaline Phosphatase Reaction

1. PBT: PBS containing 0.05% Tween 20.
2. Blocking reagent (Roche, cat. no. 1096176). Solubilization of blocking reagent at 0.2% in PBT will require heating the solution.
3. Antidigoxigenin antibody coupled to alkaline phosphatase (Roche, cat. no. 1093274).
4. Phosphatase buffer: 100 mM NaCl, 50 mM $MgCl_2$, 100 mM Tris-HCl, pH 9.5.
5. Nitroblue tetrazolium stock solution: NBT (Roche) 75 mg/mL in dimethylformamide 70%. Keep at –20°C.
6. 5-bromo-4-chloro-3-indolyl phosphate disodium salt stock solution: X-phosphate (Roche) 20 mg/mL in water. Keep at –20°C.

2.3. β-*Galactosidase Staining*

2.3.1. X-Gal Staining

1. Z Buffer: 60 mM Na$_2$HPO$_4$, 40 mM NaH$_2$PO$_4$, 10 mM KCl, 1 mM MgSO$_4$, pH 7.0. Sterilize by autoclaving. Store at 4°C.
2. Fixing solution: 3.7% formaldehyde solution, prepared by diluting ten times the commercial formaldehyde (37%) in Z Buffer. The diluted solution is stable for several days although it is better if freshly prepared.
3. Permeabilization solution: 0.1% NP40 in Z Buffer.
4. Potassium ferrocyanide, 33 mM. Store at 4°C, wrapped in aluminium foil.
5. Potassium ferricyanide, 33 mM. Store at 4°C, wrapped in aluminium foil.
6. X-Gal (5-Bromo-4-chloro-3-indolyl β-D-galactopyranoside), 20 mg/mL in dimethyl-formamide. Store at –20°C, wrapped in aluminium foil.
7. Staining solution: 5 mM potassium ferrocyanide, 5 mM potassium ferricyanide, 0.4 mg/mL X-Gal in Z Buffer. Prepare immediately before use by mixing 680 μL Z Buffer, 150 μL potassium ferricyanide stock solution, 150 μL potassium ferro-cyanide stock solution and 20 μL X-Gal stock solution per milliliter of staining solution.
8. Eosin Y (Sigma) 100 mg/mL in Z Buffer.
9. Permeabilization buffer for spores: 20% dimethyl sulfoxide (DMSO), 0.2% NP40 in Z Buffer.
10. Fixation solution for spores: 0.5% glutaraldehyde in Z Buffer.

3. Methods

3.1. Development of D. discoideum *on Nitrocellulose Filters*

1. Culture *D. discoideum* axenic cells in HL-5 media to a density of $1–3 \times 10^6$ cells/mL.
2. Under standard conditions, 2×10^7 cells are collected by centrifugation at 200g for 5 min. Cells are resuspended in 5 mL of PDF and centrifuged again at 200g for 5 min. Cells are resuspended in 800 μL of PDF to be laid on each filter. Smaller amounts of cells can also be used and resuspended in a proportional volume of PDF (for example, $1–2 \times 10_6$ cells in 100 μL of PDF).
3. Nitrocellulose filters are placed on Petri dishes as follows: place an absorbant pad on each dish and soak it with 1 mL of PDF, then place the nitrocellulose filter on top of the pad. Nitrocellulose filters are previously soaked on PDF or distilled water.
4. Cells are homogeneously distributed on the nitrocellulose filter. The PDF buffer must be absorbed by the filter. Once the filter is dry, more PDF is added to the absorbant pad until completely soaked.
5. Filters are incubated in a humidified chamber for the required periods of time.

3.2. In Situ *Hybridization*

3.2.1. Probe Labeling

1. The DNA to be labeled should be cloned into an appropriate vector containing pro-moters for SP6, T3, or T7 RNA polymerases adjacent to the polylinker (*see* **Note 6**).

2. One microgram of the plasmid per probe must be linearized with an appropriate restriction enzyme in order to generate "run off" transcripts. Both sense and antisense RNA transcripts may be synthesized from the same construct. The sense probe can be used as a control for background staining (*see* **Note 7**).

3. Linearized DNA is subsequently purified by phenol/chloroform extraction and ethanol precipitation.

4. Plasmid DNA is resuspended in 14 μL of DEPC-treated water and the following kit components added as described by the manufacturer's instructions: 2 μL of Digoxigenin RNA Labeling Mix (10X conc.); 2 μL of transcription buffer (provided with the RNA polymerases, as described in the Materials section); 2 μL of RNA polymerase (SP6, T7 or T3). Mix and incubate 2 h at 37°C. After incubation put on ice and add 2 μL of 0.2 *M* EDTA, pH 8.0, to stop the reaction.

5. Save 1 μL for determination of synthesis efficiency (*see* **step 7**). Adjust the rest of the reaction to 25 μL with water, add 25 μL of 2X carbonate buffer and incubate at 65°C for 15 min. This step is optimized for partial hydrolysis of the RNA to give fragments of shorter size (around 50–500 nucleotides) allowing better penetrability into fixed structures (*see* **Note 8**).

6. Precipitate the labeling reaction with acetate and ethanol and re-dissolve it with 100 μL of DEPC-treated water. The probe is now ready to be used and should be stored at –70°C (*see* **Note 9**).

7. Synthesis and hydrolysis efficiency should be checked afterward by simply running a standard DNA agarose gel as follows: run both the sample saved in **step 5** (before hydrolysis) and an equivalent amount of the sample after hydrolysis (10 μL) side by side, together with a DNA size marker. Besides the linearized plasmid, an additional band (sometimes two bands) corresponding to the newly synthesized RNA should appear. After hydrolysis, RNA will be seen as a smear of lower size (approximately between 50 and 500 nucleotides). Typically, the ethidium bromide intensity of the RNA band should be higher than that of the linearized template, which indicates good level of synthesis.

3.2.2. *Fixing* Dictyostelium *Structures*

1. Structures developed on nitrocellulose filters (*see* **Subheading 3.1.**) at different times of development are transferred to clean, sterilized glass or polypropylene tubes with PBS (*see* **Note 10**).

2. Remove the PBS from the bottom of the tube (most of the structures will be floating and clumping) and wash the structures twice with 2 mL of 100% methanol. At this step, the structures will form big clumps that must be dispersed (*see* **Note 11**).

3. Methanol is removed and replaced with 4% paraformaldehyde in PBS. After fixation for 2.5 h at room temperature, the structures are washed three times for 5 min each in PBS and incubated with 20 μg/mL Proteinase K in PBS for 20 to 60 min, washed in PBS for 5 min, and further fixed with in 4% paraformaldehyde-PBS solution at room temperature for 20 min (*see* **Note 12**).

3.2.3. Hybridization

1. After three washes with PBS for 5 min each, the structures are prehybridized at 50°C for 3 h in prehybridization buffer.
2. Hybridization is carried out at 50°C in 200 µL of hybridization buffer containing 1 µL of the RNA probe. After hybridizing for 20 h, remove excess probe by successive washes with 500 µL of 2X SSC, 1X SSC, 0.5X SSC, and 0.1X SSC at 50°C for 30 min each (*see* **Notes 13** and **14**).

3.2.4. Anti-DIG-Alkaline Phosphatase Reaction

1. The preparations are then equilibrated in PBT for 5 min and preincubated for 30 min at room temperature with PBT containing 0.2% blocking reagent. They are then incubated overnight at 4°C in PBT containing 0.2% blocking reagent with a 1/1000 dilution of antidigoxigenin antibody coupled to alkaline phosphatase.
2. The next day, samples are washed three times for 15 min each with PBT at room temperature, equilibrated with phosphatase buffer for 5 min, and incubated in the same solution containing 4.5 µL/mL NBT stock solution and 8.7 µL/mL 5-bromo-4-chloro-3-indolyl phosphate stock solution. Once a satisfactory signal is obtained, the reaction is stopped by washing with PBS twice, and the samples are observed and photographed under a stereomicroscope (*see* **Note 15**). An example of the results produced after hybridization of *Dictyostelium* structures with the well characterized genes *ecmA*, *ecmB* and *cotB* is shown in **Fig. 1A**.

3.3. β-Galactosidase Staining

3.3.1. Staining the Structure on Nitrocellulose Filters

1. Develop the structures over white nitrocellulose filters in PDF (*see* **Subheading 3.1.**). Filters can be split in two or more pieces if more than one developmental stage is to be analyzed. While one of the pieces is processed, the others are allowed to continue development.
2. Filters are taken from the pad and the excess of PDF removed by blotting them briefly on filter paper. Filters are placed on a clean Petri dish and structures are fixed by incubation in 3.7% formaldehyde in Z Buffer (fixing solution) for 10 min at room temperature. This and subsequent solutions must be loaded onto the filter carefully, drop by drop, to avoid detachment of the structures. A volume of 700–800 µL is enough for a filter of 47 mm in diameter. *See* **Note 16** for alternative fixation methods.
3. The formaldehyde solution is removed very carefully to avoid the detachment and loss of structures. Most of the solution can be taken off with a pipet tip, soaking from the margin of the filter. The rest of the solution can be sucked off with filter paper or a paper tissue from the margin or underneath the filters. Once the formaldehyde solution has been removed, the permeabilization solution (700–800 µL/filter) is added, dropwise, on top of the structures. Permeabilization is allowed to continue for 20 min at room temperature. *See* **Note 17** for alternative permeabilization methods.

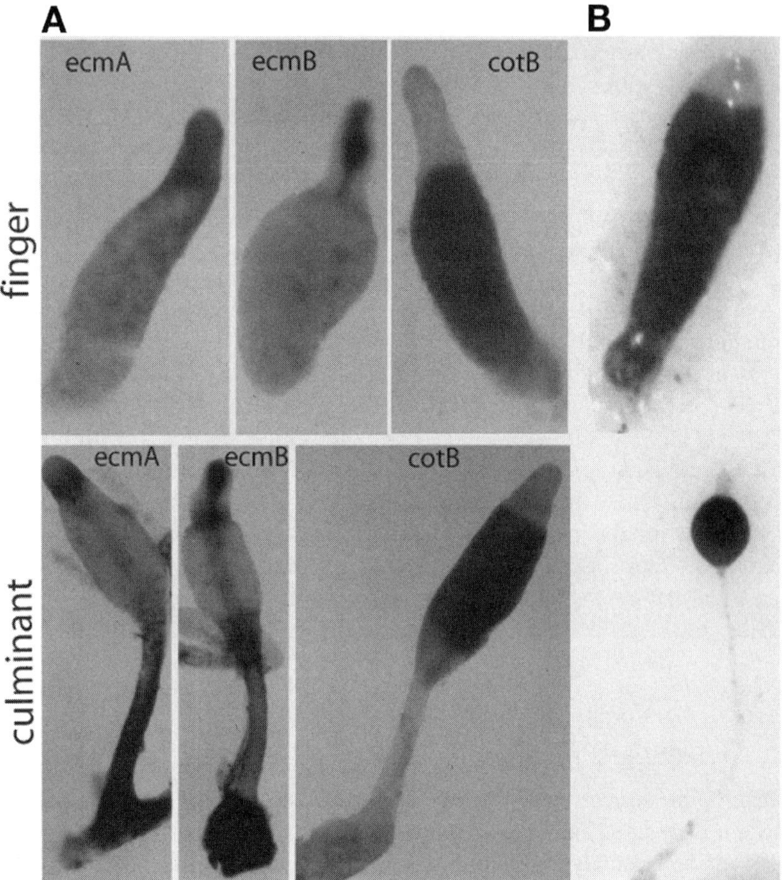

Fig. 1. *In situ* hybridization and β-galactosidase staining of *Dictyostelium* structures. (**A**) Results obtained after *in situ* hybridization of finger (upper panels) or culminant (lower panels) structures with Digoxigenin-labeled RNA probes specific for the prestalk-specific *ecmA* and *ecmB* genes or the prespore-specific *cotB* gene, as indicated on each panel. (**B**) Pattern of β-galactosidase staining in finger (upper panel) and culminant (lower panel) structures from a strain that expresses the *lacZ* gene under the control of the prespore and spore specific *sigA* promoter.

4. The permeabilization solution is carefully removed, as indicated previously, and Z Buffer is added dropwise. This step is repeated a second time to eliminate most of the permeabilization buffer.

5. The Z Buffer from the second wash is carefully removed and 700 to 800 μL of staining solution dropped on the structures. Staining is allowed to proceed as long as necessary. Ten minutes of incubation in the staining solution might be sufficient if *lac-Z* expression is driven from a strong promoter. Weaker promot-

ers may require several hours of incubation. X-Gal staining can be monitored on a stereomicroscope.

6. Structures can be stained with eosin. Morphology of the structures is difficult to observe on the white filter background. Eosin staining of the structures makes it easier to define the morphology and highlights the blue X-Gal staining. Eosin staining is obtained by incubation in 100 μg/mL of eosin in Z Buffer for a few minutes at room temperature (the X-Gal staining solution having been carefully removed). The eosin solution can be washed away with Z Buffer before observation and/or photography of the structures. β-galactosidase activity can be also determined using spectrophotometric methods (*see* **Note 18**). Examples of β-galactosidase staining are shown in **Fig. 1B**.

3.3.2. X-Gal Staining of Spores

1. Spores are collected from structures developed for 24 h on nitrocellulose filters or PDF-agar plates. There are several ways to collect the sporocarps, but one of the simplest is scraping the filter or the plate with a rubber policeman or a flat-ended spatula. This method collects stalks together with the sporocarps, which may not be desirable in some experiments. Alternatively, sporocarps can be individually collected from the filters using a pipet tip.
2. Permeabilization. Sporocarps are resuspended in permeabilization buffer for spores. A volume of 500 μL is enough for the spores coming from a single plate or filter. Structures are disrupted passing them five times through a 25-G (0.5 mm) needle. Permeabilization is allowed for 30 min at room temperature.
3. Spores are centrifuged for 5 min at 20,000*g* in a microfuge and the permeabilization buffer removed.
4. Spores are resuspended in 500 μL of fixation solution for spores and incubated for 10 min at room temperature (*see* **Note 19**).
5. Spores are centrifuged for 5 min at 20,000*g* in a microfuge and resuspended in 500 μL of Z Buffer.
6. Spores are centrifuged 5 min at 20.000*g* in a microfuge and resuspended in 500 μL of staining solution. X-Gal staining is accomplished by incubation at 37°C for the required time. *LacZ* expression driven by a strong promoter allows the detection of X-Gal staining after less than 1 h of incubation. More often, weaker promoters require overnight incubation at 37°C for X-Gal staining detection. β-galactosidase activity can be also determined using spectrophotometric methods (*see* **Note 18**).

4. Notes

1. Cloning vectors that incorporate the i-α-gal gene, coding for short-lived β-galactosidase have been developed by Detterbeck et al. *(13)*, who constructed a vector using the spore-specific promoter of the *psa* gene to direct short-lived β-galactosidase expression.
2. Several laboratories have developed vectors that are suitable for cloning promoter regions upstream of the *lacZ* reporter gene. Many of them derive from the

pDdGal vectors developed by Harwood and Drury *(19)*, which contain a multiple cloning site upstream of the *lacZ* gene. Three similar vectors (pDdGal-15, -16, and -17) were constructed that contained the *lacZ* coding sequence in the three possible reading frames in relation to the BglII site of the multiple cloning site.

3. Several reporter vectors have been constructed that express *lacZ* under the control of cell type-specific promoters, which are very useful in many experimental approaches, including the characterization of developmental mutants. A vector expressing *lacZ* from the constitutive Actin 15 promoter was constructed by Harwood and Drury *(19)*. Several constructs containing prespore-specific gene promoters have been described, including pD19lac-2 initially described by Dingermann et al. *(15)*. Some of the prespore-specific *lacZ* vectors more often used are *pspA*-gal (available from the Dicty Stock Center) or *cotB::lacZ* *(20)*. *spiA::LacZ* is a very useful spore reporter *(21)*. There are also several reporter constructs that incorporate prestalk-specific promoters. Among them, *ecmA*-gal, ecmAO-gal, *ecmB*-gal, and *ecmO*-gal are extensively used to identify the different prestalk regions, PstA, PstB, and PstO *(11)*, and are available from the Dicty Stock Center.

4. In order to prevent RNA degradation of the samples, it is important to avoid any possible contamination with RNases. A number of precautions are advisable: Wear gloves, use DEPC-treated distilled water to prepare all solutions, reserve chemicals for work with RNA, and use baked or disposable spatulas. Glassware must be cleaned thoroughly or baked at 180°C for 12 h to destroy RNases.

5. The original report describing the use of *in situ* hybridization in *Dictyostelium* made use of DNA-labeled probes *(22)*. This method has proven suitable for genes that are expressed at high levels. However, the use of RNA-antisense-labeled probes has been an important advance because its sensitivity is typically much better than that obtained with DNA probes. Variations of this basic method, both in the labeling and detection, have been described. For example, double staining to detect two different transcripts can be performed by simultaneously using digoxigenin- and fluorescein-labeled riboprobes, which can be detected by means of fluorescent-labeled anti-digoxigenin and anti-fluorescein antibodies *(4)*.

6. In order to obtain good sensitivity and low background, it is important, when possible, to use only the DNA coding regions of the gene of interest. Untranslated sequences, such as 5'- (or 3')-untranslated regions, as well as introns should be avoided. Similarly, regions coding for homopolymer runs such as those of glutamines and asparagines, which commonly occur in *Dictyostelium* proteins, must also be excluded. The simplest way may be to generate a fragment of the gene of interest by polymerase chain reaction (PCR) and to clone it by means of the T-overhangs into any of the common commercially available vectors, such as pGEM-T from PROMEGA. Subsequent sequencing will give the information required to determine clone orientation. Typically, large fragments (1–2 kb) give a higher signal than shorter ones. Small fragments (of less than 100 basepairs) might also give sufficient signal if the expression of the gene is high enough. Other vectors useful for synthesizing riboprobes from the SP6, T3, or T7 promoters are the pBluescript (Stratagene) and pSPT (Roche) plasmid vector families.

Basic methods for DNA manipulation such as PCR and cloning techniques have become routine in most laboratories and therefore are not described here (for more information, *see* **ref. 23**).

7. Although the best control for background staining should be sense probes, we have found that sometimes they give much higher background than the antisense probe. It is always advisable to carry out positive control hybridizations over a fraction of the structures with well-known markers (such as *ecmA*, *ecmB* and *cotB*) to check for integrity of the structures and state of cellular RNA. Even for an optimal probe, a certain degree of background staining can always be obtained if the color reaction is prolonged. In this connection, one cannot rule out the possibility of a low level of expression under the detection limit of the technique.

8. If the DNA template is 500 pb or shorter, this step might not be necessary.

9. We have found it unnecessary to remove the DNA remaining in the probe.

10. Taking the structures out of the filters without breaking them is the most delicate step. We have done this successfully by two different methods, but surely anyone might find better ways depending on his or her skills. One method is as follows: cut the filter in a half and put it inside a 15-mL polypropylene tube with the structures facing the air. Transfer the structures into the tube by throwing streams of buffered salt solution (PBS) upon them with a pipet. Although some of the structures will get stuck in the walls of the tube, many of them will be recovered in the following step. The filter is then removed and the other half is also recovered in the same way. Another possible way to get the structures out of the filter is to place the filter in a 10-cm Petri dish and wash the structures out with buffer as before by keeping the dish in a steep slope. After the methanol washes, the structures can be transferred from the dish to a polypropylene tube.

11. Structures should be dispersed thoroughly by repetitive pipetting with a Pasteur pipet (structures are stronger than they appear, and it is desirable to have them well dispersed even if a few of them get broken in the process). Finally, one should end up with a reasonable amount of dispersed, intact structures that sink to the bottom of the tube (sunken structures are a good sign–this means they are permeable). In the subsequent steps, the structures must be allowed to sink before the buffers are replaced from the top. A filter set up with 5×10^7 cells should be enough for one or two hybridizations.

12. The samples can be stored in methanol after the fixation procedure, but the sensitivity might decrease slightly. In this case, wash the samples twice with 100% methanol and store them in 100% methanol at –20°C.

13. Recently, we have found that another, simpler recipe for both prehybridization and hybridization buffers works fairly well (4XSSC; 60% formamide; 0.5 mg/mL denatured sonicated calf thymus DNA), but we have not tested whether the background level is equal to or higher than that obtained with the other recipe.

14. The hybridization temperature should be higher when cross-hybridization must be avoided in the case of homologous genes (for example, the markers *ecmA* and *ecmB* can cross-hybridize at 50°C, but they become specific at 60°C). Washes with SSC can also be performed at 60°C.

15. If the structures are transferred to small Petri dishes for the color reaction, the progress of the color can be easily monitored under the stereomicroscope. Photographs can be taken directly, or the structures can be mounted on glass slides for microscopy.
16. Alternative fixation methods. Glutaraldehyde has often been used to fix the structures at concentrations of 0.5–1% in Z buffer, for 10 to 30 min (*see*, for example, **refs.** *15,21,24*, and *25*).
17. Alternative structure permeabilization methods. Many authors do not include a permeabilization step in their protocols and stain the structures after fixation. In other protocols, the permeabilizing detergent (0.05-0.1 % Triton X-100 or NP-40) is included in the fixation solution (for example, *see* **ref.** *25*).
18. Determination of β-galactosidase activity. The activity of this enzyme can be determined by spectrophotometric methods using substrates that generate colored products, such as *o*-nitrophenyl-β-D-galactoside (ONPG) or chlorophenol-red-β-D-galactopyranoside (CPRG). Protocols were originally developed by Dingermann et al. *(15)*. Updated protocols for the determination of β-galactosidase activity from vegetative or differentiated cells can be found in Stevens at al. *(6)* and Thompson et al. *(10)*.
19. We have at times omitted the fixation step in preliminary assays and have still obtained good spore staining.

References

1. Escalante, R. and Vicente, J. J. (2000) *Dictyostelium discoideum*: a model system for differentiation and patterning. *Int. J. Dev. Biol.* **44**, 819–835.
2. Maeda, M., Sakamoto, H., Iranfar, N., et al. (2003) Changing patterns of gene expression in prestalk cell subtypes of *Dictyostelium* recognized by in situ hybridization with genes from microarray analyses. *Euk. Cell.* **2**, 627–637.
3. Shimada, N., Maeda, M., Urushihara, H., and Kawata, T. (2004) Identification of new modes of Dd-STATa regulation of gene expression in *Dictyostelium* by in situ hybridization. *Int. J. Dev. Biol.* **48**, 679–682.
4. Maruo, T., Sakamoto, H., Iranfar, N., et al. (2004) Control of cell type proportioning in *Dictyostelium discoideum* by differentiation-inducing factor as determined by in situ hybridization. *Euk. Cell* **3**, 1241–1248.
5. Fukuzawa, M. and Williams, J. G. (2000) Analysis of the promoter of the cudA gene reveals novel mechanisms of *Dictyostelium* cell type differentiation. *Development* **127**, 2705–2713.
6. Stevens, B. A., Flynn, P. J., Wilson, G. A., and Hames, B. D. (2001) Control elements of *Dictyostelium discoideum* prespore specific gene 3B. *Differentiation* **68**, 92–105.
7. Varney, T. R., Ho, H., Petty, C., and Blumberg, D. D. (2002) A novel disintegrin domain protein affects early cell type specification and pattern formation in *Dictyostelium*. *Development* **129**, 2381–2389.
8. Kibler, K., Nguyen, T-L., Svetz, J., et al. (2003) A novel developmental mechanism in *Dictyostelium* revealed in a screen for communication mutants. *Develop. Biol.* **259**, 193–208.

9. Wang, N., Shaulsky, G., Escalante, R., and Loomis, W. F. (1996) A two-component histidine kinase gene that functions in *Dictyostelium* development. *EMBO J.* **15,** 3890–3898.

10. Thompson, C. R. L., Fu, Q., Buhay, C., Kay, R. R., and Shaulsky, G. (2004) AbZIP/bRLZ transcription factor required for DIF signaling in *Dictyostelium*. *Development* **131,** 513–523.

11. Jermyn, K. A., Duffy, K. T., and Williams, J. G. (1989) A new anatomy of the prestalk zone in *Dictyostelium*. *Nature* **340,** 144–146.

12. Bachmair, A., Finley, D., and Varshavsky, A. (1986) In vivo half-life of a protein is a function of its N-terminal residue. *Science* **234,** 179–186.

13. Detterbeck, S., Morandini, P., Wetterauer, B., Bachmair, A., Fischer, K., and MacWilliams, H. K. (1994) The "prespore-like cells" of *Dictyostelium* have ceased to express a prespore gene: Analysis using short-lived beta-galactosidases as reporters. *Development* **120,** 2847–2855.

14. MacWilliams, H., Gaudet, P., Deichsel, H., Bonfils, C., and Tsang, A. (2001) Biphasic expression of rnrB in *Dictyostelium discoideum* suggests a direct relationship between cell cycle control and cell differentiation. *Differentiation* **67,** 12–24.

15. Dingermann, T., Reindl, N., Werner, H., et al. (1989) Optimization and in situ detection of *Escherichia coli* beta-galactosidase gene expression in *Dictyostelium discoideum*. *Gene* **85,** 353–362.

16. May, T., Kern, H., Muller-Taubenberger, A., and Nellen, W. (1989) Identification of a cis-acting element controlling induction of early gene expression in *Dictyostelium discoideum*. *Mol. Cell. Biol.* **9,** 4653–4659.

17. Howard, P. K., Ahern, K. G., and Firtel, R. A. (1988) Establishment of a transient expression system for *Dictyostelium discoideum*. *Nucleic Acids Res.* **16,** 2613–2623.

18. Yin, Y. Z., Williamson, B. D., and Rutherford, C. L. (1994) An autonomously propagating luciferase-encoding vector for *Dictyostelium discoideum*. *Gene* **150,** 293–298.

19. Harwood, A. J. and Drury, L. (1990) New vectors for expression of the *E. coli* lacZ gene in *Dictyostelium*. *Nucleic Acids Res.* **18,** 4292.

20. Fosnaugh, K. L. and Loomis, W. F. (1993) Enhancer regions responsible for temporal and cell-type-specific expression of a spore coat gene in *Dictyostelium*. *Dev. Biol.* **157,** 38–48

21. Richardson, D. L., Loomis, W. F., and Kimmel, A. R. (1994) Progression of an inductive signal activates sporulation in *Dictyostelium discoideum*. *Development* **120,** 2891–2900.

22. Escalante, R. and Loomis, W. F. (1995) Whole-mount in situ hybridization of cell-type-specific mRNAs in *Dictyostelium*. *Dev. Biol.* **171,** 262–266.

23. Sambrook, J., Frisch, E. F., and Maniatis, T. (1989) *Molecular Cloning: a Laboratory Manual*, Cold Spring Harbor Laboratory, Cold Spring Harbor, New York.

24. Haberstroh, L. and Firtel, R. A. (1990) A spatial gradient of expresssion of a cAMP-regulated prespore cell type-specific gene in *Dictyostelium*. *Genes Devel.* **4,** 596–612.

25. Powell-Coffman, J. A. and Firtel, R. A. (1994) Characterization of a novel *Dictyostelium discoideum* prespore-specific gene, PspB, reveals conserved regulatory sequences. *Development* **120,** 1601–1611.

16

Application of 2D and 3D DIAS to Motion Analysis of Live Cells in Transmission and Confocal Microscopy Imaging

Deborah Wessels, Spencer Kuhl, and David R. Soll

Summary

The chemotactic signal in *Dictyostelium* is a cAMP wave that is relayed over relatively large distances through a cell population during aggregation. Cells exhibit unique behaviors in response to the different spatial, temporal, and concentration components of the cAMP wave, suggesting that distinct signal transduction pathways are evoked in each of the various phases of the wave. For this reason, we designed a set of experimental protocols to test responses of normal and mutant *Dictyostelium* amoebae to the different components of a wave of chemoattractant. We then used computer-assisted two- (2D) and three-dimensional (3D) technologies (2D and 3D Dynamic Image Analysis System [DIAS]) for analysis of cells in the absence of a chemotactic signal (basic motile behavior) and in response to the temporal, spatial, and concentration components of the wave. As a result, we have elucidated parallel and independent pathways activated by specific phases of the cAMP wave. Likewise, human polymorphonuclear neutrophils (PMNs) respond to experimentally applied waves of the chemotactic peptide fMLP, and also exhibit discrete behavioral responses to the different phases. Using *Dictyostelium* as a paradigm, we applied our protocols to normal and diseased human PMNs and precisely defined a chemotactic defect. In this chapter, we describe methods for quantifying behaviors in *Dictyostelium* amoebae, PMNs, and other amoeboid cells using 2D and 3D DIAS. These methods are useful in the reconstruction and motion analysis of most migrating cells with transmitted and/or confocal microscopy.

Key Words: Cell motility; neutrophil chemotaxis; temporal gradient; spatial gradient; 2D motion analysis; 3D motion analysis.

1. Introduction

As an amoeboid cell translocates across or through a substratum, anterior and lateral lamellipodia expand, touch the surface, and adhere, or elevate and retract *(1,2)*. Translocation also requires coordinated detachment and retrac-

From: *Methods in Molecular Biology, vol. 346:* Dictyostelium discoideum *Protocols*
Edited by: L. Eichinger and F. Rivero © Humana Press Inc., Totowa, NJ

tion of the rear or uropod of the cell *(3,4)*. Furthermore, even in the absence of a chemotactic signal, cells such as *Dictyostelium discoideum* and human polymorphonuclear neutrophils (PMNs) *(5)* exhibit morphological polarity as a result of asymmetric intracellular distribution of a plethora of structural and signaling molecules *(6–8)* that generate basic motile behavior, i.e., motility in the absence of chemoattractant *(9)*. As complex as basic motile behavior is from a molecular standpoint, it is altered and regulated in the presence of chemotactic signals. In the case of *Dictyostelium (10)*, and possibly neutrophils *(5)* and primary tumor cells during metastasis *(11)*, that signal is not or may not be a static spatial gradient of chemoattractant. In other words, like *Dictyostelium*, at least a subset of chemotactic gradients generated in vivo must be relayed over great distances relative to the dimensions of a single cell. Furthermore, because relayed signals have spatial, temporal, and concentration components *(9,12)*, they are not entirely mimicked by the exogenous application of a uniform concentration of chemoattractant or by its delivery from a pipet tip *(9)*. For this reason, signal transduction models derived exclusively from these types of experiments may be inadequate.

With this in mind, we developed a contextual framework *(9)* for the analysis of basic motile behavior and its modulation by the chemotactic wave in normal *Dictyostelium* cells as well as in cytoskeletal and signaling mutants *(13–20)*. We then formulated a model of chemotaxis that incorporates independent and parallel signal transduction pathways activated by different phases of the chemotactic wave *(9)*. The application of this paradigm to human neutrophils *(5)* led to identification of a highly specific chemotactic defect in PMNs of children with Shwachman–Diamond Syndrome (SDS) *(21)*.

Discrimination of signal transduction pathways activated by different components of a dynamic chemotactic signal is not possible simply by qualitative comparisons of mutant cells with normal cells. Likewise, a cursory analysis of chemotaxis in SDS or other diseased neutrophils does not provide the insights necessary to pinpoint a defect and advance research toward more expedient diagnoses, improved treatment, and possible cures. For these reasons, we developed the advanced computer-assisted 2D and 3D motion analysis systems, 2D and 3D Dynamic Image Analysis System (DIAS) *(22–25)*. In this chapter, we describe protocols for the use of DIAS in the analysis of basic motile behavior and chemotaxis within the contextual framework mentioned previously (*see* **Fig. 1**) in live cell images acquired through transmission and confocal microscopy. We have also included a section on the use of the multiplatform DIAS 4.0 *(26)*, the successor to the Macintosh-based 3D DIAS, to analyze the dynamics of filopodia in basic motile behavior and in response to chemoattractant.

2. Materials

1. *Dictyostelium* BSS: 20 mM KCl, 2.5 mM MgCl$_2$, 20 mM KH$_2$PO$_4$ (pH 6.4).
2. H-HBSS: 0.01 M HEPES in Hank's balanced salt solution with CaCl$_2$ (0.10 g/L) and MgCl$_2$ (0.10 g/L), pH 7.4 (Gibco BRL, Gaithersburg, MD).
3. Heparinized vacutainer tubes (Becton-Dickinson).
4. Polymorphprep (Axis-Shield, Oslo, Norway) for separation of viable neutrophils from whole blood.
5. cAMP: 500 µM stock solution is frozen until use.
6. N-formyl peptide (fMLP): 500 µM stock solution of N-formyl-met-leu-phe (Sigma-Aldrich, Inc.) in dimethyl sulfoxide (DMSO) is frozen until use.

2.1. Preparation of Cells

2.1.1. Preparation of Dictyostelium discoideum

Growth-phase *Dictyostelium discoideum* amoebae at a density of 2×10^6 cells per mL in suspension cultures or a monolayer stage in a Petri dish are washed free of nutrient media and dispersed on filter pads saturated with *Dictyostelium* BSS *(27–29)*. At the onset of aggregation (approx 6–7 h on filter pads), cells are washed from the filter pads and diluted to a concentration of 3×10^4 cells/mL *(27–29)*.

2.1.2. Isolation of Human Polymorphonuclear Neutrophils

Neutrophils are isolated from 5 to 10 mL of whole venous blood drawn from healthy donors into a heparinized vacutainer tube. The tube is gently inverted several times and 5 mL of blood layered onto the top of 5 mL of Polymorphoprep in a 15-mL Falcon centrifuge tube according to directions provided by the manufacturer and as described in detail in **ref. 33**. Centrifugation for 30–35 min at 450g results in a distinct PMN layer. The cells are washed in H-HBSS and the final pellet resuspended in H-HBSS at a density of 1×10^6 cells/mL

2.2. 2D DIAS

1. Cell chamber for single cell imaging such as the Sykes-Moore perfusion chamber, (Bellco Glass, Inc., Vineland, NJ), a spatial gradient chemotaxis chamber *(30,31)*, or 35-mm Petri dish for analysis of behavior in natural cAMP waves.
2. Perfusion system capable of generating defined gradients such as the NE-1000 Multiphase Programmable Pumps (New Era Pump Systems, Farmingdale, NY) (*see* **Fig. 2**) or custom made microfluidic device.
3. Upright or inverted microscope (depending on the chamber) equipped with brightfield optics and 10× to 40× objectives.
4. Firewire camera (*see* **Note 1**) with a means to record and store video data such as iStopMotion for Macintosh, or with Adobe Premiere on a Windows machine. Alternatively, one can use an analog to digital converter in conjunction with iMovie and an analog charge-coupled device (CCD) camera or a framegrabber board such as the Data Translation board (Marlboro, MA) if available.

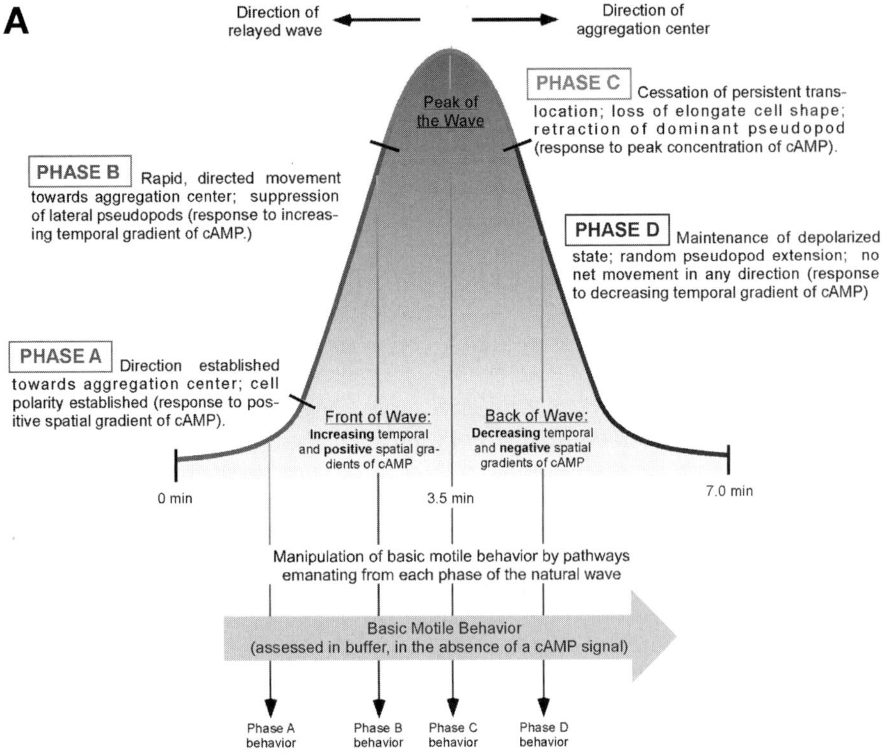

B Experimental protocols for determining the behavioral defects of S13A mutants in basic motile behavior and chemotaxis.

Tested Behavior	Experimental Protocol
1. "Basic motile behavior" in the absence of an extracellular cAMP signal.	1. Perfusion of amoebae with buffer in perfusion chamber.
2. Capacity to assess the direction of a spatial gradient of cAMP (Phase A).	2. Spatial gradient of cAMP generated on the bridge of a single cell spatial gradient chamber.
3. Behavior in the middle of the front of the wave in response to increasing temporal gradient of cAMP (Phase B), the response to the concentration of cAMP at the peak of the wave (Phase C) and the behavior in the back of the wave (Phase D).	3. Exposure to a sequence of interspersed increasing and decreasing temporal gradients of cAMP that mimic the temporal dynamics of natural waves in the absence of established spatial gradients.
4. Response to the concentration of cAMP at the peak of the wave (Phase C).	4. Rapid addition of 10^{-6}M cAMP to cells crawling in buffer.
5. Behavior in all phases (A, B, C and D) of self-generated natural waves of cAMP.	5. Aggregation in a dilute monolayer on a plastic surface.
6. Behavior in response to all phases (A, B, C and D) of natural waves generated by wild type cells.	6. Aggregation in which fluorescently stained mutant cells are mixed with unstained wild type cells in a 1 to 5 or ratio.

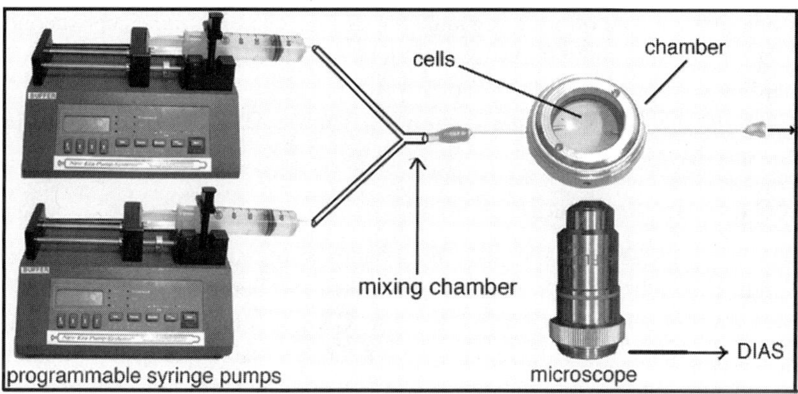

Fig. 2. Programmable syringe pump system used to generate temporal waves of chemoattractant in the perfusion chamber *(5)*. One syringe is filled with the peak concentration of chemoattractant, the other with buffer, and the solutions are mixed in the mixing chamber before entering the chamber. The flow rate and round shape of the chamber preclude establishment of a spatial gradient.

5. Macintosh G3 computer or higher with OSX, 128 MB RAM or more and 20 MB hard disk space.
6. 2D DIAS software (Soll Technologies, Inc.).

2.3. 3D DIAS

1. Short-working-distance chamber suitable for DIC optics such as the Dvorak-Stotler chamber (http://www.nevtek.com), the Bioptechs chamber (http://www.bioptechs.com), or a modified Zigmond chamber *(31)* for upright or inverted microscopes.
2. High-numerical-aperture condenser and objectives on an upright or inverted microscope equipped with DIC optics.
3. Motorized stage or external motor with controller and means to distinguish up from down slices in a z-series (*see* **Note 2**).
4. Equipment and software to record and store video data as described in **Subheading 2.2.**, **item 4**.

Fig. 1. *(opposite page)* (**A**) *Dictyostelium* cell behavior in the different spatial, temporal, and concentration components of a natural cAMP wave (phases A, B, C, D). Vertical arrows indicate regulatory pathways that emanate from each phase and target machinery embedded in the basic motile behavior, leading to cell responses that are specific to the corresponding phase *(16)*. (**B**) Experimental protocols used to measure the responses of wild-type and mutant *Dictyostelium* cells to each phase of the cAMP wave *(18)*.

5. Computer capable of grabbing and storing images at 30 frames per second.
6. Macintosh G3 computer or higher with OSX, 128 MB RAM or more and 20 MB hard disk space and 3D DIAS software.
7. For filopodia, DIAS 4.0 on a Windows XP or MacOSX computer, 2 Ghz processor speed.

2.4. 2D and 3D Confocal

1. Chambers suitable for laser-scanning confocal microscopy (LSCM) imaging of single cells labeled with a vital dye such as the lipophilic tracer dialkylcarbocyanine (DiI, Molecular Probes, Eugene, OR), or cells transformed with a green fluorescent protein (GFP)-tagged protein. For perfusion and generating temporal waves, the Sykes–Moore chamber works well. The modified Zigmond chamber with the quartz bridge is suitable for chemotaxis in a spatial gradient.
2. Gradient generating system as described in **Subheading 2.2.**, **item 2**.
3. Confocal software that can convert or export images from the native acquisition format to a series of TIFF images.
4. QuickTime Time Pro® software (*see* **Note 3**).
5. 2 D DIAS and/or 3 D DIAS.

3. Methods
3.1. 2D DIAS
3.1.1. Sample Preparation
3.1.1.1. Analysis of Basic Motile Behavior

1. Inoculate 1.1 mL of cells prepared as described under **Subheading 2.1.1.** into the Sykes–Moore perfusion chamber *(27)*. Allow cells to adhere to the round coverslip until they resume elongate morphology and translocation, a process that usually requires 5 min of incubation at room temperature.
2. Seal the chamber, connect it to the pump system (*see* **Fig. 2**), and place it on the microscope stage.
3. Fill a 60-cc syringe with BSS and set the flow rate on the pump system to 4 mL/min to continually refresh the environment *(27)*. The pumps can be programmed with instructions provided by the manufacturer. Cells may be attached to either the top or bottom coverslip for use on an inverted or upright scope. Alternatively, for a rapid shift from buffer to a constant concentration of chemoattractant, a second syringe can be filled with chemoattractant, and the pump system programmed to switch from the syringe delivering BSS to the syringe delivering the chemical.

3.1.1.2. Analysis of Cell Behavior in Temporal Waves of Chemoattractant *(29)*

1. Inoculate cells into the Sykes-Moore perfusion chamber as described previously and connected to the pump system (*see* **Fig. 2**).
2. Program the pumps to deliver a series of increasing and decreasing temporal gradients of chemoattractant that mimic the temporal dynamics of natural cAMP

waves. The periodicity of the waves is 7 min and the concentration increases from 0.01 μM cAMP in the trough to 1 μM at the peak in the case of *Dictyostelium* and from 0.001 to 0.1 μM fMLP for PMNs *(5)* (*see* **Note 4**).

3.1.1.3. ANALYSIS OF CELL BEHAVIOR IN A SPATIAL GRADIENT OF TEST CHEMOATTRACTANT

1. Streak 30 μL of cells across the center of a 22×30 mm coverslip at a density that results in about 10–20 cells per mm^2.
2. After 5 min, invert the coverslip over the 2-mm wide bridge of the Plexiglas chemotaxis chamber *(25,28)* and gently clamp it.
3. Fill one trough with buffer (sink: BSS or H-HBSS) and the other with buffer containing the appropriate concentration of test solution (source). For *Dictyostelium*, 1 μM cAMP in BSS is used in the source trough, and in the case of human neutrophils, 0.05 μM fMLP in H-HBSS is used *(5)* (*see* **Note 5**). We have found that mouse neutrophils respond more efficiently to 50 μM fMLP.

3.1.1 4. ANALYSIS OF *DICTYOSTELIUM* CELL BEHAVIOR IN NATURAL cAMP WAVES

1. Wash log-phase cells free of nutrient media and plate 5×10^6 cells on the bottom of a 35-mm Petri dish in a total volume of 2 mL of BSS *(32)*.
2. After 30 min, cells are well adhered to the bottom of the dish, which can then be placed on the stage of an inverted microscope equipped with low magnification (10–20×) brightfield objectives. The Petri dish should be covered and left undisturbed for several hours before starting image acquisition (*see* **Note 6**).

3.1.2. Image Capture and Outlining With DIAS

1. Images can be digitally captured using the equipment described above (*see* **Subheading 2.2.**, **items 4** and **5**) at a rate of 15 frames per min, depending on the rate of cell movement.
2. DIAS uses a grayscale threshold algorithm to automatically detect the edges of a cell in each frame (*see* **Note 7**). DIAS software converts the x,y coordinates of the cell perimeter to a β-spline replacement image and determines the centroid for each frame based on either the area or geometric center of that image, depending on the complexity of the image as determined by the user (*see* **Note 8**).

3.1.3. Computing Parameters

1. The standard DIAS 3.4.2 menu allows the user to select up to 40 parameters or enter other functions if desired, as described elsewhere in detail *(20,22,24,25)*. In brief, instantaneous velocity, direction of travel, and direction change are computed from the centroid position and cell shape parameters are computed from the β-spline replacement image. The "chemotactic index" is computed as the net distance moved toward the source of chemoattractant divided by the total distance traveled.
2. DIAS creates a database file and has a built-in graphics package with features such as smoothing, finding minima and maxima, linear and quadratic fits, and Fourier transforms. If desired, data can also be exported as text into spreadsheet

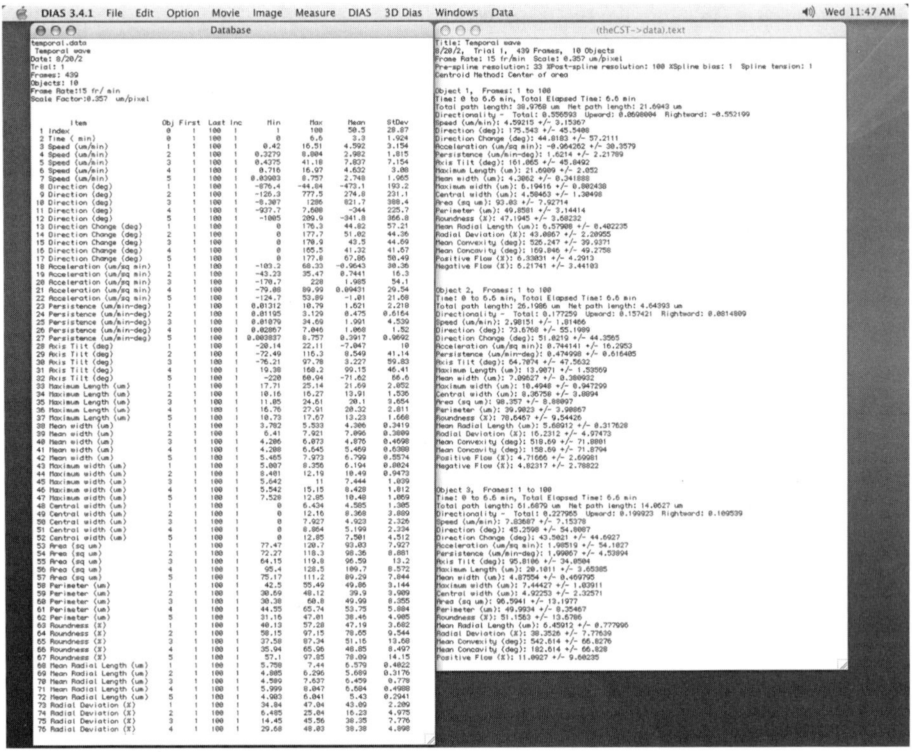

Fig. 3. Sample of parameters in tabular form computed by dynamic image analysis system (DIAS). Data can be exported as a text file for analysis with other programs.

software such as Microsoft Excel® (*see* **Fig. 3**). Cell movement is visualized by several display modes such as stacked perimeter plots (*see* **Fig. 4**) that can also be saved as QuickTime movies for presentation. Any frame from a movie can be saved as a PICT file.

3.2. 3D DIAS

3.2.1. Sample Preparation for the 3D Reconstruction of Cells Translocating in Buffer (Basic Motile Behavior) or Responding to Chemoattractant

1. *Dictyostelium* amoebae harvested at the onset of aggregation as described under **Subheading 2.1.1.**, or neutrophils prepared as described under **Subheading 2.1.2.**, are inoculated into a short-working-distance perfusion chamber, such as the Dvorak-Stotler chamber, according to methods described elsewhere in detail (*34*).
2. Cells are allowed to adhere to the coverslip for 5 min after the chamber is assembled. The chamber is then positioned on the microscope stage and imaged

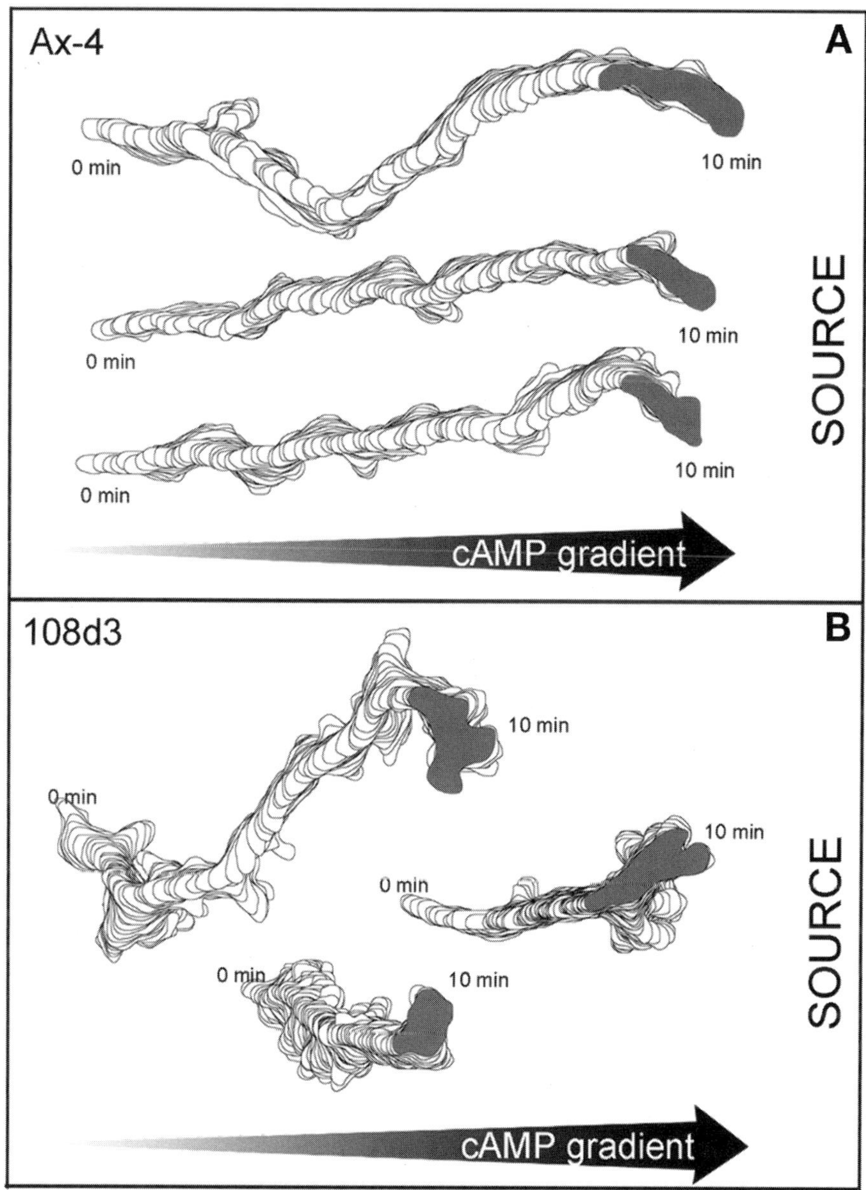

Fig. 4. Dynamic Image Analysis System (DIAS)-generated stacked perimeter plots of *Dictyostelium* cells migrating in a spatial gradient of cAMP. This display is useful for comparing turning and pseudopod dynamics of normal (**A**) and mutant (**B**) *Dictyostelium* (*16*) or normal and diseased neutrophils. The last image of the series is filled in dark gray.

with a high numerical aperture DIC objective and condenser. The chamber is perfused with buffer or media to prevent conditioning of the microenvironment.

3. For 3D reconstruction of cells migrating in a spatial gradient of chemoattractant, cells are inoculated into a spatial gradient chamber with a quartz bridge described in **ref. *31***.

3.2.2. Image Capture, Optical Sectioning, Outlining, and Reconstructing Live Cells in 3D

1. Optical sectioning of live cells is performed by programming the motor to move through the z-axis of the single cell, usually 10 to 20 μm for *Dictyostelium* and PMNs.
2. Approximately 60 optical sections should be obtained in 2–3 s, and the process repeated every 4–5 s for a minimum of 5 min.
3. Images are written directly onto a computer hard drive, using the equipment described in **Subheading 2.2.**, **items 4** and **5**, at a rate of 30 frames per second.
4. The in-focus perimeter of each optical section is automatically outlined by DIAS using a pixel complexity algorithm with user-defined threshold and smoothing. Methods for outlining and reconstructing the cell, pseudopod, and nucleus are described in detail elsewhere *(22,24,25,35)*. A representative series of 12 optical sections at 0.33-μm increments is presented in **Fig. 5A** *(26)*.
5. The use of trace slots in DIAS 3.4.2 allows outlining of multiple features within a single frame. For example, the nonparticulate cell body is outlined in one trace slot, the pseudopods in another, the nucleus in a third, and so on. The number of trace slots is virtually unlimited.
6. The trace slots are combined to yield a composite 3D reconstruction that can be rotated and viewed at any angle (*see* **Fig. 5B**). The reconstructed cell can be viewed in stereo using the Crystal Eyes® system. Alternatively, DIAS can reconstruct red/blue movies of crawling cells that can be saved as QuickTime and viewed in 3D with commercially available red/blue glasses.

3.2.3. Motility and Dynamic Morphology Parameters Computed by 3D DIAS

DIAS generates a database file of 3D parameters based on the position of the cell centroid and the 3D contours of the cell body that can be plotted and analyzed as described under **Subheading 3.1.3.** for 2D database files. 2D parameters can also be measured on any planar slice of the cell.

3.2.4. Reconstruction of Filopodia Using 3D DIAS

1. Images are acquired using the high numerical aperture DIC optics in conjunction with the motor parameters and acquisition hardware described in **Subheading 2.2.**, **items 4** and **5**. However, because of the fine structure of filopodia, video compression is avoided during processing of the movie to DIAS format. Java-based DIAS 4.0 *(26)* is used for reconstruction. This version allows use of the

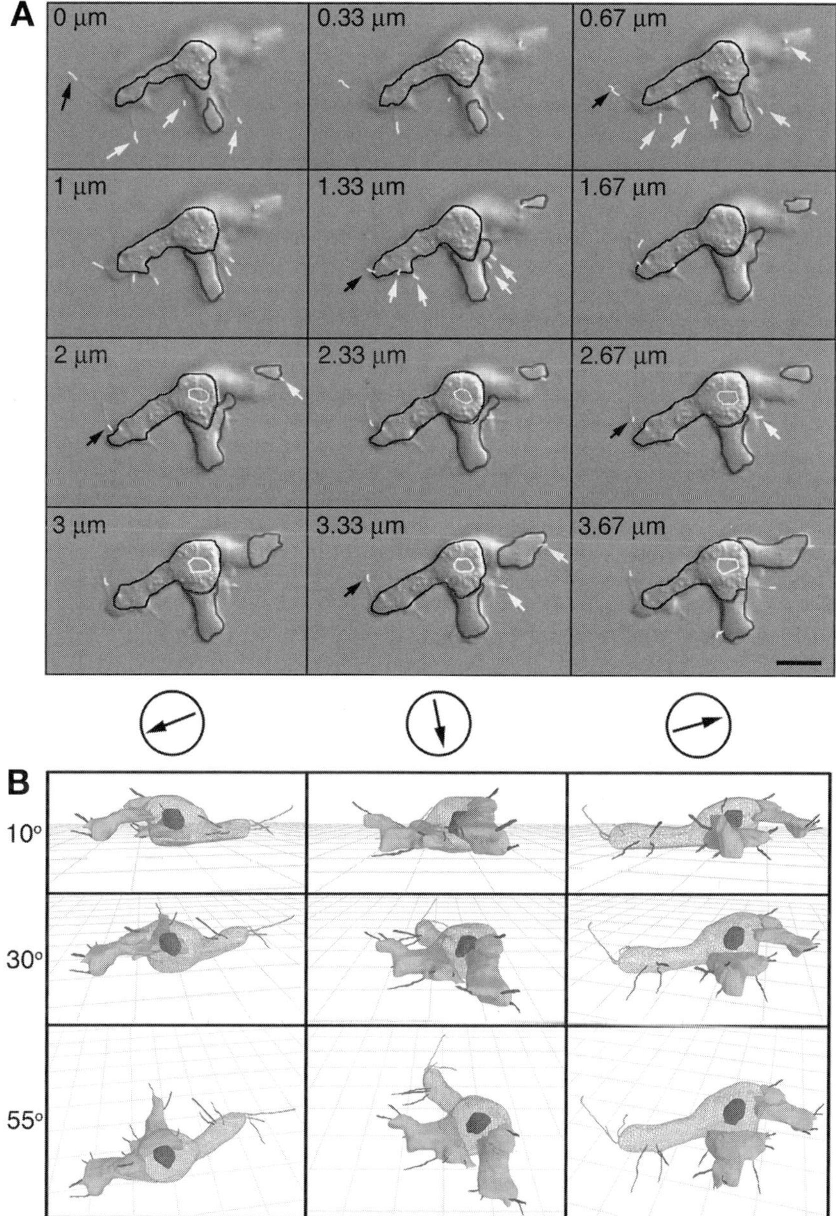

Fig. 5. (**A**) The in-focus edges of the cell body, nucleus, and pseudopods are outlined in black, white, and gray, respectively. Filopods and tail fibers are outlined in white and indicated by white and black arrows, respectively. All trace slots are superimposed. Only the first twelve optical sections are shown. (**B**) The outlined cell is reconstructed and viewed at 10°, 30°, and 55° from three rotational vantage points *(26)*.

"Swing" graphical user interface and provides the necessary speed through JVM 1.4 and IBM WebSphere implementations of the Java virtual machine.

2. As a result of the complex 3D trajectory of filopodia and their narrow diameters (<0.1 μm), filopodia are more accurately traced using the manual trace option. Dilation is used to thicken the segments to achieve continuity, and erosion is used to restore the diameters to a more realistic width. 3D reconstructions of cells with nuclei, pseudopods, and filopodia are generated as described under **Subheading 3.2.2.** and in detail elsewhere *(24–26,35)*. In addition, in order to accurately render the fine structure of filopodia, the facet limit was extended to 200,000 per object in DIAS 4.0 using object-oriented techniques and native memory management. Additional resolution up to 500,000 facets was obtained by using the Open GL (Silicon Graphics) 3D graphics language to communicate with an NVIDIA GeForce4 Ti400 video display board. The lighting, color, texture, size, and view angle of the image are controlled by DIAS in an Open GL processing station.

3. Generally, three rounds of smoothing are applied by averaging the x, y, and z coordinates of each vertex with those of the adjacent vertices, using equal weights. An example of a wild-type *Dictyostelium* cell reconstructed with filopodia in DIAS 4.0 at different angles and views is presented in **Fig. 5B**.

3.3. 2D and 3D Analysis of Confocal Images

3.3.1. Sample Preparation for Analysis of Basic Motile Behavior or Response to Chemoattractant

1. To examine localization of a GFP-tagged protein during basic motile behavior of live cells or in response to temporal waves of chemoattractant, transformed cells are inoculated into a Sykes-Moore perfusion chamber using 25-mm quartz coverslips (SPI Supplies, West Chester, PA) on both the top and the bottom of the chamber (*see* **Note 9**).

2. The chamber is connected to a pump system capable of delivering BSS or H-HBSS at a rate that results in replacement of chamber fluid every 15 s. To introduce temporal waves of chemoattractant, the Sykes-Moore chamber is connected to the NewEra NE1000 pumps capable of generating a linear gradient and programmed to deliver waves of chemoattractant as described under **Subheading 3.1.1.2.**

3. In order to examine chemotaxis in a spatial gradient, labeled cells are streaked onto the coverslip of a modified Zigmond chamber with a quartz bridge as described in more detail in **ref. *31***.

4. Mixing (chimeric) experiments are performed to examine the response of mutant *Dictyostelium* amoebae to natural cAMP waves generated by wild-type cells or, conversely, to measure the response of normal amoebae to signals generated by mutants. Growth-phase cells of either parental or mutant strains are incubated overnight in nutrient media supplemented with 50 μ*M* DiI. Labeled cells are then washed free of media in BSS, mixed with the unlabeled population at a ratio of 1:9, and 5×10^6 cells plated onto a 35-mm Petri dish in a final volume of 2 mL

BSS. Once the cells have adhered, the dish is placed on the microscope stage and viewed with a 10× or 20× objective.

3.3.2. Microscopy and Image Acquisition

1. Transmitted and fluorescent images of cells in buffer or gradients of chemo-attractant are simultaneously collected using the 488-nm argon laser and the 60× planapochromat water immersion objective (N.A. 1.2) on a BioRad Radiance 2000MP LSCM system equipped with a Nikon TE2000U microscope or comparable configuration on the user's system.
2. For rapidly crawling amoebae, 512×512 images are collected at 4-s intervals at a scan rate of 166 lines per second for 5 min. A z-series can be collected if the signal is sufficiently bright to be imaged at a rapid scan rate (500 lines per second). In that case, 10 slices are collected in 5 s at a resolution of 256×256 pixels. In all cases, the grayscale dynamic range is set between 0 and 255 to eliminate excess saturation and ensure linearity.
3. Images are batch converted to a TIFF sequence, imported into QuickTime Pro 7®, and saved. The QuickTime movie is opened in DIAS.
4. In the mixing (chimeric) experiment, transmitted and fluorescent images are acquired with the appropriate laser line for the dye in use. Optical zoom may be used if desired. Separate QuickTime movies of fluorescent and transmitted images are made from the TIFF sequence as described in **item 3** above. The movies are merged in DIAS and labeled and unlabeled cells outlined using a combination of automatic and manual outlining features. **Figure 6** is an example of graphic data and cell tracks obtained from a mixing experiment. In this case, *acaA⁻* cells in *Dictyostelium* were labeled with (DiI and responses of the mutants to waves generated by wild-type cells were examined *(13)*.

3.3.3. Outlining Images Using Trace Slots and the Analysis of Fluorescence Intensity

1. Through the use of trace slots as described in **Subheading 3.2.2.**, **items 5** and **6**, and also by employing separate threshold values, it is possible to outline and subsequently quantify different intensity levels or differentially labeled organelles in a confocal image (*see* **Fig. 7A**). In this figure, the cell body is outlined in trace slot 1, a mid-range intensity label is outlined in trace slot 2, and the highest intensity label in trace slot 3. In **Fig. 7B**, the trace slots from three frames of a z series have been combined into 3D reconstructions. Separate data files can be computed from the outlines in each trace slot.
2. DIAS also measures pixel intensity in any planar slice using a line profile or area profile method (*see* **Fig. 8**). A text file of pixel intensities is generated and can be imported into spreadsheet software and plotted.

4. Notes

1. To enhance automatic outlining, it is helpful to use a camera with contrast and brightness adjustments, but this is not essential.

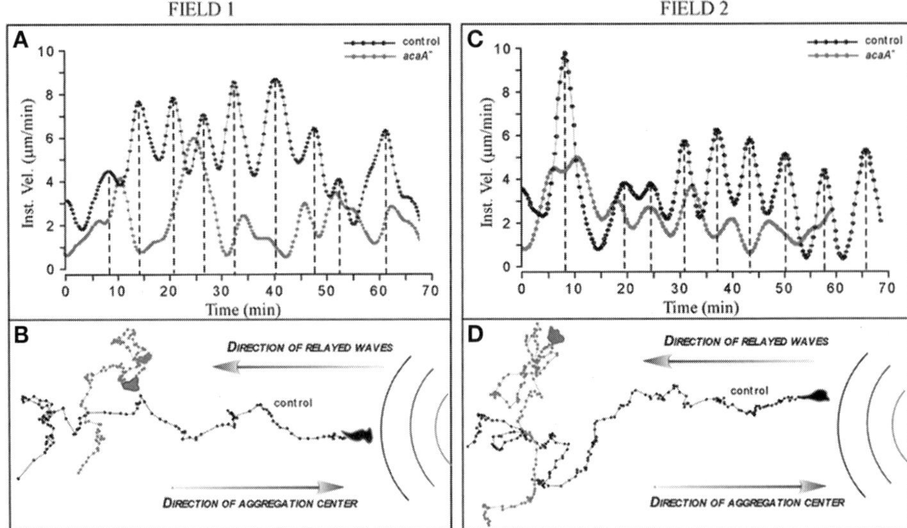

Fig. 6. Example of data acquired from two different mixing experiments (Fields 1 and 2) in which *acaA⁻ Dictyostelium* (**13**) cells were labeled with Dialkylcarbocyanine (DiI) as described in **Subheading 3.3.1., item 4** and mixed at a 1:9 ratio with unlabeled wild-type. Images were acquired with a laser-scanning confocal microscope. Velocity plots (**A,C**), cell tracks, and perimeters (**B,D**) of mutant and wild-type cells responding to cAMP waves produced by the latter were generated by Dynamic Image Analysis System (DIAS).

2. We use a Burst Electronics (Corrales, NM) micro character generator inserted into the camera video path to stamp synchronization (up/down/pause) information corresponding to motor position directly onto the captured movie.

3. It is possible to open a TIFF sequence directly in DIAS, but this depends on the TIFF compression algorithm built into the confocal software. If DIAS will open the TIFF images, then they can be saved directly as a DIAS format movie. If not, then it is necessary to first make a QuickTime movie in QuickTime Pro®.

4. The shapes of the four temporal waves can be assessed by generating gradients with a fluorescent dye such as xylene cyanole FF (Sigma-Aldrich, St. Louis, MO). Samples can be collected from the chamber outlet at 10-s intervals and absorption measured with a spectrophotometer at 615 nm. Optical density plots of these results reveal the actual shape of generated waves.

5. Neutrophil chemotaxis is enhanced if the temperature of the microscope stage is maintained at 37° C using an air curtain or stage heater. Under these conditions, it is preferable to use a sealed, modified Zigmond chamber such as that described by Shutt et al. (*31*) to prevent evaporation of solutions.

6. Incubation time varies depending on the strain and the mutation. To avoid cell death from phototoxicity, start image acquisition no sooner than 1 h prior to the anticipated onset of aggregation.

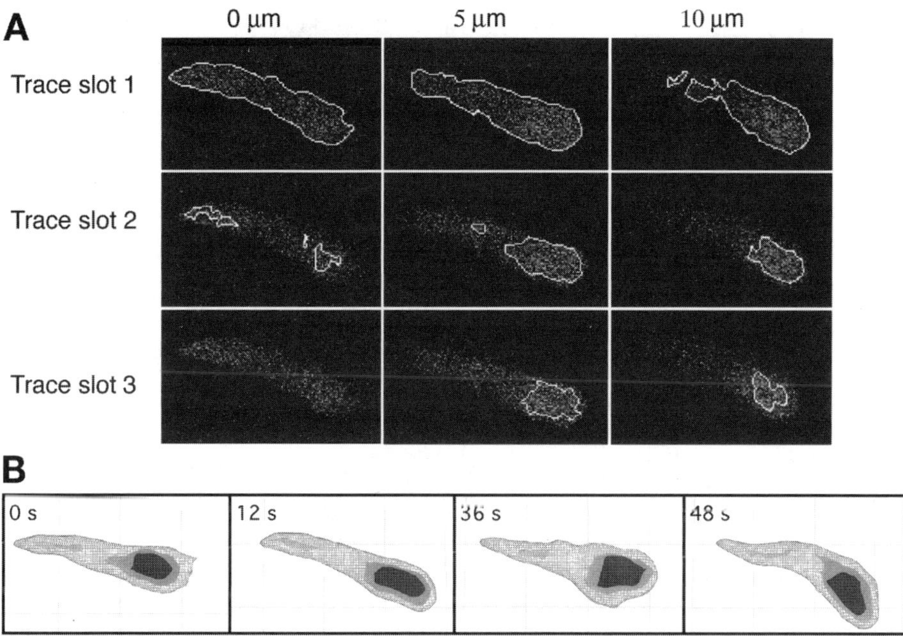

Fig. 7. (**A**) laser-scanning confocal microscope-acquired z-series of a *Dictyostelium* cell expressing a green fluorescent protein (GFP)-tagged protein and chemotaxing in a cAMP spatial gradient with the source to the right. A fast scan rate (500 lines per second) was necessary because of the rapid translocation rate. The user can define separate threshold values so that dynamic image analysis system (DIAS) automatically outlines different intensity levels in different trace slots. Trace slot 1 contains outlines of the cell body illustrated at the substrate level, 5 µm above the substrate and 10 µm above the substrate. Trace slots 2 and 3 contain outlines of mid range and high pixel intensity, respectively, at the same levels presented for the cell body. (**B**) Three-dimensional reconstruction of the cell body outlined in trace slot 1, the middle intensity GFP label as determined by user-defined threshold and outlined in trace slot 2, and the high-intensity GFP label outlined in trace slot 3. The trace slots are combined and the reconstructed cell presented at the time (seconds) indicated in the upper left of each panel.

7. Adjustment of brightness and contrast to darken the cell relative to the background is very helpful in automatic outlining with the grayscale threshold method. In addition, slight defocusing of the image may also improve connectivity. Ideally, the cell suspension should be diluted to minimize outlining errors associated with cell–cell contact. However, the manual tracing option allows the user to correct any errors that arise during automatic outlining.

8. If the outline of the image is unusually complex, i.e., exhibits a high degree of concavity and convexity, the centroid will sometimes appear outside the cell. In

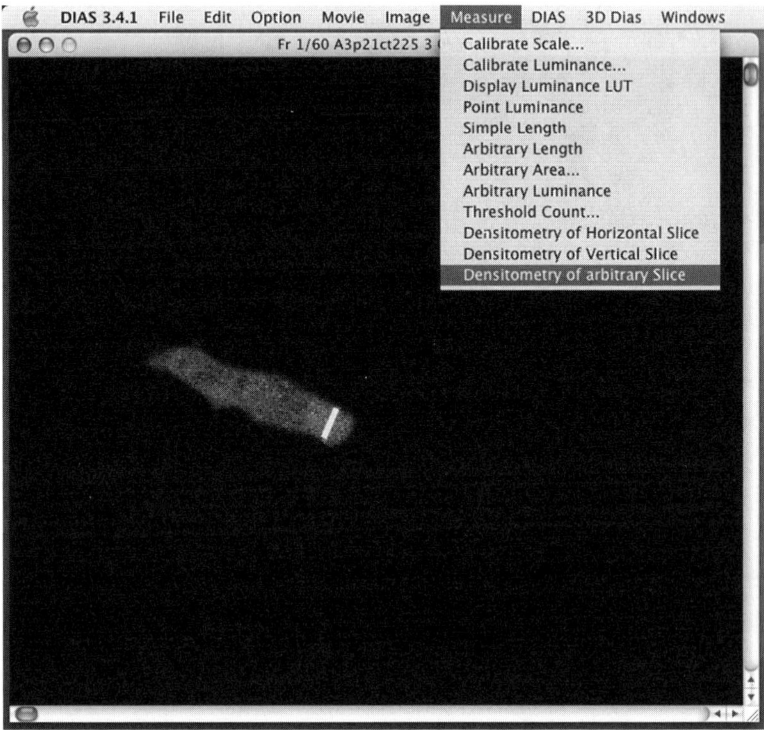

Fig. 8. Pixel intensity of a green fluorescent protein-labeled protein can be measured across a line or within a region of interest using dynamic image analysis system (DIAS), as shown in this screen shot of the DIAS program. Data can be exported into spreadsheet software.

these cases, select perimeter-based centroid from the path file header. The software then uses both the area and perimeter to determine the centroid position.

9. Quartz coverslips minimize oscillations due to pressure from pump.

Acknowledgments

The development of 2D DIAS and 3D DIAS was supported in part by grants HD18577 and AI40040 from the National Institutes of Health and a generous facility grant from the W.M. Keck Foundation.

References

1. Murray, J., Vawter-Hugart, H., Voss, E., and Soll, D. R. (1992) A three-dimensional motility cycle in leukocytes. *Cell Motil. Cytoskel.* **22,** 211–223.
2. Wessels, D., Vawter-Hugart, H., Murray, J., and Soll, D. R. (1994) Three dimensional dynamics of pseudopod formation and turning during the motility cycle of *Dictyostelium. Cell Motil. Cytoskel.* **27,** 1–12.

3. Kirfel, G., Rigort, A., Borm, B., and Herzog, V. (2004) Cell migration: mechanisms of rear detachment and the formation of migration tracks. *Eur. J. Cell Biol.* **83,** 717–724.

4. Uchida, K., Kitanishi-Yumura, T., and Yumura, S. (2002) Myosin II contributes to the posterior contraction and the anterior extension during the retraction phase in migrating *Dictyostelium* cells. *J. Cell Sci.* **116,** 51–60.

5. Geiger, J., Wessels, D., and Soll, D. R. (2003) Human polymorphonuclear leukocytes respond to waves of chemoattractant, like *Dictyostelium. Cell Motil. Cytoskel.* **56,** 27–44.

6. Firtel, R. and Chung, C. (2000) The molecular genetics of chemotaxis: sensing and responding to chemoattractant gradients. *Bioessays* **22,** 603–615.

7. Devreotes, P. and Janetopoulos, C. (2003) Eukaryotic chemotaxis: distinctions between directional sensing and polarization. *J. Biol. Chem.* **278,** 20,445–20,448.

8. Parent, C. (2004) Making all the right moves: chemotaxis in neutrophils and *Dictyostelium. Curr. Opin. Cell Biol.* **16,** 14–23.

9. Soll, D.R., Wessels, D., Heid, P., and Zhang, H. (2003) A contextual framework for characterizing motility and chemotaxis mutants in *Dictyostelium discoideum. J. Muscle Res. Cell Motil.* **23,** 659–672.

10. Alcantara, E. and Monk, M. (1974) Signal propagation in the cellular slime mould *Dictyostelium discoideum. J. Gen. Microbiol.* **84,** 321–334.

11. Wyckoff, J., Wang, W., Lin, E., et al. (2004). A paracrine loop between tumor cells and macrophages is required for tumor cell migration in mammary tumors. *Cancer Res.* **64,** 7022–7029.

12. Wessels, D., Murray, J., and Soll, D. R. (1992) Behavior of *Dictyostelium* amoebae is regulated primarily by the temporal dynamic of the natural cAMP wave. *Cell Motil. Cytoskel.* **41,** 225–246.

13. Stepanovic, V., Wessels, D., Daniels, K. Loomis, W. F., and Soll, D. R. (2005) Intracellular role of adenylyl cyclase in regulation of lateral pseudopod formation during *Dictyostelium* chemotaxis. *Euk. Cell* **4,** 775–786.

14. Wessels, D., Brincks, R., Kuhl, S., et al. (2004) RasC plays a selective role in the transduction of temporal gradient information in the cAMP wave of *Dictyostelium. Euk. Cell* **3,** 646–662.

15. Kumar, A., Wessels, D., Daniels, K., Alexander, H., Alexander, S., and Soll, D. R. (2004) Sphingosine-1-phosphate plays a role in the suppression of lateral pseudopod formation during *Dictyostelium discoideum* cell migration and chemotaxis. *Cell Motil. Cytoskel.* **59,** 227–241.

16. Zhang, H., Heid, P., Wessels, D., et al. (2003) Constitutively active protein kinase A disrupts motility and chemotaxis in *Dictyostelium. Euk. Cell* **2,** 62–75.

17. Falk, D., Wessels, D., Jenkins, L., et al. (2003) Shared, unique and redundant functions of three members of the class I myosins (MyoA, MyoB and MyoF) in motility and chemotaxis in *Dictyostelium. J. Cell Sci.* **116,** 3985–3999.

18. Zhang, H., Wessels, D., Fey, P., Daniels, K., Chisholm, R., and Soll, D. R. (2002) Phosphorylation of the myosin regulatory light chain plays a role in cell motility and polarity in *Dictyostelium* chemotaxis. *J. Cell Sci.* **115,** 1733–1747.

19. Wessels, D., Zhang, H., Reynolds, J., et al. (2000) The internal phosphodiesterase *RegA* is essential for the suppression of lateral pseudopods during *Dictyostelium* chemotaxis. *Mol. Biol. Cell* **11**, 2803–2820.

20. Wessels, D. and Soll, D. R. (1998) Computer-assisted characterization of the behavioral defects of cytoskeletal mutants of *Dictyostelium discoideum*, in *Motion Analysis of Living Cells* (Soll, D. R. and Wessels, D., eds.), John Wiley, Inc., New York: pp. 101–140.

21. Stepanovic, V., Wessels, D., Goldman, G., Geiger, J., and Soll, D. R. (2004) The chemotaxis defect of Shwachman-Diamond syndrome leukocytes. *Cell Motil. Cytoskel.* **57**, 158–174.

22. Soll, D. R. and Voss, E. (1998) Two and three-dimensional computer systems for analyzing how cells crawl, in *Motion Analysis of Living Cells* (Soll, D. R. and Wessels, D., eds.), John Wiley, Inc., New York: pp. 25–52.

23. Soll, D. R. (1999) Computer-assisted three-dimensional reconstruction and motion analysis of living, crawling cells. *Comp. Med. Imag. Graph.* **23**, 3–14.

24. Soll, D. R., Voss, E., Johnson, O., and Wessels, D. (2000) Three dimensional reconstruction and motion analysis of living crawling cells. *Scanning* **22**, 249–257.

25. Soll, D. R., Wessels, D., Voss, E., and Johnson, O. (2000) Computer-assisted systems for the analysis of amoeboid cell motility, in *Methods in Molecular Biology: Cytoskeleton Methods and Protocols* (Gavin, R. H., ed.), Humana, Totowa, NJ: pp. 45–58.

26. Heid, P., Geiger, J., Wessels, D., Voss, E., and Soll, D. R. (2005) Computer assisted analysis of filopod formation and the role of myosin II heavy chain phosphorylation in *Dictyostelium*. *J. Cell Sci.* **118**, 2225–2237.

27. Varnum, B. and Soll, D.R. (1984). Effect of cAMP on single cell motility in *Dictyostelium*. *J. Cell Biol.* **99**, 1151–1155.

28. Varnum-Finney, B., Voss, E., and Soll, D. R. (1987) Frequency and orientation of pseudopod formation of *Dictyostelium discoideum* amoebae chemotaxing in a spatial gradient: further evidence for a temporal mechanism. *Cell Motil. Cytoskel.* **8**, 18–26.

29. Varnum, B., Edwards, K., and Soll, D. R. (1985) *Dictyostelium* amoebae alter motility differently in response to increasing versus decreasing temporal gradients of cAMP. *J. Cell Biol.* **101**, 1–5.

30. Zigmond, S. (1978) A new visual assay of leukocyte chemotaxis, in *Leukocyte Chemotaxis: Methods, Physiology and Clinical Implication* (Gallin, J. I. and Quie, P. G., eds.), Raven, New York: pp. 57–66.

31. Shutt, D., Jenkins, L., Carolan, E., et al. (1998) T cell syncytia induced by HIV release T cell chemoattractants: demonstration with a newly developed single cell chemotaxis chamber. *J. Cell Sci.* **111**, 99–109.

32. Escalante, R., Wessels, D., Soll, D. R., and Loomis, W. F. (1997) Chemotaxis to cAMP and slug migration in *Dictyostelium* both depend on MigA, a BTB protein. *Mol. Biol. Cell* **8**, 1763–1775.

33. Geiger, J., Wessels, D., Lockhart, S., and Soll, D. R. (2004) Release of a potent polymorphonuclear leukocyte chemoattractant is regulated by white-opaque switching in *Candida albicans*. *Infect. Immunl.* **72**, 667–671.

34. Wessels, D., Schroeder, N., Voss, E., Hall, A., Condeelis, J., and Soll, D. R. (1989) cAMP mediated inhibition of intracellular particle movement and actin reorganization in *Dictyostelium*. *J. Cell Biol.* **109,** 2841–2851.
35. Wessels, D., Voss, E., von Bergen, N., Burns, R., Stites, J., and Soll, D. R. (1998) A computer-assisted system for reconstructing and interpreting the dynamic three-dimensional relationships of the outer surface, nucleus, and pseudopodia of crawling cells. *Cell Motil. Cytoskel.* **41,** 225–246.

17

Using Quantitative Fluorescence Microscopy and FRET Imaging to Measure Spatiotemporal Signaling Events in Single Living Cells

Xuehua Xu, Joseph A. Brzostowski, and Tian Jin

Summary

The mechanisms that mediate how migratory eukaryotic cells amplify a shallow, extracellular chemoattractant gradient into a steep intracellular gradient of signaling components to guide chemotaxis remains unknown. To unravel these mechanisms, it is essential to quantitatively measure the spatiotemporal patterns of chemoattractant gradients, the dynamic movement of intracellular signaling pathway molecules, and the localized activation of these molecules in single living cells. Recent developments in live-cell fluorescence microscopy have permitted direct visualization and quantitative measurement of signal transduction events with high temporal and spatial resolution. Here, we outline fluorescence imaging methods to simultaneously visualize and quantitatively measure spatiotemporal changes in chemoattractant concentration by using the fluorescent tracer dye Alexa 594. Next, we provide a method to correlate the dynamic changes in ligand to the spatiotemporal changes in the second messenger phosphatidylinositol 3,4,5-triphosphate (PIP_3) along the inner surface of the plasma membrane in live cells. Finally, we describe a fluorescence resonance energy transfer (FRET) method to determine the extent of heterotrimeric G protein activation in single living cells in response to various chemoattractant fields.

Key Words: Confocal fluorescence microscopy; FRET; G protein; spatiotemporal dynamics.

1. Introduction

Chemotactically competent cells have the ability to rapidly and precisely remodel the components of intracellular signal transduction networks to respond to changes to a variety of environmental cues *(1–4)*. When activated by extracellular stimuli, multimolecular complexes transiently form at specific loca-

From: *Methods in Molecular Biology, vol. 346:* Dictyostelium discoideum *Protocols*
Edited by: L. Eichinger and F. Rivero © Humana Press Inc., Totowa, NJ

tions within the cell, which in turn leads to a diverse array of cellular behaviors *(5–8)*. Traditional biochemical approaches that analyze the behavior of cell populations or examination of the static spatial distribution of signaling network components in fixed cells do not fully reveal the spatiotemporal aspects that mediate intracellular response pathways. However, recent improvements in live-cell imaging have made it an essential approach in dissecting the dynamic nature of signal transduction events.

It has been realized with practice that many proteins can be genetically fused with green fluorescent protein (GFP) or its color variants without compromising the native function of the protein or the viability of the cell *(5–9)*. By expressing such GFP fusion proteins in living cells, dynamic changes in protein localization can be visualized by tracking intensity changes using fluorescence microscopy. Recent developments in our laboratory have allowed us to simultaneously visualize and quantitatively measure both changes in extracellular stimuli and the intracellular distribution of signaling proteins in single living cells. These methods have provided a means to reveal the relationship between dynamic patterns of intracellular signaling components and spatiotemporal changes in external stimuli *(10)*. In addition, advances in fluorescence resonance energy transfer (FRET) imaging techniques have allowed us to go beyond simple localization events to examine protein–protein interactions and conformational changes in live cells *(10–13)*. The development and application of such fluorescence imaging techniques to advance our understanding of the molecular mechanisms of chemotaxis represents one of the success stories for the model system *Dictyostelium discoideum*.

D. discoideum amoebae migrate directionally toward an increasing concentration of the extracellular chemoattractant cAMP, which is perceived by the heterotrimeric G protein-coupled receptor cAR1 *(5)*. Activation of cAR1 induces the dissociation of the associated $G\alpha 2$ and $G\beta\gamma$ subunits *(6,10,12)*, which in turn regulate a key process in gradient sensing—the localized balance of activity of the phosphoinositide kinase PI3K and the phosphatase phosphatase and tensin homolog (PTEN) to control the local level of the membrane lipid phosphatidylinositol 3,4,5-triphosphate (PIP_3) along the inner surface of the plasma membrane *(7,8)*. PIP_3 acts as binding site for certain plekstrin homology (PH) domain proteins that go on to regulate cell polarity and movement among other downstream processes *(8–10)*. Here, we describe fluorescence imaging methods that directly visualize and quantitatively measure cAMP-triggered signaling events in single living *D. discoideum* cells. We measure the relative external concentration of an exogenously applied cAMP stimulus and relate receptor occupancy to the changes of intracellular distribution of PIP_3 by using the PIP_3-reporter PH_{Crac}-GFP. Finally, we show the degree of G protein activation by measuring the dissociation of the $G\alpha$ and $G\beta\gamma$ subunits using FRET imaging. These methods can be easily modified to visualize and measure various signaling events in other cell types *(14)*.

2. Materials

2.1. Cell Culture

1. D3-T Growth Media (KD Medical, Columbia, MD).
2. Penicillin (10,000 U/mL) and streptomycin (10 mg/mL) solution (GIBCO, Grandland, NY) is frozen at –20°C in aliquots; add 3 mL to 1 L of D3-T before use.
3. Geneticin (Sigma, Steinheim, Germany). A 1000X stock concentration at 50 mg/mL is made in water and frozen at –20°C in aliquots.
4. Blasticidin S (Invitrogen, Carlsbad, CA). A 1000X stock concentration at 10 mg/mL is made in water and frozen at –20°C in aliquots.
5. 10-cm, round tissue culture dishes are used for stationary growth (Falcon, Franklin Lakes, NJ).
6. 125- and 250-mL plastic Erlenmeyer flasks (Corning, Corning, NY) are used for shaking culture.
7. Cell lines: Ax2 expressing PH_{Crac}-GFP *(10)*; Myc2 co-expressing $G\alpha$ cyan fluorescent protein (CFP) and yellow fluorescent protein (YFP)$G\beta$ *(10,12)*.

2.2. Cell Development

1. Cells are developed to a chemotactically competent stage in development buffer (DB): 1.34 g Na_2HPO_4, 0.68 g KH_2PO_4 per liter with 0.2 mM $CaCl_2$ and 2 mM $MgSO_4$. For convenience, concentrated stock solutions can be made for each component.
2. cAMP (Sigma, Steinheim, Germany). A 0.01 M stock is made in water and frozen at –20°C. The final pulse concentration is 75 nM.
3. ChronTrol XT programmable timer (ChronTrol Corp, San Diego, CA).
4. Miniplus 3 peristaltic pump (Gilson, Middletown, WI).
5. Platform rotary shaker with accurate rpm control.
6. 125- and 250-mL plastic Erlenmeyer flasks.
7. Caffeine (Sigma, Steinheim, Germany). A 100 mM stock is made in water and frozen at –20°C in aliquots. The working concentration is 2.5 mM.

2.3. Cell Stimulation

1. Latrunculin B (Molecular Probes, Eugene, OR). A 1 mM stock is made in dimethylsulfoxide (DMSO); it is stored in aliquots at –20°C and is used fresh. The final concentration in experiments is 2.5 μM.
2. Alexa 594 (Molecular Probes, Eugene, OR). A 1mg/mL stock is made in water and frozen at –20°C in aliquots. The working concentration is 0.1 mg/mL.
3. cAMP (Sigma, Steinheim, Germany). A 0.01 M stock is made in water and frozen at –20°C. Its working concentration can range (*see* below).

2.4. Confocal Microscopy

1. Single- and four-well Lab-Tek II coverglass chambers (Nalge Nunc International, Naperville, IL).
2. DB (*see* **Subheading 2.2.**, **item 1**).

3. FemtoJet microcapillary pressure supply (Eppendorf, Germany).
4. TransferMan NK2 micromanipulator (Eppendorf, Germany).
6. LSM 510 META or equivalent with either a 40× 1.3 NA or 60× 1.4 NA oil differential interference contrast (DIC) Plan-Neofluar objective lens.

3. Methods
3.1. Cell Culture and Development

1. *D. discoideum* cells are grown at 22°C in stationary on 10-cm, round tissue-culture plates in 10 mL of D3-T media and are fed weekly with fresh D3-T. The antibiotics penicillin and streptomycin from a frozen stock are always added to growth media to a final concentration of 3.3 U/mL and 3.3 µg/mL, respectively, before use. Strains carrying a blasticidin or geneticin resistance expression cassette are cultured with 10 µg/mL and 50 µg/mL of drug, respectively. The drugs are always added when cells are fed or expanded for shaking culture.

2. To grow cells for experiments, old media is removed from the stationary plate, 10 mL of fresh D3-T containing the appropriate drug(s) is added, and cells are removed from plates by pipetting. The suspended cells are then added to a 125-mL, sterile Erlenmeyer flask containing 25 mL of D3-T. The flask is mixed and a 10-mL aliquot is returned to the stationary plate. Cells are grown to log phase (1–5×10^6 cells/mL) in the 125-mL flask at 22°C on a platform rotary shaker at 200 rpm. Expect a 9- to 10-h doubling time for wild-type cells. Shaking cultures can be expanded to 50 mL in the 125-mL flask or 100 mL in a 250-mL flask. In general, 100 mL of cells grown to 2–3×10^6 cells/mL is more than adequate for these experiments.

3. To develop cells (*see* **Note 1**) to a chemotactically competent stage, 2×10^8 cells are harvested from shaking flasks by centrifugation at 1200*g* for 2–3 min at room temperature and washed once with 100 mL of DB and repelleted. Cells are suspended in 10 mL of DB for a final concentration of 2×10^7/mL and are shaken at 100 rpm at 22°C (*see* **Note 2**). After 1 h, a 100-µL aliquot of cAMP is delivered every 6 min to achieve a final concentration of 70 n*M* cAMP in the flask. The concentration of the 100-µL cAMP aliquot must be calculated according to the volume of cells in the shaking flask. To deliver the cAMP pulses, thin tubing is taped to the mouth of the flask and is connected through a peristaltic pump to a reservoir of cAMP in DB. An external programmable timer controls the pump (*see* **Subheading 2.2.**).

4. After 4–5 h (*see* **Note 3**) of pulsing, cells are centrifuged at 1200*g*, suspended in 20 mL of DB with 2.5 m*M* caffeine (*see* **Note 4**), and shaken at 200 rpm at 22°C for 20 min.

5. To plate pulse-developed cells in Lab-Tek II coverslip single- or four-well chambers:
   - After caffeine treatment, remove an aliquot of cells and centrifuge at 500*g* for 3 min.
   - Remove buffer and dilute cells to 5×10^5 cells/mL with fresh DB containing caffeine.

- Apply 1 mL of cell suspension to a single-well chamber or 0.4 mL to each well of a four-well chamber. Allow cells to adhere for 10 min, carefully pipet off the buffer to remove unattached cells, and replace with the same volume (*see* **Note 5**).

6. Cells are incubated with 2.5 μM (final concentration) Latrunculin B for 10 min prior to use in experiments. Note that Latrunculin B is not washed from cells before stimulation. If numerous stimulations are to be performed, Latrunculin B treatment is staggered to maintain consistency (*see* **Note 6**).

3.2. Generation and Monitoring of Exogenously Applied Chemoattractant

Chemoattractant binding to receptors initiates a rapid re-organization and activation of intracellular signaling components that polarize the cell and coordinate directional movement toward the source of chemoattractant. Remarkably, cells can determine the direction of a chemoattractant source within a gradient that differs by as little as 2% across the cell body. Therefore, it becomes important to relate the amount of ligand that surrounds the cell surface and how it changes in a quantitative manner to the spatiotemporal changes that occur to intracellular signaling networks as a consequence of ligand binding. Here, we show a method to indirectly visualize the spatiotemporal changes of an applied cAMP stimulus around cells by measuring the changes in fluorescence intensity of the hydrophilic dye Alexa 594 mixed with the cAMP ligand.

1. Backfill a micropipet (we use Eppendorf Femtotips II; *see* **Note 7**) using the manufacturer's supplied microcapillary tip with a freshly prepared 30-μL solution of 1 μM cAMP and Alexa 594 at 0.1 μg/μL in DB (*see* **Notes 8** and **9**).
2. Attach the Femtotip to a micropipet holder and connect the tubing to a pressure supply apparatus. We use the Eppendorf FemtoJet system (*see* **Note 10**).
3. Attach the micropipet assembly to a micromanipulator. We use the Eppendorf TransferMan NK2 motorized micromanipulator.
4. Mount a one-well LabTek chamber filled with 6 mL of DB over a 40× oil lens on a confocal microscope. Using brightfield optics, center the Femtotip into the field of view. Using the Z-control, bring the tip of the micropipet to lie just above the bottom of the chamber (*see* **Note 11**).
5. Turn on the pressure supply and set the compensation pressure (Pc) for 70 hPa to establish a gradient of the cAMP/Alexa 594 mixture (*see* **Note 12**).
6. Image Alexa 594 fluorescence by excitation with a 543-nm laser line. Use a 580 to 650 band-pass filter and set the Z-axis resolution <3 μm to eliminate unwanted fluorescent signal from other focal planes. Results and quantification are shown in **Fig. 1A,B**. See figure legend for details on quantification.

3.3. Simultaneous Imaging of cAMP Stimulation and Cell Response in Single Living Cells

An increase in cAMP receptor occupancy activates a G protein-dependent signaling network, leading to a rapid change in PIP$_3$ levels in the plasma

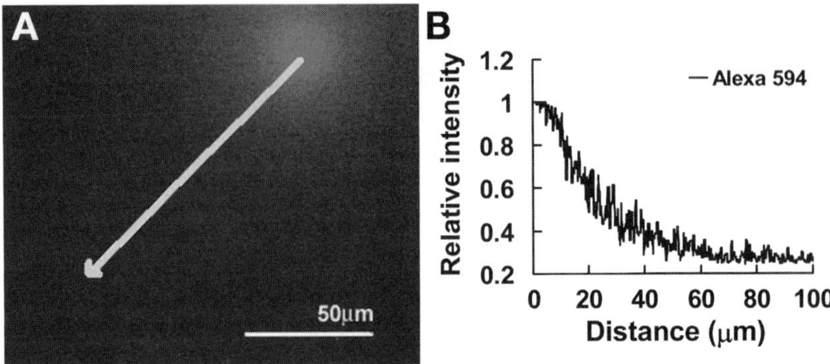

Fig. 1. Generation and monitoring of exogenously applied chemoattractant. (**A**) Shown is an image of a circular cAMP gradient. A micropipet was linked to a Femtojet with constant pressure (Pc = 70 and Pi = 70) to release a small, constant volume of an Alexa 594/cAMP mixture into a single-well chamber containing 6 mL of buffer. The micropipet tip is in the center of the gradient and is not imaged. When the micropipet is moved from one position to another, the gradient remains nearly constant (not shown). (**B**) The intensity change of fluorescence is normalized along the white line in **A** in the direction of the arrow and is plotted against distance (μm). The normalized Relative Intensity equals $(I_T\text{-}I_B)/(I_C\text{-}I_B)$, where I_T is the maximum fluorescence, I_B is the background intensity value measured in a field without a gradient, and I_C is the intensity value measured along the white line. I_T equals I_C at the start of the line; therefore, the maximum relative intensity begins at one.

membrane. The dynamic changes in PIP_3 can be monitored by measuring changes in the membrane association of PH_{Crac}-GFP (**9,10**). To determine the relationship between the relative degree of cAMP receptor occupancy over the cell surface and the temporal changes of PIP_3 in these regions, we simultaneously image by confocal microscopy the relative cAMP concentration with Alexa 594 and the dynamics of PH_{Crac}-GFP in single living cells. We analyze the spatiotemporal PIP_3 response in two ways: by either providing cells a rapid uniform increase of cAMP around the cell periphery or by moving cells rapidly into a cAMP gradient.

The general outline for both procedures is as follows:

1. Plate pulse-developed cells (*see* **Subheading 3.1.**) in a four-well LabTek chamber for uniform stimulation experiments or in a one-well LabTek chamber for gradient stimulation.
2. Allow cells to adhere to the coverslip for 10 min and then cover with 400 μL or 6 mL DB buffer for the four-well chamber or one-well chamber, respectively.
3. Image cells on a confocal microscope using a 40× objective lens and use a Z-axis resolution <3 μm.

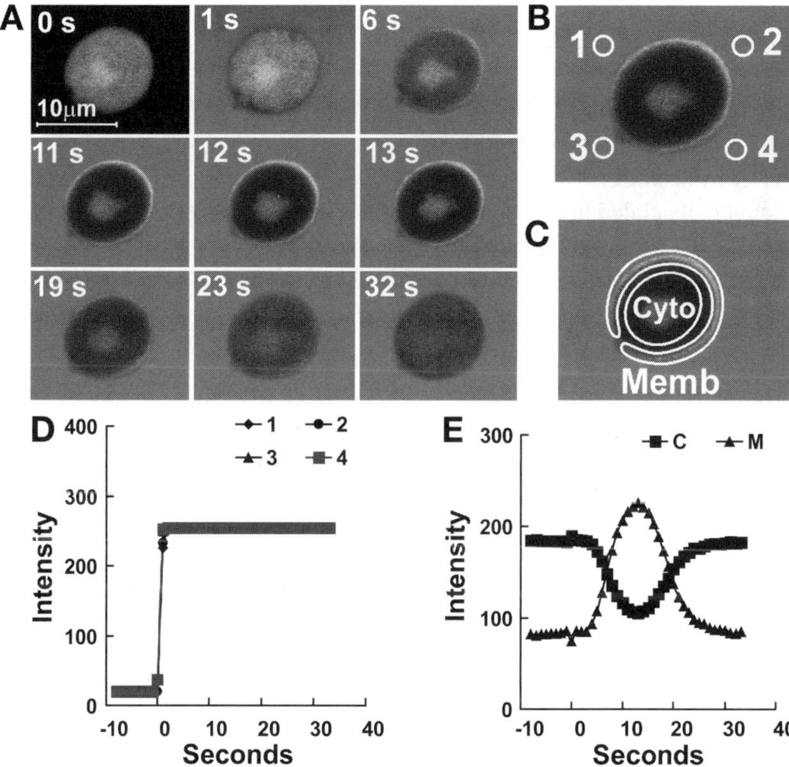

Fig. 2. Simultaneously visualizing a uniform cAMP stimulation and PH_{Crac}-green fluorescent protein (GFP) translocation. **(A)** A cell expressing PH_{Crac}-GFP was uniformly stimulated with 1 μ*M* cAMP mixed with 0.1 μg/μL Alexa 594 (as indicated by fluorescence intensity change from black at 0 s to grey at 1 s outside the cell). Translocation of PH_{Crac}-GFP to membrane associated phosphatidylinositol 3,4,5 triphosphate (PIP_3) peaks at approx 13 s (as indicated by the white fluorescence intensity peak along the inner membrane surface). Time (seconds) is shown in the upper left corner of each image. **(B)** Four circular regions of interest (ROIs) were selected in the time series stack (one image of the stack is shown) to calculate the intensity change of the Alexa 594 signal over time. The plot is shown in **D**. **(C)** Two ROIs were drawn to calculate the change of intensity of the PH_{Crac}-GFP signal over time in the plasma membrane and the cytoplasm. The plot of membrane (M) and cytoplasmic (C) intensity values is shown in **E**.

4. To monitor the relative cAMP concentration and PH_{Crac}-GFP localization, excite the specimen with two laser lines: 543 nm for Alexa594 and 488 nm for GFP.
5. Simultaneously record time-lapse images in three channels. In channel one, collect the GFP (green) signal with a 505- to 530-nm band-pass filter; in channel

two, use a 580- to 650-nm band-pass filter for the orange emission from Alexa 594; and in channel three (not shown), collect the transmitted image (*see* **Fig. 2**).

3.3.1. Acquisition of Images of Cells Exposed to a Uniform Stimulation

1. Prepare a fresh mixture of 1 μM cAMP (*see* **Note 13**) and Alexa 594 (0.1 mg/mL) in DB from stocks. The solution is made each time to ensure maximum efficacy.
2. Mount cells plated in a four-well LabTek chamber over a 40× oil lens and image cells. Image cells that have adhered well to the coverslip.
3. Remove a 100-μL aliquot of the cAMP/Alexa 594 mixture with a standard 200-μL pipetter and set aside.
4. Begin a time-lapse acquisition. We usually use a three second interval (*see* **Note 14**).
5. After acquiring several images of unstimulated cells, position the 200-μL pipetter directly over the cells using the spot of laser light as a guide. Pipet the cAMP/Alexa 594 mixture over the cells being imaged with a quick and even flow (*see* **Note 15**). Results are shown in **Fig. 2**.

3.3.2. Acquisition of Images of Cells Suddenly Exposed to a cAMP Gradient

1. Backfill a Femtotip with a freshly prepared 30-μL solution of 1 μM cAMP (*see* **Note 16**) and Alexa 594 at 0.1 mg/mL in DB.
2. Attach the Femtotip to a micropipet holder and connect tubing to the FemtoJet.
3. Attach the micropipet assembly to the Eppendorf TransferMan NK2 (*see* **Note 17**). Three positions can be stored in memory by using the Pos. buttons 1, 2, or 3.
4. Mount cells plated in a one-well LabTek chamber over a 40 or 60× oil lens and image cells with bright-field optics.
5. Using brightfield optics, move the Femtotip into the middle of the field with the pressure supply off and position the tip within 100 μm from the cells of interest and just above the bottom of the chamber. Store the position into memory (Pos. 1).
6. Move the Femtotip laterally, about 1000 μm from the cells, and store the position into memory (Pos. 2).
7. Turn on the FemtoJet and apply a positive pressure of 70 hPa to establish the gradient of the cAMP/Alexa 594 mixture.
8. Begin a time-lapse with a 3-s interval and acquire several images of unstimulated cells. Press the Pos. 1 button to move the gradient into place. Results are shown in **Fig. 3**.

3.4. Spectrally-Resolved FRET Imaging Detects Dynamic Protein–Protein Interaction in Single Living Cells

Dynamic changes in protein conformation or protein–protein interaction can be monitored in single living cells in real time using FRET imaging methodologies. A basic understanding of donor/acceptor fluorophore energy transfer is assumed here. Briefly, FRET is a nonradiative process in which energy from an excited fluorophore (donor) is transferred to an acceptor fluorophore. FRET detection is dependent on the distance between the donor and acceptor

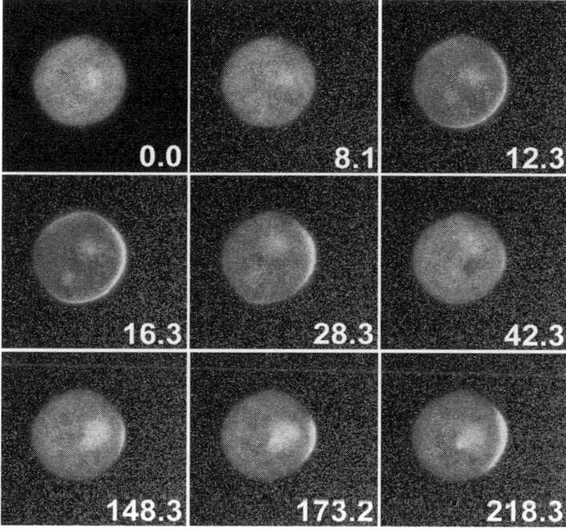

Fig. 3. Simultaneously visualizing an imposed cAMP gradient and PH_{Crac}-green fluorescent protein (GFP) translocation. A cell expressing PH_{Crac}-GFP was suddenly exposed to a cAMP gradient from a micropipet containing 1 μM cAMP mixed with 0.1 $\mu g/\mu L$ Alexa 594 (as indicated by speckled grey values outside the cell). The gradient source (not shown) is at the lower right corner of the image. Translocation of PH_{Crac}-GFP (as indicated by the white fluorescence intensity peak along the inner membrane surface) to the plasma membrane occurs in the direction of the gradient source. Time (seconds) is shown in the lower right corner of each image.

fluorophores, which ranges between 10 and 100 Å. In the context of live-cell imaging, FRET measurements are usually made with CFP (donor) and YFP (acceptor) pairs. The fluorescent proteins can be fused to the respective ends of a single protein, requiring a change in protein conformation to bring the fluorophores into proximity to achieve FRET, or can be engineered onto separate proteins that form a complex. In general, useful FRET fluorophore pairs require sufficient separation of respective excitation spectra, whereby the wavelength of light used to excite the donor fluorophore does not significantly excite the acceptor fluorophore. To achieve efficient energy transfer, the donor emission spectrum should overlap by >30% with the acceptor fluorophore's excitation spectrum. Excellent reviews of the power of this technology and its inherent caveats are available *(10–13)*. As implied, imaging and quantifying FRET can be technically challenging. In general, FRET signal intensities are weak. In addition, resulting acceptor emission spectra caused by FRET significantly overlaps with donor emission spectral wavelengths (see overlap centering at approx 520 nm in **Fig. 4B**). Therefore, it can be difficult to discriminate a true FRET acceptor signal from the bleed-through of donor emission into the FRET image. Here, we describe a new FRET technique using the Lambda Stack Acquisition mode of the Zeiss LSM 510 META. The system employs a specialized photomultiplier tube (PMT) that collects a series of emission spectra that are recorded by up to 32 separate detectors rather than the typical single intensity measurement obtained through selected emission band-pass filters with a single PMT. In the Lambda Stack mode, fluorescence intensities of the donor and acceptor fluorophore are calculated based on their individual emission spectra, therefore allowing FRET to be precisely measured without interference from donor emission bleed-through. Here, we describe a method for using this technique to study the interaction of heterotrimeric G protein Gα2 and Gβ subunits in living cells.

3.4.1. Acquisition of Emission Fingerprint of Individual Fluorophores

An emission fingerprint of each fluorophore must be obtained in cells expressing Gα2CFP alone and YFPGβ alone (*see* **Note 18**).

1. Pulse-develop Gα2CFP- and YFPGβ-expressing cell lines (*see* **Subheading 3.1.**).
2. Plate each cell line separately in one well of a four-well LabTek chamber, mount over a 40× lens, and treat cells with Latrunculin B (*see* **Note 19**).
3. Set the system to the Lambda Stack Acquisition mode and apply the following settings: choose the 458 laser line (*see* **Note 20**); choose the 458 main dichroic beam splitter; adjust the pinhole size for an optical slice of 3 μm; define the spectral range from 464 to 624 nm with a 10-nm width and collect the Lambda Stack (**Fig. 4A**) for *each* fluorophore. This selected spectral range will collect images from 16 of the 32 META detectors (*see* **Note 21**).
4. To create the emission fingerprint, enter the Mean ROI mode and choose a region of interest (ROI) for the images obtained for each fluorophore; this action will create an intensity plot over the spectral range collected in the Lambda Stack (*see*

Fig. 4. *(opposite page)* Spectrally resolved fluorescence microscopy detecting intensity changes in cyan fluorescent protein (CFP) and yellow fluorescent protein (YFP). (**A**) Lambda Stack acquisition of multi fluorescence signals using a Zeiss Confocal Laser Scanning Microscope 510 META. A cell co-expressing Gα2CFP and YFPGβ was excited with a 458-nm laser line using a Plan-Neofluar differential interference contrast (DIC) 40× NA 1.3 oil lens. The spectral emissions of every pixel in the image were recorded in 16 channels from 464 nm to 624 nm. (**B**) Emission spectra of the cells expressing CFP or YFP only were obtained using the same scanning conditions, and were used as the fingerprint references for the Linear Unmixing Function. (**C**) The resulting CFP (direct emission) and YFP (emission via fluorescence resonance energy transfer [FRET]) images are shown after the Linear Unmixing function was applied to the Lambda Stack acquired in **A**. Two regions of interest (ROIs) (R1, and R2) were drawn and the intensities were plotted in **B**. (**D**) The resulting images after the Linear Unmixing function

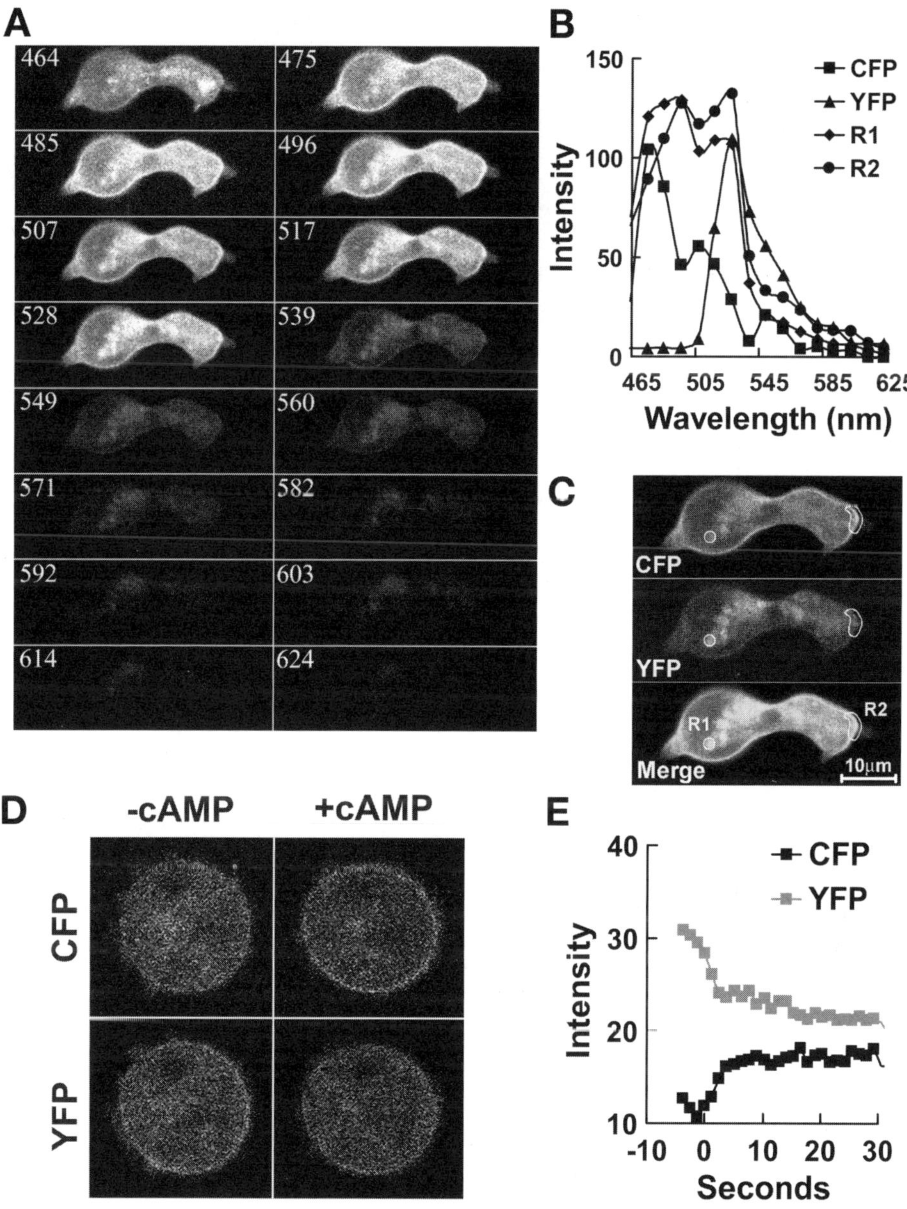

Fig. 4 *(continued)* was applied to a Lambda Stack (not shown) acquired over time for a Latrunculin B treated cell co-expressing Gα2CFP and YFPGβ. The cell was stimulated with 10 µ*M* cAMP (+) and the resulting increase in CFP signal and concomitant decrease in YFP signal due to loss of FRET is readily apparent by eye along the inner membrane surfae and is plotted as a change in intensity over time (seconds) in **E**.

Note 22). Save each plot to the Spectral Database; they will be used as reference curves for the Linear Unmixing function described next (**Fig. 4B**).

3.4.2. Acquisition of a Lambda Stack From Cells Co-Expressing Fluorophores

To acquire a Lambda Stack for cells co-expressing Gα2CFP and YFPGβ and to separate the CFP and YFP fluorescent signals:

1. Develop and plate cells co-expressing Gα2CFP and YFPGβ as described in **steps 1** and **2** and acquire the Lambda Stack as in **Subheading 3.4.1., step 3** (*see* **Notes 23** and **24**).
2. In the Process window, select the Unmix function and choose the saved CFP and YFP emission fingerprints from **step 4**. Apply the fingerprints to the Lambda Stack acquired from co-expressing cells (**Fig. 4B,C**). The Linear Unmixing function mathematically calculates the contribution of each fluorophore in the Lambda Stack acquired from co-expressing cells on a pixel-by-pixel basis over the spectral window to separate the CFP and YFP into individual channels. The total emission signal is therefore separated into the weighted contributions of the CFP and YFP signal based on the knowledge of their individual emission fingerprints saved in **Subheading 3.4.1., step 4**.

3.4.3. Observation of Loss of FRET

To observe a loss of FRET between Gα2CFP and YFPGβ after receptor stimulation:

1. Prepare co-expressing cells as above (*see* **Subheading 3.4.1.**). To significantly increase acquistion speed, define the spectral range from 464 to 544 nm and step with 10-nm increments.
2. Begin a time lapse with a 2-s interval and acquire three to five Lambda Stacks of unstimulated cells.
3. Stimulate cells with a uniform dose of cAMP as described under **Subheading 3.3.1.** and acquire 30 stacks.
4. To analyze the FRET change, apply the Linear Unmixing function to the Lambda Stack time series. Select ROI and plot intensity change over time (**Fig. 4D,E**).

4. Notes

1. Developing cells with exogenous cAMP pulses shortens the period of time for cells to reach a chemotactically competent stage as compared with natural development on a solid substrate; it is a standard technique used by many *D. discoideum* researchers.
2. Smaller volumes of cells can be pulse-developed. Keep the volume of cells between 5 and 15 mL in 125-mL flasks for pulse-development.
3. Slight variations of time are observed between mutant strains, and the best duration of pulsing must be determined empirically. Also, day-to-day variability can occur with any strain if the room temperature is not finely controlled.

4. Caffeine is used to inhibit the endogenous production of cAMP, bringing cells to a quiescent or basal signaling state. The point of action of the drug is unknown.

5. Make sure that a thin film of buffer remains on top of the cells as you remove the majority of it from the well. Quickly replace with fresh buffer at an edge of the well, taking care not to disturb the attached cells. If the well is sucked dry, cells will lyse. We have noticed a strange lot-to-lot variability in chambers where the bottom of the well will not remain wet when the buffer is pipetted away. In this situation, we do not exchange buffer, leaving the unstuck cells to float around.

6. Latrunculin disrupts the actin cytoskeleton, causing polarized, chemotactically competent cells to round up. These rounded cells can still sense the direction of an extracellular cAMP gradient and respond to a global stimulation *(6,9,10)*. Because these cells are immobile, the accuracy of spatiotemporal fluorescent measurements within the cell is improved.

7. Occasionally, an air bubble will form at the end of the tip after backfilling; this can be removed after attaching the micropipet to the pressure supply and pressing either the "injection" or "clear" button on the apparatus.

8. If a gradient does not establish from the needle tip after several attempts to clear with pressure, particulate matter may be in the cAMP/Alexa 594 solution. Centrifuging at full speed for several minutes and reloading a new micropipet usually solves the problem.

9. We chose Alexa 594 for several practical reasons: it is water-soluble; it does not readily cross the plasma membrane; chemotaxis is unaffected by its presence; and it emits in the orange spectra (peaking at 615 nm), thereby allowing simultaneous detection of shorter wavelength GFP fusion proteins *(10)*.

10. The FemtoJet (Eppendorf) pressure supply is a stable, small ($22 \times 28 \times 17$ cm) apparatus that does not require an external gas tank, making it easily portable between microscopes.

11. For the novice, we suggest centering the micropipet tip at low magnification with brightfield optics. If the lamp intensity is raised, the resulting circle of light above the specimen can be used to roughly center the tip. Manipulate the tip to enter the buffer of the chamber. Once the tip is centered in the light, lower the intensity, look into the oculars, and focus on the tip to make sure it is finely centered. Bring the tip to a point about several cell heights above the bottom of the chamber by slowly lowering the height while adjusting the focus to follow the tip down. Change to the 40× oil lens, refocus the tip, and lower it to a point where the tip is just above the bottom of the chamber. The tip should be straight. If there is a slight bend, then it is touching the bottom. If it is touching, the tip is likely to break after movement. Some practice is required.

12. Note that the pressure given is a good starting point. Occasionally, variability in optimal compensation pressure (Pc) has been observed between micropipet lots. A stable gradient should establish within about 10 s after turning on the pressure.

13. A range of cAMP concentrations of up to 10 μM, which is saturating for cAR1, can be used in these experiments.

14. Make sure that the scan time does not exceed the time interval set for the time lapse. We tend to use a 512×512 resolution, scan in one direction, acquire in

8-bit mode, and set the scan speed as fast as possible without compromising image quality. Image quality can be improved by averaging, but keep in mind that scan time will increase.

15. We cut the end off of a standard 200-μL pipet tip to produce a wider bore-size; this provides an even, rapid flow. Note that loosely attached cells can be washed away after stimulation. Although annoying, this happens on occasion. Tap the stage with a finger to evaluate how well cells are adhered to the coverslip.

16. 1 μ*M* cAMP provides a robust response, but the cAMP concentration can be varied from 0.001 μ*M* to 10 μ*M* to study such effects.

17. Manually operated manipulators can be used if travel in the *x–y* plane is tuned well enough so that the micropipet can be moved without the need to visualize the movement with standard optics and without vibration of the instrumentation.

18. The emission fingerprint of each fluorophore serves as a reference curve to mathematically separate the contribution of the CFP and YFP signal in images of cells co-expressing the fluorophores. The separation is performed by the Linear Unmixing function.

19. Incubation with Latrunculin B immobilizes otherwise motile pulse-developed cells. Minimizing cell movement increases the accuracy of FRET measurements in specific regions of the cell such as the plasma membrane.

20. Although a shorter wavelength laser would provide an excitation line closer to the optimal excitation wavelength of CFP, effective FRET measurements can be made using the 458-nm line from the Argon laser.

21. The META is designed to collect eight images over the chosen spectral range from eight detectors simultaneously. If a large enough spectral range is chosen, all 32 META detectors are used and four scans are required to collect all the wavelengths. Keep in mind that collecting "unnecessary" wavelengths will impede time-lapse experiments if speed is required.

22. It is of particular importance that the peak height of the intensity plot for CFP at 475 nm and YFP at 530 nm be equal, so as to not bias the calculation of the Linear Unmixing function for each fluorophore's contribution in an image of cells co-expressing CFP and YFP fusion proteins (*see* **Fig. 4B**).

23. Choose cells for FRET analyses that have a balanced CFP/YFP signal. Make adjustments for relative signal strength of each fluorophore depending on the limitations of the available filter sets on your system. For example, we visually inspect for strong CFP signal with the mercury lamp using a CFP (or 4',6-diamidino-2-phenylindole [DAPI]) filter set. We use a GFP filter set to inspect for the cells co-expressing YFP and choose those that have a slightly higher YFP signal relative to the CFP signal for analysis.

24. For the Linear Unmixing function to accurately separate the signal of each fluorophore in images of cells co-expressing CFP and YFP, acquisition of the Lambda Stack of co-expressing cells *must* be identical to the acquisition made for each individual fluorophore in the creation of the emission fingerprints; that is, all laser, filter, mode, and objective lens settings are to be the same.

Acknowledgments

The authors would like to thank the members of the Chemotaxis Signal Section, Xuanmiao Jiao, Jimmy Chen, Lauren Nelson, Jun Fang, and Ling Yi, for their contributions and thoughtful discussions. This research was supported by the Intramural Research Program of the National Institutes of Health (NIH), National Institute of Allergy and Infectious Disease (NIAID).

References

1. Parent, C. and Devreotes, P. N. (1999) A cell's sense of direction. *Science* **284**, 765–770.
2. Devreotes, P. N. and Janetopoulos, C. (2003) Eukaryotic Chemotaxis: Distinctions between directional sensing and polarization. *J. Biol. Chem.* **278**, 20,445–20,448.
3. Servant, G., Weiner, O. D., Herzmark, P., Balla, T., Sedat, J. W., and Bourne, H. R. (2000) Polarization of chemoattractant receptor signaling during neutrophil chemotaxis. *Science* **287**, 1037–1040.
4. Zigmond, S. H., Levitsky, H. I., and Kreel, B. J. (1981) Cell polarity: an examination of its behavioral expression and its consequences for polymorphonuclear leukocyte chemotaxis. *J. Cell Biol.* **89**, 585–592.
5. Xiao, Z., Zhang, N., Murphy, D. B., and Devreotes, P. N. (1997) Dynamic distribution of chemoattractant receptors in living cells during chemotaxis and persistent stimulation. *J. Cell Biol.* **139**, 365–374.
6. Jin, T., Zhang, N., Long, Y., Parent, C. A., and Devreotes, P. N. (2000) Localization of the G protein βg complex in living cells during chemotaxis. *Science* **287**, 1034–1036.
7. Iijima, M. and Devreotes, P. N. (2002) Tumor suppressor PTEN mediates sensing of chemoattractant gradients. *Cell* **109**, 599–610.
8. Funamoto, S., Meili, R., Lee, S., Parry, L., and Firtel, R. A. (2002). Spatial and temporal regulation of 3-phosphoinositides by PI 3-kinase and PTEN mediates chemotaxis. *Cell* **109**, 611–623.
9. Parent, C., Blacklock, B., Froelich, W., Murphy, D., and Devreotes, P. N. (1998) G protein signaling events are activated at the leading edge of chemotactic cells. *Cell* **95**, 81–91.
10. Xu, X., Meier-Schellersheim, M., Jiao, X., Nelson, L. E., and Jin, T. (2005) Quantitative imaging of single live cells reveals spatiotemporal dynamics of multi-step signaling events of chemoattractant gradient sensing in *Dictyostelium. Mol. Biol. Cell* **16**, 676–688
11. Miyawaki, A. (2003) Visualization of the spatial and temporal dynamics of intracellular signaling. *Dev. Cell* **4**, 295–305
12. Janetopoulos, C., Jin, T., and Devreotes, P. N. (2001) Receptor mediated activation of heterotrimeric G-proteins in living cells. *Science* **291**, 2408–2411.

13. Sekar, R. B. and Periasamy, A. (2003) Fluorescence resonance energy transfer (FRET) microscopy imaging of live cell protein localization. *J. Cell Biol.* **160,** 629–633.

14. Jiao, X., Zhang, N., Xu, X., Oppenheim, J. J., and Jin, T. (2005) Ligand-induced partitioning of human CXCR1 chemokine receptors with lipid-raft microenvironments facilitates G-protein-dependent signaling. *Mol. Cell. Biol.* **25,** 5752–5762.

18

Visualizing Signaling and Cell Movement During the Multicellular Stages of *Dictyostelium* Development

Dirk Dormann and Cornelis J. Weijer

Summary

Time-lapse microscopy provides a powerful tool with which to study cell behavior during *Dictyostelium* development. On a macroscopic level, the overall cell movement patterns that give rise to the complex multicellular structures such as slugs and fruiting bodies can be studied together with the signal waves that coordinate cell movement. Using green fluorescent protein fusion proteins, it is also possible to visualize the cytoskeleton or signal transduction processes at high resolution in single cells in their multicellular environment.

Key Words: Imaging; cell movement; wave propagation; morphogenesis; image processing.

1. Introduction

Dictyostelium discoideum is not only an ideal organism in which to study the cell and molecular biology of processes such as cell motility, chemotaxis, phagocytosis, and cytokinesis, but it is also an excellent system in which to study the cellular basis of multicellular morphogenesis and the signals that control this process. Multicellular development of *Dictyostelium* is initiated by the aggregation of hundreds to thousands of cells into multicellular mounds followed by complex morphogenetic movements leading to the formation of slugs and finally fruiting bodies. Using fluorescent labels like green fluorescent protein (GFP) cell behavior can be studied in the context of large populations during the entire developmental process using time-lapse microscopy *(1,2)*. Combined with cell type-specific expression of GFP variants, movement patterns of different cell types can be followed simultaneously *(3)*. Even signal transduction events or cytoskeletal structures can be visualized at high resolution

From: *Methods in Molecular Biology, vol. 346:* Dictyostelium discoideum *Protocols*
Edited by: L. Eichinger and F. Rivero © Humana Press Inc., Totowa, NJ

during development in order to gain insight into the mechanisms that govern cell behavior in a multicellular environment *(4,5)*. The effect of mutations in signaling or cytoskeletal components can be studied in synergy experiments in which mutant cells are allowed to co-aggregate with wild-type cells *(6)*. The analysis of the relative movement of mutant and wild-type cells can be useful in establishing whether a mutation plays a functional role during development. In this chapter, we describe how to label cells chemically with a fluorescent dye if GFP-expressing cells are unavailable and how to perform synergy experiments. We explain how to obtain the different developmental structures how they are prepared for fluorescence time-lapse microscopy to generate high-resolution images, and describe basic methods of image analysis and visualization.

2. Materials

1. KK2 buffer: 20 mM KH$_2$PO$_4$/K$_2$HPO$_4$, pH 6.8.
2. 1% H$_2$O agar plates: 1% (w/v) agar (e.g., Bacto-Agar, Difco) in H$_2$O dest.; 7 mL per Petri dish (9 cm diameter).
3. 1% KK2 agar plates: 1% (w/v) agar (e.g., Bacto-Agar, Difco) in KK2 buffer; 1 mL per Petri dish (3.5 cm diameter).
4. Low-viscosity silicon oil (e.g., Dow Corning 200/20cs or AR20 from Wacker Chemie).
5. Microscope observation chamber (e.g., Attofluor Cell Chamber, Invitrogen/ Molecular Probes).
6. CellTracker™ Green CMFDA (5-chloromethylfluorescein diacetate, Invitrogen/ Molecular Probes): prepare 10 mM stock solution in dimethylsulfoxide (DMSO) and freeze in 20-μL aliquots at –20°C.

3. Methods

3.1. Preparation of Synergy Experiments: Cell Labeling

1. *Dictyostelium* cells growing in suspension culture at 22°C are harvested during exponential growth (cell density 2–4 × 10^6 cells/mL; *see* **Note 1**)
2. Cells are pelleted by centrifugation at 800g for 2 min, washed once in KK2 buffer, and pelleted again.
3. Adjust cell density with KK2 buffer to 1 × 10^7 cells/mL and incubate on a shaker at 160 rpm at 22°C. If cells are already labeled because they express markers such as GFP, proceed to **step 8**.
4. Place 1 mL of the cell suspension in a 1.5-mL microcentrifuge tube and add 2 μL of the 10 mM CellTracker stock solution (1:500 dilution). Wrap tube with tinfoil, as CellTracker is light sensitive and mount horizontally on a shaker (160 rpm) at 22°C.
5. Incubate for 45 min. After 10 to 15 min, the staining solution should turn yellow-green, indicating the processing of the CellTracker dye by the cells as esterases cleave off acetate groups, turning the membrane-permeable nonfluorescent dye into the fluorescent form.

6. In order to remove unprocessed and secreted dye, labeled cells are washed twice in KK2 by pelleting the cells (5 s at maximum speed of microcentrifuge) and resuspending them in KK2 buffer.

7. Resuspend cells after final wash in 1 mL KK2 (cell density 1×10^7 cells/mL).

8. Mixing of labeled and unlabeled cells. For synergy/sorting experiments one would make a 10% mixture of labeled cells by mixing 100 µL of labeled cells with 900 µL of unlabeled cells. For experiments requiring the tracking of individual cells at the multicellular stages of development, the percentage of labeled cells must be reduced to 1 or 2%.

9. Generate the following combinations of cells:

> 10% labeled mutant cells + 90% unlabeled wild-type cells
> 10% labeled mutant cells + 90% unlabeled mutant cells (control)
> 10% labeled wild-type cells + 90% unlabeled mutant cells
> 10% labeled wild-type cells + 90% unlabeled wild-type cells (control)

9. The following steps depend on the type of experiments. In order to study cell movement and sorting during the aggregation phase, samples are prepared as described under **Subheading 3.2.** In order to study the slug stage, follow the steps described under **Subheading 3.3.**

3.2. Preparation of Aggregation Streams and Mounds

1. Carefully dissociate cell clumps into single cells by pipetting cell suspension up and down 10 to 20 times using a 1000-µL micropipet. This step is important if one is to achieve a homogenous cell layer on the agar plate.

2. Pipet cell suspension on a small KK2 agar plate at a density of about 5.2×10^5 cells/cm². For example, on a 3.5-cm diameter agar plate, add 500 µL of the cell suspension (=5 × 10⁶ cells). Spread the cell suspension evenly over the entire agar surface by tilting and rotating the agar plate.

3. After 10 min, the cells have attached to the agar and the supernatant can be removed by carefully tilting the Petri dish and blotting the excess fluid from the edge of the dish with tissue paper. Sometimes cells or cell clumps detach from the substrate as well when the plate is tilted; this can happen if the cells have not been dissociated properly.

4. Keeping the plates tilted for another 2 min allows more excess fluid to accumulate at the bottom edge of the plate, which is blotted again with tissue paper.

5. Put the plates down horizontally and open the lid to allow the remaining fluid to evaporate completely. This takes about 10 to 15 min.

6. Close lids and incubate plates inverted at 22°C. Darkfield waves should be visible within 4 to 5 h, followed by aggregation streams and mounds (6 to 9 h, depending on strains used).

3.3. Preparation of Slugs

1. The cells are washed once in H_2O and resuspended in H_2O to a density of 3×10^7 cells/mL. The KK2 buffer is removed, as aggregates tend to culminate immediately on buffered agar plates.

2. Carefully dissociate cell clumps by pipetting cell suspension up and down 10 to 20 times using a 100-µL micropipet.
3. Place four to five drops of 40 µL of the cell suspension on 1% H_2O agar plates as shown in **Fig. 1A**.
4. After 20 min, the cells have settled on the agar. Tilt the plate carefully and blot off the excess fluid from the edge of the Petri dish with tissue paper.
5. Open the lid of the Petri dish and let the remaining fluid evaporate. This should take about 10 to 15 min.
6. Close the lid and place the Petri dishes upright in a light-tight container, e.g., cardboard box, and incubate at 22°C. Slugs form after 16 to 18 h and migrate away from the cell drops (*see* **Fig. 1B**). As light triggers fruiting body formation, it is important that the slugs be kept in the dark as much as possible.

3.4. Microscopy

3.4.1. Sample Preparation to Study Cell Movement

In order to study cell behavior in a large cell population, as during the early stages of development (streams, loose mounds), the agar plates can be directly mounted on the stage of an inverted microscope for imaging through the agar. Low magnification (5× to 20×) is usually enough to obtain sufficient detail of individual cells in a large field of view. However, following aggregation, the optical properties of the multicellular structures deteriorate rapidly as a result of increased light scattering and reflection on the surface (*see* **Fig. 1C**).

This can be reduced by submersing the structures in silicon oil, which has a similar refractive index *(7)*: a small rubber ring (1 mm thick, 5 mm diameter) is placed around the structure, e.g., a slug, and slightly pushed against the agar surface to prevent leaking of the oil onto the surrounding agar. The ring is then

---

Fig. 1. *(opposite page)* Sample preparation. (**A**) To obtain slugs, four to five drops of the cell suspension are placed on a water agar plate as indicated. (**B**) The cells aggregate at this position (indicated by the dashed line) and form slugs that migrate outwards as shown. These slugs are used for further experiments. (**C**) Brightfield and fluorescence image of a slug containing 10% CellTracker™-labeled cells recorded on a dissecting microscope at 10× magnification. Individual labeled cells are clearly visible. Light scattering and reflection on the slug surface—especially at the slug tip, which is usually raised above the substrate—can give the impression of increased fluorescence at the tip. (**D**) Improved optical conditions due to the addition of silicon oil to the same slug. The inset shows a macroscopic view of the rubber ring that has been placed around the slug and filled with oil. Note especially the effect on the slug tip, where individual cells are now visible. The fluorescent cells are uniformly distributed throughout the slug. (**E**) To obtain a side view, an agar block containing, for example, a slug is placed on a coverslip and rotated by 90°. (**F**) Aggregating cells and streams can be sandwiched between agar and coverslip (schematic side view). A scalpel and a

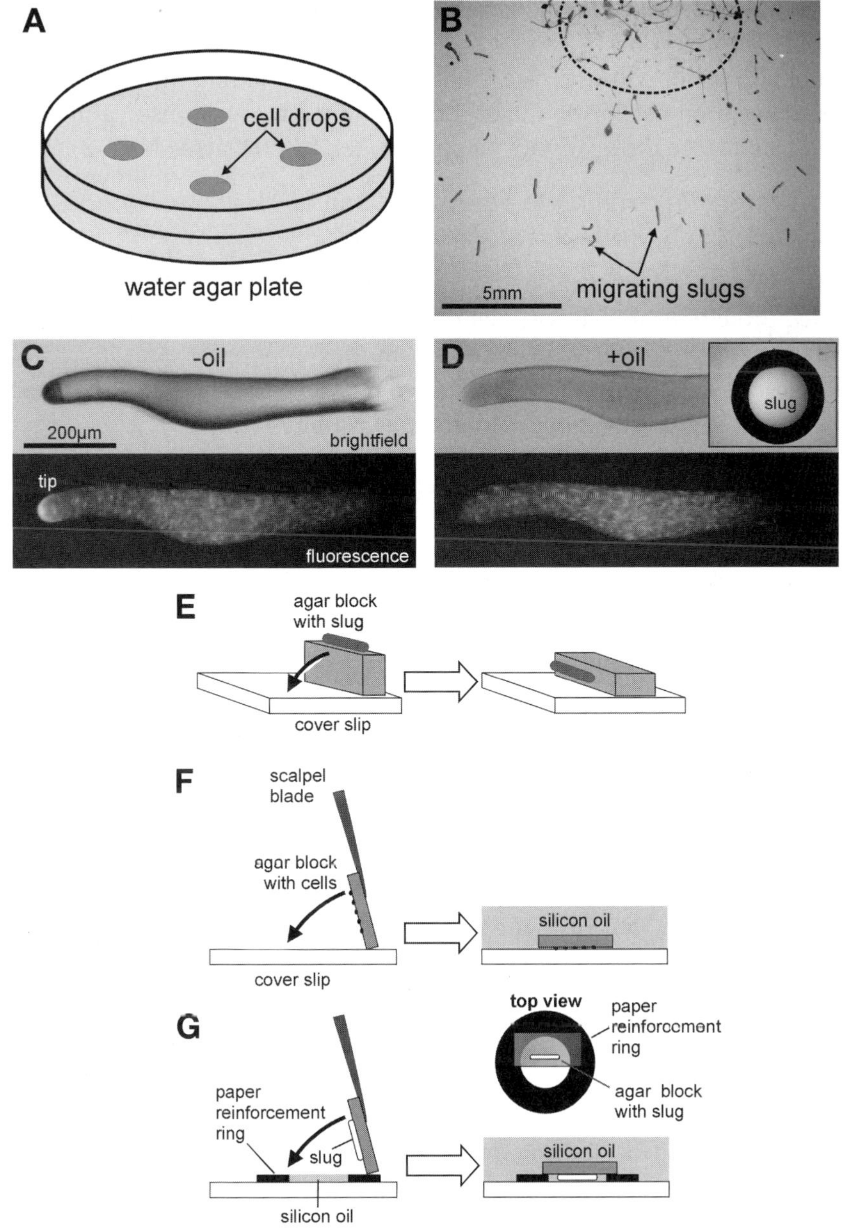

Fig. 1 *(continued)* syringe needle are required for the handling of the agar. Subsequently, the agar is submersed in silicon oil to prevent desiccation. **(G)** Larger structures such as slugs require a spacer to avoid squashing (schematic side view). A paper reinforcement ring is sufficient for this purpose. The agar block should rest on the reinforcement ring as indicated (top view), allowing the slug to move freely.

filled with a few drops of silicon oil until the structure is completely covered (*see* inset **Fig. 1D**).

To obtain a side view, excise the agar piece containing the sample with a scalpel, turn it on its side on the coverslip of a microscope observation chamber, and submerge the agar with silicon oil (*see* **Fig. 1E**) (*see* **Note 2**) *(8)*.

3.4.2. Sample Preparation to Study the Intracellular Distribution of Fluorescent Probes During Development

Because of the short working distance of the high (40× to 100×) NA objectives required to resolve the cellular distribution of fluorescent probes, it is not possible to image the distribution in the multicellular aggregates in sufficient detail through the agar layer on an inverted microscope. The samples are therefore inverted as described in the following section *(4,5)*.

3.4.2.1. Aggregation Stage Cells

1. Cut out a piece of agar (up to 10×10 mm) containing the structure of interest (dark field wave stage cells, aggregation streams, mounds) with a scalpel. Very large slices of agar are difficult to handle and break easily.
2. Lift up the agar piece with the scalpel and transfer it to the coverslip of the microscope observation chamber.
3. Use a syringe needle to push the agar from the blade of the scalpel so that the agar block is standing on the coverslip on its side with the tip of the scalpel still supporting the top of the agar block (*see* **Fig. 1F**).
4. Slowly lower the agar piece onto the coverslip with the scalpel and the syringe needle so that the cells are sandwiched between the agar and the coverslip. Although this may destroy some delicate structures such as aggregation streams, they form again very quickly.
5. Cover the agar with silicon oil to prevent desiccation and mount chamber on microscope (*see* **Fig. 1F**) (*see* **Note 3**).

3.4.2.2. Slugs and Culminates

1. As extensive squashing would disrupt slug migration, a paper reinforcement ring (commercially available self-adhesive reinforcement rings for notebook paper, thickness about 90 to 100 µm) is attached to the center of the coverslip and filled with a drop of silicon oil. The reinforcement ring acts as a spacer between agar and coverslip so that the slug can move freely.
2. A rectangular piece of agar, large enough to rest on the reinforcement ring, is cut out and put on its side on the reinforcement ring (*see* **Fig. 1G**).
3. The agar is carefully lowered so that it lies on the reinforcement ring as shown in **Fig. 1G**. Depending on the diameter of the slug and the distance of the slug from the paper ring, the slugs are either just touching the surface of the coverslip or completely surrounded by the silicon oil.
4. Cover the agar block with silicon oil to prevent desiccation (*see* **Note 3**).

3.4.3. Time-Lapse Recordings

Time series can be recorded on any camera or confocal laser scanning system. However, for synergy experiments, it can be sufficient to record only snapshots of the different cell mixtures (*see* **Fig. 2A**). To view CellTracker Green-labeled cells, a standard fluorescein filter set is required. Depending on the system, it may be possible to use more than one fluorophore/GFP. This can be particularly useful for synergy experiments in which the wild-type cells are labeled with one type of GFP and the mutant cells with another GFP variant or when using cell type specific expression of different GFPs. Useful combinations of GFPs are wild-type GFP and the red-shifted variant (S65T), GFP(S65T) together with the red mRFPmars *(9)*, or cyan fluorescent protein (CFP) and yellow fluorescent protein (YFP). CellTracker Green-labeled cells can also be mixed with cells expressing wild-type GFP or mRFPmars.

It is paramount to reduce exposure to the fluorescence excitation light as much as possible, even during the setup of the experiment while adjusting exposure time, gain, and so on, as the cells are easily damaged by excessive light. Slug tips, for example, are very light-sensitive; excessive excitation light will damage the tip within seconds and cause it to stop moving. For experiments requiring high NA objectives (63×/NA 1.4 or 100×/NA 1.4 lenses) for the study of protein translocation or cytoskeletal components, direct visual inspection of the sample should be avoided altogether, as even a few seconds' exposure to the fluorescence excitation light is detrimental; snapshots taken by the imaging system should be used instead. For recordings, the excitation light intensity should generally be reduced to a minimum. It is better to reduce the excitation light intensity with neutral density filters and to increase the exposure time than to use high intensity light and short exposure times. On laser scanning systems, it can be advantageous to use a higher scanning frequency at a given laser intensity (e.g., 800 Hz instead of 400 Hz) to reduce the pixel dwell time and then to average two scans. Time-lapse sequences can be recorded at room temperature (about 20–25°C); higher temperatures should be avoided, because slug movement, for example, will slow down drastically at 29°C and stop at even higher temperatures.

3.4.3.1. High-Resolution Imaging of Slug Cells Using a Confocal Microscope

1. Prepare slugs on a spacer as described under **Subheading 3.4.2.2.** However, for this preparation, the slug surface should just touch the coverslip, as the optical conditions deteriorate if the slug surface is separated from the coverslip by a layer of silicon oil.
2. Use a high NA objective, e.g., 63×/NA 1.4 or 100×/NA 1.4, and the appropriate optimal pinhole setting.
3. Because the slugs move rapidly, set up a fast scan speed (e.g., 800 Hz). This may also exclude the use of line averaging methods to reduce noise. Increase laser

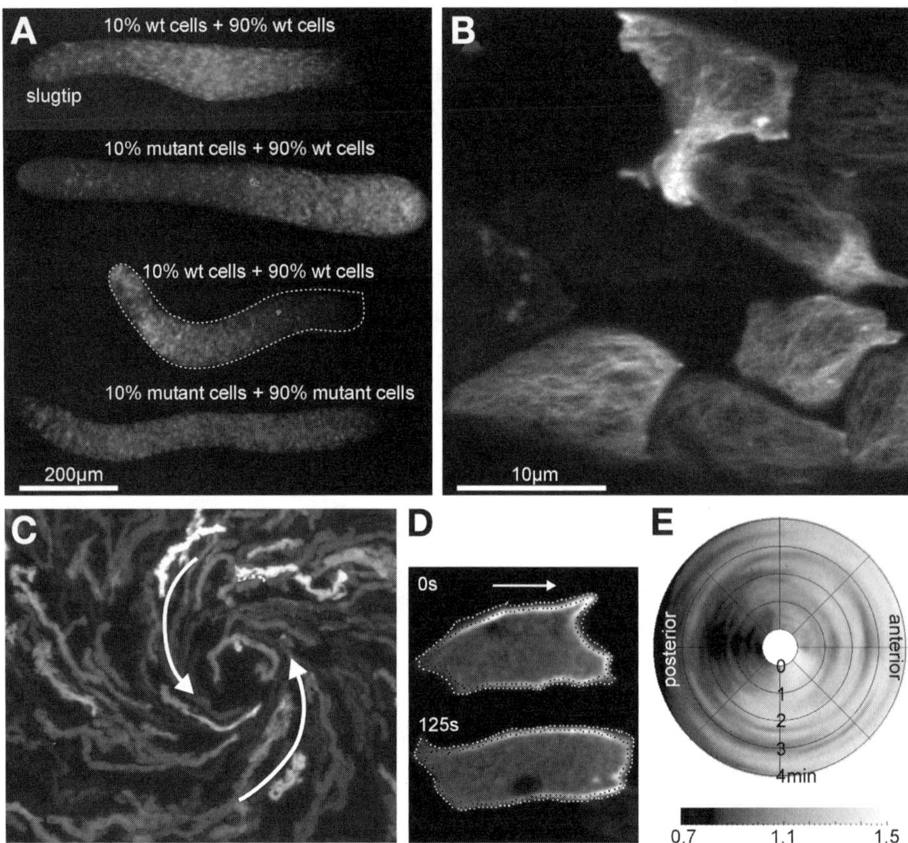

Fig. 2. Data visualization. (**A**) Fluorescence images of slugs from a synergy experiment using Ax2 wild-type cells (wt) and a *paxB* knockout strain (mutant). Ten percent of cells are labeled with CellTracker™. The control experiments (wt/wt and mutant/mutant) show a uniform distribution of fluorescent cells, whereas the mutants are restricted to the slug posterior when mixed with wild-type cells. Ax2 cells sort to the front of the slug in competition with mutant cells (the dashed line indicates the outline of the slug). (**B**) Cells in the posterior part of the slug expressing a F-actin specific probe, the actin binding domain of ABP120 fused to green fluorescent protein (GFP) *(18)*. The filamentous actin cytoskeleton can clearly be resolved. (**C**) Cell traces marking the movement of GFP-expressing cells toward a small aggregation center where movement is coordinated by a propagated spiral wave. The direction of cell movement is indicated by arrows. Maximum projection over 150 images (= 25 min). (**D**) Measuring the spatial distribution of GFP fusion proteins along the plasma membrane. Two snapshots of a slug cell expressing GRP1-PH-GFP, which binds specifically to PtdIns(3,4,5)P_3 in the membrane *(5)*. The direction of movement is indicated by the arrow. The membrane-binding of the probe is highest at the leading edge and decreases gradually toward the cell posterior. Active contours were applied to auto

     intensity to compensate for the fast scanning, as an increase in the photomultiplier gain can cause excessive noise.

4. In order to resolve, for example, individual F-actin or myosin filaments/bundles, the zoom factor must be set very high (e.g., 4× for a 1024 × 1024 pixel image or 5× to 6× for a 512 × 512 pixel image). Slight oversampling can sometimes be useful to improve the image quality.

5. When studying cytoskeletal GFP probes in the cell cortex, start focusing on the cell/coverslip interface of the outer cell layer of the slug and then focus slightly further (in the range of nanometers) into the sample until structures in the cell cortex become visible (*see* **Fig. 2B**). Probably as a result of spherical aberrations, it is not possible to resolve individual actin filaments in cells two or three layers deep inside the slug.

6. Record the time-lapse sequence. The time interval depends on the dynamics of the process being studied, but frequent scanning will rapidly bleach the sample.

3.4.3.2. Visualizing Optical Density Waves During Development

A simple and convenient way to study cell–cell signaling during aggregation and in mounds is to record the so-called optical density or darkfield waves. These waves occur as a result of the coordinated changes in cell shape and cell movement in response to the propagated extracellular cAMP waves *(10–12)*.

1. Use plates with aggregation-stage cells prepared as described under **Subheading 3.2.** Any type of microscope can be used for this purpose. A 10× objective should be sufficient.

2. In order to improve the visibility of the optical density waves, it is important to create an oblique illumination. This can be achieved by moving the annular stop slider of the condensor or the condensor turret into an intermediate position between the brightfield and the phase contrast position. The sample should appear slightly brighter on one side and have a slight shadow on the other side (*see* **Fig. 3A**).

3. Record images every 10 to 20 s.

4. Use image subtraction to improve visibility of wave bands (*see* **Fig. 3B**) (*see* **Subheading 3.5.3.**).

Fig. 2 *(continued)* matically detect the cell outline (white dots) and the plasma membrane (black dots) in the image sequence and to measure the fluorescence intensity along the entire membrane *(16)*. (**E**) Fluorescence intensities along the membrane are visualized using polar plots, in which the intensities are plotted on circles and arranged like tree rings progressing from inside to outside. The plot shows a constant, polarized membrane-binding of the GRP1-PH-GFP probe, with strong binding at the leading edge and reduced association with the cell posterior *(16)*.

3.5. Data Analysis and Visualization

There are many available software packages with which to analyze and visualize time-series data. The freely available ImageJ software (http://rsb.info. nih.gov/ij/), however, can be sufficient for many purposes.

3.5.1. Generating Cell Tracks

For the simple visualization of cell movement patterns in a cell population, it is useful to combine many individual fluorescence images of a sequence into a single image using a maximum projection, in which each pixel contains the maximum gray value over all the images of the entire sequence at that pixel location. This leads to cell traces that mark the movement of the fluorescent cells during that time interval (*see* **Fig. 2C**) *(13)*.

1. Start ImageJ. Import the image series (from the main menu, select: File ♦ Import ♦ Image sequence) as image stack.
2. To generate the cell traces, select Image ♦ Stacks ♦ Z Project…
3. In the ZProjection dialog box, select the range of images to calculate the tracks and the Projection type. For fluorescence images, select the "Max Intensity" (maximum intensity) option and press "OK." A new window containing the cell traces is created.

In order to obtain quantitative data on cell velocity, directionality of movement, and so on, individual cells must be tracked *(1,14)*. A suitable ImageJ plugin for manual cell tracking is available (http://rsb.info.nih.gov/ij/plugins/ manual-tracking.html).

3.5.2. Measuring Fluorescence Intensity

For the analysis of fluorescent probes that change their cellular localization, the changes in fluorescence intensity in different parts of the cell can be quantified by measuring the average gray levels at these locations (membrane, nucleus, cytosol) with small windows (e.g., 5×5 or 10×10 pixel) over time. Using this methodology, we were able to measure the periodic translocation of PH-domain GFP fusion proteins to the leading edge of cells in response to the natural cAMP waves in aggregation streams and mounds and the cAMP-induced nuclear translocation of STATa-GFP in slugs *(4,5,15)*. These measurements can also be done in ImageJ.

1. Import image sequence as described above (*see* **Subheading 3.5.1.**).
2. Select Analyze ♦ Set Measurements, tick the "Mean Gray Value" option in the dialog box, and press "OK." This enables the measurement of the mean gray intensity in the selected area.
3. Create a small measuring window using the "Rectangular selections" tool.
4. Move the window to where the fluorescence is going to be measured, e.g., the plasma membrane.

5. Select Analyze ♦ Measure to make the measurement. The result is immediately indicated in the "Results" window.
6. Select the next image of the sequence and move the measuring window accordingly if required.
7. Repeat **steps 5** and **6** until all the measurements have been made. Save the data from the "Results" window as text file.

The analysis of GFP fusion proteins that shuttle between cytosol and plasma membrane can be significantly extended to allow the measurements of fluorescence intensities over the entire length of the membrane and the entire cytosol, allowing detailed analysis of both spatial and temporal changes in protein localization (*see* **Fig. 2D** and **E**). ImageJ plug-ins are available for this analysis as well *(16)*.

3.5.3. Image Subtraction

In order to enhance the visibility of the optical density waves, which appear as very faint wave bands in brightfield movies and are usually not visible in still images, image subtraction can be used *(11)*. It enhances the differences between two images, e.g., the rapidly propagated optical density waves that appear at different positions in subsequent images. The following steps describe the image subtraction procedure using ImageJ.

1. Open two brightfield images from the sequence, e.g., image 1 and image 2.
2. Select Process ♦ Image Calculator… from the menu.
3. In the Image Calculator dialog box, select image 1 in the "Image 1" pull-down menu and image 2 in the "Image 2" menu. For the image subtraction, choose "Subtract" from the "Operation" menu. Image 2 will be subtracted from image 1. Tick the "Create new window" box so that the subtracted image is displayed in a new window and tick the "32-bit Result" box. This is necessary, as otherwise all negative values would be set to 0 (=black). Press "OK" and a new window with the subtracted image appears.
4. The structures that have moved between two images, e.g., the wave fronts, appear as bright and dark bands. Structures that have not moved have an intermediate gray value (*see* **Fig. 3B**). If the wave pattern is not clear enough, use images taken at longer time intervals (e.g., image 1 and image 5), as the wave fronts might not have traveled far enough between subsequent images.

For a more detailed description on the further analysis of wave propagation during development, *see* **refs. *10,11*,** and ***17*.

4. Notes

1. Growth conditions and cell density (e.g., stationary cells vs exponentially growing cells) affect the synergy experiments, as they can already cause cell sorting. Growth conditions must be carefully controlled.

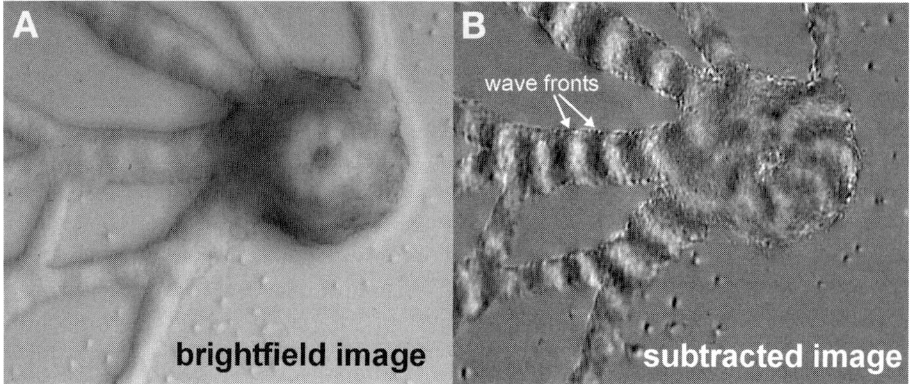

Fig. 3. Use of image subtraction to enhance optical density waves. (**A**) Brightfield image of a mound with aggregation streams under oblique illumination. (**B**) Subtracted image of the same mound shows the propagated wave fronts as bright and dark bands. For the subtraction, two images taken 25 s apart were used, followed by contrast enhancement. The arms of the multi-arm spiral wave rotating around the center of the mound are also visible.

2. The agar of the KK2 plates (1 mL per 3.5 cm dish) is too thin for this procedure; the agar volume should, therefore, be increased to 2.5 mL to allow easy excision and handling of the agar block.
3. The silicon oil can be collected at the end of the experiment and reused.

References

1. Dormann, D. and Weijer, C. J. (2001) Propagating chemoattractant waves coordinate periodic cell movement in *Dictyostelium* slugs. *Development* **128,** 4535–4543.
2. Rietdorf, J., Siegert, F., and Weijer, C. J. (1996) Analysis of optical density wave propagation and cell movement during mound formation in *Dictyostelium discoideum. Dev. Biol.* **177,** 427–438.
3. Dormann, D., Vasiev, B., and Weijer, C. J. (2000) The control of chemotactic cell movement during *Dictyostelium* morphogenesis. *Phil. Trans. R. Soc. Lond. B* **355,** 983–991.
4. Dormann, D., Weijer, G., Parent, C. A., Devreotes, P. N., and Weijer, C. J. (2002) Visualizing PI3 kinase-mediated cell-cell signaling during *Dictyostelium* development. *Curr. Biol.* **12,** 1178–1188.
5. Dormann, D., Weijer, G., Dowler, S., and Weijer, C. J. (2004) In vivo analysis of 3-phosphoinositide dynamics during *Dictyostelium* phagocytosis and chemotaxis. *J. Cell Sci.* **117,** 6497–6509.

6. Sucgang, R., Weijer, C. J., Siegert, F., Franke, J., and Kessin, R. H. (1997) Null mutations of the *Dictyostelium* cyclic nucleotide phosphodiesterase gene block chemotactic cell movement in developing aggregates. *Dev. Biol.* **192,** 181–192.

7. Siegert, F. and Weijer, C. J. (1992) Three-dimensional scroll waves organize *Dictyostelium* slugs. *Proc. Natl. Acad. Sci. USA* **89,** 6433–6437.

8. Dormann, D., Siegert, F., and Weijer, C. J. (1996) Analysis of cell movement during the culmination phase of *Dictyostelium* development. *Development* **122,** 761–769.

9. Fischer, M., Haase, I., Simmeth, E., Gerisch, G., and Muller-Taubenberger, A. (2004) A brilliant monomeric red fluorescent protein to visualize cytoskeleton dynamics in *Dictyostelium. FEBS Lett.* **577,** 227–232.

10. Siegert, F. and Weijer, C. (1989) Digital image processing of optical density wave propagation in *Dictyostelium discoideum* and analysis of the effects of caffeine and ammonia. *J. Cell Sci.* **93,** 325–335.

11. Siegert, F. and Weijer, C. J. (1995) Spiral and concentric waves organize multicellular *Dictyostelium* mounds. *Curr. Biol.* **5,** 937–943.

12. Dormann, D., Kim, J. Y., Devreotes, P. N., and Weijer, C. J. (2001) cAMP receptor affinity controls wave dynamics, geometry and morphogenesis in *Dictyostelium. J. Cell Sci.* **114,** 2513–2523.

13. Yang, X., Dormann, D., Munsterberg, A. E., and Weijer, C. J. (2002) Cell movement patterns during gastrulation in the chick are controlled by positive and negative chemotaxis mediated by FGF4 and FGF8. *Dev. Cell* **3,** 425–437.

14. Rietdorf, J., Siegert, F., Dharmawardhane, S., Firtel, R. A., and Weijer, C. J. (1997) Analysis of cell movement and signalling during ring formation in an activated Gα1 mutant of *Dictyostelium discoideum* that is defective in prestalk zone formation. *Dev. Biol.* **181,** 79–90.

15. Dormann, D., Abe, T., Weijer, C. J., and Williams, J. (2001) Inducible nuclear translocation of a STAT protein in *Dictyostelium* prespore cells: implications for morphogenesis and cell-type regulation. *Development* **128,** 1081–1088.

16. Dormann, D., Libotte, T., Weijer, C. J., and Bretschneider, T. (2002) Simultaneous quantification of cell motility and protein- membrane-association using active contours. *Cell Motil. Cytoskel.* **52,** 221–230.

17. Sawai, S., Thomason, P. A., and Cox, E. C. (2005) An autoregulatory circuit for long-range self-organization in *Dictyostelium* cell populations. *Nature* **433,** 323–326.

18. Pang, K. M., Lee, E., and Knecht, D. A. (1998) Use of a fusion protein between GFP and an actin-binding domain to visualize transient filamentous-actin structures. *Curr. Biol.* **8,** 405–408.

19

Under-Agarose Chemotaxis of *Dictyostelium discoideum*

David Woznica and David A. Knecht

Summary

In the vegetative state, *Dictyostelium* amoebae are chemotactic toward pterins released by bacteria, whereas during multicellular development, they become chemotactic to endogenously produced cAMP. A variety of assays have been used to visualize and quantify chemotactic movement. Under-agarose chemotaxis provides a simple and flexible assay that permits high-resolution imaging and quantification of the motility behavior of individual cells and populations by both transmitted light and fluorescence microscopy. The assay requires cells to deform a solid but flexible matrix; therefore, it also provides a way to measure defects in the ability of mutant cells to move in these restrictive conditions.

Key Words: Chemotaxis; *Dictyostelium*; cAMP; folate.

1. Introduction

The ability of *Dictyostelium* to move up a chemoattractant gradient is important for the organism's successful existence in both the vegetative and aggregative states, enabling *Dictyostelium* to detect either prey or aggregation centers. In the vegetative state, *Dictyostelium* possesses chemotactic sensitivity for bacterially released pteridines such as folic acid *(1)*. After a 7- to 9-h period of starvation, this sensitivity is lost and replaced by sensitivity for cAMP, which is released by aggregating *Dictyostelium* and relayed front to rear by each individual cell *(2)*. The nature of the attractant signaling in the developmental state has some interesting differences from that of the vegetative state. In the vegetative state, *Dictyostelium* is likely to perceive a steady level of attractant that increases as cells approach the source, but might decrease or increase over time depending on what the bacteria are doing. In contrast, developmental signaling in *Dictyostelium* has a wavelike characteristic, with waves occurring less than every 13 min at the start of aggregation, and

From: *Methods in Molecular Biology, vol. 346:* Dictyostelium discoideum *Protocols*
Edited by: L. Eichinger and F. Rivero © Humana Press Inc., Totowa, NJ

"

increasing in frequency to every 4 min by the completion of aggregation *(3)*. The two developmental states of *Dictyostelium* are not only characterized by distinct chemosensitivities, but also by differing physical morphologies. *Dictyostelium* in the vegetative state have a rounded shape and protrude from almost any part of the periphery, whereas developing *Dictyostelium* are highly elongated with protrusion taking place almost exclusively at the front.

Dictyostelium's developmental state notwithstanding, the capacity to move toward a chemoattractant is governed by the cell's ability to detect a chemoattractant, determine the spatial distribution of attractant in the extracellular environment, and produce the forces necessary for motion. The chemotactic behavior of *Dictyostelium* is of interest because of its many parallels with mammalian cells such as leukocytes. The detection of chemoattractant is made possible by G protein-coupled cell-surface receptors present in *Dictyostelium*'s membrane *(4)*. In the vegetative state, *Dictyostelium* cells respond chemotactically to many pteridines, such as folic acid, pterin, aminopterin, and methotrexate *(2)*. Although there is evidence for the existence of at least two different pterin receptors, none of the genes or proteins have been identified *(5)*. In the starved state, *Dictyostelium* responds to cAMP through four different serpentine receptors, cAR1-cAR4, each coupled to the same heterotrimeric G protein complex *(6)*. Although a folic acid receptor has not been isolated, it has been shown to be ubiquitously distributed on the cell membrane *(7)*, and appears to couple to the same G protein response system as the cAR receptors *(8)*. Both folic acid and cAMP can be degraded by their respective deaminases or phosphodiesterases and released or endocytosed along with the receptor.

Because cAMP receptors are uniformly distributed along the cellular surface, *Dictyostelium* must have a method of determining the distribution of attractant in order for appropriate cell polarization and pseudopod protrusion to occur. In the vegetative state, *Dictyostelium* will perceive both a positive spatial gradient and increasing temporal gradient as it approaches a bacterial food source. If oriented toward the source, the cell will have a higher level of attractant at its front than at its rear, and the amount detected in both locations will increase with time as *Dictyostelium* approaches the source. During the developmental state, each wave of cAMP has a time period during which the cell perceives an increasing temporal gradient of cAMP followed by a decreasing gradient of cAMP. The increasing temporal gradient at the front of the wave brings about an increase in motility, whereas the decreasing gradient is either ineffectual or reduces motility *(9)*. Furthermore, the cAMP concentration at the peak of the developmental wave is inhibitory to motion. However, it has been shown that cells can respond to spatial as well as temporal gradients of cAMP *(10)*.

The detection of chemoattractant results in downstream signaling and cytoskeletal rearrangement. The establishment of cell polarity, i.e., the formation of an asymmetric cell with anterior and posterior portions, is a direct outcome of these rearrangements. The anterior portion of the cell is characterized by molecules associated with cytoskeletal assembly, such as actin and actin binding proteins. Conversely, the posterior of the cell is associated with molecules involved in uropodial retraction, such as actin and myosin II.

A number of models have been developed to explain the internal dynamics that result in cell directionality and polarization in response to chemoattractant, but none are able to account for cell behavior under all experimental conditions. A recent and arguably the most encompassing model is the local excitation–global inhibition model, which states that cells respond to alterations in receptor occupancy *(11)*.

1.1. Chemotactic Assays

A variety of different assays have been developed to study the phenomenon of chemotaxis, many having their roots in leukocyte research. The bridge visual assay, commonly known as the Zigmond chamber, allows the observation of individual cells within a relatively stable spatial gradient of chemoattractant. Although it allows live cell observation, the assay is unable to hold a gradient for an extended period of time, can be difficult to set up, and is unable to address questions concerning force production. Several modifications to the chamber assay have been developed to increase the ease of assembly and the duration of the assay *(12,13)*. The Millipore, or Boyden, filter assay permits a quantitative assessment of a cell population's migration as a result of a gradient. However, the filter must be fixed prior to visualization, thus only a population measurement is obtained because one cannot observe the cells directly. Furthermore, the gradient across the filter is steep and short in duration, and the ability to discern chemokinesis from chemotaxis in the filter assay is very much conjectural *(14)*.

The use of agarose as a medium for chemotaxis results from its ease of preparation, compatibility with living cells, and ability to stabilize the chemoattractant gradient over distance and time *(15)*. Many *Dictyostelium* assays have been developed using an on-agarose approach, in which a droplet of cells is positioned on an agarose surface and a droplet of chemoattractant is placed a small distance away *(7,16)*. Alternatively, the agarose is made with a uniform concentration of chemoattractant and the gradient is formed by the cells themselves secreting the degrading enzyme *(17,18)*. This enzyme destroys much of the nearby chemoattractant, while largely unaffecting the levels of distant chemoattractant, thereby creating a gradient that increases as the cells move away from the droplet area. Although these assays work well to show population behavior and are simple to do, the quantitative motility data are limited because it is difficult to analyze individual cell movement or do high-resolution imaging.

When using an inverted microscope, images must be collected by focusing through the dish and the agarose. When using an upright microscope, the dish must be uncovered, and the cells rapidly become dehydrated.

Under-agarose chemotaxis was first developed for leukocytes using a circular well assay *(19)* and later modified to a linear well format *(20)*. We have modified this assay to make it suitable for monitoring *Dictyostelium* chemotaxis *(21)*. Most of our work has focused on folate chemotaxis of vegetative cells; however, under-agarose cAMP chemotaxis is also possible. The assay is simple to set up and carry out, and the cellular response can be visualized on anything from an inexpensive inverted tissue-culture microscope to a multi-wavelength confocal microscope. Although routine assays can be carried out in plastic Petri dishes, fluorescence imaging is best carried out in special dishes that have a glass coverslip insert glued into the bottom of the dish. Both phase-contrast and differential interference contrast (DIC) imaging are aided by the flattening of the cells that occurs as a result of the pressure from the agarose. The higher the percentage of the agarose, the greater the flattening that occurs. This process is related to the agarose overlay flattening technique developed by Fukui to reduce the three-dimensionality of cells for widefield fluorescence microscopy of fixed *Dictyostelium* cells *(22)*. Unlike Fukui's technique, the agarose is not dehydrated to artificially flatten the cells; rather, the cells flatten themselves as they move out of the trough and create space underneath the agarose. In the environment of the under-agarose chemotaxis assay, the cell is in contact with two surfaces, top and bottom, and this results in some redistribution of cytoskeletal components *(23)*, but it may also lead to a situation more akin to the natural soil environment.

The under-agarose folate chemotaxis assay has a number of advantages for analyzing cell behavior. First, even if one is not studying motility, it is often advantageous to examine protein localization in polarized cells rather than cells that are randomly moving (or not moving) and continuously changing direction. The establishment of cell polarity may reveal aspects of protein localization that are not evident in randomized cells. Second, because the cells must push their way under the agarose to move, defects in cortical structure or force production can be assessed. By using different concentrations of agarose, the relative ability of different mutants to deform the agarose and move can be measured *(24)*. Third, because the cells are flattened, a larger percentage of the cell volume can be visualized at one time and more processes occur in the plane of focus. The reduced three-dimensionality makes phase-contrast images sharper and localization of proteins to structures clearer. Fourth, cell movement can be analyzed on either a population or individual basis. The low-magnification image of a moving population of cells provides a measure of average cell speed, whereas higher-magnification movies can be used to quantify individual cell movement behavior. Last, the localization of fluorescent fusion proteins in cells can be easily assessed in cells moving in a polarized fashion up a folate gradient.

2. Materials

1. SM medium: 10g Difco Bacto-Peptone, 10 g glucose, 1 g yeast extract, 1.9 g KH_2PO_4, 0.6 g K_2HPO_4, 0.43 g $MgSO_4$. Dissolve in 1 L H_2O and adjust pH to 6.5. Autoclave and store at 4°C.
2. 50 mM folic acid stock: dissolve 0.22 g folic acid in 10mL 0.1 N NaOH. This is stable for 1 yr if stored in the dark at 4°C. Working solutions are made fresh weekly by diluting the stock into appropriate media or buffer.
3. 1 mM folic acid working solution: Add 100 µL of 50 mM folic acid stock solution to 5 mL of SM medium. Store at 4°C for about 1 wk.
4. PDF buffer: 2.51 g K_2HPO4, 1.80 g KH_2PO_4, 1.49 g KC, 0.147 g $CaCl_2$, 0.301 g $MgSO_4$, in 1 L H_2O. Adjust to pH 6.4 and filter-sterilize.
5. MCPB buffer: 1.42 g Na_2HPO_4, 1.36 g KH_2PO_4, 0.19 g $MgCl_2$, 0.03 g $CaCl_2$, 0.5 g dihydrostreptomycin sulfate per 1 L H_2O. Adjust to pH 6.5 and filter sterilize.
6. cAMP: Make 10 mM stock in distilled water or development buffer, store at –20°C.
7. Agarose, e.g., Seakem GTG (FMC Corp.; *see* **Note 1**)
8. Inverted microscope and digital image acquisition system or laser scanning confocal microscope.

3. Methods

3.1. Folate Chemotaxis

3.1.1. Preparation of the Agarose Gel

1. Combine 300 mg of agarose and 30 mL of SM medium in a **loosely** covered 125-mL flask (*see* **Notes 1–3**). A glass screw-capped flask works well for visibility, but care must be taken to not tighten the cap and cause an explosion during boiling.
2. Heat the mixture in a microwave oven and stop when it first begins to boil (*see* **Note 4**). Hold the flask with a heat-protective glove or pad and swirl gently to mix. Return the flask to the microwave and repeat brief pulses, swirling in between until all the agarose is melted. The small crystals of agarose can be hard to see if not in a glass flask.
3. For low-resolution imaging, add 20 mL of agarose solution to a standard 100-mm Petri dish or 8 mL to a 60-mm dish (*see* **Note 5**). For high-resolution imaging, add sufficient solution to a glass bottom dish (Willco Wells, Inc., or MatTek Inc.) to produce an agarose gel 4–5 mm deep.
4. Allow agarose dishes to harden on a level surface. Plates can be poured the day before or just before the experiment. We have not tried storing plates for longer times.

3.1.2. Formation of Cell and Chemoattractant Troughs

1. Repeatedly move a thin spatula between the edge of the agarose and the walls of the dish until agarose visibly turns. This breaks tight contacts between the agarose and the bottom of the dish.

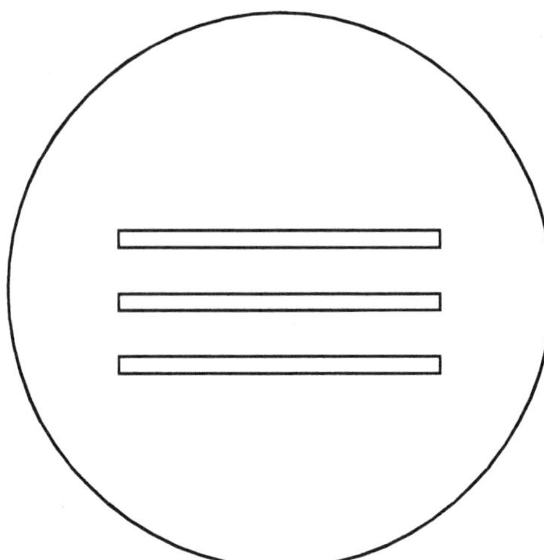

Fig. 1. Trough-cutting template. Lay the dish on top of a copy of this template and use it as a guide to cut the troughs. The troughs should be 2 mm wide and 5 mm apart.

2. Place the Petri dish on top of a template with the trough pattern (*see* **Fig. 1**). It is important to make clean vertical cuts with a razor blade and not to use a sawing motion, which can score the plastic surface and hinder movement. Using a standard single-edged, 39-mm razor blade, insert the blade several times into the agarose in an area of the dish near the edge that will not be used. This wets the blade and makes it easier to cut. Place the blade on the agarose using the template as a guide and then press gently, but firmly, straight down so that the blade cuts through to the bottom. Cut three sets of troughs 2 mm wide, 5 mm apart, and 39 mm long (*see* **Note 6**).

3. After cutting the lengths of the troughs, use a pointed spatula or chisel-point knife to cut the ends of each trough. Pry out the agarose slab by lifting one end with the spatula and then grabbing it with your fingers. Try not to disturb the surrounding agarose, especially the strip that will form the bridge between the troughs.

3.1.3 Cell Preparation and Trough Filling

1. The order of addition of chemoattractant and cells is not critical because chemotaxis continues for at least 9 h. We routinely add the chemoattractant first and then prepare the cells, but both can be done at the same time. Depending on the thickness of the agarose and the width of the trough, it will hold about 100–200 µL of solution. Using a pipetter and standard tip, add 1 m*M* folic acid (*see* **Notes 7** and **8**) to the center trough until it is nearly full, but do not overfill and spill over the top. Then begin preparing the cells.

2. Axenically grown cells should be in log phase, less than 1×10^6 cells/mL, for optimal chemotaxis. Harvest cells from the plate or flask, take a small sample to titer on a hemocytometer, and centrifuge the rest at $200g$ for 5 min. Resuspend the cells in a volume of SM that brings the titer to 5×10^6–1×10^7 cells/mL. If bacterially grown cells are to be used, three cycles of centrifugation should be performed in order to wash the cells free of most bacteria.

3. Add 100–200 μL of the cell suspension to each cell trough, vortexing the cells before pipetting each aliquot (*see* **Notes 8–10**). Add cells with the same care as for the chemoattractant. Cells may take up to 30 min to attach, so transport the dish carefully or not at all during this time. Cells generally begin to move out underneath the agarose within an hour. They will continue to move all the way to the chemoattractant trough. Cells continue to move out of the trough throughout the assay, so the leading cells are the first cells to exit, not the fastest cells. There is a zone near the trough edge where the cells appear to not be as flattened, but once they are beyond about 100 μm, they remain comparably flattened for the duration of the experiment. For long-term experiments, keep watch on the troughs and, if the level of fluid begins to drop as a result of evaporation, SM can be added so that they do not dry out.

4. Place the dish on the stage of the microscope for time-lapse imaging or, alternatively, place the plate in a humidified incubator and take images periodically using the trough edge as a reference point (*see* **Note 11**). As soon as the cells have attached, check the troughs with an inverted microscope. Make sure that no cells have leaked out from the trough—this will happen if the agarose bridge is disturbed during cell loading. Once cells begin to move, check along the edge of the cell trough to ensure uniformity of movement. Sometimes, many cells exit from a small region, with the majority of the cells lagging behind. These regions are ignored in the analysis.

3.2. Chemotaxis to cAMP

Chemotaxis using cAMP as a chemoattractant and aggregation-stage cells requires an adaptation of the assay. This assay is not as well optimized as the folate assay, but has been used successfully.

3.2.1. Preparation of the Agarose Gel

1. The agarose matrix must be made using development buffer (i.e., PDF, MCPB) instead of SM medium.

2. Agarose concentrations under 1.5% are difficult to work with. Buffered agarose binds very tightly to dishes, and these contacts must be broken. It may be more difficult to rotate buffered agarose than to roll it up and lay it back down.

3.2.2. Cell Preparation and Trough Filling

1. Harvest vegetative cells 8 h prior to setting up the assay. Wash twice in development buffer and resuspend cells to a titer of 5×10^6 to 1×10^7 cells/mL. Shake the

Fig. 2. An image of cells moving up the folate gradient 2.5 h after starting the assay.

cells in a flask at 150 rpm on a platform shaker for 6–8 h to allow expression of the cAMP receptor.

2. Add about 200 μL of cells to the outer troughs and 200 μL of 5 m*M* cAMP to the center trough.

3.3. Imaging and Analysis

3.3.1. Population Analysis

1. For population analysis, images are taken every hour with a 10× (or lower) objective. The edge of the trough is included in the image as a reference point (*see* **Fig. 2** and **Note 11**).
2. Once the leading front of cells is beyond the field of view, a series of overlapping images starting at the trough edge are captured at each time point. These images can be combined into a single larger view to measure the distance of the leading cells from the trough.
3. To merge the images, open the individual images in an image processing program such as Photoshop (Adobe Inc.). Change the canvas size of the first image to increase the length to a size that will accommodate the merged set of images. Increase the transparency of the image so that the next image can be seen through it and then copy and paste the second image into the enlarged canvas. Drag the pasted image and use the overlapping cells as a guide to align the two images. Repeat this process until all images are merged (*see* **Fig. 3**).
4. Calibrate the microscope and camera system by capturing an image of a stage micrometer. Open the image in an image analysis program such as the freeware

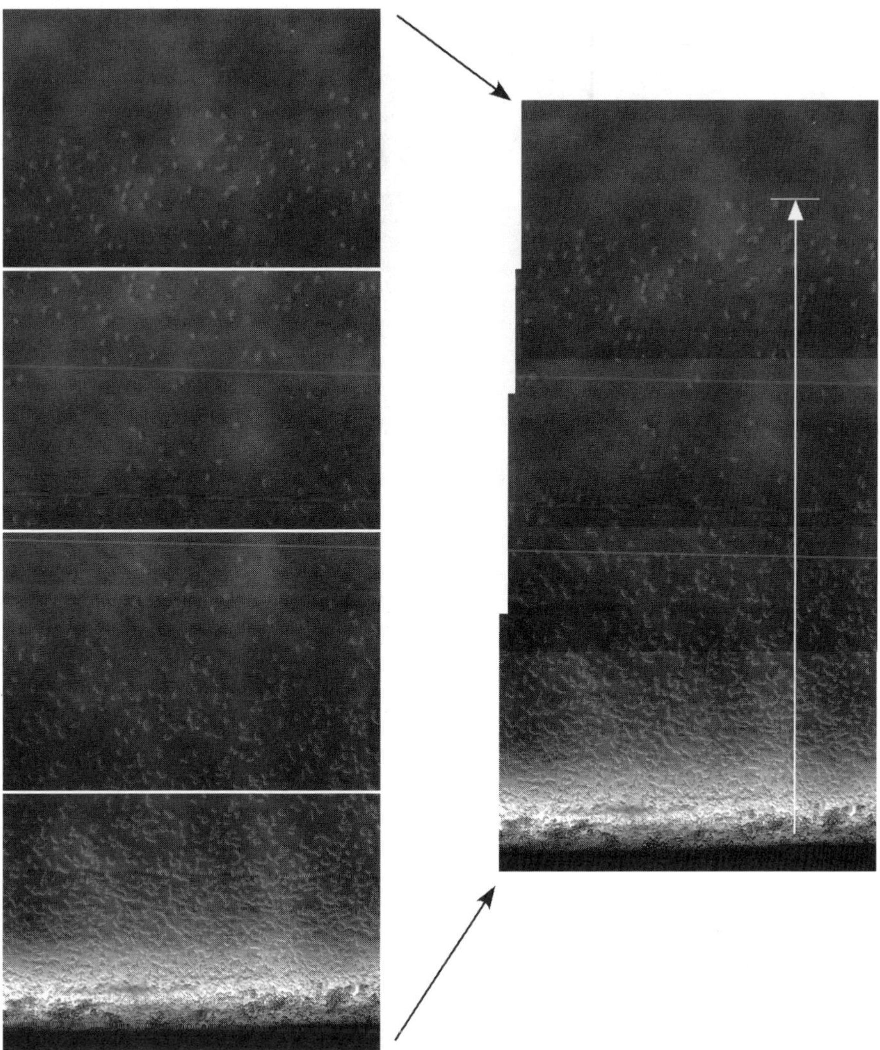

Fig. 3. Four hours after starting the assay, the cells have moved beyond one field of view with a 10× objective. Overlapping images were merged into a single image to allow measurement of the distance cells had moved from the trough (line with arrow).

open source program, ImageJ (http://rsb.info.nih.gov/ij/). Draw a line of known distance on the image of the micrometer and then use the Analyze/Set Scale command to enter the calibration so that the program converts pixel distances to microns.

5. Open the images and, for each, measure the distance from the trough edge to the leading cells. This can be done either as the leading cell in each field of view or

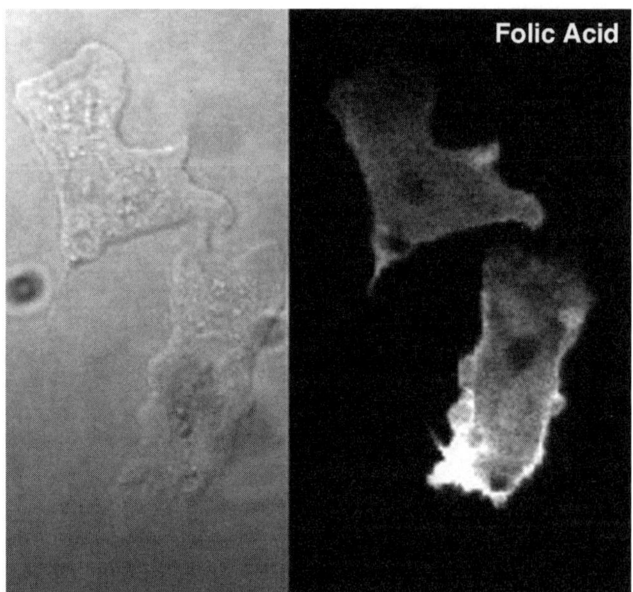

Fig. 4. Cells expressing a GFP-ABD120 fusion protein imaged simultaneously with differential interference contrast and fluorescence on a Leica SP2 confocal microscope during under agarose chemotaxis.

the average of the leading 5–10 cells in each field, depending on the density of moving cells. Divide these distances by total time to get an average speed of the population.

3.5.2. Individual Cell Analysis

1. For individual cell speed and directionality measurements, capture images every 2–5 s for a total of approx 30 min using a 10–20× phase -contrast objective (*see* **Note 12**).
2. For high-resolution imaging in glass-bottom dishes, we routinely use a confocal microscope with a 100× oil-immersion objective to capture both fluorescence and DIC images simultaneously (*see* **Fig. 4**). A 3-inch by 2-inch glass slide is placed over the top of the dish for DIC imaging (plastic depolarizes the light) and to prevent the agarose from drying out.
3. Image sequences (movies) can also be quantified using ImageJ. A number of plug-ins are available to make the process easier, including Multitracker and Manual Tracking (*see* **Fig. 5**). Instructions for using the ImageJ plugins are available online in the same location as the plug-in. Briefly, a number of individual cells are marked in each frame of a movie. The program then calculates the distance the object moves between frames. Knowing the frame rate of data collection allows one to calculate the speed of movement or to plot the speed over time.

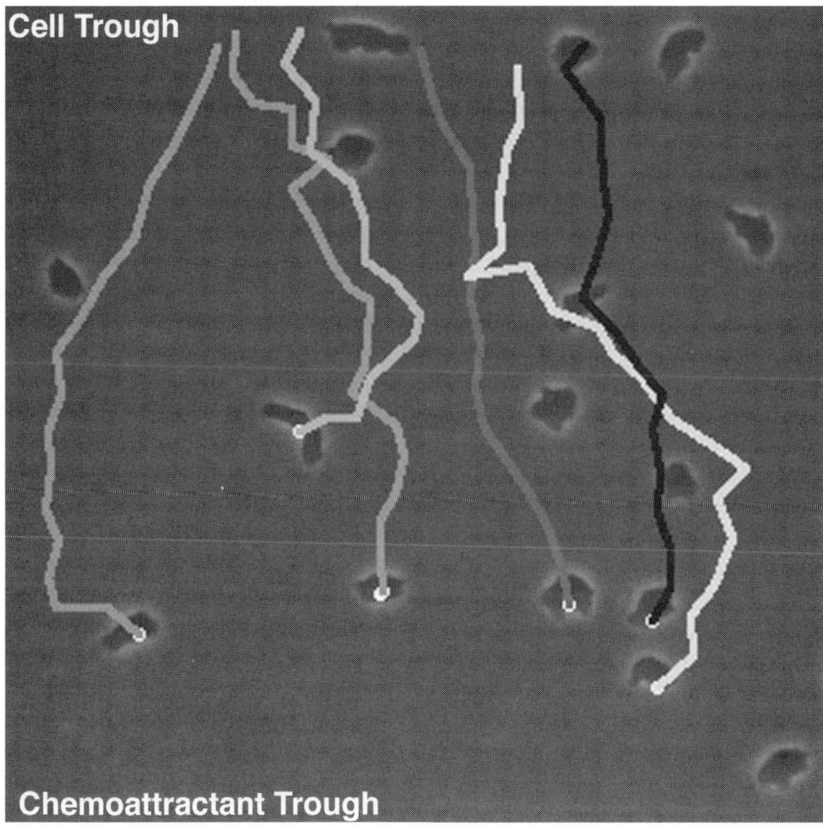

Fig. 5. The paths of cells as determined by the ImageJ Manual Tracking plug-in.

More sophisticated analysis of cell motility behavior can be accomplished using commercial programs such as Dynamic Image Analysis System (DIAS; Solltech Inc.).

4. Notes

1. We have used many different types of agarose for the assay. Agarose intended for gel DNA electrophoresis seems to be generally suitable. Different types of agarose differ somewhat in their stiffness, so the minimum workable concentration for cutting wells and the maximum concentration that inhibits cell movement should be titrated for a given batch or type of agarose. FMC Seakem GTG is particularly easy to work with at concentrations below 1%.
2. There are a variety of media in which the assay could be carried out. HL5 growth medium does not produce robust chemotaxis, presumably because it contains about 0.2 μM folate *(25)*. We routinely use SM medium because it gave good results in early testing. Nonnutrient development buffers did not work very well, presumably because the cells are starving and so progressing toward development

as the assay proceeds. It is possible that a buffer containing amino acids to suppress development, defined FM media lacking folate *(25)*, or a reduced nutrient buffer would work as well or better, but these conditions have not been tested.

3. Varying the percent composition of the agarose drastically changes the environment that the cells must traverse in order for successful chemotaxis to occur. We typically use agarose gels ranging in concentration from 0.5% to 3%. The lower end is permissive and hinders cell migration the least. At such a level, the earliest and most plentiful migration is obtained, although it is common to see premature leakage of cells under both walls of the trough unless care is taken during cell addition. If not excessive, this artifact can be disregarded as long as it is noted by looking at the troughs under the microscope immediately after the cells have settled. Another caveat of using low agarose concentrations is that the physical manipulation of the agarose gel will be difficult. As the agarose concentration increases, the ability of cells to exit the trough is reduced and their subsequent movement speed is hindered. Most cells will not move under agarose at concentrations higher than 3%. An advantage of testing chemotaxis at a range of agarose concentrations between 0.5% and 3% is the ability to discern mutated abilities of force production in comparison to wild-type cells. Typical standard concentrations are 1–1.5%. Under these conditions, the gel is easy to handle and the cells move rapidly, with only a modest flattening.

4.   The agarose can be melted in a steam oven or briefly in an autoclave as well.

5.   Usable volumes for a 100-mm Petri dish range from 12 to 25 mL and from 5 to 10 mL for a 60-mm Petri dish. Larger volumes generally do not interfere with the assay, and are more easily handled. The thickness of the agarose does not appear to affect cell movement, as the deformation of the agarose enabling the cells to move underneath is a local deformation. Therefore, aim for a 4- to 5-mm deep agarose layer regardless of the dish size.

6.   To cut shorter wells in smaller dishes, carefully break off part of a razor blade with a pair of pliers. The same trough-spacing pattern is used as for larger dishes.

7.   Cells respond to a range of folic acid concentrations. We have traditionally used 0.1 m*M* folic acid, but have recently used as high as 5 m*M* folic acid.

8.   The precise volume of chemoattractant is not critical and cannot be precisely stated because the width and volume of the well will depend on the exact width of the cuts. Therefore, we usually add chemoattractant to nearly fill the well without spilling over. The cells are added to target about 5×10^5 cells per trough. The goal is to have a near monolayer of cells in the trough, so that once the cells have settled, more cells may be added if the density appears too low.

9.   We typically add chemoattractant to the center trough and two different cell lines to the outer troughs to compare mutant and wild-type chemotaxis in the same plate. Alternatively, chemoattractant can be added to the outside troughs and cells to the center trough to get a replicate view of the same cells on opposite sides.

10. For visualizing a good cell front, cells should be at 1×10^7/mL. For a less cluttered analysis of individual cells, 1×10^6/mL to 5×10^6/mL is appropriate.

11. When imaging near the edge of the trough, there is an optical artifact that sometimes creates light and dark zones that make it hard to see the cells in both the

Fig. 6. Imaging near the edge of the trough sometimes generates these light/dark zones. A coverslip overlay will usually eliminate this problem (*see* **Note 11**).

trough and the agarose at the same time (*see* **Fig. 6**). To avoid this problem, move the field of view so that the trough edge is just out of view. If you need to image the trough and agarose simultaneously to visualize the exit of cells from the trough, a coverslip overlay will eliminate this artifact (thanks to Robert Insall for suggesting the solution). To do this, remove most of the supernatant medium after the cells have attached and place a glass coverslip (22 mm × 22 mm) over the trough, laying it down gently and not allowing it to reach the chemoattractant trough. Tilt the plate and slowly replace the liquid in the trough, allowing it to fill underneath the coverslip without air bubbles. It is wise to also remove some of the chemoattractant from its trough to avoid spilling. It is difficult to move the coverslip after placement, because the surface tension will move the agarose bridge and the cells.

12. For individual cell behavior, it is best to use the leading edge cells for quantifying movement. Cells behind the front have less directional movement, presumably because the chemoattractant gradient is altered as a result of the action of secreted and cell surface folate deaminase.

References

1. Pan, P., Hall, E. M., and Bonner, J. T. (1975) Determination of the active portion of the folic acid molecule in cellular slime mold chemotaxis. *J. Bacteriol.* **122,** 185–191.

2. Nandini-Kishore, S. G. and Frazier, W. A. (1981) [3H]Methotrexate as a ligand for the folate receptor of *Dictyostelium discoideum*. *Proc. Natl. Acad. Sci. USA* **78,** 7299–7303.

3. Alcantara, F. and Monk, M. (1974) Signal propagation during aggregation in the slime mould *Dictyostelium discoideum*. *J. Gen. Microbiol.* **85,** 321–334.

4. Devreotes, P. and Janetopoulos, C. (2003) Eukaryotic chemotaxis: distinctions between directional sensing and polarization. *J. Biol. Chem.* **278,** 20,445–20,448.

5. Rifkin, J. L. (2002) Quantitative analysis of the behavior of *Dictyostelium discoideum* amoebae: stringency of pteridine reception. *Cell Motil. Cytoskeleton* **51,** 39–48.

6. Parent, C. A. and Devreotes, P. N. (1996) Molecular genetics of signal transduction in *Dictyostelium*. *Annu. Rev. Biochem.* **65,** 411–440.

7. Rifkin, J. L. (2001) Folate reception by vegetative *Dictyostelium discoideum* amoebae: distribution of receptors and trafficking of ligand. *Cell Motil. Cytoskeleton* **48,** 121–129.

8. Xiao, Z. and Devreotes, P. N. (1997) Identification of detergent-resistant plasma membrane microdomains in *Dictyostelium*: enrichment of signal transduction proteins. *Mol. Biol. Cell* **8,** 855–869.

9. Wessels, D., Murray, J., and Soll, D. R. (1992) Behavior of *Dictyostelium* amoebae is regulated primarily by the temporal dynamic of the natural cAMP wave. *Cell Motil. Cytoskeleton* **23,** 145–156.

10. Mato, J. M., Losada, A., Nanjundiah, V., and Konijn, T. M. (1975) Signal input for a chemotactic response in the cellular slime mold *Dictyostelium discoideum*. *Proc. Natl. Acad. Sci. USA* **72,** 4991–4993.

11. Ma, L., Janetopoulos, C., Yang, L., Devreotes, P. N., and Iglesias, P. A. (2004) Two complementary, local excitation, global inhibition mechanisms acting in parallel can explain the chemoattractant-induced regulation of PI(3,4,5)P3 response in *Dictyostelium* cells. *Biophys. J.* **87,** 3764–3774.

12. Dunn, G. A. and Zicha, D. (1993) Long-term chemotaxis of neutrophils in stable gradients: preliminary evidence of periodic behavior. *Blood Cells* **19,** 25–39.

13. Shutt, D. C., Jenkins, L. M., Carolan, E. J., et al. (1998) T cell syncytia induced by HIV release. T cell chemoattractants: demonstration with a newly developed single cell chemotaxis chamber. *J. Cell Sci.* **111(Pt 1),** 99–109.

14. Wilkinson, P. C. (1988) Micropore filter methods for leukocyte chemotaxis. *Methods Enzymol.* **162,** 38–50.

15. Lauffenburger, D. A., Tranquillo, R. T., and Zigmond, S. H. (1988) Concentration gradients of chemotactic factors in chemotaxis assays. *Methods Enzymol.* **162,** 85–101.

16. Konijn, T. M. (1970) Microbiological assay of cyclic 3',5'-AMP. *Experientia* **26,** 367–369.

17. Bonner, J. T., Kelso, A. P., and Gillmor, R. G. (1966) A new approach to the problem of aggregation in the cellular slime molds. *Biol. Bull.* **130,** 28–42.

18. Tillinghast, H. S. and Newell, P. C. (1987) Chemotaxis towards pteridines during development of *Dictyostelium*. *J. Cell Sci.* **87(Pt 1),** 45–53.

19. Cutler, J. E. and Munoz, J. J. (1974) A simple in vitro method for studies on chemotaxis. *Proc. Soc. Exp. Biol. Med.* **147,** 471–474.
20. Lauffenburger, D., Rothman, C., and Zigmond, S. H. (1983) Measurement of leukocyte motility and chemotaxis parameters with a linear under-agarose migration assay. *J. Immunol.* **131,** 940–947.
21. Laevsky, G. and Knecht, D. A. (2001) Under-agarose folate chemotaxis of *Dictyostelium discoideum* amoebae in permissive and mechanically inhibited conditions. *Biotechniques* **31,** 1140–1149.
22. Fukui, Y., Yumura, S., Yumura, T. K., and Mori, H. (1986) Agar overlay method: high-resolution immunofluorescence for the study of the contractile apparatus. *Methods Enzymol.* **134,** 573–580.
23. Neujahr, R., Heizer, C., Albrecht, R., et al. (1997) Three-dimensional patterns and redistribution of myosin II and actin in mitotic *Dictyostelium* cells. *J. Cell Biol.* **139,** 1793–1804.
24. Laevsky, G. and Knecht, D. A. (2003) Cross-linking of actin filaments by myosin II is a major contributor to cortical integrity and cell motility in restrictive environments. *J. Cell Sci.* **116,** 3761–3770.
25. Franke, J. and Kessin, R. (1977) A defined minimal medium for axenic strains of *Dictyostelium discoideum. Proc. Natl. Acad. Sci. USA* **74,** 2157–2161.

20

Optimized Fixation and Immunofluorescence Staining Methods for *Dictyostelium* Cells

Monica Hagedorn, Eva M. Neuhaus, and Thierry Soldati

Summary

Recent years have seen a powerful revival of fluorescence microscopy techniques, both to observe live cells and fixed objects. The limits of sensitivity, simultaneous detection of multiple chromophores, and spatial resolution have all been pushed to the extreme. Therefore, it is essential to improve in parallel the quality of the structural and antigenic preservation during fixation and immunostaining. Chemical fixations are broadly used but often lead to antigenicity loss and severe membrane damages, such as organelle vesiculation. They also must be followed by membrane permeabilization by detergents or solvents, which can lead to extensive extraction and cytosol leakage. Fixation with solvents bypasses the need for permeabilization, but when carried out at "high" temperatures, leads to severe extraction of soluble proteins and lipids and cytosol wash-out, and has therefore been used routinely to visualize the cytoskeleton. Here, we describe a few modifications to the common aldehyde fixation protocol that help decrease the usual artifacts induced by chemical fixation. Alternatively, new techniques have now been established that are based on rapid freezing using a variety of coolants followed by fixation in solvents at low temperature. We present detailed protocols and notes that allow the achievement of optimal preservation and permeabilization for both light and electron microscopy.

Key Words: Immunofluorescence; microscopy; rapid freezing; fixation; detergents; preservation; structure; antigenicity; *Dictyostelium*.

1. Introduction

Recent years have seen a powerful revival of fluorescence microscopy techniques *(1)*, both to observe live cells and fixed objects. The limits of sensitivity, simultaneous detection of multiple chromophores, and spatial resolution have all been pushed to the extreme *(2)*. Nevertheless, very often, preparation of cells or tissues for immunocytochemistry, whether for light or electron microscopy

From: *Methods in Molecular Biology, vol. 346:* Dictyostelium discoideum *Protocols*
Edited by: L. Eichinger and F. Rivero © Humana Press Inc., Totowa, NJ

(EM), takes its inspiration from cooking and witchcraft. All the recipes have in common the aim of solving the three major problems of preservation of antigenicity, preservation of native structure, and accessibility of the antigen. Despite about half a century of mostly empirical approaches, no optimal solution has been uncovered; instead, painstaking trial-and-error procedures are needed to determine the technique "best" suited to each particular case. It is not within the scope of this chapter to present an exhaustive theoretical and experimental list of the possible approaches and strategies, but the reader is advised to seek wisdom in such textbooks as the one from Gareth Griffiths *(3)* and further work from the authors. No panacea can be offered, but only a guide to the steps that can be optimized to reach the best compromise between the three parameters mentioned previously.

Briefly, in classical methods, fixation is responsible for the degree of preservation of both structure and antigenicity. Very often the improvement of one parameter is inversely proportional to the loss of the other: use of increased concentrations of aldehydes leads to better preservation of structure, but chemical modification of proteins (and other molecules) destroys the antibody binding sites. The same is generally true when switching from formaldehyde to glutaraldehyde, and most of the additives used in EM (osmium tetroxyde and other empirically selected chemicals) are deleterious to antibody recognition. Even when the optimal compromise between antigenicity and structure has been defined, accessibility remains a hurdle that is most often overcome by the use of detergents that either selectively or generically solubilize membrane lipids and create pores in the membranes. There is an empirical gradation from mild detergents (often specific to some lipids, e.g., sterols, such as digitonin and saponin) to stronger ones, such as Triton X100 or similar nonionic detergents. Depending on the strength of the fixation method used, detergents can remove or solubilize proteins and other small molecules, generating (sometimes desired) additional artifacts.

An alternative strategy has slowly emerged that bypasses the need for chemical denaturation of proteins and the use of detergents, namely the immobilization by freezing. Obviously, this approach has some drawbacks, mainly in the more sophisticated equipment needed, but offers serious advantages. Because the immobilization of all biological processes and biochemical machines happens almost instantaneously (at least one to two orders of magnitude faster compared with chemical fixations), there is little time for modification of the native structure. The major enemy is the formation and growth of ice crystals that leads to the appearance of artifactual hexagonal patterns inside the cell or organelles. Therefore, the ultimate goal of all the methods used—slamming against supercooled copper blocks, high-pressure freezing, and freezing on coverslips—is to generate amorphous ice. Whereas the first two methods can lead to several 100-µm thick, crystal-free samples, the rapid freezing on cover-slips is limited to cell monolayers of up to 10–15 µm thickness. The use of coolants such as liquid ethane has been shown to lead to excellent structure preservation when followed either by freeze substitution in solvents such as acetone or methanol or by resin embedding, but is rarely compatible with antigen detection. The most popular method for immuno-EM is likely the Tokuyashu method (*4*, and references therein), in which frozen sections are thawed and incubated with antibodies before visualzation, leading to remarkable antigenicity preservation; however, with this method, structure preservation is usually minimal. Recently, alternatives have been developed that take advantage of rapid cooling, either in liquid ethane for EM fine structure preservation or in methanol for light microscopy level preservation (*5*), and which are followed by fixation by coagulation or precipitation as a consequence of replacing the water by solvent around proteins and other molecules. This fixation results in excellent preservation of antigenicity, and precise temperature control can lead to graded levels of lipid extraction and thus of permeabilization, making the use of detergents unnecessary (*5,6*).

In summary, depending on the degree of structure and/or antigenicity preservation needed, the choice of the appropriate technique must be carefully tailored. Here, we present a robust, rapid-freezing, methanol fixation-permeabilization method and an alternative chemical fixation that is well adapted to *Dictyostelium* (an illustration of the results obtained can be found in **Figs. 2** and **3**, discussed later). It is also worth noting that optimization of the method is dependent on the structrure and the antigen that one uses to judge the improvements. We strongly recommend labeling with a contractile vacuole, as this is arguably the most complex and beautiful organelle in *Dictyostelium*. The preservation of its delicate structure, made of interconnected bladders and a reticular tubular network, is an excellent "standard meter" with which to monitor the efforts. Finally, use of 4',6-diamidino-2-phenylindole (DAPI) to stain nuclear DNA allows one to monitor the formation of hexagonal ice that might result from suboptimal freezing conditions inside the otherwise homogeneously stained nucleus.

2. Materials

2.1. Buffers, Fixatives, and Other Reagents

1. Soerensen buffer (SB): 15 mM KH$_2$PO$_4$, 2 mM Na$_2$HPO$_4$, pH 6.0.
2. Soerensen/Sorbitol buffer (SSB): Soerensen buffer containing 120 mM sorbitol.
3. Phosphate-buffered saline (PBS): 140 mM NaCl, 2.7 mM KCl, 10 mM Na$_2$HPO$_4$, 1.8 mM KH$_2$PO$_4$.
4. Quenching buffer: PBS with 100 mM glycine
5. Blocking buffer: PBS with 0.2% gelatin and 0.1% Triton X100.

6. PIPES buffer: Piperazine-1,4-bis(2-ethanesulfonic acid) (PIPES) 20 mM buffered to pH 6.0 with NaOH.

7. Preparation of the saturated picric acid solution: dissolve 3 g of solid picric acid in 1 L of double-distilled (dd)-H$_2$O and warm up to 80°C overnight. Leave to cool down to room temperature (crystals may build up and precipitate), set the pH to 6.0, and store at 4°C.

8. Preparation of the picric acid/paraformaldehyde fixative: always prepare fresh, in a 50 mL Falcon tube.

   a. Weigh 0.4 g of paraformaldehyde (PAF) (powder, kept at 4°C). Caution should be exerted with PAF: do not inhale it, and always weigh it under a fume cabinet.

   b. Add 10 mL 20 mM PIPES buffer.

   c. Microwave in brief pulses until it dissolves (takes only a few seconds), then cool immediately (on ice) to room temperature.

   d. Add 7 mL of dd-H$_2$O (or accordingly less if you add 0.5–1.0 mL of 2.5 M sucrose in order to enhance preservation of filopodia).

   e. Add 3 mL of a saturated picric acid solution (stock prepared in advance and kept at 4°C (*see* **Subheading 2.1.1.**).

2.2. Equipment

1. Coverslips: The choice of coverslip type is crucial. For conventional chemical fixation, standard coverslips (12 mm diameter, grade 1, approx 145 μm thick, http://www.hecht-assistent.de) can be used, whereas for fixation in ultracold methanol, thinner glass coverslips (12 mm diameter, grade 0, approx 100 μm thick) must be used. For rapid-freezing techniques using liquid ethane and destined to ultrastructural observations by EM, highest heat conductance is necessary, and thus 50-μm thick sapphire coverslips are used (3 or 6 mm diameter, Groh & Ripp, Germany, http://www.groh-ripp.de).

2. Liquid ethane manipulation: The ethane or propane can be directly condensed into a cold vessel (aluminium cup) set down on a small stage at the bottom of a styrofoam container filled with liquid nitrogen to the height of the stage. This should keep the environment around the liquid coolant purged of air. It may be necessary to refill the liquid nitrogen from time to time. If it happens that the coolant freezes, it can be thawed either by adding small volumes of liquid coolant or by touching the aluminium cup with a warm metal rod. Static electricity may be a real problem for liquid propane/ethane. Therefore, it is safer to use metal container and needles instead of glass or plastic (*see also* Chapter 21).

3. Freezing Dewar (FH Cryotec, Instrumentenbedarf Kryoelektronenmikroskopie, Plankstadt, Germany; schematically presented in **Fig. 1**): Alternatively, a styrofoam box with aluminium cups and racks can be used, similar to the one used in **item 1**. It should be placed in a –80°C freezer, taken out only briefly before plunging the coverslips (*see* **Subheading 3.3.**), and left at room tempera-

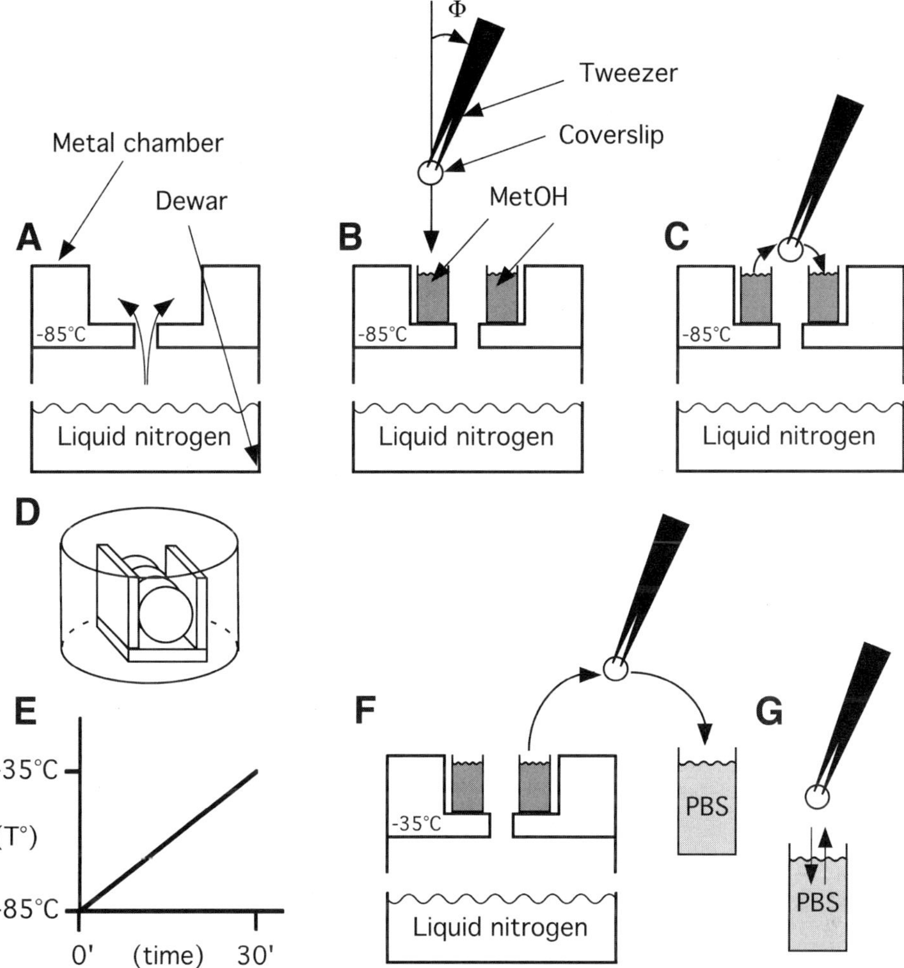

Fig. 1. Schematic drawing of the Dewar setup used for rapid freezing and ultracold-methanol fixation (*see* **Subheading 3.3.**). A metal chamber with an opening at the bottom is placed over a Dewar filled with liquid nitrogen. The temperature is controlled by varying the flux of liquid nitrogen vapor through the opening (**A**). The chamber is cooled to –85°C with the methanol-filled cups and racks (**B**) before the coverslips, with cells on top, are plunged at an angle of 15° (**Φ**) and transferred to the rack (**C**). Details of the coverslip rack are shown schematically in **D**. Next, the temperature is raised to –35°C in about 30 min (**E**) and the coverslips are transferred to PBS at room temperature (**F**). The methanol is diluted by moving the coverslip through the air–water interface (**G**).

ture until the methanol reaches –35°C (*see also* Chapter 21).

4. Humid chamber: a humid chamber for immunostainings is easily made out of a large (20-cm diameter) Petri dish with a glass lid. The bottom is covered with clean Parafilm before each incubation, and paper towels soaked in water are pressed at the periphery. A dark bench coat facilitates visualization of coverslips, and a light-tight cover can be placed on top to prevent photobleaching.

3. Methods

A comparison of the results achieved using the different fixation procedures described are shown in **Figs. 2** and **3**.

3.1. Preparation of Coverslips and Cell Plating

Coverslips are usually cleaned as follows:

1. Place 100 glass coverslips (or fewer if expensive sapphire is used) in a glass beaker, with a glass dish as a lid to prevent splashing.
2. Immerse in a solution of 50% nitric acid in dd-H_2O and incubate at room temperature for 2 h under a fume hood, swirling gently from time to time.
3. Decant the acid solution (can be re-used) and replace with an ample volume of dd-H_2O to rinse beaker and coverslips; swirl gently for a few seconds and decant.
4. Repeat 5–10 times and follow with two similar rinses in pure ethanol.
5. Dry the coverslips in a microwave at full power until the coverslips are completely dry. Use a glass dish as a lid to prevent coverslips from "popping" out of the beaker.
6. Place the coverslips in a box, between layers of paper tissues. With this treatment, we usually do not need to sterilize the coverslips by autoclaving before plating the cells. Also, autoclaving sometimes induces coverslip bending.
7. Place up to eight coverslips at the bottom of a 6-cm diameter plastic dish, and plate cells at an adequate density so as to reach 70–80% confluency after overnight growth.
8. Before fixation, cells can be treated as necessary, for example by feeding fluid phase markers or particles.

3.2. Rapid Freezing in Liquid Ethane

This step is necessary for achieving the highest degree of structure preservation, and is useful for observation of delicate and/or transient structures. This is

---

Fig. 2. *(opposite page)* Effects of fixation on structure preservation. Fluorescence microscopy of *Dictyostelium discoideum* cells stained against the vacuolar H^+-ATPase in order to visualize the contractile vacuole system. Each row of images represents three maximum intensity projections of equal number of optical sections through one cell. The total number of sections for **A–C** was 21; for **D–F** and **G–I** it was 18. Stack pictures were taken using a Leica, AS MDW-widefield microscope equipped with a 100× objective on a piezo z-positioner and a charge-coupled device camera. Raw

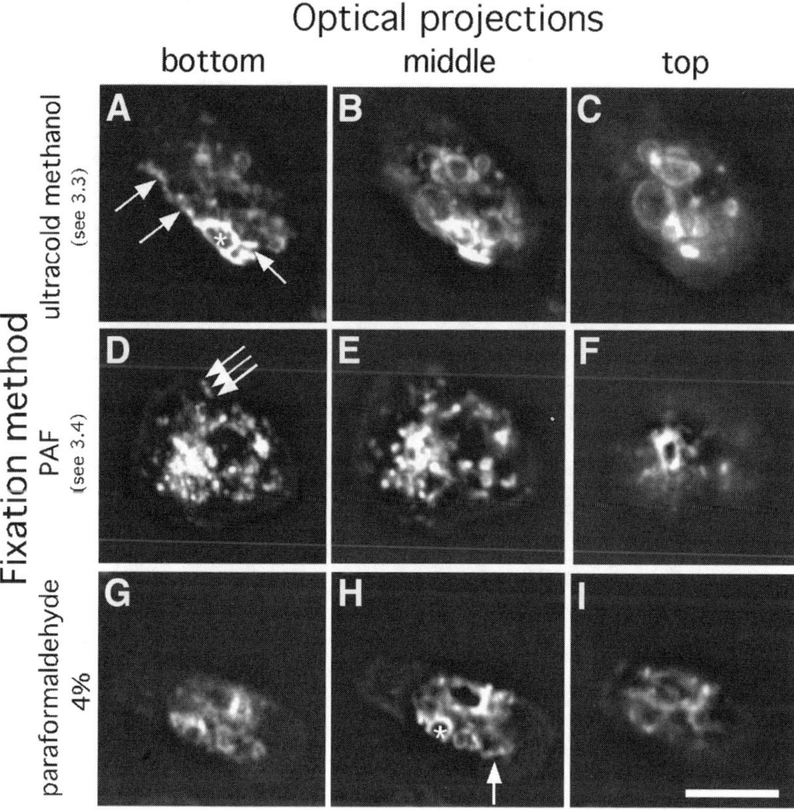

Fig. 2 *(continued)* fluorescence images were deconvoluted (Leica Deblur software, Blind Deconvolution, 20 iterations) and processed to projections using ImageJ V1.32j. The specimens were prepared and treated as follows: Cells were grown on coverslips (**A–C**, grade 0; **D–I**, grade 1), and three different fixation protocols were applied. For images **A–C**, coverslips were plunged into ultracold methanol (*see* **Subheading 3.3.**); in images **D–F**, cells were fixed using the paraformaldehyde (PAF) fixative as described under **Subheading 3.4.**; in images **G–I**, cells were fixed with 4% PAF in PBS for 45 min at room temperature. After blocking with 2% fetal calf serum and permeabilization (only for chemical fixations [**D–I**], 0.1% Triton X100 in blocking solution), the monoclonal, primary anti-VatA antibody was used at a dilution of 1:10 and, subsequently, the secondary antibody, anti-mouse IgG Alexa 594 (Molecular Probes), was diluted 1:1000. All three fixation approaches show similar antigenicity preservation but clear differences in structural conservation of the contractile vacuole system. In the specimen fixed in ultracold methanol, a main vacuole (indicated by "*") can be distinguished from a tubular (arrows) and vesicular system. In the other fixation approaches, the contractile vacuole network has mainly vesiculated (parallel arrows in **D** indicate vesiculated tubule) and poor structural preservation is demonstrated. Scale bar = 5 µm.

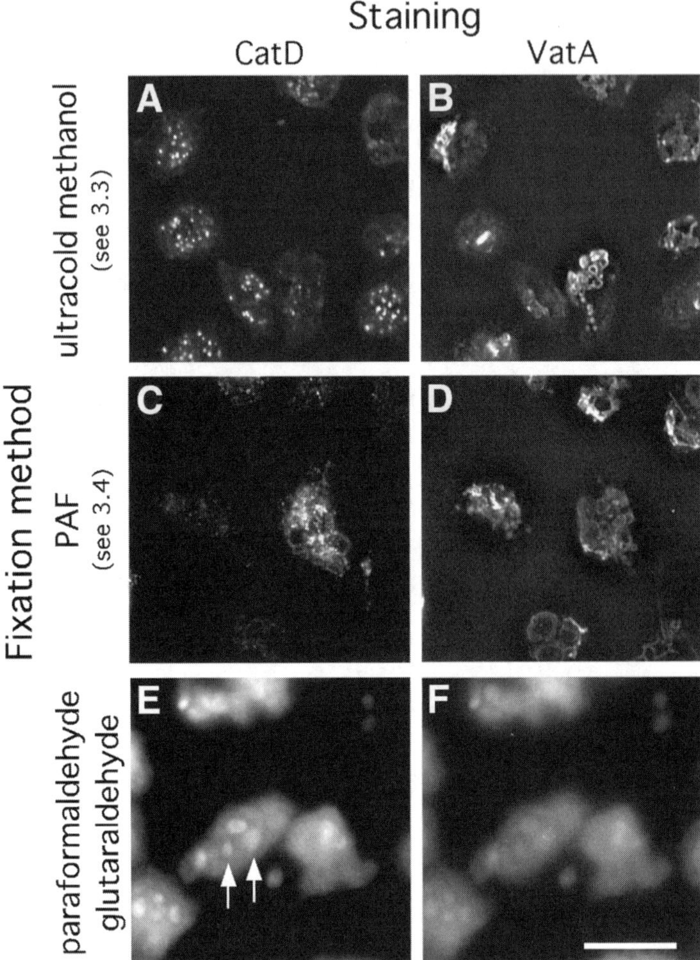

Fig. 3. Effect of fixation on antigenicity preservation. Fluorescence microscopy of *Dictyostelium discoideum* cells stained against cathepsin D and the vacuolar H$^+$-ATPase illustrates the various levels of antigenicity preservation obtained after different fixation techniques. Each image represents a maximum intensity projection through the complete height of cells. Image acquisition and processing was carried out as described for **Fig. 2**. Specimens were prepared as follows: cells were grown on coverslips (**A,B**, grade 0; **C–F**, grade 1), fed with yeasts for 45 min (two ingested yeasts are indicated by arrrows in **E**), and fixed by methanol fixation (**A–B**) or paraformaldehyde (PAF) (**C–D**) or 4% PAF and 0.12% glutaraldehyde. Permeabilization, blocking, and staining were performed as for **Fig. 2**. Antibodies were diluted as follows: anti-VatA at 1:10, anti-CatD at 1:500, and anti-mouse IgG-Alexa594 and anti-rabbit IgG Alexa488 at 1:1000. The different fixations show varying degrees of antigenicity preservation. Fixation with glutaraldehyde masks antigens for both antibodies (**E,F**) and

usually not necessary for standard immunofluorescence microscopy, but is essential for EM-level immunocytochemistry.

1. Sapphire or glass coverslips, grade 0, are either held with a tweezer or at the tip of a pipet microtip that has been lightly dipped in silicone grease.
2. Plunge coverslip in liquid ethane with the cell layer facing up, and wait a few seconds, keeping the coverslip under the ethane.
3. Transfer to ultracold methanol for fixation (*see* **Subheading 3.3.**, **step 2**) and/or freeze-substitution.

3.3. Fixation in Ultracold Methanol

1. After the desired treatments (*see* **Subheading 3.1.**, **step 8** and/or **Subheading 3.2.**), the cells can be rinsed briefly in fresh growth medium or SB (for example, to remove excess uningested particles) if necessary. Coverslips are held with a tweezer and delicately tipped vertically on a tissue, then laid down for a second to blot the back free of liquid. Do not overdry, as cells are rapidly damaged by surface tension if the liquid film evaporates completely.
2. Plunge vertically in ultracold (–85°C) methanol, holding the coverslip with at a 15° angle to the vertical, cell layer up, so as to maximize the flux of coolant on the back of the coverslip (*see* **Fig. 1** for an illustration of **steps 2–6**).
3. Hold under methanol for a few seconds and transfer rapidly to a rack placed in a second ultracold methanol container.
4. Repeat until all of the coverslips are processed.
5. Start the warming and monitor until the temperature reaches –35°C, in approx 30–60 min (*see* **Note 1**).
6. Transfer the coverslips one by one into PBS at room temperature. Grab the coverslips with a tweezer, lift them from the rack, and immediately plunge them vertically into a beaker of PBS. Rapid warming and removal of excess methanol is produced by five to eight quick but gentle movements through the air–buffer interface, monitoring progressive wetting of the coverslip. Do not move laterally, as it detaches the cell layer.
7. The cells can either be processed immediately or stored in PBS, for example in a multiwell dish.

Fig. 3 *(continued)* produces massive autofluorescence (this can be quenched by incubating with 1 mg/mL of NaBH$_4$ in phosphate-buffered saline for 15 min, prior to blocking). The PAF fixative results in better antigenicity preservation (**C,D**). However, the anti-CatD labeling is very weak, and longer exposure times were necessary, thereby increasing visibility of unspecific background signals (**C**). Best antigenicity preservation was achieved by fixation in ultracold methanol. The anti-CatD staining reveals the strong, punctuate lysosomal staining. Note that VatA staining is also optimal in these conditions (**B**). Scale bar = 10 µm.

3.4. Alternative Fixation in Picric Acid-Paraformaldehyde

This method, first referred to in **ref. 7**, is a reasonably good alternative to methanol fixation. It is routinely used by many labs and was made popular by the laboratory of Dr. G. Gerisch.

1. Wash the cells briefly in 1X SB (be aware that SB has an osmolarity of about 50 miliOsm and the medium has an osmolarity of around 200 miliOsm; therefore, one can use SSB instead) (*see* **Note 2**).
2. Plunge the coverslips in PAF fixative or, alternatively, gently pipet the fixative over the cells, and incubate for 30 min at room temperature (*see* **Note 3**).
3. Plunge the coverslips briefly in PBS (to get rid of most of the fixative).
4. Incubate the cells for 10 min in quenching buffer to quench the free aldehyde groups.
5. Block the cells for 30 min in blocking buffer with 0.1% Triton X100 (*see* **Note 4**).
6. From this point, the protocol described under **Subheading 3.5.**, **step 2** can be followed, using 0.1% Triton X100 in all buffers (*see* **Note 5**).

3.5. Immunofluorescence Staining

An illustration of results achieved with this procedure is shown in **Figs. 2** and **3**.

1. Incubate fixed cells in blocking buffer for 30 min in a multiwell dish (*see* **Notes 5** and **6**).
2. Cut some parafilm and place at the bottom of a humid chamber for antibody incubations.
3. Dilute the antibody in blocking buffer and deposit a 50-µL drop on the parafilm (*see* **Note 7**).
4. Use tweezers to place the coverslip, with cells facing down, on the drop (drain excess fluid from the coverslips beforehand, in order not to dilute the antibody further).
5. Incubate for 60 min at room temperature.
6. Gently lift the coverslips and rinse for a few seconds by gently plunging repeatedly in a beaker of PBS, then incubate twice for 5 min in 3 mL PBS (in a multiwell plate).
7. Place a 50-µL drop of secondary antibody diluted in blocking buffer on a CLEAN parafilm.
8. Use tweezers to place the coverslip, with cells facing down, on the drop (drain excess fluid from the coverslips beforehand, in order not to dilute the antibody further).
8. Incubate 60 min at room temperature .
9. Repeat washes as in **step 6** (*see* **Notes 8** and **9**).
10. Dip the coverslip once in water, drain excess fluid, and mount in a SMALL drop (5–10 µL) of Prolong Antifade (Molecular Probes) on a clean glass slide (mount

one by one on a "fresh" convex drop of mounting medium). Alternatively, mount in Fluoromount G (Southern Biotechnology Associates, USA) (*see* **Note 10**).

11. Wait until the mounting medium has hardened (optimally overnight, but can be accelerated by placing under an air flux).
12. Have fun watching the cells. The slides can be kept for months in a cold (4°C) and dark place.

4. Notes

1. The end T° dictates the degree of extraction and permeabilization; if less extraction is needed (e.g., staining of plasma membrane proteins facing the extracellular space), the coverslip can be taken out at –45°C or –40°C. If more extraction is required (e.g., visualization of cytoskeleton), prolong the incubation at –35°C or –30°C for some minutes. Presence of detergents is not required for subsequent immunodetection.
2. If no sucrose is added to the fixation solution, the cells can round up and many will be lost; therefore, use relatively confluent cell layers. If sucrose is added, the cells show many spikes, which do not look very natural; however, up to this point, they look healthy and stay on the coverslip.
3. The fixation should be carried out in a Petri dish (or multiwell plate, with wells at least 2 cm in diameter, to ease the manipulation of the coverslips).
4. Alternatively, block with 1X PBS containing 2% fetal calf serum and 0.1% Triton (*see also* below).
5. As a result of the presence of Triton X100 in nearly all solutions, surface tension is very much decreased and incubation "on a drop" might be difficult. Alternatively, it can be performed in small volumes in microtiter plates.
6. The type of blocking is dependent on the antibody, the cells, and the subsequent technique (EM or immunofluorescenece) used. Two percent fetal calf serum or a cocktail of 0.5% bovine serum albumin and 0.045% fish skin gelatin are alternatives.
7. Primary antibody incubations can be performed at room temperature or 37°C for various times—each antibody is different. It is a good idea to use hybridoma supernatants, NOT undiluted but diluted at least 1:2 in blocking buffer (buffered and blocked).
8. We have never heard of an indirect fluorescent antibody that had been over-washed, be patient!
9. Staining with DAPI (Sigma) can easily be performed between the last two washes. Incubate the coverslips for 5 min on a drop of PBS with about 1–5 µg/mL of DAPI, and continue with the last wash. This is very useful for focusing on the cell layer when setting the microscope, in addition to being very informative with regard to cell damage during freezing. Indeed, the nuclear DAPI staining should be homogeneous (except for the dark nuclear caps of heterochromatin), and a honeycomb pattern will indicate that ice crystals have formed and, hence, structural damage has occurred.
10. Other mounting media are usable, just use one that hardens with time (e.g., Mowiol-based). Also, if photobleaching of the chromophore is a problem, include

1,4-diazabicyclo[2,2,2]octane (DABCO; Sigma) at 100 mg/mL (dissolve powder directly in the desired small amount of mounting medium, mix by inverting for some minutes, and use the same day). Some people seal the coverslip on the slide with nail polish, making it tight to microscopy oil, which otherwise can damage the preparation.

Acknowledgments

This work was mainly carried out at the Max-Planck Institute for Medical Research in Heidelberg and was supported by a grant from the Deutsche Forschungsgemeinschaft. We also acknowledge support from The Wellcome Trust, the UK Biotechnology and Biological Sciences Research Council (BBSRC), and the Swiss National Science Foundation. A big "thank you" goes to all of the lab members who have participated in the establishment of the protocols and have suggested amendments and improvements.

References

1. Griffiths, G., Parton, R. G., Lucoq, J., et al. (1993) The immunofluorescence era of membrane traffic. *Trends Cell Biol.* **3,** 214–219.
2. Garini, Y., Vermolen, B. J., and Young, I. T. (2005) From micro to nano: recent advances in high-resolution microscopy. *Curr. Opin. Biotech.* **16,** 3–12.
3. Griffiths, G. (1993) *Fine Structure Immunocytochemsitry.* Springer-Verlag, Berlin/Heidelberg.
4. Tokuyasu, K. T. (1978) A study of positive staining of ultrathin frozen sections. *J. Ultrastruct. Res.* **63,** 287–307.
5. Neuhaus, E., Horstmann, H., Almers, W., Maniak, M., and Soldati, T. (1998) Ethane-freezing/methanol-fixation of cell monolayers. A procedure for improved preservation of structure and antigenicity for light and electron microscopies. *J. Struct. Biol.* **121,** 326–342.
6. Neuhaus, E. M., Almers, W., and Soldati, T. (2002) Morphology and dynamics of the endocytic pathway in *Dictyostelium discoideum. Mol. Biol. Cell* **13,** 1390–1407.
7. Humbel, B. M. and Biegelmann, E. (1992) A preparation protocol for postembedding electron microscopy of *Dictyostelium discoideum* cells with monoclonal antibodies. *Scanning Microsc.* **6,** 817–825

21

Cryofixation Methods for Ultrastructural Studies of *Dictyostelium discoideum*

Mark J. Grimson and Richard L. Blanton

Summary

Cryopreservation methods, including rapid freezing, freeze-substitution, and low-temperature embedment, lead to superior ultrastructural preservation compared with traditional fixation procedures. This is particularly true for the multicellular stages of *Dictyostelium discoideum*, in which the hydrophobic sheath that surrounds the structures causes delayed penetration by the already slow-acting aqueous chemical fixatives, resulting in cell shape changes, loss of cell–cell contacts, and changes in cell–matrix interactions. The surface tension effects of traditional fixation methods can also result in disruption of the delicate structures. Depth of freezing is often greater than expected because of the relatively dehydrated state of the multicellular structures. Variations in freeze-substitution solvents and embedment media can be employed to allow for antigenic preservation. Commercial instruments exist for most of the procedures, but excellent results can be obtained using inexpensive hand-crafted apparatus.

Key Words: Ultrastructure; transmission electron microscopy; cryofixation; cryopreservation; rapid-freezing; freeze-substitution; *Dictyostelium*.

1. Introduction

One of the greatest challenges facing the electron microscopist is to prepare specimens that reflect accurately the living state of the cell. Specimen preparation involves a series of processes, each capable of introducing artifacts and requiring optimization for a particular cell type or tissue. The initial fixation step is particularly critical: the ideal would be a procedure that preserves cellular structure exactly as it is at the moment of fixation. In fact, penetration of the fixative and its subsequent cross-linking action are very slow. Cellular processes continue, cells change shape, and cell–cell and cell–matrix interactions change during the course of fixation, resulting in a fixed specimen that may be

From: *Methods in Molecular Biology, vol. 346:* Dictyostelium discoideum *Protocols*
Edited by: L. Eichinger and F. Rivero © Humana Press Inc., Totowa, NJ

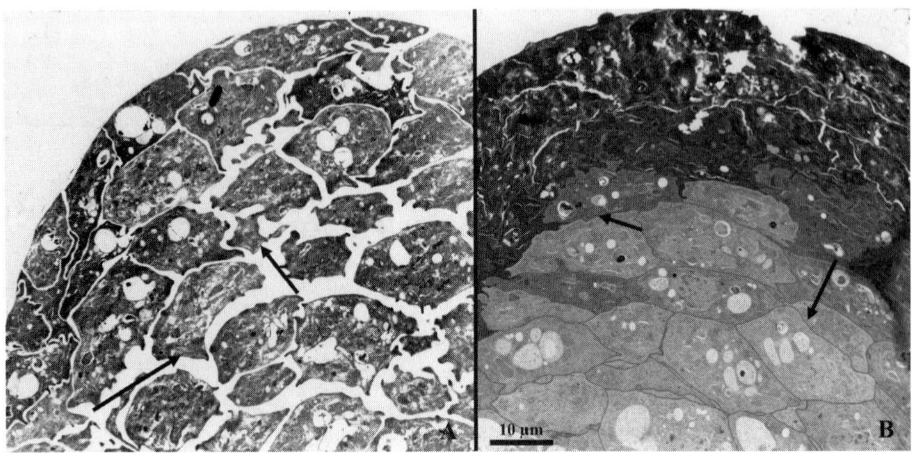

Fig. 1. Transmission electron micrographs of the tip of a culminant prepared by conventional aqueous chemical fixation methods (**A**) and plunge-freezing/freeze-substitution (**B**). Note how the cells have pulled apart from one another in **A**, but maintain both spatial and structural integrity in **B** (arrows). In the rapidly frozen culminant (**B**), a layer of electron dense cells can be observed just beneath the outer hydrophobic sheath (the peripheral layer). The slow aqueous fixation in **A** has caused the cells in this layer to rehydrate and enlarge. The preservation of internal fine cell structure is also superior in **B**. Notice that in **A**, the cells all have a similar appearance, whereas in **B**, there is a gradual change in cell size and shape moving down from the top.

very different from reality. The multicellular stages of *Dictyostelium discoideum* are surrounded by a hydrophobic sheath, which further interferes with penetration of fixatives. In addition, the surface tension changes that occur during attempts to get these structures to sink in the fixative can often result in the destruction of the structures.

Cryofixation (alternatively, cryoimmobilization) offers the potential to come much closer to the fixation ideal, i.e., to immobilize cells so quickly that dynamic cellular processes and delicate cellular structures and interactions are preserved. Its particular advantage with *Dictyostelium* is that the surface sheath does not interfere with the freezing process (*see* **Fig. 1** for a comparison of chemically fixed and cryopreserved specimens). The methodology has its own challenges, including optimization of freezing methods (for discussions of the physics of cryofixation, *see* **refs.** *1–4*) and limitations to the depth of optimal freezing in multicellular specimens. Cryofixation is also just the first step of a

Table 1
Commercially Available Cryofixation and Freeze-Substitution Devices

Freezing/preparation technique	Manufacturer(s)[a] and models	Approximate cost	Depth of adequate freezing Type of sample
Plunge-freezing[b]	Leica EM CPC	$15,000	10–20 μm
	EMS 002	$6000	Single/multicellular
Spray-freezing	None	NA	10–20 μm
			Single cells
Slam-freezing	Leica EM CPC	$15,000	15–30 μm
(metal-mirror	Leica MM80 E	$8000	Single cells
freezing)	RMC MF 7000	$15,000	
Propane-Jet freezing	BAL-TEC 030	$15,000	10–20 μm
			Single cells
High-pressure freezing	Leica EM PACT	~$100,000	≤200 μm
	BAL-TEC HPM 010		Single/multicellular
Freeze-substitution	Leica EM AFS	$15,000	Single/multicell
	RMC FS 7500	$15,000	

[a]Leica Microsystems, www.em-preparation.com; BAL-TEC/RMC, www.baltec-rmc.com/; Electron Microscopy Sciences, www.emsdiasum.com

[b]The two commercially available plunge-freezing devices can also perform metal-mirror freezing (also known as slam-freezing).

multistep process. For each step there is a range of options, the choice of which depends on the specimen and the ultimate use of the material, i.e., for immuno-cytochemistry or for high-resolution structural observations.

In this chapter, we describe cryopreservation methods that we have used to prepare *Dictyostelium* cells and multicellular stages for transmission electron microscopy. Space does not allow us to present all of the potential alternatives to these methods (*see* **refs.** *1–7* for detailed discussions of these). We also assume knowledge of the fundamentals of electron microscopy (*see* **refs.** *8–11* for background). The chapter will first describe cryofixation by spray-freezing and plunging, both of single cells and multicellular stages. It will then describe freeze-substitution, and conclude with infiltration and embedment. We describe how to construct apparatus to perform cryofixation and freeze-substitution (*see* **Table 1** for a listing of commercially available equipment for these techniques).

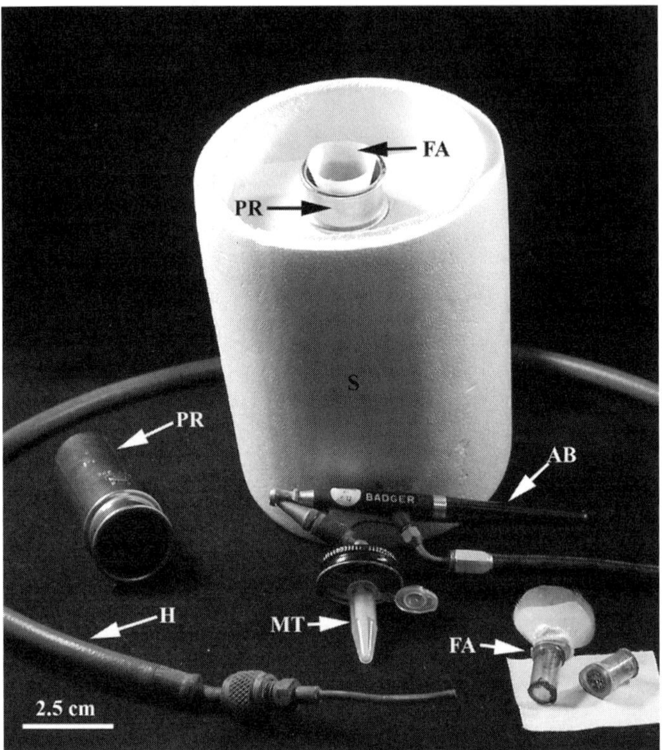

Fig. 2. The spray-freezer consists of the Styrofoam cylinder (S), that holds the liquid nitrogen with the propane reservoir (PR) supported in the center by a custom-fitted piece of Styrofoam. An artist's airbrush (AB) connected to a nitrogen gas tank (not shown) is used to aerosolize and propel cells in suspension from a microcentrifuge tube (MT) into the Nitex funnel assembly (FA) (*see* **Fig. 3**) that is submerged in the liquefied propane. The propane hose (H) attached to the propane source used to fill the reservoir shows the copper fill tube attached.

2. Materials

2.1. Cryoimmobilization

2.1.1. Spray-Freezing (see **Fig. 2**)

1. Badger Model 350 airbrush (Badger Air-Brush Co.) connected to a standard gas tank regulator and a tank of gaseous nitrogen.
2. Liquid nitrogen reservoir: a Styrofoam shipping container that can hold at least 1 L of liquid nitrogen (we use the cylindrical containers used to ship 1 L glass bottles).
3. Propane reservoir: a steel clinical centrifuge test tube holder (3 cm diameter × 10 cm height) or a copper pipe to which a bottom has been silver-soldered. The 3-cm

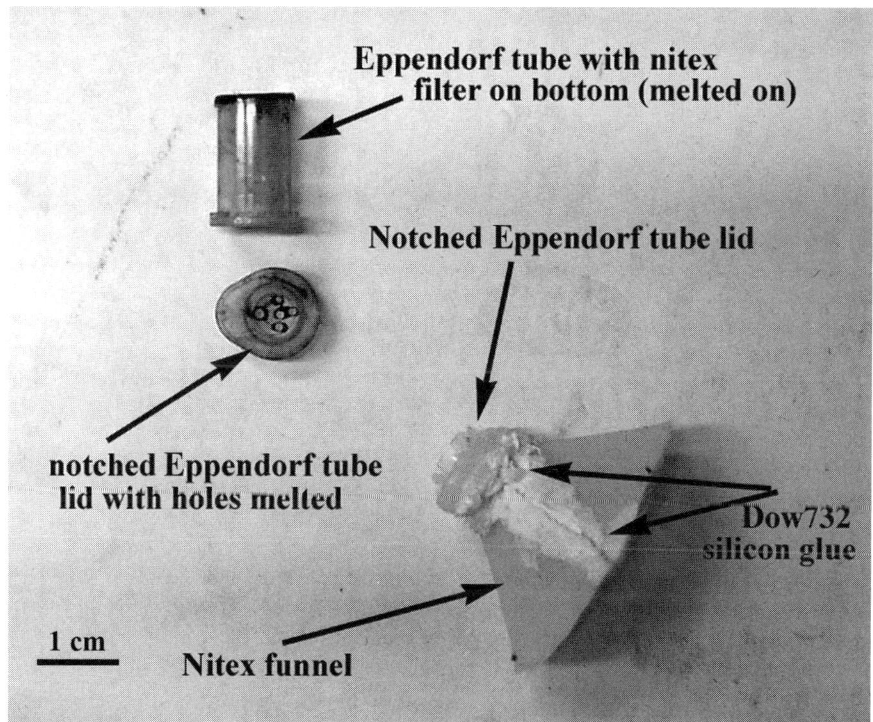

Fig. 3. Details and notes on making the Nitex funnel assembly used in the spray-freezer.

diameter is important, but the depth is not, although the advantage of a shallower reservoir is the lower quantity of propane required to fill it.

4. Propane reservoir holder: a support to suspend the propane reservoir in the liquid nitrogen reservoir. A piece of scrap Styrofoam, about 3 cm thick, cut to press against the sides of the foam container is sufficient for this purpose. The foam holder should be cut to allow liquid nitrogen to be added from the sides without the chance of pouring any into the propane. A hole is made in the center of this foam piece to hold the steel propane reservoir by first heating the tube on a Bunsen burner and then using the tube to melt through the foam, producing a solid support for the tube (the tube is surrounded by liquid nitrogen on all sides).

5. Sample collection filter-funnels (*see* **Fig. 3**), made by transferring the funnel template (*see* **Fig. 4**) to Nitex filter material (5 µm mesh or smaller; Sefar America). The cut filters are glued at the edges with Silastic 732 RTV Adhesive/Sealant (Dow Corning) to form a funnel, and allowed to dry overnight while being held together with a paperclip. A second coat of adhesive on the seams will extend the life of the funnels. It is advisable to glue a piece of a larger mesh (e.g., 250 µm) Nitex filter to the outside of the 5-µm mesh filter funnel; the stiffness of the former provides support for the latter.

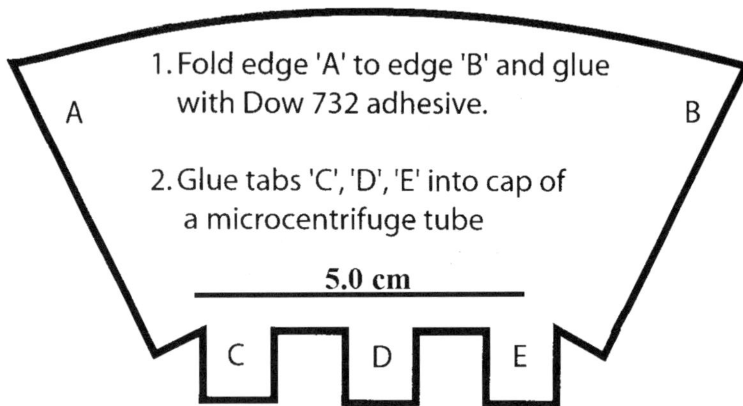

Fig. 4. Template of the Nitex funnel for the spray-freezer reduced 4×. The funnel may need to be trimmed, particularly the tabs that hold it to the cap.

The funnel base is fashioned from the cap of a 1.5-mL Eppendorf microcentrifuge tube (*see* **Note 1**). Cut the cap from a tube and melt out the center of the cap with a heated metal rod of the appropriate diameter, leaving the rim of the cap and the flange that extends down into the tube intact. Use a razor blade to cut three evenly spaced, "V-shaped" slots around the circumference of the rim; these cuts help to keep the rim flexible in liquid nitrogen and enable the caps to be placed on the microcentrifuge tubes while cold. Use Silastic 732 RTV Adhesive/Sealant to glue the tabs at the base of the Nitex funnel into the hole on the top of the microcentrifuge cap. Turn the funnel over and carefully place a small weight on the bottom of the cap to insure that the cap and funnel will be well attached. After drying overnight, apply another layer of adhesive around the attachment point.

6. Sample transfer/storage containers, made from the uncapped Eppendorf tubes (*see* **Note 1**) resulting from filter funnel construction. Cut the bottom of the tube off with a razor blade, about 1.5 cm from the top. Lay a piece of 5-µm Nitex filter on the lab bench adjacent to a Bunsen burner. Heat a flat spatula in the burner until it is glowing slightly. Use the hot spatula to melt the cut surface of the tube bottom. Quickly slide the tube off of the spatula and push the melted plastic immediately onto the Nitex. If done quickly enough, the filter will fuse with the bottom of the microcentrifuge tube, making a tight seal. Cut the excess filter off the bottom with scissors, and use a heated needle to smooth the filter around the bottom of the tube.

Using the cap from another microcentrifuge tube (*see* **Note 1**), cut the V-shaped slots into the rim as described in **item 5** above. Using a hot needle, carefully melt about six holes in the center of the cap to allow free exchange of the freeze-substitution media with the sample.

7. Wooden applicator sticks (Fisher Scientific).

2.1.2. Plunge-Freezing

1. A homemade plunger (*see* **Fig. 5**), following the description of **ref. *12***. This is a simple, robust, inexpensive (US $75 for parts; *see* **Note 2** for list and sources), and readily made device. Its fabrication requires access to a machine shop (for metal-cutting, bending, and silver-soldering).
2. Micro-serrefine surgical clamps (Fine Science Tools).
3. Styrofoam shipping container, as used for 1-L glass bottles. Line the inside with a polypropylene beaker (approx 9 cm diameter). Trim the top of the beaker to match the height of the Styrofoam container.
4. 20 mL disposable syringe barrel cut from the plunger end so that it will be level with the top of the cryogen reservoir (this is used to catch the plunged specimens; *see* **Fig. 5H**). Drill or melt several holes in the barrel wall.
5. Round (5 mm diameter) glass No. 1 coverslips (Esco, Erie Scientific) or Thermanox (Ted Pella) coverslips punched out with a standard paper punch (5 mm diameter). Treat with 0.05–0.1% (w/v) poly-L-lysine solution (Sigma, P-8920) to facilitate the cells sticking to coverslips when wicking off excess liquid. The glass coverslips should first be washed in either chromic-sulfuric acid (e.g., Fisher cleaning solution, #SC88-500) or 2.4 *M* HCl for 15 min, followed by several washes in deionized H_2O. Attach coverslips to double-sided tape in a Petri dish, place a drop of the poly-L-lysine solution on each coverslip for 3–5 min, wick off the excess, and dry with a stream of nitrogen gas. The coated coverslips should be used within several days.
6. Copper planchets (specimen carriers; Bal-Tec RMC, #BU 012-057-T).
7. Filter paper (9 cm Whatman No. 1), cut into wedges (8 wedges per circle).

2.2. Freeze-Substitution

2.2.1. Apparatus

1. Freeze-substitution module (*see* **Fig. 6**): The freeze-substitution module is constructed starting with a solid cylindrical block of brass (10 cm diameter × 14 cm height; obtainable from most metal supply businesses). Brass is preferred because it is easy to machine, suitable for repeated use in ultralow temperatures, and resistant to corrosion from condensate and the freeze-substitution chemicals. The block is cut transversely to create a 6-cm-height top and an 8-cm-height bottom. Six cavities are drilled in the lower block, each 2.2 cm diameter × 3.5 cm deep, and numbers are embossed adjacent to each to allow identification of samples. Cavities of the same dimension are drilled in the upper block, positioned so that they will align with those in the lower block. To secure the two block halves together, a small hole is drilled through the center of the top and partially into the bottom, the hole in the bottom is threaded, and a long machine bolt is passed through the top and screwed into the bottom.
The cavities are sized to hold and allow easy placement and removal (particularly at low temperatures) of 10-mL polypropylene sample vials (Fisher #03-338-1E). The sample vial caps are modified to allow for easy removal for

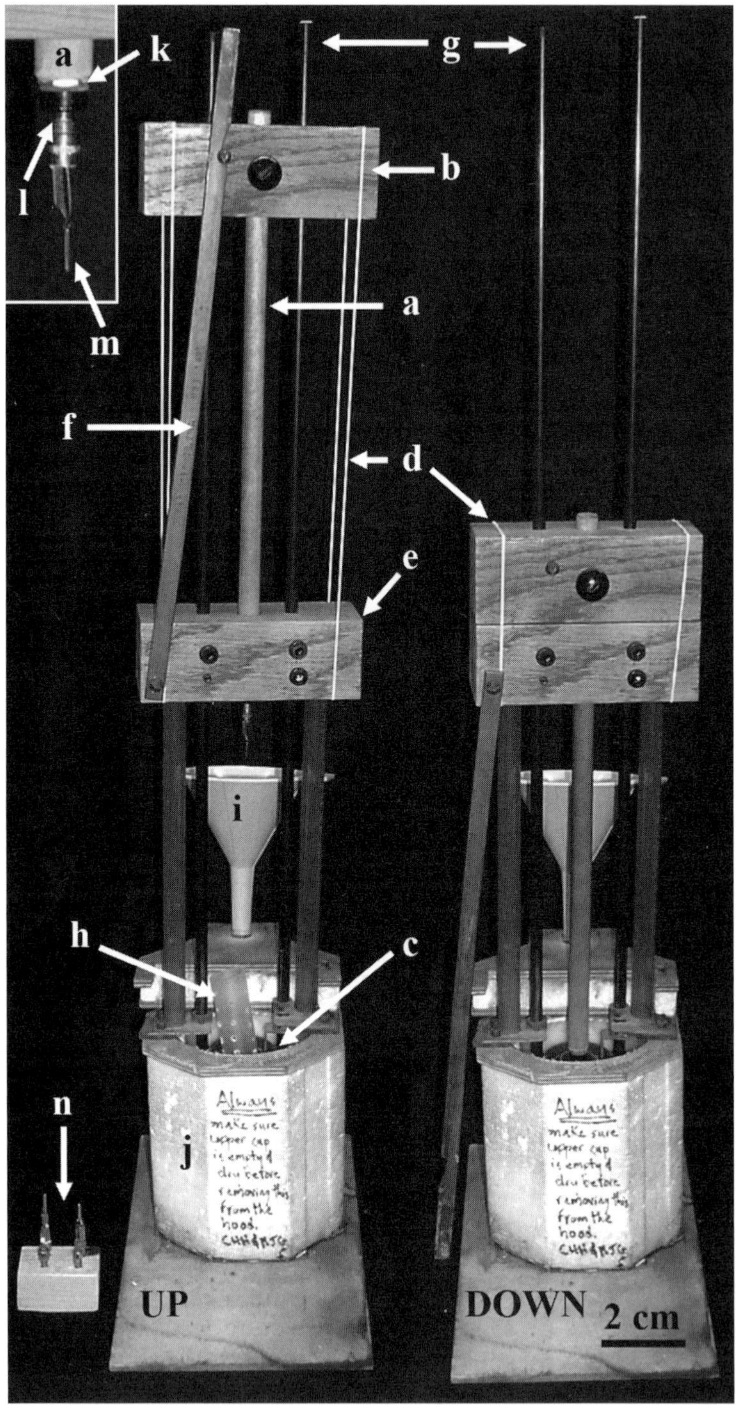
a
k
l
m
b
g
a
f
d
e
i
h
c
n
Always
make sure
upper cap
is empty of
dm before
removing this
from the
hood.
CHEERS!!
j
UP
DOWN
2 cm

exchange or addition of substitution fluids without removing the vials from the block (*see* **Fig. 6**). The cap is severed from the vial, a hole drilled in its center, and a handle created by inserting a small brass or stainless steel machine screw through the hole. Nuts threaded on the screw on either side of the cap hold it in place and at the desired length.

To facilitate transfer of the freeze-substitution module, obtain a Styrofoam shipping container with a cavity of sufficient size and depth to contain the freeze-substitution module (when its two halves are separated) and two or three of the solvent bottles used to store the freeze-substitution solutions. Line the cavity with aluminum foil. The foam box lid should always be in place during transfers between incubation conditions; remove the lid once the freeze-substitution module is in place. The freeze-substitution module should be kept in the foam box while exchanging fluids on the bench-top or in the fume hood to help maintain its temperature (*see* **Note 3**).

2. Freezers and refrigerators: Freeze-substitution requires pre-equilibration of solvents at various temperatures and also the transfer of the freeze-substitution module

---

Fig. 5. *(opposite page)* The plunge-freezer in its loading and prefiring condition (UP) and in its fired position (DOWN; i.e., in this position the sample is immersed and frozen in liquid propane). The specimen to be frozen is attached to the plunging dowel via a sample carrier clamp (*see* caption for inset below). The plunging mechanism consists of a plunging dowel (**A**), which is inserted and tightly held in the moveable upper hardwood block (**B**), which in turn is propelled toward the liquefied propane reservoir (**C**) by rubber bands (**D**) stretched between the stationary bottom hardwood block (**E**) and the upper moveable hardwood block after the cocking mechanism (**F**) is released. The lower block is used to stop the plunging motion of the upper block. A hole slightly larger than the diameter of the dowel has been drilled in the lower block, in alignment with the plunging dowel, to guide it and to allow the dowel to move freely during its plunge. The carbon composite guide rods (**G**) support the bottom block, while the upper block slides along these guides as the sample plunges into the propane. The 20-mL specimen-collection syringe barrel (**H**), which is normally seated in the liquefied propane reservoir, can be seen leaning on the propane reservoir in (UP) and in place in the reservoir in (DOWN). The funnel (**I**) in the back of the device is used to add liquid nitrogen to the bottom of the Styrofoam container (**J**) so the liquid nitrogen does not get into the propane reservoir. Generally, the sample will fall off into the collection syringe after the plunge, but it can also be knocked off with pre-cooled forceps.

The inset (upper left) shows a detail of the end of the plunging dowel (**a**) with a female banana plug connector (**k**) inserted and glued into its drilled end. The male banana plug (**l**) has been modified to hold coverslips or planchets by using Epoxy to glue a vascular clamp (**m**) to its end. The banana plug system makes sample loading and subsequent plunging very fast, so that the specimen is not subjected to drying. Extra specimen carriers are held upright in a foam base so they can be warmed between runs (**n**).

Fig. 6. The freeze-substitution module consists of a solid brass cylinder cut transversely to produce a top and bottom. The bottom has six holes drilled to hold the 10-mL sample chambers, whereas the top has six matching, although shallower, holes to make room for the chamber tops. The top of each polypropylene sample chamber has been modified by attaching a stainless steel machine screw that serves as a handle to facilitate opening of the chamber so substitution fluids and samples can be added without removing the chamber from the module. The hole in the center of the bottom half of the freeze-substitution module is threaded so the bolt running through the top half can firmly attach the two parts together. The entire device is precooled at $-80°C$ overnight prior to freeze-substitution. The microscope slide in the bottom is an example of a flat-embedded sample.

through a graded temperature series (from $-80°C$ to $-20°C$ to $4°C$ to $22°C$). The procedure involves the use of inflammable solvents (ethanol, acetone, methanol); hence, explosion-proof freezers at $-80°C$ and $-20°C$ and an explosion-proof refrigerator at $4°C$ should be used. Consultation with your institution's safety committee is advised.

2.2.2. Reagents

Use electron microscopy-grade fixatives (e.g., glutaraldehyde, osmium tetroxide [OsO_4], tannic acid), available from a number of vendors (e.g., Electron Microscopy Sciences, Ted Pella). Familiarize yourself with the hazards of these fixatives and use appropriate precautions.

All solutions are kept over a molecular sieve to remove water from the exchange media both before and during the substitution procedure. Prior to use, the molecular sieve must be baked at 300–400°C for several hours. All containers at subambient temperatures must only be opened for the minimum time possible to reduce the potential hydration of the dried solutions by condensate formation. Freeze-substitution solutions can be unstable, but the commonly used OsO_4 is relatively stable at ultralow temperatures for extended periods. It is prudent to prepare all solutions in small quantities just prior to use. The container volumes should be such that there is not substantial air space above the liquid. When making these solutions, the solvent (e.g. methanol, acetone, ethanol) should always be at the ultralow temperature prior to adding the fixatives, which can react rapidly with the organic solvent at room temperature and thereby be ruined. It is important that all substitution solutions are precooled by overnight incubation at their respective temperatures prior to the start of a freeze-substitution procedure.

2.3. Specimen Infiltration

1. Specimen rotator (The Cold Rotator, EBSciences, #ER-22C).
2. Embedding resins: Epon or Spurr's low-viscosity Epoxy resin *(13)* or LR White acrylic resin (all from Electron Microscopy Sciences).
3. For low-temperature infiltration with LR White, an appropriate freezer (–20°C) or refrigerator (4°C) is needed. If centrifugation of specimens is required, a microcentrifuge that will maintain those temperatures or operate contained in the freezer or refrigerator.
4. Insect pin (#00, Carolina Biological, #65-4300).

2.4. Embedment

1. Appropriate temperature conditions for resin. Spurr's or Epon require a 70°C oven. LR White requires a Freezer (–20°C) or refrigerator (4°C).
2. Ultraviolet (UV) lamp (Blak-Ray Ultraviolet Lamp, 365 nm long-wave, Ted Pella, #23061).
3. Small binder clips (no larger than the width of the slide).
4. Instantly acting glue (e.g., Super Glue).

2.4.1. Liquid Release Agent-Treated Slides

1. Clean standard microscope slides with 95% (v/v) ethanol.
2. Glue two 22 × 22 mm glass coverslips, one at each end of the slide, with Permount (Fisher).
3. Incubate the slides on a slide warmer at 40–50°C for at least 1 d.
4. Clean the same number of slides, but do not glue coverslips to them.
5. Fill a Coplin jar with the Liquid Release Agent, (Electron Microscopy Sciences, EMS, #70880).

6. Dip the slides (both those with and without attached coverslips) into the jar.
7. Lean the slides against the back of the bench on top of some paper towels and allow them to dry for 2 h to overnight.
8. Place the slides in empty Coplin jars and dry them in a 40–70°C oven.
9. Store slides covered in the Coplin jars. A large number of these slides may be prepared at one time and stored for extended periods.

2.5. Cryogens and Cryogen Storage

1. Liquid nitrogen, the primary cryogen (i.e., the one used to liquefy the secondary cryogen and for storage of specimens), as obtained from any gas supplier. Safety: wear appropriate protective equipment (e.g., cryogenic gloves, cryogenic apron or coat, goggles, face shield) and follow your institution's procedures for working with liquid nitrogen.
2. Liquid propane, the secondary cryogen (i.e., the one responsible for cryoimmobilization). Gaseous propane (as used for barbeque grills) is liquefied by releasing it into a vessel that is surrounded by liquid nitrogen (*see* individual procedures below). A bulk propane adapter (Coleman), high-pressure extension hose (Coleman), and 5-cm-diameter copper filler tube (a specialized gas and welding supply company will have a selection of fittings for propane hoses) are required to fill the liquid propane container. Safety: propane is explosive. All procedures in which it is used should be performed inside a fume hood operating at the highest range of exhaust speed. Ideally, the hood will be dedicated solely for this purpose. The use of propane in a basement laboratory should be avoided. Wear appropriate protective equipment. Avoid open flames and sparks. Care should be taken to insure that all liquid propane has evaporated before removing the apparatus from the fume hood. Seek approval of your propane protocols from your institution's safety committee.
3. Cryogen Dewars, including a 5-L transfer Dewar (for filling apparatus) and a long-term-storage Dewar (e.g., Barnstead-Thermolyne Thermo 30) for storage of frozen specimens.
4. Short-term sample storage box: a shallow Styrofoam box (as received with dry ice shipments from chemical vendors) with an aluminum plate (available from a machine shop) cut to cover its bottom interior.
5. Storage tubes for frozen specimens. These are fashioned from disposable 50-mL conical screw-cap centrifuge tubes (Falcon). Holes are punched through the cap and the tube sides. A Teflon-coated lead weight (made by wrapping a lead fishing weight in Teflon pipe-wrapping tape) is placed in the bottom of the tube and a cotton ball or wadded-up tissue is stuffed into the tube to hold the weight in place. Specimens contained in sample/transfer storage containers (*see* **Subheading 2.1.1.**, **item 6**) or as monolayers on coverslips are placed in the tube, another cotton ball or tissue is stuffed into the top of the tube, and the cap screwed on. A wire of sufficient length to emerge from the top of the storage Dewar and yet allow the tube to sink fully should be attached to the cap.

2.6. Specimen Transfer and Manipulation

1. Fine-tip forceps with O-rings that can hold the forceps in the closed position (Dumont #5, Fine Science Tools).
2. Scalpel (fine point; e.g., Feather, No. 10 or No. 11).

3. Methods

Specimens are first cryoimmobilized, followed by infiltration with an embedding resin. The resin is then polymerized and the specimens can be sectioned for light or electron microscopy. For each stage there are a number of alternatives, the choice of which depends upon the specimen, the nature of the study (i.e., high-resolution structure, immunolocalization), and equipment availability.

3.1. Cryoimmobilization

We describe two cryoimmobilization methods: spray-freezing and plunge-freezing. Other methods include slam-freezing (also known as metal mirror freezing), in which the specimen is rapidly propelled into a precooled (with liquid nitrogen or liquid helium), mirror-polished block; jet-freezing, in which the specimen contained within copper specimen planchets is rapidly frozen by application of high-pressure jets of cryogen; and high-pressure freezing, in which the specimen is momentarily subjected to high pressure prior to freezing with a jet of liquid nitrogen. Refer to **Table 1** for the commercial instruments available for each of these methods.

A critical consideration is depth of freezing. Most methods will produce good freezing only to a depth of 10–30 μm, but depths of 250 μm have been attained with high-pressure freezing *(14)*. It is also possible to extend good freezing to greater depths by use of cryoprotectants. We have not included cryoprotectants in this chapter and refer the reader to **ref.** *2* for a complete discussion of their use. It should be noted that the depth of freezing that we obtained in *D. discoideum* developing culminants often exceeded expectations, perhaps because of the dehydrated state of those stages.

3.1.1. Spray-Freezing

The spray-freezing technique has been developed for particulate samples, including cell suspensions. Spraying has the advantage of reducing the size of the sample into aerosol droplets (from 10 μm to 30 μm) and, therefore, the time required to freeze it *(15)*. Freezing also occurs from all sides of the sample rather than from one direction as in plunging. Spraying can be done in either a liquid cryogen or onto a precooled metal block. One of its disadvantages is potential damage from the shear forces placed on the sample leaving the spray device. The freeze-sprayer that we describe here sprays the aerosol generated by an artist's airbrush into a Nitex funnel assembly that is submerged in liquid propane.

3.1.1.1. Assembly and Testing of the Airbrush

1. Screw the sprayer/airbrush regulator on the nitrogen gas tank and tighten with a wrench. Turn the tank on and adjust the secondary regulator to 20 psi (1.4×10^5 Pa; adjust the regulator while pushing the silver button on the airbrush). Place the airbrush in the hood.
2. Look at the spray orifice on the airbrush (where the spray comes out). This generally has salt build-up from the last spray no matter how well it was cleaned, so it must be cleaned again. Unscrew the spray orifice by backing off the lock nut immediately below it. Rinse the orifice in distilled H_2O. Look for other salt build-up and clean as required.
3. Reassemble the airbrush. Adjust the spray orifice such that the fluid needle is just beginning to emerge from the orifice. Spray some distilled H_2O through the airbrush against a piece of paper to make sure that the aerosol is working correctly. Adjust the orifice setting as necessary to achieve a fine aerosol. Be sure to perform this test prior to the freezing run.

3.1.1.2. Assembly of the Cryogen Container

1. Place the Styrofoam sprayer container in the fume hood. Place the propane reservoir in its holder with the top about 2–3 cm below the top of the container.
2. Fill the container with liquid nitrogen. It is acceptable to allow liquid nitrogen to get into the steel tube at this time: it will help to cool the tube more quickly and will boil off when the warm propane gas is added.
3. After the foam box and steel tube are cooled down and the liquid nitrogen has been refilled to below the top of the metal tube, place the copper spout of the propane fill tube into the **bottom** of the steel propane cup. Slowly open the main valve on the propane tank until a high-pitched whine is heard. This initial flush of gas will boil off significant quantities of liquid nitrogen as it liquefies. Once a boiling sound is heard from the propane in the cup, close the propane tank and refill the device with liquid nitrogen. Take care not to fill liquid nitrogen above the level of the steel propane cup after liquefied propane is in the cup. Liquid nitrogen in the liquid propane reservoir will cause the propane to pop and spatter, causing a safety hazard.
4. Return to filling the cup with propane. Once liquid propane has formed in the bottom of the cup, filling will be relatively rapid if the propane spout is kept slightly under the level of the liquid propane. Once the steel cup is filled to the top with liquid propane, maintain the liquid nitrogen level at just below the top of the tube.

3.1.1.3. Sample Preparation and Spraying

1. Push one of the funnels onto the mesh-covered microcentrifuge collecting tube (note the sample collecting tube number for your sample reference). Drop the

assembly into the propane cup. The funnel top should be below the level of the propane. Top off the propane as required.

2. Do not take more than 10 min to spray and remove the specimens, because the propane will solidify and freeze the sample holder into the cup (*see* **Note 4**).

3. Fill the short-term sample storage box with liquid nitrogen. After its aluminum plate is cooled, add enough liquid nitrogen to just cover the plate's surface.

4. Place the cell suspension (no more than 0.75 mL) in a microcentrifuge tube from which the cap has been removed.

5. Place the plastic siphon tube of the airbrush into the suspension. Tape a piece of paper to the wall of the fume hood for testing the sprayer. Hold the airbrush/tube up to the test paper and push the silver button on the airbrush until the suspension is pulled up from the microcentrifuge tube into the airbrush siphon tube. Use short bursts with the sprayer and wait until the suspension starts to appear on the paper.

6. Hold the airbrush/sample at about a 45° angle immediately above the propane cup. Use short bursts of about a second each to shoot the sample into the funnel. The level of sample should drop in the sample microcentrifuge tube. If the sample level in the tube does not change, the siphon tube or spray orifice might be clogged. Transfer the airbrush to a tube filled with water and spray it against the test paper until it is cleared.

7. Try to avoid spraying the sample into the same spot in the funnel after each burst. While keeping the sprayer at 45°, rotate the tip of the sprayer so the suspension will hit a new place on the funnel with each burst.

8. Repeat the burst spraying into the propane. The spraying of the sample will cause some of the liquid propane to blow out of the propane reservoir. Before each burst, let the propane settle back down. The 0.75-mL sample should be exhausted after five to seven bursts.

9. After the spraying is complete (the sample microcentrifuge tube is empty), obtain a wooden stick (the end trimmed flat, into the shape of a spatula is useful) and use it to gently push the sample that remains on the funnel down into the attached micro-centrifuge tube. Make sure the stick is precooled prior to touching the samples.

10. Lift the sample funnel assembly out of the propane with pliers or forceps (these must be precooled by liquid nitrogen) and place it in the sample storage box. You have about 10 s to accomplish this transfer without risk of damage to the speci-mens. (If the funnel is frozen in the propane, do not force it; *see* **Note 4**.)

11. Remove the funnel from the sample collection tube and seal the tube with one of the perforated microcentrifuge tube caps (*see* **Subheading 2.1.1.**, **item 6**). Leave the sample collection tube in the Styrofoam box for temporary storage until all of the freeze runs are completed. If chunky parts of the frozen sample fall out of the container, they can be picked up and replaced in the container prior to putting the cap on. Be sure to always use precooled forceps. Once the freezing session is concluded, the samples may be stored (*see* **Note 5**).

12. Drop the funnel into a beaker of deionized H_2O (*see* **Note 6**). Shake excess water off and allow it to dry before using it again (a hair dryer can be used to speed this

process). Do not drop the funnel back in the propane while it is wet (the propane will freeze the water onto the mesh, and propane will not be able to get into the funnel during the spraying process).

13. To repeat the spraying process for other samples: refill the liquid nitrogen and propane as required (remember that it is important to avoid getting liquid nitrogen into the propane cup). Do not add propane while samples are in the propane cup, because the warm propane can melt the samples. Spray distilled H_2O through the airbrush to clean the last sample out of the apparatus. Spray nitrogen gas to clear the airbrush of the H_2O.

3.1.1.4. SHUTDOWN OF THE SPRAYER

1. Spray distilled H_2O through the airbrush to clean any leftover sample and media (salts and so on) from the apparatus. Spray nitrogen gas to clear the airbrush of the H_2O prior to storage.
2. Shut off the nitrogen gas at the primary regulator and press the spray button on the airbrush to bleed off excess gas.
3. Do not remove the sprayer (airbrush and all containers) from the hood until all of the propane has evaporated. Do not attempt to speed the process by using a hair dryer or other heating device. A prudent practice is to leave the sprayer in the fume hood overnight.

3.1.2. Plunge-Freezing

The major advantages of plunge-freezing are its applicability to a wide variety of specimens *(2)* and repeatable results *(1)*. In addition, a freeze-plunging device can be easily and inexpensively made *(12)* (*see* **Subheading 2.1.2.**, **item 1**). Successful plunge-freezing will satisfy the following criteria: (1) the specimen is as small as possible; (2) the specimen supports are small and low mass; (3) the secondary cryogen is at its lowest possible temperature while still remaining liquid; (4) the entry rate of the specimen into the coolant is maximized; (5) the specimen is the first part of the injection assembly to enter the cryogen; and (6) the secondary cryogen is of sufficient depth to allow the specimen to keep moving at the entry speed over at least part of the critical cooling phase (0°C to –100°C). Care must be taken to avoid sample precooling (*see* **Note 7**), including keeping the cryogen reservoir at full volume so the specimen does not first pass through a cooling gas layer.

Freeze-plungers are available commercially (*see* **Table 1**). It is possible to hand-plunge, but it is more difficult to obtain consistency of results and avoid precooling. The handmade device we describe meets all of the criteria listed above and has yielded excellent and consistent results.

3.1.2.1. PREPARING THE PLUNGE-FREEZER

1. Depending on the specimen, either a clip or flat-end mount can be used, and each requires adjustment of the plunging depth. For clips: place a specimen clip into

the banana plug at the end of the plunging rod. Lower the rod into the propane cup. The lip of the cup should be even with the place on the clip where the tangs cross over one another. For flat-end mounts: place a flat-end mount into the banana plug at the end of the plunging rod. Lower the rod into the plunging cup. The lip of the cup should be even with the part of the mount that narrows down from the mount portion to the plug portion. Plunging depth for this mount is not as deep as for the clip.

To adjust the height, remove the plunging arm from the guides and loosen the two plastic screws on the wooden block. Push the rod one way or the other to achieve the correct height for the clip (put it back on the guides to check). Retighten the screws and return the plunging mechanism to the guides.

2. Carefully place the plunger in the hood with the plunging mechanism positioned on the graphite guide rods. Lock the plunging mechanism in the "UP" position by hooking the notch on the copper rod onto the screw head on the mechanism.

3. Stretch the rubber band accelerator from the right-hand side of the "Stop-Block" to the right-hand side of the plunging mechanism. Repeat for the left-hand side.

4. Attach the fill hose with the copper fill spout to the "quick-release" connection on the hose coming from the propane tank.

5. Fill the plunger with liquid nitrogen from the Dewar through the funnel system in the back of the plunger (make sure the Tygon tube from the funnel is in the hole in the plunger). Do not fill the nitrogen above the level of the brass propane cup. Wait 1–2 min for the liquid nitrogen to cool down the interior of the plunger and refill the device with liquid nitrogen.

6. After the interior is cooled down and the liquid nitrogen has been refilled in the device, place the copper spout of the propane fill tube into the bottom of the brass propane cup. Slowly open the main valve on top of the propane tank until a high-pitched whine is heard. This initial flush of gas will boil off significant quantities of liquid nitrogen as it liquefies. Once a boiling sound is heard from the propane in the cup, close the propane tank and refill the device with liquid nitrogen. Do not fill liquid nitrogen above the level of the brass propane cup.

7. Return to filling the cup with propane. Once liquid propane has formed in the bottom of the cup, filling will be relatively rapid if the propane spout is kept just under the level of the liquid propane. Once the cup is full, refill the plunger's liquid nitrogen reservoir.

8. Drop the sample collector tube (a 20-mL syringe barrel with holes melted through the sides; *see* **Subheading 2.1.2.**, **item 4** and **Fig. 5H**) into the support ring in the middle of the propane cup and allow it to cool. Top off the propane in the reservoir if the level has dropped during use.

3.1.2.2. Sample Preparation

One advantage of the plunger is the variety of specimens that it can freeze: cell suspensions, cell monolayers on coverslips, and intact multicellular structures on filters or agar. For cell suspensions, use a copper planchet, held in a surgical clamp attached to a banana plug. Place a drop of the suspension on the planchet and let it sit for 30–40 s. Wick off as much excess media and cells as possible, using a filter paper wedge. To obtain the largest number of well frozen cells, it is important to have as thin a layer of cells and medium as possible.

For cells grown on (or suspension cells allowed to settle onto) glass or Thermanox coverslips, remove the coverslips from the medium, attach the clamp/banana plug, and wick off as much medium as possible. For multicellular stages on filters, use flat-ended banana plugs. These allow the samples to plunge "head first" into the cryogen rather than sideways, which might result in the stages being washed from the mount. Put the banana plug mounts in the aluminum holder. Place a **very small** drop of mucilage or other water-based adhesive (glue) on the flat-ended mount (too much mucilage will not allow the samples to fall off by themselves after plunging). Cut a piece of the filter with a scalpel and gently press the filter onto the mucilage on the mount. In considering how to mount specimens, remember that good freezing will only be obtained to a depth of 10–15 μm on the side facing the cryogen. Therefore, orientation of the sample is important.

1. Do not allow samples to air-dry! Allow as little time to lapse from the time that a coverslip, planchet, or filter is mounted to the plunge-freezing. In a quick and practiced move, take the male clip or mount with the specimen and push it into the female plug at the end of the plunging rod. Slide the copper locking rod off of the screw on the plunging mechanism, thereby allowing the specimen to plunge into the cryogen. The specimens usually detach from the clip and fall to the bottom of the plastic tube; those attached to the flat mounts usually do not release. In either case, lift the plunging device just high enough out of the cryogen to determine whether the sample has been released. If not, carefully knock it off into the collecting tube with a pair of cooled forceps (clamp-held samples) or pry it off with a cooled pointed scalpel (flat-mount samples).

2. Lift the plunging mechanism up and cock it again with the copper rod. Remove the banana plug.

3. Repeat for as many samples as required. If you re-use a clip or flat mount, make certain it is warmed to room temperature before using again.

4. The propane level will remain high enough for about 30 min of use (time enough for 15–20 samples). The liquid nitrogen reservoir may need to be refilled once or twice during this period. Propane should not be added while frozen samples are in the collector tube, because the warm gas may adversely affect them. If the propane level drops more than 1.5 cm from the top of the propane cup, remove the specimens (*see* **Subheading 3.1.2.3.**), refill with liquid nitrogen and propane as required, and start the process over.

3.1.2.3. Sample Retrieval

1. If plunging has taken a prolonged period (i.e., over 20 min), the collector tube will probably be frozen in the propane (*see* **Note 4**).

2. Fill the sample storage box with liquid nitrogen. With a pair of forceps, partially lift the collector tube out of the propane and swirl it around to cause any samples frozen against the sides to break free. Using a gloved hand (latex gloves are generally sufficient), lift the collector out of the propane while swirling. In a practiced motion, quickly invert the collector over the storage box and tap it so the samples fall out. Return the collector to the propane. There is a window of approx 10 s in which to accomplish this task without risking damage to the specimens. Look into the collector to determine whether any specimens remain. If they do, swirl the tube again and repeat the above procedure. If samples remain stuck, obtain a long stick or spatula and, after precooling, gently scrape the samples off of the walls of the tube and repeat the above procedure.

3. For sample storage, *see* **Note 5**.

3.1.2.4. Shutdown of the Plunger

1. Do not remove the plunger from the hood until all of the propane has evaporated. Leaving it overnight is the best way to ensure that this has happened.
2. Once the plunger can be removed from the hood, remove the plunging mechanism from the guides and store them together.

3.2. Freeze-Substitution

The objective of freeze-substitution is to slowly substitute the ice in a cryoimmobilized specimen at temperatures below the recrystallization point with an anhydrous liquid that often, but not always, contains a chemical fixative such as OsO_4 or glutaraldehyde *(16)*. Advantages of freeze-substitution over traditional chemical fixation include superior structural preservation *(17)* and applicability to specimens that are sensitive to fixation or dehydration, such as those that cannot be stabilized by chemical cross-linking and which thereby collapse or aggregate upon fixation *(7)*.

Efficient water removal from the exchange media is an important consideration. Excess water in and around the cells can cause ice crystals to grow during the warming process, in turn causing damage to a potentially well frozen specimen *(1)*. Water can be removed by including a molecular sieve in all solutions and incubation chambers and/or by making certain that the volume of exchange media is at least 1000× greater than that of the sample *(16)*. The time course of the warming process during substitution is also important. The samples should be kept in a metal block that can slowly warm as it is moved to progressively higher temperatures. To slow the warming process, the block can be moved to progressively warmer cold cabinets, where it is allowed to equilibrate before moving to a warmer cabinet (e.g., –85°C to –40°C to –20°C to 4°C to room temperature).

The ideal exchange medium would be able to dissolve ice below its recrystallization temperature (–150°C), but the rate of exchange of media with that ability is too slow (on the order of weeks or months) for routine use *(16)*. However, little recrystallization actually occurs until much higher temperatures are reached (around –75°C), and there are several liquids that will dissolve ice at this temperature in a reasonable amount of time (we describe using methanol and acetone; for a list of exchange media, *see* **ref. *2***).

We describe two freeze-substitution protocols, one optimized for excellent structural preservation and the other for immunolocalization. For reviews of theory and practice of freeze substitution, *see* **refs. *2, 17***, and ***18***.

3.2.1. Preparation of the Freeze-Substitution Module

1. Examine the freeze-substitution module (stored between procedures at room temperature) for moisture, particularly in the cavities and chamber vials. Bake the assembly at 70°C to ensure its dryness.
2. Pour just enough baked-out molecular sieve to cover the bottom of the required number of chambers (one per sample, maximum of six). Put the freeze-substitution module cover on and screw it down.
3. Place the freeze-substitution module in the foil-lined foam box with lid and place this in the –80°C freezer. Remove the lid from the foam box and allow the freeze-substitution module to equilibrate overnight.

3.2.2. Placing Samples in the Cooled Freeze-Substitution Module

1. Fill a short-term sample storage box (*see* **Subheading 2.5.**, **item 4**) with liquid nitrogen. Samples not already in sample transfer/storage containers (*see* **Subheading 2.1.1.**, **item 6**) must be transferred into one that has been precooled in liquid nitrogen. Loose specimens in storage tubes (*see* **Subheading 2.5.**, **item 5**) should be poured into the short-term sample storage box and transferred with precooled forceps to the sample transfer/storage container. Precool the perforated container lid and snap it on. Use a flat-sided instrument to push it tightly down.
2. Remove foam box containing the freeze-substitution module (cooled to –80°C) from the freezer. Place a bottle of the required precooled freeze-substitution solvent (for structural studies: 1% w/v OsO_4 in dry acetone; for immunolocalization: 100% dry acetone) in the box. Place the lid on the box (*see* **Note 8**). Remove the freeze-substitution module top, placing it to one side inside the foam box (*see* **Note 3**).
3. Working as quickly as possible, open a chamber and solution bottle and use a Pasteur pipet to fill the chamber with freeze-substitution solvent to just below the rim (the volume of each chamber is 9 mL, equivalent to 3-3.5 pipetfuls. Quickly re-cover the chamber and go to the next one.
4. After the chambers are filled and their covers replaced, return the stocks to the –80°C freezer.
5. Remove the top from a chamber. Using precooled forceps, lift a sample transfer/ storage container with specimens out of the liquid nitrogen and sink it into the

chamber. Hold it down while the residual propane frozen onto the specimens boils off. Note the level of the exchange media after the boil-off, and top the chamber off with more from the freezer if required. Replace the chamber cover.

6. After all of the samples are in their respective chambers, obtain the dry ethanol stock from the freezer. Ethanol is used to ensure good thermal contact between the block and the chambers during the warming process. Lift each chamber out of its slot, and using a Pasteur pipet, fill the slots with the ethanol. Several pipetfuls are required. Ethanol should overflow the slots when the chamber is pushed back down.

3.2.3. Initial Temperature Series

The temperature series regime is the same for both protocols until after the –20°C step, at which point they diverge.

1. Return the freeze-substitution module to the –80°C freezer and leave it for 3–4 d. During this time, the H_2O in the specimens will exchange with the substitution medium.
2. Transfer the freeze-substitution module in the foam box (covered) to the –20°C freezer. Remove the foam box cover and leave the freeze-substitution module for 2–3 d. If using the 1% (w/v) OsO_4 substitution medium, it is during this incubation that the OsO_4 fixes the cell contents.
3. Remove the freeze-substitution module from the freezer and transfer it (still in its foam box) to a fume hood. Wash the samples with three changes of –20°C dry acetone (*see* **Note 3** for precautions to follow). Continue with **Subheading 3.2.4.** for structural studies or **Subheading 3.2.5.** for immunolocalization.

3.2.4. Temperature Series Continuation for Structural Studies

1. Transfer the freeze-substitution module to the 4°C refrigerator. Leave overnight.
2. Transfer to room temperature. In this case only, leave the lid in place on the foam box to slow the warming event. Leave overnight.
3. The samples are now ready for infiltration, described under **Subheading 3.3.1.**

3.2.5. Temperature Series Continuation for Immunolocalization

Resin infiltration and embedding should be carried out at the lowest temperature possible, either –20°C or 4°C, whichever is more convenient. The heat generated by the polymerization of the resin can cause denaturation of sensitive proteins, leading to loss of antigenic determinants. Low-temperature infiltration and embedding helps to avoid this problem. The samples are now ready for infiltration, described under **Subheading 3.3.2.**

3.2.6. Postprocedure Cleanup

1. Wash the brass block, cover, and bolt with water. Wash out the chambers. Discard the molecular sieve. Rinse the sample transfer/storage containers. Air-dry or dry in a 50–70°C oven. Store at room temperature.

2. Clean out the freezers and refrigerator and discard any resin or solvents into the proper waste containers.

3.3. Infiltration of Specimens with Embedment Resin

The freeze-substituted specimens must now be infiltrated with an embedment resin. As with all of the other steps, there are several choices of resins and infiltration conditions (refer to **refs. *6,7*** for options). We continue with our presentation of procedures optimized for structural studies and immunolocalization.

3.3.1. Epoxy Resin Infiltration for Structural Studies

The specimens for structural studies are in 100% acetone at room temperature and contained within a sample transfer/storage container (*see* **Subheading 3.2.4.**) Infiltration is with Spurr's *(13)* or Epon Epoxy resins.

1. Pour a small quantity of dry acetone into a 60-mm glass Petri dish. Empty the contents of a sample transfer/storage container into the dish. Use an insect pin to gently scrape off samples on planchets, filters, or glass coverslips.
2. For single cells, use a plastic pipet to transfer specimens into microcentrifuge tubes. Use the lowest speed on a microcentrifuge to pellet the cells. Remove the supernatant. For intact multicellular stages, simply transfer them by pipet into the microcentrifuge tube (i.e., centrifugation should not be required).
3. Infiltration should proceed through a graded series of resin:dry acetone of 1:3, 1:1, and 3:1. For each step, allow infiltration to proceed for 12–24 h at room temperature with the tubes being rotated. For each change, pellet single cells as before, carefully removing as much of the supernatant as possible before resuspending the pellet in the next mix. For multicellular stages, allow them to settle and then carefully remove the resin mix (alternatively, transfer the stages by pipet to a new tube containing the next mix). Continue until the specimens have had 2–3 changes of 100% resin (*see* **Note 9**).
4. Proceed to embedment (*see* **Subheading 3.4.**)

3.3.2. Acrylic Resin Infiltration for Immunolocalization

The specimens for immunolocalization are in 100% acetone at –20°C or 4°C and contained within a sample transfer/storage container (*see* **Subheading 3.2.5.**) Low-temperature infiltration methods help to retain antigenicity. We describe a procedure using LR White, but other resins are available and have been used by us successfully, including LR Gold and Unicryl.

Follow the same steps as in **Subheading 3.3.1.**, except that all steps must be performed at the appropriate below-ambient temperature (steps carried out in a cold-room, freezer, or incubator) and all resins, dishes, pipets, forceps, and so on should be precooled to the appropriate temperature (*see* **Note 10**). At the conclusion of the infiltration, proceed to **Subheading 3.4.**

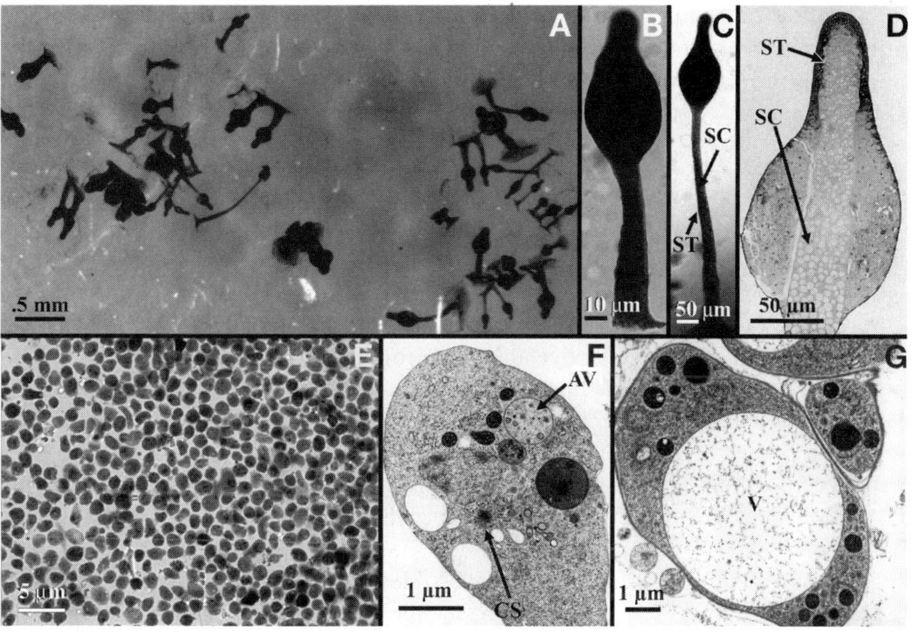

Fig. 7. Examples of cryo-immobilized *Dictyostelium discoideum* specimens. (A–D) Plunge-frozen culminants. (A) A group of culminants on a polymerized flat-embedded slide before being cut out and mounted for sectioning, showing their similar orientations. These can be cut and mounted on a blank in this orientation for longitudinal sections, or turned upright and glued into a slot cut in a blank for transverse sectioning. (B,C) Unsectioned late culminants fixed with OsO_4 showing the overall quality of preservation of the delicate stalk tubes (ST). Note the stalk cells (SC) in the stalk tube in C. (D) Light microscope image of a culminant section from a cryo-prepared flat-embedded sample showing the quality of internal preservation. (E–G) Spray-frozen *D. discoideum* cells. (E) A mass of unsectioned amoebae on a flat-embedded slide showing the large numbers of similarly oriented cells that can be generated with this method. (F,G) Transmission electron microscopy of a prestalk cell (F) and vacuolating stalk cell (G) developed as monolayers. Note the quality of fine structural preservation of delicate internal structure such as the centrosome (CS), and autophagic vacuoles (AV), and large internal vacuoles (V).

3.4. Embedment of Infiltrated Specimens using Flat-Embedding Techniques

Specimens can be embedded in block, but flat-embedding allows the initial identification in the light microscope of specific stages of interest (*see* **Fig. 7**). Embedment proceeds with the specimens spread in a thin layer between two glass slides (*see* **Fig. 6**). Once the resin is polymerized (cured), one of the glass slides can be removed and the remaining resin/slide combination can be examined with a compound microscope. Stages of interest can be cut from the resin, mounted in the desired orientation, and sectioned for transmission electron microscopy in the usual manner.

1. Place several drops of the specimen/resin suspension in the center of a Liquid Release Agent-treated slide with coverslip spacers (*see* **Subheading 2.4.1.**). Place another treated slide (without coverslip spacers) on top and clip each end of the slide sandwich with a binder clip. Fill any air space around the edge of the slide sandwich with resin.

2. Cure Epoxy resins by incubation for 8–16 h at 70°C. For LR White, curing can occur at the same low temperature as the infiltration (–20°C or 4°C). Place the slides on a sheet of reflective aluminum foil. Place a long-wave UV lamp 20 cm above the samples and place another sheet of aluminum foil between the sample and the bulbs of the lamp, thereby blocking the direct irradiation of the samples (curing by slightly indirect UV irradiation is optimum). Curing should be completed within 24 h. However, it is necessary to ensure that the cure is complete. To do so, transfer the specimens to a benchtop at room temperature and provide indirect UV irradiation for several hours (*see* **Note 11**).

3. Remove the binder clips and pry the slides apart; the resin will usually remain with one slide. Place that slide resin-side up on a compound microscope and determine whether the specimens are close to the exposed surface or against the slide (important information enabling one to mount the specimen so that the cells will be sectioned immediately rather than having to section through empty resin to reach the cells).

4. Search for the cells or stage of interest with the compound microscope (for multi-cellular stages, a dissecting microscope may serve this purpose). Mark cells/ stages of interest with a razor blade.

5. Locate the area of interest under the dissecting microscope. Use a new razor blade to cut through the resin on all sides down to the slide surface. This can be a relatively large piece, which will be easier to handle and can be trimmed later.

6. Place a small drop of Super Glue on the tip of a blank resin block (e.g., Spurr's Epoxy resin; *13*). Gently "rough up" the surface of the blank with a razor blade. Place the resin piece with the specimen centered on the tip of the Spurr's block and closest to the top. Press gently. Allow to dry overnight.

7. Trim the block as usual for ultramicrotomy. Sectioning, collection of sections on grids, poststaining, and observation in the transmission electron microscope should proceed by standard procedures (*see* **refs.** *8–11* for details).

4. Notes

1. Always use the same tubes to construct the filter funnels and sample transfer and storage containers so that the parts will be interchangeable. Microcentrifuge tubes from Eppendorf tend to perform the best for these purposes at liquid-nitrogen temperatures.

2. The following parts are required for construction of the Mollenhauer *(12)* plunger. From a model airplane supply store: 4 Nylon wing bolts (1/4-20 × 2"; Du-Bro #142); 4 Nylon oval head wing mount bolts (1/4-20 × 2"; Rocket City Specialties); 8 Nylon wing bolts (10-32 × 2"; Du-Bro #164); 10 plated brass dura-collars (1/4"; Du-Bro #244); 10 threaded inserts (8-32; Du-Bro #393); 2 fiberglass pushrod systems (82 cm; Dave Brown Products, Inc., #5400); 4 8-32 × 3/4 socket head screws (steel) (Carl Goldberg Models, Inc., #525); 1 oak dowel rod (32 × 1 cm); 1 sheet of 1/4" × 12" × 24" birch plywood; 15-min "Mid-Cure" Epoxy (two-part Epoxy kit, although any two-part Epoxy will work); Southern's sorghum glue (for gluing wood to foam; Dave Brown Products). From an electronics supply store: Banana Jack connector (fits 8.2-mm hole; Radio shack # 274-725B); solderless banana plugs (1 for each sample holder; Radio Shack # 274-721c). From a quality wood supply store: one block of oak wood (at least 5.5" × 4" × 2"). From a metal supply store: one copper or brass tube (7 cm long × 4 cm wide); one brass or copper tube (15 cm long × 0.5 cm wide); one brass tube (1 cm long × 3 cm wide); one sheet of brass or copper to silver-solder on to the bottom of the 7-cm long tube and some extra to form the guide supports for the injector rod assemblies.

3. Working within the confines of the foam box is important to the success of this procedure. The freeze-substitution module chills the air within the cavity, which excludes the warm air of the room and minimizes condensation. If prechilled transfer pipets are kept within the cavity, condensation on them is reduced. All solution exchanges should be performed as quickly as possible.

4. If the sample holder freezes and traps your specimens, or the filter assembly becomes frozen in the propane, you will need to wait until the propane melts (30–45 min). Do not try to force the filter assembly out. As long as there is liquid propane in the cup, the samples will remain well frozen. Liquid propane will take about 90 min to boil completely away, but check it often. After the propane is melted, the filter assembly can be easily removed or the samples retrieved as outlined in **Subheading 3.1.1.3., items 10** and **11**.

5. Frozen specimens can be stored in a liquid nitrogen storage Dewar. Specimens are placed in the storage tubes (*see* **Subheading 2.5., item 5**). Storage tubes must be precooled in liquid nitrogen; this is most conveniently accomplished in the short term sample storage box (*see* **Subheading 2.5., item 4**) in which the specimens are being collected, but take care to cool the tube in an area of the box away from the specimens. It is important that the tubes each have unique numbers and accurate records are kept concerning what specimens are in a particular tube and the location of a particular tube in the storage tube. This minimizes the time taken to retrieve samples, an important consideration in reducing the possibility of sample melting.

6. Filter-funnels, transfer tubes, and caps can be reused indefinitely if they are soaked in water immediately after each use. Repair as necessary with the silicon glue.

7. It is important to avoid circumstances that might lead to precooling of specimens (e.g., placement of a culture near or above cryogens or near chilled implements

or surfaces, using chilled implements to handle filters, coverslips, or planchets, or re-using specimen clips and mounts without first allowing them to warm to room temperature). It is also important to avoid circumstances that might lead to melting of specimens (e.g., taking too much time to transfer a sample from the primary cryogen to the storage cryogen, handling specimens with implements that have not been prechilled, neglecting to prechill storage tubes, pipets, or handling instruments).

8. It is possible to have different freeze-substitution solutions in each cavity of the freeze-substitution module, although it is necessary that they all have the same processing times.

9. In our experience, we have found that freeze-substituted samples require longer infiltration times than chemically fixed ones.

10. Single cells that were spray-frozen will have their original medium frozen around them. If the cells were freeze-substituted in acetone, the salts in the medium may have precipitated and will obscure the samples in low-temperature embedment. The salts can be dissolved by following the –20°C stage of freeze-substitution (when the samples will still be in acetone; *see* **Subheading 3.2.3.**, **item 3**) with one wash of 1:1 acetone:methanol and two washes of 100% methanol (all at –20°C). Perform the LR White infiltration series with resin:methanol rather than resin:acetone mixes.

11. If OsO_4 has been used in the substitution protocol, UV light cannot be used to cure the resin because the black samples will block the UV light.

References

1. Echlin, P. (1992) *Low-temperature Microscopy and Analysis*. Plenum, New York.
2. Robards, A. W. and Sleytr, U. B. (1985) *Low Temperature Methods in Biological Electron Microscopy*. Elsevier, New York.
3. Malecki, M. and Roomans, G. M. (eds.) (1996) *The Science of Biological Specimen Preparation for Microscopy. Scanning Microscopy Supplement no. 10*. Scanning Microscopy International, Chicago.
4. Plattner, H. (1989) *Electron Microscopy of Subcellular Dynamics*. CRC, Boca Raton, Florida.
5. Polak, J. M. and Van Noorden, S. (1984) *An Introduction to Immunocytochemistry: Current Techniques and Problems. Royal Microscopical Society Handbook no. 11*. Oxford University Press, Oxford, UK.
6. Polak, J. M. and Priestley, J. V. (1992) *Electron Microscopic Immunocytochemistry: Principles and Practice*. Oxford University Press, Oxford, UK.
7. Newman, G. R. and Hobot, J. A. (1993) *Resin Microscopy and On-section Immunocytochemistry*. Springer-Verlag, Berlin, Germany.
8. Bozzola, J. J. and Russell, L. D. (1999) *Electron Microscopy: Principles and Techniques for Biologists*. Jones and Bartlett, Sudbury, MA.
9. Dykstra, M. J. (1992) *Biological Electron Microscopy: Theory, Techniques and Troubleshooting*. Plenum, New York.

10. Dykstra, M. J. and Reuss, L. E. (2003) *Biological Electron Microscopy: Theory, Techniques and Troubleshooting.* Kluwer Academic/Plenum, New York.
11. Watt, I. M. (1997) *The Principles and Practice of Electron Microscopy.* Cambridge University Press, Cambridge, UK.
12. Mollenhauer, H. (1999) Simple apparatus for plunge-freezing of biological material: some design considerations. *Microsc. Res. Tech.* **44,** 195–200.
13. Spurr, A. R. (1969) A low-viscosity Epoxy resin embedding medium for electron microscopy. *J. Ultrastruct. Res.* **26,** 31–43.
14. Dahl, R. and Staehelin, L. A. (1989) High pressure freezing for the preservation of biological structure: theory and practice. *J. Electron Microsc. Techn.* **3,** 305–335.
15. Roos, N. and Morgan, A. J. (1990) *Cryopreparation of Thin Biological Specimens for Electron Microscopy: Methods and Applications.* Oxford University Press, New York.
16. Steinbrecht, R. A. and Muller, M. (1987) Freeze-substitution and freeze-drying, in *Cryotechniques in Electron Microscopy* (Steinbrecht, R. A. and Zierold, K., eds.). Springer-Verlag, Berlin, Germany: pp. 149–172
17. Hobot, J. A., Villiger, W., Escaig, J., Maeder, M., Ryter, A., and Kellenberger, E. (1985) Shape and fine structure of nucleoids observed on sections of ultrarapidly frozen and cryosubstituted bacteria. *J. Bacteriol.* **162,** 960–971.
18. Harvey, D. W. (1981) Freeze substitution. *J. Microsc.* **127,** 209–221.

IV

Dictyostelium as Model Organism

22

Analysis of Signal Transduction

Formation of cAMP, cGMP, and Ins(1,4,5)P₃ In Vivo and In Vitro

Peter J. M. Van Haastert

Summary

This chapter describes biochemical assays with which to analyze signal transduction from surface cAMP receptors via G proteins to the effector enzymes adenylyl cyclase, guanylyl cyclase, and phospholipase C. The cAMP-mediated formation of the second messengers cAMP, cGMP, and Ins(1,4,5)P$_3$ in cells will provide information on the entire pathway; here, details on isotope-dilution assays, through which the concentrations of these second messengers can be determined, are offered. Subsequently, several assays that analyze specific parts of the sensory transduction pathway are described. The assays include equilibrium and nonequilibrium binding to surface cAMP receptors on cells and on isolated membranes. The interaction between the cAMP receptor and G proteins can be measured as GTPγS-mediated inhibition of cAMP-binding to the receptor, or as cAMP-mediated stimulation of GTPase activity or GTPγS-binding to G proteins. Finally, assays for the *in vitro* G protein-mediated activation of the effector enzymes adenylyl cyclase, guanylyl cyclase, and phospholipase C are described.

Key Words: cAMP; cGMP; Ins(1;4;5)P$_3$; adenylyl cyclase; guanylyl cyclase; phospholipase C; G protein; receptor; GPCR.

1. Introduction

The eukaryotic microorganism *Dictyostelium discoideum* provides a convenient model with which to investigate signal transduction. The sensory transduction mechanism of extracellular cAMP resembles hormone signal transduction in mammalian cells *(1)*. cAMP is detected by surface receptors, which activate several G proteins and second messenger enzymes including adenylyl cyclase, guanylyl cyclase, phospholipase C, and phosphatydyl inositol 3-kinases *(2,3)*. The completely sequenced genome *(4)* reveals many genes that potentially transduce extra- or intracellular signals to chemotaxis, morphogenesis, or cell

From: *Methods in Molecular Biology, vol. 346:* Dictyostelium discoideum *Protocols*
Edited by: L. Eichinger and F. Rivero © Humana Press Inc., Totowa, NJ

differentiation. These genes include at least 55 G protein-coupled receptors, 11 G-protein α-subunits, 2 G protein Gβ and 1 G protein Gγ subunit *(4)*. In addition, three adenylyl cyclases, two guanylyl cyclases, one phospholipase C, and at least six type I phosphatydyl inositol 3-kinases have been identified. The produced second messengers are degraded by a multitude of enzymes and usually fulfill their action by activating multiple target enzymes. cAMP and cGMP are degraded by seven cyclic nucleotide phosphodiesterases that have different cellular localization and substrate specificities *(5)*. cAMP activates a relatively classical protein kinase A (PKA) *(6,7)*, whereas two unusual target proteins for cGMP have been identified that contain a multitude of domains including Ras, Ras-GEF, and MAPKKK *(8,9)*. The product of phospholipase C, Ins(1,4,5)P$_3$, is degraded by a series of phosphatase reactions *(10,11)* and activates an Ins(1,4,5)P$_3$ receptor to release Ca^{2+} ions from the endoplasmatic reticulum *(12)*. The product of phosphatydyl inositol 3-kinases, PtdIns(3,4,5)P$_3$, may be degraded by the 3-phosphatase PTEN *(13)* and by several 5-phosphatases *(14)*. Areas of membranes enriched in PtdIns(3,4,5)P$_3$ form binding regions for proteins with pleckstrin homology (PH) domains *(15)*, of which there are at least 40 recognized in the genome. These second messengers may regulate a multitude of other proteins, such as kinases, guanine nucleotide exchange factors, actin binding proteins, transcription factors, and so on. In *Dictyostelium,* inactivation of a gene is easily achieved by homologous recombination. This has been done for a number of the genes listed above, which may allow the identification of that proteins' function in signal transduction.

Biochemical assays are indispensable tools for the analysis of both wild-type and mutant cells. This chapter is an update on previously published assays *(16,17)* for the analysis of signal transduction from surface cAMP receptors via G proteins to the effector enzymes adenylyl cyclase, guanylyl cyclase, and phospholipase C.

2. Materials

2.1. General Materials

1. [³H]cAMP (34 Ci/mmol = 1.26 TBq/mmol, Amersham).
2. [³H]cGMP (25 Ci/mmol = 0.91 TBq/mmol, Amersham).
3. [³H]Ins(1,4,5)P$_3$ (40 Ci/mmol = 1.48 TBq/mmol, Amersham).
4. [³⁵S]GTPγS (1320 Ci/mmol = 48.8 TBq/mmol, New England Nuclear).
5. [γ-³²P]GTP (43 Ci/mmol = 1.59 TBq/mmol, New England Nuclear).
6. cAMP, 2'deoxy-cAMP, (Sp)-cAMPS, cGMP, ATP, GTP, AppNHp, ATPγS, GTPγS, GDPβS (Boehringer).
7. Dithiothreitol (DTT): 1 *M* stock solution stored at −20°C, made fresh each week.
8. Silicon oils AR 20 and AR 200 (Wacker-Chemie, Munich).
9. Scintillator 299 (Packard).

10. Nucleopore polycarbonate filters, 3 μm pore size, mounted in filter holders (Whatman).

2.2. Dictyostelium *Culture and Preparation of Membranes*

1. HG5 medium (1 L): 14.3 g peptone, 7.15 g yeast extract, 10 g glucose, 0.49 g KH_2PO_4, and 1.36 g $Na_2HPO_4\cdot2H_2O$, autoclaved. HG5 medium: an economic version of HL5 medium *(18)*.
2. Solid SM medium (1 L): 15 g agar, 3.3 g glucose, 3.3 g peptone, and 4.5 g KH_2PO_4 and 1.5 g $Na_2HPO_4\cdot2\ H_2O$, autoclaved.
3. PB: 10 mM Na_2HPO_4/KH_2PO_4 buffer, pH 6.5.
4. Nonnutrient agar (1 L): 15 g agar in PB, autoclaved.
5. LB^- buffer: 40 mM HEPES/NaOH, 0.5 mM EDTA, pH 7.7.
6. LB^+ buffer: LB^- supplemented with 250 mM sucrose.

2.3. *Isotope-Dilution Assays for cAMP, cGMP, and Ins(1,4,5)P₃*

1. cGMP assay buffer: 200 mM K_2HPO_4/KH_2PO_4, 10 mM EDTA, 10 mM EGTA, pH 7.0; store at 4°C.
2. cAMP assay buffer: 100 mM K_2HPO_4/KH_2PO_4, 10 mM EDTA, 2 mg/mL bovine serum albumin (BSA), 3 mM NaN_3, pH 7.0; store at 4°C.
3. Ins(1,4,5)P₃ assay buffer: 100 mM Tris-HCl, 4 mM BSA, 4 mM EDTA, pH 9.0; store at 4°C.
4. Tracers: 100,000 cpm cGMP and Ins(1,4,5)P₃ or 400,000 cpm cAMP per mL in assay buffer, prepare fresh.
5. 60% saturated ammonium sulfate precipitate of calf serum, dissolved in cGMP assay buffer to the original volume; store at 4°C.
6. cGMP antiserum, diluted in the calf serum preparation; prepare fresh (*see* **Subheading 3.3.1.**).
7. cAMP-binding protein, diluted in cAMP assay buffer; prepare fresh (*see* **Subheading 3.3.1.**).
8. Ins(1,4,5)P₃-binding protein, undiluted (*see* **Subheading 3.3.1.**).
9. Charcoal (5% w/v): 1.25 g of activated charcoal and 0.5 g of BSA in 25 mL cAMP assay buffer; store at 4°C.

2.4. *cAMP Binding Assays*

1. Aggregation competent cells at 10^8 cells/mL in PB
2. Radioactive binding mixture containing per mL: 4 μL of [³H]cAMP stock, 50 μL of 1 M DTT, and 946 μL of PB; prepare fresh from stock solutions that are stored at –20°C.
3. Mixture of silicon oils. The mixture for 20°C is AR20:AR200 = 2:1; store at 22°C. The mixture for 0°C is 11:4; store at 4°C.
4. 90% saturated ammonium sulphate in PB. Prepare saturated ammonium sulfate in PB at room temperature, place on ice until equilibrated, and dilute the clear solution with PB; store at 4°C.
5. BSA in H_2O at 1 mg/mL; store at –20°C.

6. Sucrose, 20 % (w/v) in PB; store at 4°C.
7. Acetic acid 0.1 *M*; store at 22°C.

2.5. cAMP-Induced Second Messenger Responses

1. Aggregation-competent cells are suspended at 5×10^7 cells/mL in PB, and air is bubbled through the cell suspension for 10 min at a rate of about 15 mL air per mL suspension per min.
2. Stimulus solution: 5 µ*M* cAMP in PB for the cGMP and Ins(1,4,5)P$_3$ response, and 50 µ*M* 2'-deoxy-cAMP and 50 m*M* DTT in PB for the cAMP response; prepare fresh from 1 m*M* stock solutions that are stored at –20°C.
3. 3.5% (v/v) perchloric acid (PCA); store at 22°C.
4. KHCO$_3$ (50% saturated at 22°C); store at 22°C.

2.6. GTP-Inhibition of cAMP Binding

1. Membranes resuspended in PB to a density equivalent to 10^8 cell/mL, prepared as described under **Subheading 3.2.**; membranes are used within 1 h after preparation.
2. Radioactive binding mixture containing per mL: 4 µL of [^{3}H]cAMP stock, 50 µL of 1 *M* DTT, and 946 µL of PB; prepare fresh from stock solutions that are stored at –20°C.
3. 1% sodium dodecyl sulfate (SDS); store at 22°C.
4. 1 m*M* cAMP; store at –20°C.
5. 0.3 m*M* GTPγS; store at –20°C.
6. Acetic acid 0.1 *M*; store at 22°C.

2.7. cAMP-Stimulation of GTPγS Binding

1. Membranes resuspended in PB to a density equivalent to 10^8 cells/mL, prepared as described in **Subheading 3.2.**; membranes are used within 1 h after preparation.
2. Binding mix: 2 n*M* [^{35}S]GTPγS, 1 m*M* ATP, 30 m*M* MgCl$_2$, and 20 m*M* PB; prepare fresh from 1 mM stock solutions that were stored at –20°C.
3. 1% SDS; store at 22°C.
4. 1 m*M* cAMP; store at –20°C.
5. 1 m*M* GTP; store at –20°C.
6. Acetic acid 0.1 *M*; store at 22°C.

2.8. cAMP-Stimulation of GTPase Activity

1. Membranes are prepared as described under **Subheading 3.2.**, except that the membrane pellet is washed once with 10 m*M* triethanolamine-HCl, pH 7.4 (TAA) containing 0.5 m*M* EDTA, and the final pellet is resuspended in TAA to the equivalent of 1×10^8 cells/mL.
2. [γ-^{32}P]GTP, 0.1 µCi/assay (60,000 cpm/assay).
3. 0.1, 1, and 500 µ*M* GTP; store at –20°C.

4. 100 μ*M* cAMP; store at –20°C.

5. Activated charcoal: 5% (w/v) in 50 m*M* sodium phosphate buffer, pH 2.0; store at 4°C.

6. Reaction mixture, the content per mL is prepared from stock solutions as follows: 125 μL of 1 *M* TAA, 50 μL of 10 m*M* AppNHp, 25 μL of 1 *M* DTT, 25 μL of 10 m*M* EGTA, 50 μL of 100 m*M* MgCl$_2$, 25 μL of 10 m*M* ATPγS, 250 μL of BSA (20 mg/mL), 0.4 mg of creatine kinase, 4 mg of creatine phosphate, 450 μL of H$_2$O; prepare fresh from 1 m*M* stock solutions that are stored at –20°C.

2.9. GTPγS-Stimulation of Adenylyl Cyclase In Vitro

1. Aggregation competent *Dictyostelium* cells. Starve cells on nonnutrient agar for 16 h at 6°C and shake for an additional h at 22°C at a density of 10⁷ cells/mL in PB. Collect cells by centrifugation and adjust to a density of 1.5 × 10⁸ cells/mL in ice-cold PB.

2. Lysis buffer: 20 m*M* Tris-HCl, pH 8.0, 4 m*M* MgCl$_2$; store at 4°C.

3. 1 mL assay mixture: 40 μL of 0.5 *M* DTT, 10 μL of 0.1 *M* ATP, 510 μL of lysis buffer, 440 μL of H$_2$O; prepare fresh from 1 m*M* stock solutions that were stored at –20°C.

4. 1 m*M* GTPγS; store at –20°C.

5. 1 m*M* 2'deoxy-cAMP; store at –20°C.

6. 0.5 *M* DTT; store at –20°C.

7. 0.1 *M* ATP; store at –20°C.

8. 0.1 *M* EDTA, pH 8.0; store at 22°C.

2.10. GTPγS-Regulation of Phospholipase C

1. Aggregation-competent *Dictyostelium* cells. Starve cells for 4 h by shaking at 150 rpm in PB 22°C at a density of 10⁷ cells/mL. Collect cells by centrifugation, wash, and resuspend in 40 m*M* HEPES-NaOH, pH 6.5, at a density of 5 × 10⁷ cells/mL.

2. 59 m*M* CaCl$_2$; store at 4°C.

3. 118 m*M* EGTA; store at 4°C.

4. 3.5% (v/v) PCΛ; store at 22°C.

5. 20 μ*M* cAMP; store at –20°C.

6. 1 m*M* GTPγS; store at –20°C.

2.11. GTPγS-Stimulation of Guanylyl Cyclase

1. Aggregation-competent *Dictyostelium* cells. Starve cells on nonnutrient agar for 16 h at 6°C and shake for an additional hour at 22°C at a density of 10⁷ cells/mL in PB. Collect cells by centrifugation; wash and resuspend the cells in 40 m*M* HEPES-NaOH, pH 7.0 to a density of 1.5 × 10⁸ cells/mL.

2. Lysis buffer: 40 m*M* Hepes-NaOH, 6 m*M* MgSO$_4$, 2 m*M* EGTA, 0.2 m*M* AppNHp, pH 7.0; prepare fresh from 1 m*M* stock solutions that are stored at –20°C.

3. Assay mixture (1 mL): 20 µL of 0.5 M DTT, 60 µL of 0.01 M GTP, 920 µL of H_2O; prepare fresh from 1 mM stock solutions that are stored at –20°C.
4. 0.5 M DTT; store at –20°C.
5. 0.01 M GTP; store at –20°C.
6. 1 mM GTPγS; store at –20°C.
7. PCA 3.5 % (v/v); store at 22°C.

3. Methods

Under **Subheadings 3.1–3.3.**, general assays for preparation of *Dictyostelium* cells and membranes, as well as isotope dilution assays for determining the concentration of cAMP, cGMP, and Ins(1,4,5)P$_3$, are described. Under **Subheadings 3.4–3.11.**, assays with which to investigate the interactions between the cAMP receptor, G protein, and effector enzymes are described. These interactions are transient in cells and unstable in vitro. It is important to complete the experiments as soon as possible after preparation of the cells or lysates. Therefore, each assay is divided into three parts: (1) preparations that should be implemented before the cells are lysed, (2) the assay itself, which should be completed within a short period, and (3) procedures that are carried out afterward in order to finish the experiment.

3.1. Dictyostelium *Culture Conditions*

The strains that have been used in G protein studies are wild-type NC4, axenic strains AX2 or AX3, several chemically mutagenized strains derived from NC4, and transformants derived mainly from AX3. Axenic strains are grown at 22°C in liquid HG5 medium. Nonaxenic strains are grown in co-culture with *Klebsiella aerogenes* on solid SM medium. Axenically grown strains are harvested in the late logarithmic phase and bacterially grown strains just before clearing of the bacterial lawn. Cells are washed three times in PB by repeated centrifugation at 300g for 3 min and resuspension in PB.

These vegetative cells acquire aggregation competence by starvation. Cells are starved either by shaking in suspension for 4–6 h (10^7 cells/mL in PB at 150 rpm and 22°C), or by incubation on nonnutrient agar (2.5×10^6 cells/cm^2). These plates are incubated at 22°C for 4–6 h or at 8°C for 16 h.

3.2. *Preparation of Membranes*

This is a general procedure for the rapid preparation of lysates or membranes that preserve the interaction between G proteins and surface receptors or effector enzymes . For some assays, other buffers are used.

1. After starvation, *Dictyostelium* cells are washed twice with PB, once with LB⁻ buffer, and resuspended in LB⁺ buffer to a density of 1–2 $\times$ 10^8 cell/mL. Cells and solutions are kept at 0°C during the procedure.

2. Homogenization of cells is performed by pressing the cell suspension through a Nuclepore filter (pore size 3 μm).
3. The lysate is centrifuged at 14,000g for 5 min at 4°C.
4. The pellet is washed once with PB, and the final pellet is resuspended in PB to a density equivalent to 2×10^8 cells/mL. The membranes are kept on ice during the experiment, which is completed within 1 h after membrane preparation (*see* **Note 1**).

3.3. Isotope-Dilution Assays for cAMP, cGMP, and Ins(1,4,5)P₃

Three isotope-dilution assays are used for the determination of cAMP, cGMP, or Ins(1,4,5)P₃ levels, respectively. These assays are based on the competition between a fixed quantity of radioactive tracer and the unlabeled second messenger from the sample for binding to a specific antibody or receptor. The amount of labeled tracer bound to the binding protein is inversely related to the amount of second messenger in the sample. Commercial assay kits are available for each ligand. We have used the cGMP radioimmunoassay and the cAMP- and Ins(1,4,5)P₃-binding protein kits from Amersham. For optimal use, we have modified the protocols for these commercial kits as follows: the reagents are dissolved as described by the manufacturer, but in the assay all volumes are divided by five. When many assays are to be performed, we suggest preparing your own assay kits as outlined below.

3.3.1. Outline for the Preparation of Isotope Dilution Assay Kits

The main ingredients in all kits are a radioactive tracer of high specific activity (>25 Ci/mmol; >1TBq/mmol) and a protein that binds the radioactive tracer with high specificity and affinity. The binding proteins are diluted in the appropriate buffers such that in the absence of an unlabeled competitor, about 30% of the radioactive tracer is bound (*see* **Note 2**).

1. The anti-cGMP antiserum is prepared in rabbits by immunization of BSA-coupled cGMP as described in **ref. *19***. The concentration of cGMP that induces half-maximal displacement of tracer cGMP should be below 10 nM. Specificity is analyzed using different concentrations of cAMP, GTP, GDP, GMP, and ATP. A good 20-mL bleed provides sufficient materials for about 40,000 assays.
2. The regulatory site of cAMP-dependent protein kinase type I is used as the cAMP-binding protein, and is isolated from bovine muscle as described *(20)*. In 3 d, the binding protein is isolated from 500 g of fresh beef shoulder sufficient for 200,000 assays.
3. The Ins(1,4,5)P₃-binding protein is isolated from bovine liver *(21,22)*. Bovine liver (500 g fresh from the slaughterhouse, cut in small pieces) is homogenized in 1 L of 20 mM NaHCO₃. The homogenate is centrifuged for 10 min at 500g at 4°C to remove larger materials. The supernatant is centrifuged again for 20 min at 20,000g at 4°C, and the pellet is resuspended in 15 volumes (about 400 mL) of 20 mM Tris-HCl, pH 7.5. The protein concentration is now approx 15 mg/mL.

Store as 1-mL aliquots at –80°C. In 4 h, 500 g of liver can be processed for 20,000 assays.

3.3.2. Procedure for cGMP

1. Incubate 20 μL of tracer, 20 μL of H_2O, standard or sample, and 20 μL of cGMP antiserum at 0°C in 1.5-mL tubes.
2. Terminate the incubation after 2 h by addition of 0.5 mL of 60% saturated ammonium sulfate. Incubate for 5 min.
3. Centrifuge tubes for 2 min at 14,000g; aspirate the supernatant.
4. Finish by adding 100 μL of H_2O. Dissolve the pellet and add 1.3 mL of scintillator.

3.3.3. Procedure for cAMP

1. Incubate 20 μL of tracer, 20 μL of H_2O, standard or sample, and 20 μL of cAMP-binding protein at 0°C in 1.5-mL tubes.
2. Terminate the incubation after 2 h by addition of 60 μL of charcoal suspension. Incubate for 1 min.
3. Centrifuge tubes for 2 min at 14,000g.
4. Transfer 90 μL of the supernatant to a scintillator vial. Finish by adding 2 mL of scintillator.

3.3.4. Procedure for Ins(1,4,5)P₃

1. Incubate 20 μL of tracer, 20 μL of H_2O, standard or sample, and 20 μL of Ins(1,4,5)P_3 binding protein at 0°C in 1.5-mL tubes.
2. Terminate the incubation after 10 min by centrifugation of the tubes for 2 min at 14,000g.
3. Aspirate the supernatant.
4. Finish by adding 100 μL of H_2O. Dissolve the pellet and add 1.3 mL of scintillator.

3.3.5. Calculation

The binding of radioactive tracer is determined in the absence of cold ligand (C_0), in the presence of excess cold ligand (Bl), and in the presence of known amounts of ligand or unknown samples (C_x). The relationship between pmole ligand and measured C_x is given by the following equation:

$$\text{pmole} = X*[(C_0 - Bl)/(C_x - Bl) - 1]$$

The value of X is determined from a standard curve with known amounts of ligand in pmole at the abscissa and the observed $(C_0\text{-Bl})/(C_x\text{-Bl})-1$ at the ordinate. Then, X is used to calculate the amounts of unknown ligand in the samples.

3.3.6. Results

A summary of the primary data for the three isotope dilution assays is presented in **Table 1**. Each assay has a characteristic ratio of maximal/minimal

Table 1
Primary Data for Isotope-Dilution Assays, Receptor-Stimulated Responses In Vivo, and Receptor- and G Protein-Stimulated Enzyme Activities In Vitro

	cGMP		Ins(1,4,5)P$_3$		cAMP	
	cpm	pmol	cpm	pmol	cpm	pmol
Standard curve						
Input radioactivity	2000		2000		8000	
0 pmol standard (C$_0$)	620 ± 31		767 ± 8		2417 ± 224	
0.1 pmol standard	511 ± 23		ND		2012 ± 186	
0.25 pmol standard	403 ± 32		635 ± 12		1726 ± 163	
0.5 pmol standard	310 ± 20		545 ± 16		1346 ± 129	
1 pmol standard	227 ± 19		463 ± 15		1007 ± 98	
2 pmol standard	135 ± 11		369 ± 20		675 ± 63	
4 pmol standard	148 ± 8		315 ± 14		461 ± 53	
10 pmol standard	ND		248 ± 7		303 ± 26	
blank (Bl)	95 ± 7		196 ± 10		178 ± 16	
X	0.376		0.895		0.572	
Cell stimulation						
Unstimulated	424 ± 21	0.22 ± 0.04	572 ± 53	0.54 ± 0.22	2056 ± 192	1.1
Stimulated cells	194 ± 11	1.56 ± 0.21	528 ± 38	0.72 ± 0.15	591 ± 56	22.4
Enzyme activities						
Lysate t$_0$	425 ± 11	0.21 ± 0.02	568 ± 43	0.53 ± 0.20	2525 ± 227	0
Basal activity	320 ± 18	0.49 ± 0.07	272 ± 9	6.50 ± 0.90	1530 ± 148	9.7
Plus GTPγS	233 ± 15	1.02 ± 0.15	251 ± 5	9.36 ± 1.75	515 ± 49	82.6
Plus cAMP	ND		246 ± 3	10.25 ± 0.65	1145 ± 103	19.1
Plus cAMP and GTPγS	ND		247 ± 6	10.10 ± 1.81	362 ± 34	162

ND, not determined.

binding; these values determine the range of ligand concentrations that can be determined accurately. The cAMP and cGMP assays allow one to determine the concentration of unlabelled cAMP or cGMP at a wide range of concentrations, whereas the range for Ins(1,4,5)P$_3$ assay is more narrow. Each assay has also a characteristic affinity for its ligand, which determines the absolute concentrations that can be determined. The cGMP and Ins(1,4,5)P$_3$ assays are more sensitive than the cAMP assay. The accuracy of all assays is sufficient to result in a standard deviation of determined concentrations of less than 10%.

3.4. cAMP Binding Assays

3.4.1. Principle

Dictyostelium cells possess surface receptors that bind cAMP with high specificity and affinity. These receptors may have different kinetic forms, which are probably related to the interaction with G proteins. The receptor–cAMP complex dissociates very fast, with half-times as short as 1 s *(23)*. Therefore, the binding assay requires the separation of bound and free cAMP without washing of the cells (*see* **Note 3**). This can be achieved by either pelleting of the cells and aspiration of the supernatant, or centrifugation of the cells though silicon oil. The latter method is based on the density of the oil, which must be denser than the buffer, but less dense than the cells. After centrifugation, the oil separates the cells in the pellet and the unbound cAMP in the supernatant (*see* **Notes 4–6**).

cAMP receptors are a heterogeneous mixture of different forms, not only with respect to the interaction with G proteins and the state of phosphorylation, but also because a substantial portion of the receptors is not assessable for cAMP binding. A fraction of these receptors is cryptic and can be exposed by polyvalent ions. Another fraction of these receptors is sequestered as an intermediate during cAMP-induced downregulation of the receptors *(24)*. We have observed that in nearly saturated ammonium sulfate, binding of cAMP to cells is increased substantially and shows a more homogeneous population. Interactions with G proteins are lost, and cryptic as well as sequestered receptors bind cAMP. In addition, ammonium sulfate largely retards the dissociation of the receptor/cAMP complex. This not only increases the affinity of the receptor, but also allows the extensive washing of cells *(25)*. Three cAMP-binding assays will be described and compared.

3.4.2. Procedure

3.4.2.1. Phosphate Buffer Pellet Assay

The cells are pelleted at the end of the binding reaction. This method is a fast and simple method to detect equilibrium binding of exposed cAMP receptors.

1. Prepare. Label 1.5-mL plastic tubes and add 10 µL of radioactive binding mixture and 10 µL of H_2O (or unlabeled 1 mM cAMP; *see* **Note 7**).
2. Assay. Add 80 µL of the cell suspension. Incubate for 30–45 s at room temperature or for 2 min at 0°C. Centrifuge the tubes for 2 min at 14,000g. Aspirate the supernatant.
3. Finish. Resuspend the pellets in 100 µL of 0.1 M acetic acid. Add 1.3 mL of scintillator and determine radioactivity.

3.4.2.2. AMMONIUM SULFATE PELLET ASSAY

The binding reaction is performed in nearly saturated ammonium sulfate. The cells are pelleted at the end of the binding reaction. This method is a fast and simple method to detect the total number of cAMP receptors.

1. Prepare. Label tubes and add 10 µL of radioactive binding mixture, 10 µL of H_2O or unlabeled 1 mM cAMP, and 880 µL of 90% saturated ammonium sulfate. Place tubes on ice.
2. Assay. Add 80 µL of cells and 20 µL of BSA (*see* **Note 8**). Incubate for 5 min at 0°C. Centrifuge the tubes for 2 min at 14,000g. Aspirate the supernatant.
3. Finish. Resuspend the pellets in 100 µL of 0.1 M acetic acid. Add 1.3 mL of scintillator and determine radioactivity.

3.4.2.3. SILICON OIL ASSAY

The binding reaction occurs in phosphate buffer, and cells are centrifuged through silicon oil. This method is more laborious, but has lower nonspecific binding and allows the determination of nonequilibrium binding.

1. Prepare. Label 1.5-mL plastic tubes and add 10 µL of radioactive binding mixture, and 10 µL of H_2O or unlabeled 1 mM cAMP. Label a second series of tubes; add 10 µL of sucrose and 200 µL of silicon oil mixture.
2. Assay. Add 80 µL of the cell suspension to the tubes containing radioactive cAMP. Transfer the incubation mixture to a tube containing the silicon oil and centrifuge the tube at the desired time for 20 s at 14,000g.
3. Finish. Place tubes at –20°C until frozen or longer. Cut the tube through the layer of silicone oil and transfer the tip of the tube containing the sucrose and cell pellets to a scintillation vial; this is done most easily with a scalpel or a dog nail clipper. Add 100 µL of H_2O and 2 mL of scintillator.

3.4.3. Results

The binding data for a typical experiment are presented in **Table 2**. The ammonium sulfate assay provides high binding with relatively low nonspecific binding, resulting in the highest ratio of specific to nonspecific binding. This assay also has the lowest standard deviation. The phosphate buffer pellet assay and the silicon oil assay have approximately the same level of specific binding. However, the pellet assay has a significantly higher level of nonspecific binding.

Table 2
Primary Data of cAMP and GTPγS-Binding to Cells and Membranes

Assay/condition	Binding (cpm)		Ratio
	Nonspecific	Specific	
cAMP-binding to cells			
Input 40,000 cpm			
Phosphate buffer (PB) pellet assay	640 ± 91	2163 ± 311	3.38
PB-silicon oil assay	247 ± 25	1846 ± 209	7.47
Ammonium sulfate assay	605 ± 55	5195 ± 127	8.59
cAMP-binding to membranes			
Input 20,000 cpm			
Control	224 ± 12	1014 ± 47	4.53
100 μ*M* GTPγS	231 ± 20	314 ± 38	1.36
GTPγS-binding in membranes			
Input 80,000 cpm			
Control	841 ± 56	8551 ± 232	10.17
1 μ*M* cAMP	847 ± 72	$12,715 \pm 319$	15.01

Each assay has different applications. The ammonium sulfate assay measures the total number of receptors, irrespective of their functional status (exposed vs cryptic or sequestered). The method is very accurate and sensitive. We have noticed that receptor levels determined by ammonium sulfate binding or by Western blots are similar. The phosphate buffer pellet assay is a fast and convenient assay for determining the level of exposed and functional receptors. The assay is reasonably accurate and sensitive. The phosphate buffer silicon oil assay is more annoying, but has the advantage of immediate separation of bound and unbound cAMP, thereby allowing to analyze nonequilibrium binding (*see* **Notes 6** and **7**).

The ammonium sulfate assay has the additional advantage that bound cAMP dissociates very slowly, thereby allowing the washing of the cells to further reduce nonspecific binding. This property has been exploited for photoaffinity labeling and purification of the receptor *(26,27)*.

3.5. cAMP-Induced Second Messenger Responses

3.5.1. Principle

Stimulation of aggregation-competent *Dictyostelium* cells with cAMP leads to the activation of several effector enzymes and the formation of the second messengers cAMP, cGMP, and Ins(1,4,5)P$_3$. Three nearly identical protocols allow the determination of these responses. Stimulated cells are lysed at the desired time by PCA (*see* **Note 9**). After neutralization, the levels of the second messengers are determined by specific isotope-dilution assays. For cAMP-induced cAMP accumulation, we use the analogue 2'-deoxy-cAMP as stimulus, because this compound has high affinity for the surface receptor and low affinity for cAMP-dependent protein kinase, which is used as the cAMP-binding protein in the isotope-dilution assay (*see* **Note 10**).

3.5.2. Procedure

3.5.2.1. CGMP AND INS(1,4,5)P$_3$ RESPONSE

1. Prepare. Label 1.5-mL tubes and add 20 µL of stimulus solution. Add 100 µL of PCA to the t$_0$ samples.
2. Assay. Add 80 µL of the cell suspension to the tubes. Stop the reaction after 3, 6, 9, 12, 15, 20, and 30 s by the addition of 100 µL of PCA. Shake and place samples on ice for about 10 min or store at –20°C.
3. Finish. Neutralize by adding 50 µL of KHCO$_3$ solution. Let stand to allow CO$_2$ to escape (shake carefully). Centrifuge 2 min at 14,000g. Use 20 µL of the supernatant in the isotope-dilution assays (*see* **Subheading 3.3.**).

3.5.2.2. CAMP RESPONSE

The assay is performed as described above, except for the stimulus solution, which contains 2'deoxy-cAMP and DTT, and the times of the reaction, which are 0, 0.5, 1, 1.5, 2, 3, and 5 min.

3.5.3. Results

The primary data for a typical experiment are presented in **Table 1**. The magnitude of the responses are very different for cAMP, cGMP, and Ins(1,4,5)P$_3$, with small responses and high basal levels for Ins(1,4,5)P$_3$ and large responses and low basal levels for cAMP and cGMP. All assays have a small standard deviation (about 6% in cpm at half-maximal inhibition). These assays have been used to determine the cGMP, cAMP, and the Ins(1,4,5)$_3$ responses under a variety of conditions in wild-type and mutant cell lines *(10,28,29)*.

3.6. GTP-Inhibition of cAMP Binding

3.6.1. Principle

The effect of guanine nucleotides on agonist binding to the surface receptor is a useful indicator of receptor–G protein interaction. Addition of guanine nucleotides reduces the apparent affinity, but not the number of cAMP receptors *(30)*. *Dictyostelium* membranes are incubated with subsaturating concentrations of [^{3}H]cAMP in the presence of GTPγS. Bound [^{3}H]cAMP is separated from free [³H]cAMP by centrifugation, and radioactivity associated with membrane pellets is measured. This assay is a very convenient and accurate assay for the interaction from G protein to receptor.

3.6.2. Procedure

1. Prepare. Label 1.5-mL tubes and add 10 µL of radioactive binding mixture, 10 µL of H₂O or GTPγS, and 10 µL of H₂O or excess cAMP. Place tubes on ice.
2. Assay. Make membranes (*see* **Subheading 3.2.**). Add 70 µL of membranes to tubes and incubate for 5 min at 0°C. Centrifuge for 2 min at 14,000*g*. Aspirate the supernatant.
3. Finish. Add 100 µL of 0.1 *M* acetic acid and mix until pellet is dissolved. Add 1.3 mL of scintillator.

3.6.3. Results

Primary data are presented in **Table 2**. Binding of 5 n*M* [³H]cAMP to membranes is higher than binding to the equivalent number of cells. This increased binding is due to an enhanced affinity and not to an increase of the number of binding sites. GTP, GDP, GTPγS, and GDPβS reduce cAMP binding, as a result of a decrease of the affinity of the receptor for cAMP *(30)*. This assay has been used extensively to investigate the effects of guanine nucleotides on the transition of different kinetic forms of the cAMP receptor. In these experiments, the rate of association and dissociation of the receptor was determined in the absence or presence of guanine nucleotides using the silicon oil assay with a microfuge swing-out rotor as described under **Subheading 3.4.2.3.** *(23)*.

3.7. cAMP Stimulation of GTPγS Binding

3.7.1. Principle

G proteins are activated by the exchange of bound GDP for GTP. In *Dictyostelium*, cAMP binding to cell surface receptors promotes release of bound GDP and permits binding of GTP. Receptor stimulation of G proteins can be measured as the cAMP-stimulated binding of GTP or GTPγS to G proteins *(31)*. Membranes are incubated with [³⁵S]GTPγS in the presence of cAMP. Bound [³⁵S]GTPγS is separated from free [³⁵S]GTPγS by centrifugation, and radioactivity associated with membrane pellets is measured. This assay is a very convenient and accurate assay for the interaction from receptor to G protein.

3.7.2. Procedure

1. Prepare. Label 1.5-mL tubes and add 10 µL of radioactive binding mixture, 10 µL of H₂O or cAMP, and 10 µL of H₂O or GTP. Place tubes on ice.
2. Assay. Prepare membranes (*see* **Subheading 3.2.**). Add 70 µL of membranes to

    tubes and incubate for 30 min at 0°C. Centrifuge for 3 min at 14,000*g*. Aspirate the supernatant.

3. Finish. Add 100 µL of 0.1 *M* acetic acid; mix until the pellet is dissolved. Add 1.3 mL of scintillator.

3.7.3. Results

The primary data for the [^{35}S]GTPγS-binding to membranes are shown in **Table 2**. Equilibrium binding is enhanced 40–80% by cAMP. Detailed analysis *(31)* has revealed that [^{35}S]GTPγS-binding is relatively slow, with half-maximal association at 2 n*M* [^{35}S]GTPγS after 10 min at 0°C; equilibrium is reached after 30 min of incubation. Scatchard analysis of [^{35}S]GTPγS-binding showed two forms of binding sites with respectively high (K$_d$ =0.2 µM) and low (K$_d$ = 6.3 µ*M*) affinity. cAMP does not affect the rate of binding, but enhances the affinity and number of the high-affinity sites, whereas the low-affinity sites are not affected by cAMP *(31)*.

3.8. cAMP-Stimulation of GTPase Activity

3.8.1. Principle

G proteins have low intrinsic GTPase activity. Surface receptors promote the exchange of GDP for GTP in G proteins. The enhanced occupancy of the G protein with GTP consequently results in stimulated GTPase activity. This can be demonstrated in crude membranes from *Dictyostelium* when high-affinity GTPase activity is measured under appropriate conditions *(32)*. [γ-^{32}P]GTP is used at submicromolar substrate concentrations. Nonspecific nucleotide triphosphatases are inhibited by the ATP analog AppNHp. Redistribution of radioactivity among guanine and adenine dinucleotides by nucleoside diphosphate kinase is prevented by a nucleoside triphosphate regeneration system and by ATPγS. Under these conditions, the release of [^{32}P]P$_i$ from [γ-^{32}P]GTP is suppressed to 8–12% of added [γ-^{32}P]GTP and stimulation by cAMP becomes detectable (*see* **Note 11**).

3.8.2. Procedure

1. Prepare. Label tubes and place on ice. Add 40 µL of reaction mixture, 10 µL of [γ-^{32}P]GTP, and 20 µL of GTP, cAMP, or H$_2$O.
2. Assay. Prepare membranes as described under **Subheading 3.2.** Preincubate tubes for 5 min at 25°C. Start the reaction by addition of 30 µL of membranes, vortex, and conduct the assay for 3 min at 25°C. Terminate the reaction by the addition of 0.6 mL of ice-cold, activated charcoal in sodium phosphate buffer; place tubes on ice.
3. Finish. Centrifuged the tubes at 4°C for 5 min at 14,000*g* and take 0.4 mL of the supernatant for the determination of radioactivity.

Table 3
Receptor-Stimulated GTPase Activity in Membranes

Condition	$[\gamma\text{-}^{32}P]$GTP hydrolyzed (cpm)		
	0.01 µ*M* GTP	50 µ*M* GTP	Difference
No enzyme	900 ± 300		
Enzyme	4980 ± 285	1896 ± 43	3084 ± 319
Enzyme + 1 µ*M* cAMP	6724 ± 427	2040 ± 174	4684 ± 461

Input 60,000 cpm $[\gamma\text{-}^{32}P]$GTP.

3.8.3. Results

A typical experiment involving the release of $[^{32}Pi]$ from $[\gamma\text{-}^{32}P]$GTP is presented in **Table 3**. Total GTPase in this experiment produces 4980 cpm $[^{32}Pi]$. In the presence of 50 µ*M* GTP, low-affinity GTPase is detected, amounting to 1896 cpm. The difference, 3084 cpm, represents high-affinity GTPase. The release of $[^{32}P_i]$ is routinely measured at 3 min of incubation. The relationship between membrane protein and GTP hydrolysis is linear, in the range of 10–40 µg membrane protein per assay for the incubation at 25°C for 3 min.

3.9. GTPγS-Stimulation of Adenylyl Cyclase In Vitro

3.9.1. Principle

Adenylyl cyclase in *Dictyostelium* lysates is activated by the surface cAMP receptor and G protein. This assay combines two described methods. Lysis is performed according to Theibert and Devreotes *(33)*. Enzyme activity is measured using the method of Van Haastert et al. *(34)*. Receptor agonist 2'deoxy-cAMP and GTPγS are present during lysis. The lysate is subsequently incubated for 5 min at 0°C. Adenylyl cyclase is assayed using nonradioactive ATP. The reaction is terminated by adding excess of EDTA and boiling of the samples. The produced cAMP is determined by isotope dilution assay (*see* **Subheading 3.3.**).

3.9.2. Procedure

1. Prepare. Label 1.5-mL tubes. Add 20 µL of assay mixture to all tubes; add 10 µL of EDTA to control tubes (t_0 incubations). Prepare 1-mL syringes with Nuclepore membrane adjusted between the gauge and the needle.
2. Assay. Mix 100 µL of cells with 100 µL of lysis buffer at 0°C (*see* **Note 12**). Immediately lyse the cells by pressing them through a Nuclepore filter with 3-µm pores. Collect the lysates in tubes at 0°C and keep on ice for 5 min (*see* **Note 13**). To measure adenylyl cyclase, add 20 µL of lysate to the tubes containing 20 µL of assay mixture. Incubate for 5 min at 20°C. Terminate the reaction by

addition of 10 µL of 0.1 *M* EDTA, pH 8.0. At the end of the experiment, all samples are boiled for 2 min.

3. Finish. Assay cAMP levels using the isotope-dilution assay (*see* **Subheading 3.3.**; note that samples do not have to be neutralized).

3.9.3. Results

Primary data for the adenylyl cyclase assay are presented in **Table 1**. Basal adenylyl cyclase activity is easily detected. GTP and GTP analogs produced significant (up to 17-fold) activation of adenylyl cyclase in lysates of *Dictyostelium*. Activation is enhanced two- to fourfold by cAMP. Detailed experiments *(33,34)* have shown that maximal activation occurred when GTPγS was present in the lysate immediately after cell lysis and the lysate was preincubated with GTPγS for 5 min prior to assay. Stimulation by cAMP is optimal when added 1 min before lysis.

Dictyostelium possesses three adenylyl cyclases: the 12-transmembrane ACA, which is similar to mammalian adenylyl cyclases *(35)*, the 1-transmembrane ACG, which is more similar to membrane-bound guanylyl cyclases *(35)*, and the soluble ACB encoded by the *acrA* gene, which is similar to bacterial adenylyl cyclases *(36)*. This and related assays have been used to show that only ACA is regulated by surface receptors and G proteins *(35)*. ACG appears to be activated by osmotic stress *(37)*, and ACB possibly by a two-component regulatory system *(29,36)*. Furthermore, activation of ACA requires a soluble protein defective in mutant *synag7 (33,34,38)*, later identified as CRAC *(39)*, by the Map kinase ERK2 *(40)*, and by pianissimo *(41)*.

3.10. GTPγS-Regulation of Phospholipase C

3.10.1. Principle

The phospholipase C assay is based on the fact that the enzyme is inactive in the absence of Ca^{2+} ions. Cells are lysed in the presence of EGTA and, subsequently, a fixed amount of Ca^{2+} is added to the lysate for a fixed amount of time *(42)*. Upon addition of Ca^{2+} to the lysate, Ins(1,4,5)P₃ is produced from endogenous substrate PtdIns(4,5)P₂. The produced Ins(1,4,5)P₃ is determined by isotope-dilution assay (*see* **Subheading 3.3.**). The G protein activator GTPγS and/ or the receptor activator cAMP are added before lysis (*see* **Note 14**).

3.10.2. Procedure

1. Prepare. Label 1.5-mL tubes. Add 5 µL of $CaCl_2$ to all tubes; add 50 µL of PCA to control tubes (t₀ incubations). Prepare 1-mL syringes with Nuclepore membrane adjusted between the gauge and the needle.
2. Assay. Mix 135 µL of cells with 7.5 µL of EGTA and 7.5 µL of stimulus (cAMP, GTPγS, or buffer for control). Transfer the suspension to a syringe and lyse the cells by pressing them through a Nuclepore filter with 3-µm pores. At 10 s after

lysis, transfer 50 µL of lysate to tubes containing $CaCl_2$. Terminate the reaction after 20 s by addition of 50 µL of PCA.

3. Finish. Assay Ins(1,4,5)P_3 levels using the isotope dilution assay (*see* **Subheading 3.3.**).

3.10.3. Results

The basal Ins(1,4,5)P_3 level of a lysate (t_0 sample) is essentially identical to that of cells, indicating that phospholipase C is inactive in EGTA. Addition of Ca^{2+} ions leads to a strong increase of the Ins(1,4,5)P_3 level within 20 s of incubation. This phospholipase C activity is stimulated about twofold by cAMP or by GTPγS (**Table 1**).

3.11. GTPγS-Stimulation of Guanylyl Cyclase

3.11.1. Principle

Guanylyl cyclase in *Dictyostelium* lysates can be activated by GTPγS. Because guanylyl cyclase has a catalytic site for GTP, it is *a priori* difficult to prove that GTPγS-stimulation of guanylyl cyclase is mediated by a G protein. Evidence has been obtained by the antagonizing effect of GDPβS *(43)*, and by mutation of the guanylyl cyclase GCA to an adenylyl cyclase retaining GTPγS-stimulation of now-cAMP production *(44)*. The assay follows the same principle as for the adenylyl cyclase assay, but the lysate is not pre-incubated on ice (*see* **Note 13**). Cells are lysed in the presence of GTPγS, and enzyme activity is measured with unlabeled GTP, using a radioimmunoassay to quantify the formed cGMP.

3.11.2. Procedure

1. Prepare. Label 1.5-mL tubes. Add 20 µL of assay mixture to all tubes; add 20 µL of PCA to control tubes (t_0 incubations). Prepare 1-mL syringes with Nuclepore membrane adjusted between the gauge and the needle.
2. Assay. Mix 100 µL of cells with 100 µL of lysis buffer at 0°C (*see* **Note 15**). Immediately lyse the cells by pressing them through a Nuclepore filter with 3-µm pores. Collect the lysates in tubes at 0°C. Start the guanylyl cyclase assay at 30 s after lysis by addition of 20 µL of lysate to the tubes containing 20 µL of assay mixture. Incubate for 1 min at 22°C. Terminate the reaction by the addition of 20 µL of PCA.
3. Finish. Assay cGMP levels using the radioimmuno assay (*see* **Subheading 3.3.**).

3.11.3. Results

Primary data for the guanylyl cyclase assay are presented in **Table 1**. Basal enzyme activity is easily detected, and GTPγS produces significant activation of guanylyl cyclase in lysates of *Dictyostelium* (*see* **Note 16**).

Dictyostelium possesses two guanylyl cyclases: the 12-transmembrane GCA, which is similar to mammalian membrane-bound adenylyl cyclases *(45)*, and the soluble sGC, which is not the homolog of mammalian soluble guanylyl cyclase, but is more similar to soluble adenylyl cyclase involved in sperm maturation *(46)*. The assay described above was used to characterize these two guanylyl cyclases. Both enzymes appear to be activated by cAMP and GTPγS, but activation of sGC is generally stronger than activation of GCA *(28)*.

4. Notes

1. Sucrose helps to keep the organelles intact (which is especially relevant for lysosomes). However, cell lysis becomes increasingly more difficult when cells are resuspended in sucrose-containing buffer for prolonged periods of time (more than 2 min).
2. Samples derived from experiments are generally lysed by PCA and neutralized to pH 7.0 with potassium bicarbonate (*see* **Subheading 3.5.**). The pH of the samples should not be above pH 7.5 for cAMP and cGMP assays, and not be below pH 6.5 for the Ins(1,4,5)P$_3$ assay.
   Stock solutions are made in H$_2$O; cAMP is stable in phosphate and HEPES buffers, but is deaminated in Tris buffers after storage for a few weeks at –20°C. cGMP does not show this problem.
3. cAMP induces the activation of adenylyl cyclase and subsequent secretion of synthesized cAMP. This should be prevented, because it dilutes the radioactive cAMP. One method is to complete the binding reaction before secretion starts, which is 45 s at 20°C and about 2 min at 0°C; binding equilibrium is reached within 30 s. Another method is to inhibit adenylyl cyclase activation with 5 mM caffeine, which is included as 50 mM in the radioactive reaction mixture. cAMP secretion does not occur in the ammonium sulfate assay.
4. During centrifugation, the temperature of the silicon oil may increase, thereby decreasing its density. This may result in the floating of the silicon oil on top of the buffer. This problem is especially important for incubations at 0°C with a centrifuge operating at room temperature. Careful preparation of the silicon oil mixture and short centrifugation times should eliminate the necessity of performing the assay in a cold-room.
5. Membranes pass through silicon oil more slowly than cells, leading to longer centrifugation times (about 30 s). Accurate tuning of the silicon oil mixture and centrifugation time allows one to centrifuge tubes at room temperature.
6. For rapid kinetics, the workshop has made a swing-out rotor that fits in an Eppendorf microfuge.
7. The incubation mixture contains 10 µL of H$_2$O, which can be replaced by different compounds. For nonspecific binding, 10 µL of 1 mM cAMP is used. For the determination of the affinity and number of binding sites by Scatchard analysis, different concentrations of radioactive and cold cAMP are used. The standard conditions described previously contain 10 µL of 100 nM [^{3}H]cAMP. Generally, we use 10 µL of radioactive [^{3}H]cAMP at 20, 50, 100, and 300 nM, and 300 nM

[³H]cAMP with additional 700, 1700, 4700, 9700, and 19,700 nM unlabeled cAMP. Compared with these concentrations, the final concentrations in the binding reaction are 10-fold lower in the phosphate buffer pellet and silicon oil assays, and 100-fold lower in the ammonium sulfate assay.

8. The BSA in the ammonium sulfate assay serves to glue the cells to the wall of the tube, which facilitates the aspiration of the supernatant. It slightly increases nonspecific binding.

9. The assay described uses independent stimulations. This procedure is optimal for the determination of the magnitude of the response. In experiments in which the time course is more important than the magnitude of the response, we stimulate 900 µL of cells with 100 µL of stimulus (twice as concentrated as described), and at the desired times 100-µL samples are transferred to tubes containing 100 µL of PCA. In these experiments, the t_0 sample is taken just before stimulation. In order to catch the early time points, cells are stimulated while being vortexed, and the same pipet is used to add the stimulus and to draw the samples.

10. The protocols will yield total levels of second messenger. For the cAMP response, it is often relevant to distinguish between intra- and extracellular levels. Just before termination of the reactions, cells are centrifuged for 5 s at 14,000g, the supernatant is transferred to a tube containing 100 µL of PCA, and 100 µL of PCA is added to the pellet.

11. Total GTPase is detected in the absence of added nonradioactive GTP, whereas low-affinity GTPase is determined in the presence of 50 µM GTP. High-affinity GTPase is defined as the difference between total GTPase and low-affinity GTPase activity.

12. Stimulated adenylyl cyclase is measured by adding 30 µM GTPγS or 50 µM 2'deoxy-cAMP (final concentrations) to the cells just prior to cell lysis.

13. The incubation of the lysate at 0°C for 5 min is essential in order to observe strong activation of adenylyl cyclase activity by GTPγS. Apparently, the GTPγS-activated state of the adenylyl cyclase is formed slowly in vitro, in contrast to phospholipase C and guanylyl cyclase, which are immediately activated by GTPγS in vitro.

14. Because the phospholipase C assay uses endogenous PtdIns(4,5)P$_2$ as substrate, a change in the concentration or availability of PtdIns(4,5)P$_2$ during the assay should be avoided. We noticed that Ins(1,4,5)P$_3$ production is decreased when the phospholipase C assay is started more than 30 s after cell lysis; control experiments with [³²P]PtdIns(4,5)P$_2$ suggest that this is not due to degradation of substrate or diminished phospholipase C activity, but probably occurs because endogenous PtdIns(4,5)P$_2$ becomes unavailable for the enzyme. Therefore, we start the phospholipase C assay at exactly 10 s after cell lysis.

15. Stimulated guanylyl cyclase is measured by adding 100 µM GTPγS (final concentration) to the cells just prior to cell lysis.

16. GTPγS-stimulated guanylyl cyclase activity is detectable only with magnesium, not with manganese as a bivalent cofactor. The magnesium-dependent activity is strongly inhibited by calcium ions; therefore, the lysis buffer should contain sufficient EGTA to chelate all the calcium ions *(28,43)*.

References

1. Devreotes, P. N. and Zigmond, S. H. (1988) Chemotaxis in eukaryotic cells: a focus on leukocytes and *Dictyostelium*. *Annu. Rev. Cell Biol.* **4,** 649–686.
2. Van Haastert, P. J. M. and Devreotes, P. N. (2004) Chemotaxis: signalling the way forward. *Nat. Rev. Mol. Cell Biol.* **5,** 626–634.
3. Manahan, C. L., Iglesias, P. A., Long, Y., and Devreotes, P. N. (2004) Chemoattractant signaling in *Dictyostelium discoideum*. *Annu. Rev. Cell Dev. Biol.* **20,** 223–253.
4. Eichinger, L., Pachebat, J. A., Glockner, G., et al. (2005) The genome of the social amoeba *Dictyostelium discoideum*. *Nature* **435,** 43–57.
5. Bosgraaf, L. and Van Haastert, P. J. M. (2002) A model for cGMP signal transduction in *Dictyostelium* in perspective of 25 years of cGMP research. *J. Muscle Res. Cell Motil.* **23,** 781–791.
6. Mutzel, R., Lacombe, M. L., Simon, M. N., de Gunzburg, J., and Veron, M. (1987) Cloning and cDNA sequence of the regulatory subunit of cAMP-dependent protein kinase from *Dictyostelium discoideum*. *Proc. Natl. Acad. Sci. USA* **84,** 6–10.
7. Mann, S. K., Yonemoto, W. M., Taylor, S. S., and Firtel, R. A. (1992) DdPK3, which plays essential roles during *Dictyostelium* development, encodes the catalytic subunit of cAMP-dependent protein kinase. *Proc. Natl. Acad. Sci. USA* **89,** 10,701–10,705.
8. Bosgraaf, L., Russcher, H., Smith, J. L., Wessels, D., Soll, D. R., and Van Haastert, P. J. M. (2002) A novel cGMP-signaling pathway mediating myosin phosphorylation and chemotaxis in *Dictyostelium*. *EMBO J.* **21,** 4560–4570.
9. Goldberg, J. M., Bosgraaf, L., Van Haastert, P. J. M., and Smith, L. (2002) Identification of four candidate cGMP targets in *Dictyostelium*. *Proc. Natl. Acad. Sci. USA* **99,** 6749–6754.
10. Drayer, A. L., Van der Kaay, J., Mayr, G. W., and Van Haastert, P. J. M. (1994) Role of phospholipase C in *Dictyostelium*: formation of inositol 1,4,5-trisphosphate and normal development in cells lacking phospholipase C activity. *EMBO J.* **13,** 1601–1609.
11. van Haastert, P. J. M. and van Dijken, P. (1997) Biochemistry and genetics of inositol phosphate metabolism in *Dictyostelium*. *FEBS Lett.* **410,** 39–43.
12. Traynor, D., Milne, J. L., Insall, R. H., and Kay, R. R. (2000) Ca(2+) signalling is not required for chemotaxis in *Dictyostelium*. *EMBO J.* **19,** 4846–4854.
13. Iijima, M. and Devreotes, P. (2002) Tumor suppressor PTEN mediates sensing of chemoattractant gradients. *Cell* **109,** 599–610.
14. Loovers, H. M., Veenstra, K., Snippe, H., Pesesse, X., Erneux, C., and van Haastert, P. J. M. (2003) A diverse family of inositol 5-phosphatases playing a role in growth and development in *Dictyostelium discoideum*. *J. Biol. Chem.* **278,** 5652–5658.
15. Parent, C. A., Blacklock, B. J., Froehlich, W. M., Murphy, D. B., and Devreotes, P. N. (1998) G protein signaling events are activated at the leading edge of chemotactic cells. *Cell* **95,** 81–91.

16. Snaar-Jagalska, B. E. and Van Haastert, P. J. M. (1994) G-protein assays in *Dictyostelium. Methods Enzymol.* **237,** 387–408.

17. Bominaar, A. A. and Van Haastert, P. J. M. (1994) Phospholipase C activity in *Dictyostelium discoideum* using endogenous nonradioactive phosphatidylinositol 4,5-bisphosphate as substrate. *Methods Enzymol.* **238,** 207–218.

18. Sussman, R. and Sussman, M. (1967) Cultivation of *Dictyostelium discoideum* in axenic medium. *Biochem. Biophys. Res. Commun.* **29,** 53–55.

19. Steiner, A. L., Parker, C. W., and Kipnis, D. M. (1972) Radioimmunoassay for cyclic nucleotides. I. Preparation of antibodies and iodinated cyclic nucleotides. *J. Biol. Chem.* **247,** 1106–1113.

20. Gilman, A. G. and Murad, F. (1974) Assay of cyclic nucleotides by receptor protein binding displacement. *Methods Enzymol.* **38,** 49–61.

21. Baukal, A. J., Guillemette, G., Rubin, R., Spat, A., and Catt, K. J. (1985) Binding sites for inositol trisphosphate in the bovine adrenal cortex. *Biochem. Biophys. Res. Commun.* **133,** 532–538.

22. Guillemette, G., Balla, T., Baukal, A. J., Spat, A., and Catt, K. J. (1987) Intracellular receptors for inositol 1,4,5-trisphosphate in angiotensin II target tissues. *J. Biol. Chem.* **262,** 1010–1015.

23. van Haastert, P. J. M., de Wit, R. J., Janssens, P. M., Kesbeke, F., and DeGoede, J. (1986) G-protein-mediated interconversions of cell-surface cAMP receptors and their involvement in excitation and desensitization of guanylate cyclase in *Dictyostelium discoideum. J. Biol. Chem.* **261,** 6904–6911.

24. Van Haastert, P. J. M., Wang, M., Bominaar, A. A., Devreotes, P. N., and Schaap, P. (1992) cAMP-induced desensitization of surface cAMP receptors in *Dictyostelium*: different second messengers mediate receptor phosphorylation, loss of ligand binding, degradation of receptor, and reduction of receptor mRNA levels. *Mol. Biol. Cell* **3,** 603–612.

25. Van Haastert, P. J. M. (1985) The modulation of cell surface cAMP receptors from *Dictyostelium discoideum* by ammonium sulfate. *Biochim. Biophys. Acta* **845,** 254–260.

26. Theibert, A., Klein, P., and Devreotes, P. N. (1984) Specific photoaffinity labeling of the cAMP surface receptor in *Dictyostelium discoideum. J. Biol. Chem.* **259,** 12,318–12,321.

27. Klein, P., Knox, B., Borleis, J., and Devreotes, P. (1987) Purification of the surface cAMP receptor in Dictyostelium. *J. Biol. Chem.* **262,** 352–357.

28. Roelofs, J. and Van Haastert, P. J. M. (2002) Characterization of two unusual guanylyl cyclases from *Dictyostelium. J. Biol. Chem.* **277,** 9167–9174.

29. Meima, M. E. and Schaap, P. (1999) Fingerprinting of adenylyl cyclase activities during *Dictyostelium* development indicates a dominant role for adenylyl cyclase B in terminal differentiation. *Dev. Biol.* **212,** 182–190.

30. Van Haastert, P. J. M. (1984) Guanine nucleotides modulate cell surface cAMP-binding sites in membranes from *Dictyostelium discoideum. Biochem. Biophys. Res. Commun.* **124,** 597–604.

31. Snaar-Jagalska, B. E., De Wit, R. J., and Van Haastert, P. J. M. (1988) Pertussis toxin inhibits cAMP surface receptor-stimulated binding of [35S]GTPγS to *Dictyostelium discoideum* membranes. *FEBS Lett.* **232**, 148–152.

32. Snaar-Jagalska, B. E., Jakobs, K. H., and Van Haastert, P. J. M. (1988) Agonist-stimulated high-affinity GTPase in *Dictyostelium* membranes. *FEBS Lett.* **236**, 139–144.

33. Theibert, A. and Devreotes, P. N. (1986) Surface receptor-mediated activation of adenylate cyclase in *Dictyostelium*. Regulation by guanine nucleotides in wild-type cells and aggregation deficient mutants. *J. Biol. Chem.* **261**, 15,121–15,125.

34. Van Haastert, P. J. M., Snaar-Jagalska, B. E., and Janssens, P. M. (1987) The regulation of adenylate cyclase by guanine nucleotides in *Dictyostelium discoideum* membranes. *Eur. J. Biochem.* **162**, 251–258.

35. Pitt, G. S., Milona, N., Borleis, J., Lin, K. C., Reed, R. R., and Devreotes, P. N. (1992) Structurally distinct and stage-specific adenylyl cyclase genes play different roles in *Dictyostelium* development. *Cell* **69**, 305–315.

36. Soderbom, F., Anjard, C., Iranfar, N., Fuller, D., and Loomis, W. F. (1999) An adenylyl cyclase that functions during late development of *Dictyostelium*. *Development* **126**, 5463–5471.

37. van Es, S., Virdy, K. J., Pitt, G. S., et al. (1996) Adenylyl cyclase G, an osmosensor controlling germination of *Dictyostelium* spores. *J. Biol. Chem.* **271**, 23,623–23,625.

38. Snaar-Jagalska, B. E. and Van Haastert, P. J. M. (1988) *Dictyostelium discoideum* mutant synag 7 with altered G-protein-adenylate cyclase interaction. *J. Cell Sci.* **91**, 287–294.

39. Lilly, P. J. M. and Devreotes, P. N. (1994) Identification of CRAC, a cytosolic regulator required for guanine nucleotide stimulation of adenylyl cyclase in *Dictyostelium*. *J. Biol. Chem.* **269**, 14,123–14,129.

40. Segall, J. E., Kuspa, A., Shaulsky, G., et al. (1995) A MAP kinase necessary for receptor-mediated activation of adenylyl cyclase in *Dictyostelium*. *J. Cell Biol.* **128**, 405–413.

41. Chen, M. Y., Long, Y., and Devreotes, P. N. (1997) A novel cytosolic regulator, Pianissimo, is required for chemoattractant receptor and G protein-mediated activation of the 12 transmembrane domain adenylyl cyclase in *Dictyostelium*. *Genes Dev.* **11**, 3218–3231.

42. Bominaar, A. A., Kesbeke, F., and Van Haastert, P. J. M. (1994) Phospholipase C in *Dictyostelium discoideum*. Cyclic AMP surface receptor and G-protein-regulated activity in vitro. *Biochem. J.* **297**, 181–187.

43. Janssens, P. M., De Jong, C. C., Vink, A. A., and Van Haastert, P. J. M. (1989) Regulatory properties of magnesium-dependent guanylate cyclase in *Dictyostelium discoideum* membranes. *J. Biol. Chem.* **264**, 4329–4335.

44. Roelofs, J., Loovers, H. M., and Van Haastert, P. J. M. (2001) GTPγS regulation of a 12-transmembrane guanylyl cyclase is retained after mutation to an adenylyl cyclase. *J. Biol. Chem.* **276**, 40,740–40,745.

45. Roelofs, J., Snippe, H., Kleineidam, R. G., and Van Haastert, P. J. M. (2001) Guanylate cyclase in *Dictyostelium discoideum* with the topology of mammalian adenylate cyclase. *Biochem. J.* **354,** 697–706.
46. Roelofs, J., Meima, M., Schaap, P., and Van Haastert, P. J. M. (2001) The *Dictyostelium* homologue of mammalian soluble adenylyl cyclase encodes a guanylyl cyclase. *EMBO J.* **20,** 4341–4348.

23

Assaying Chemotaxis of *Dictyostelium* Cells

Michelle C. Mendoza and Richard A. Firtel

Summary

Both prokaryote and eukaryote cells can sense and move up chemical concentration gradients (chemotax). As a means of finding food sources during vegetative growth, *Dictyostelium discoideum* naturally chemotaxes toward chemicals released by bacteria. As part of its developmental life cycle, *D. discoideum* chemotaxes towards cAMP. This chapter describes protocols for using *Dictyostelium* to understand the cell biology behind and the signaling events necessary for eukaryotic amoeboid chemotaxis. The chapter includes analyses of random cell motility, directed motility up chemical gradients, cellular responses to uniform chemoattractant exposure, and the utility of fluorescent probes for chemotaxis signaling events. Random cell motility in the absence of chemoattractant is analyzed to decipher the properties of self-organizing pseudopodia extension and retraction. Monitoring chemotaxis toward cAMP and folate allows the determination of signaling events required for sensing a chemical gradient and moving in a directed, persistent manner up the gradient. Uniform chemoattractant exposure is employed to elucidate the immediate intracellular responses to chemoattractant stimulation. Finally, analyzing cells expressing fluorescent fusion proteins is vital to elucidating the location of signaling events during chemotaxis.

Key Words: cAMP; amoeba; chemotaxis; *Dictyostelium*; folate; fluorescent probes.

1. Introduction

Dictyostelium discoideum chemotaxes by F-actin-mediated pseudopodia protrusion and myosin-II-mediated posterior retraction. This type of movement resembles the amoeboid chemotaxis undergone by human lymphocytes (reviewed in **refs. *1,2***). Thus, researchers have exploited *Dictyostelium's* genetic manipulability and developed the organism into a model system for studying amoeboid movement up chemical gradients.

Vegetative *Dictyostelium* cells randomly protrude pseudopodia in the absence of chemoattractant and in the absence of key signaling proteins required for

From: *Methods in Molecular Biology, vol. 346:* Dictyostelium discoideum *Protocols*
Edited by: L. Eichinger and F. Rivero © Humana Press Inc., Totowa, NJ

chemotaxis (reviewed in **ref. *3***). Analyzing strains undergoing random motility will determine whether they are defective in the core signaling components required for pseudopod extension and retraction. In this assay, vegetatively growing cells are washed and seeded onto a chambered coverslip in starvation buffer. Cell movement is then recorded by acquiring a series of phase-contrast or differential interference contrast (DIC) images.

Dictyostelium chemotaxis toward cAMP under starvation conditions is widely used to assess chemotaxis competence. Analyses of chemotaxis up cAMP gradients were fundamental to discovering the essential role of phosphatidylinositol-3 kinase (PI3K), its downstream effector protein kinase B (PKB), and the antagonistically acting phosphatase and tensin homolog (PTEN) phosphatase in sensing chemoattractant gradients and directing cellular movement up such gradients *(4–8)*. Starvation of *Dictyostelium* induces a developmental program in which the cells secrete oscillatory pulses of cAMP waves every 6 min. These cAMP waves activate receptors at the cell surface and lead to upregulation of signaling proteins involved in sensing and chemotactically responding to extracellular cAMP. The waves also create a cAMP gradient and induce the cells to move toward the areas of highest cAMP concentration. After approx 9 h of starvation, the chemotactic aggregation of up to 10^5 cells leads to the formation of multicellular mounds (reviewed in **refs. *9,10***). For chemotaxis analysis, *Dictyostelium* is put into a starvation buffer and exposed to cAMP pulses for 5 h to mimic the intrinsic oscillatory pulses of cAMP that occur when *Dictyostelium* undergoes normal differentiation on a substratum. cAMP pulsing causes the cells to induce expression of early genes, including the cAMP receptor and other signaling proteins necessary for chemotaxis, which are normally expressed after 5–6 h of development *(11)*. This chapter describes two common methods for developing *Dictyostelium* in preparation for chemotaxis to cAMP. The cells are then exposed to a cAMP gradient emitted from a micropipet and their chemotaxis recorded by acquiring phase-contrast or DIC images.

Problematically, some mutants exhibit defects in chemotaxis to cAMP after 5 h of pulsing, but not after 7 h of pulsing. Such discrepancies are widely attributed to *Dictyostelium* cell polarity being gradually strengthened as the cells become further developed. When highly polarized cells are exposed to a change in the orientation of the cAMP gradient, they U-turn into the newly positioned gradient. Less developed, and therefore less polarized, cells disassemble their old leading edges and reassemble new pseudopodia in the direction of the new gradient. This complication, in which the developmental program sometimes affects the chemotaxis phenotype, can be overcome by analyzing vegetative cell chemotaxis toward folic acid. During vegetative growth, *D. discoideum* chemotaxes toward folate and other nutrients released by bacteria *(9,10)*. Thus, folic acid can be used as a chemoattractant in much the same way as cAMP is used to create a chemoattractant gradient or uniform chemoattractant concentration *(12)*. In this assay, vegetative cells are analyzed in starvation buffer containing 20% growth medium, which reduces the amount of ungraded folic acid in the milieu without inducing the developmental program.

For rigorous comparison of chemotaxis phenotypes, the cell polarity, directionality, and speed of centroid movement during chemotaxis should be quantitated. Computer-assisted analysis of cell movement and shape change is commonly performed using image analysis software, such as dynamic image analysis system (DIAS) software (Solltech, Oakdale, IA) *(13)*, as described in Chapter 16. Generally, the perimeter (shape) and centroid position for 4–10 cells in frames 100–200 of a 300-frame movie are marked. The speed of centroid movement and changes in cell shape between different frames are calculated and averaged for a single amoeba moving for 10 min.

Analyzing cellular responses to uniform chemoattractant concentration ("global stimulation") allows the determination of whether a strain is able to sense and adapt to chemoattractant. In response to uniform chemoattractant exposure (<2% difference in concentration across the length of the cell), Ras, PI3K, and pleckstrin homology (PH) domain-containing proteins immediately translocate from the cytoplasm to the cell membrane (2–10 s poststimulation). The cells then adapt to the new chemoattractant level, causing PI3K-signalling proteins to be released from and PTEN to associate with the entire membrane (20–40 s poststimulation) *(4–8,14,15)*. After adaptation, PI3K and its effectors re-associate with the membrane in patches that are the sites of pseudopodia extension and random motility behavior. Fluorescently labeled signaling molecules are key to monitoring protein localization and activity in global stimulation analyses as described later.

Fluorescent probes can be used in combination with any of the chemotaxis assays mentioned previously to assess protein localization and activation during random motility, chemotaxis, and global stimulation. In these experiments, proteins of interest are made as fusions with fluorescent proteins, such as green fluorescent protein (GFP), yellow fluorescent protein (YFP), cyan fluorescent protein (CFP), or red fluorescent protein (RFP). The fusion proteins are then expressed in *Dictyostelium* and used to determine the localization of the protein of interest during chemotaxis in vivo. GFP fusions of PI3K and PTEN have been critical in determining that these proteins translocate to the leading edge and posterior of chemotaxing cells, respectively (*see* **Fig. 1**) *(4,5,7)*. The localization of fluorescent fusion proteins should always be confirmed by other methods, such as indirect immunofluorescence with antibodies to endogenous epitopes, to ensure that the large fluorescent protein tag does not cause an artificial localization. Further, as *Dictyostelium* is an experimental system in which gene knock-

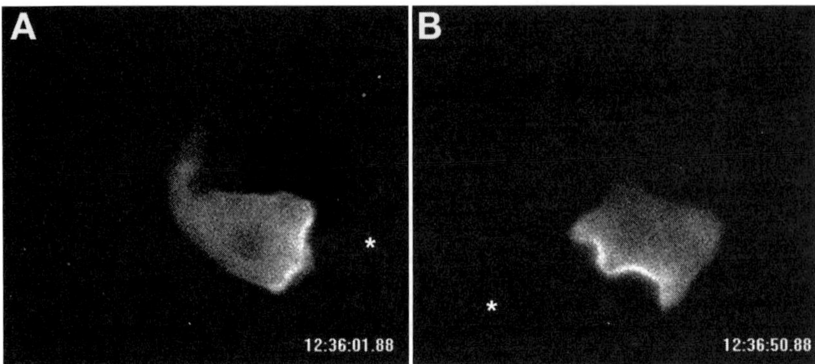

Fig. 1. Monitoring protein localization and activation using green fluorescent protein (GFP) fusion proteins. (**A**) A femtotip containing cAMP is placed on the right side of a *Dictyostelium* cell expressing GFP-PI3K. As the cell sensing the cAMP gradient moves towards the needle source (white asterisk), GFP-PI3K localizes to the leading edge. (**B**) If the needle is moved to the lower left side of the cell, (white asterisk), the *Dictyostelium* re-orients and creates a new leading edge with GFP-PI3K at the new front.

outs and gene replacements can be readily made, it should be demonstrated that the fluorescent fusion protein complements the null mutation or does not cause an abnormal phenotype when used to replace the endogenous copy of the gene.

Fluorescent protein fusions with protein domains are also used to assess the occurrence and location of protein activation in vivo. GFP fusions of the PH domains of PKB or cytosolic regulator of adenylyl cyclase (CRAC), which bind phosphatidylinositol 3,4,5 triphosphate (PIP3), are used to assess PIP3 accumulation and, therefore, PI3K and PTEN activity *(5–8,16)*. A GFP fusion of the Ras binding domain (RBD) of human Raf1 has been used to detect biologically active, GTP-loaded Ras *(15)*. The fluorescence intensity across the cells is quantitated using either ImageJ (National Institutes of Health [NIH], Bethesda, MD), Metamorph (Molecular Devices Corporation, Downingtown, PA), or IPLab-Spectrum (Scanalytics, Fairfax, VA) software. Simultaneous use of GFP and RFP fusion proteins will allow the more precise detection of the order of translocation and activation events (Sasaki and Firtel, unpublished data).

2. Materials

2.1. Buffers and Equipment

1. Na/K phosphate starvation buffer—one of two types.
    a. Na/K phosphate starvation buffer A: 12 m*M* Na/K phosphate buffer: 2.5 m*M* Na$_2$HPO$_4$, 9.5 m*M* KH$_2$PO$_4$, pH 6.1. Autoclave to sterilize. Store at 22°C. Stable for 1 yr.

   b. Na/K phosphate starvation buffer B: 10 mM Na/K phosphate buffer: 5 mM Na$_2$HPO$_4$, 5 mM NaH$_2$PO$_4$, 2 mM MgSO$_4$, 200 μM CaCl$_2$, pH 6.2. Autoclave to sterilize. Store at 22°C. Stable for 1 yr *(8)*.

2. LabTek chambered coverslip (Nunc, Naperville, IL). Alternatively, chambered coverslips can be made by hand using sterile (tissue-culture treated) polystyrene cell culture dishes with at least 8 cm^2 surface area (35 mm × 10 mm culture dishes), a medium-duty hole punch that can punch a 0.5-inch (1.3 cm) hole, a 15-mL conical tube, silicone vacuum grease, 18 × 18 mm glass coverslips, and 70% ethanol. Use the medium-duty hole punch to punch a 0.5-inch hole in the center of the bottom of a culture dish. Dip a glass coverslip in 70% ethanol and let dry. Dip the top of a 15-mL conical tube into the vacuum grease and then place the inverted tube on top of the hole to create a ring of grease around the hole. Use tweezers to place the ethanolized coverslip over the grease and press on the coverslip to seal it over the hole.

3. Pump with a variable speed drive for accurately dispensing 0.02–1 mL fluid per minute every 6 min, such as the high-precision multi-channel IPC-N 8 pump (ISMATEC, IDEX Corporation, Glattbrugg). Attach 1-mL pipets with 20- to 200-μL pipet tips to the output tubing.

4. 30 mM cAMP, pH 6.1. Dissolve cAMP in either of the above Na/K phosphate starvation buffers to achieve a buffered cAMP solution at pH 6.1–6.2. Store in aliquots at –20°C. Stable for 1 yr. When diluting 30 mM cAMP for downstream applications, always dilute into one of the starvation buffers to maintain the pH.

5. Micromanipulator (Eppendorf-Netheler-Hinz GmbH, Hamburg, Germany). For experiment to experiment reproducibility, commercial micropipets are used (Eppendorf femtotips, New York).

6. Inverted microscope equipped with DIC or phase-contrast optics, a 40× objective lens, and a mercury lamp. If imaging GFP or other fluorescent proteins, the microscope requires filter sets for the relevant wavelengths and either a 63 or 100× oil-immersion objective lens.

7. Charge-coupled device (CCD) camera for acquiring images.

8. Computer with software for controlling the CCD camera and recording images of cell movement, such ImageJ (NIH, Bethesda, MD), IPLab Spectrum (Scanalytics, Fairfax, VA), or Metamorph (Molecular Devices Corporation, Downingtown, PA).

9. 25 mM folic acid, pH 7.0. Dissolve folic acid in either of the above starvation buffers and adjust the pH with NaHCO$_3$ to help it dissolve. Store in aliquots at –20°C. Stable for 1 yr.

2.2. Culture and Preparation of *D.* discoideum *Cells*

For all chemotaxis applications, *D. discoideum* cells should be taken from axenic exponentially growing or "log-phase" cultures of 1×10^4 to 2×10^6 cells/mL. This can be obtained by shaking cells in nutrient medium, in an autoclaved Erlenmeyer flask, at 150 rpm, 22°C. Determine the cell density by counting using a hematocytometer and calculate how much to dilute the cultures in order to have the necessary number of cells the next day. Alternatively, cells can be expanded in sterile tissue-culture dishes. Because there can be differences in random motility and chemotaxis behavior after growth in shaking culture vs in dishes, it is best to compare the two types of growth when doing chemotaxis assays.

3. Methods

3.1. Analyzing Random Cell Motility

The following protocol is for analyzing unstimulated vegetative cells randomly moving in the absence of chemoattractant.

1. Grow 1×10^6 cells in exponential growth phase in nutrient medium (*see* **Note 1**).
2. Count cell density using a hematocytometer.
3. Seed cells at low density (5×10^4 cells/mL or 2×10^4 cells/cm$^2 \approx 1.5 \times 10^5$ cells in a 35 mm $\times$ 10 mm, 8 cm^2 chamber) in nutrient medium onto a chambered coverslip (*see* **Note 2**).
4. Allow cells to adhere for 10 min.
5. Wash cells three times in either Na/K phosphate starvation buffer A or B by aspirating the medium/buffer and carefully pipetting buffer onto an inside corner of the dish. Be careful not to disturb the attached cells.
6. Add 3–4 mL of fresh Na/K phosphate starvation buffer (the same buffer used in **step 5**) to cover the cells and let sit for 1 h (*see* **Note 3**).
7. View the cells through a 40× objective (63× or 100× oil-immersion objective if analyzing fluorescent fusion protein localization during random cell motility) on an inverted microscope.
8. Capture images of the cells every 6–12 s for 30–60 min (~300 images) using imaging software and a CCD camera (*see* **Note 4**).

3.2. Analyzing Chemotaxis Toward cAMP

3.2.1. Analyzing Chemotaxis Toward cAMP:
Method 1, Developed by the Firtel Laboratory

The following protocol is for analyzing developed cells chemotaxing up a cAMP gradient.

1. Grow 5×10^7 cells in exponential growth phase in nutrient medium (*see* **Note 1**).
2. Count cell density using a hematocytometer.
3. Harvest 5×10^7 cells by centrifuging 5 min at 400g (*see* **Note 5**).
4. Wash cells twice by centrifuging 5 min at 400g in 40 mL of 12 mM Na/K phosphate buffer A (*see* **Note 6**).
5. Resuspend the cells at 5×10^6 cells/mL in 10 mL of 12 mM Na/K phosphate buffer A, in a 125-mL Erlenmeyer flask.
6. Make 50 mL of 7.5 μM cAMP by diluting 30 mM cAMP into 12 mM Na/K phosphate buffer A.

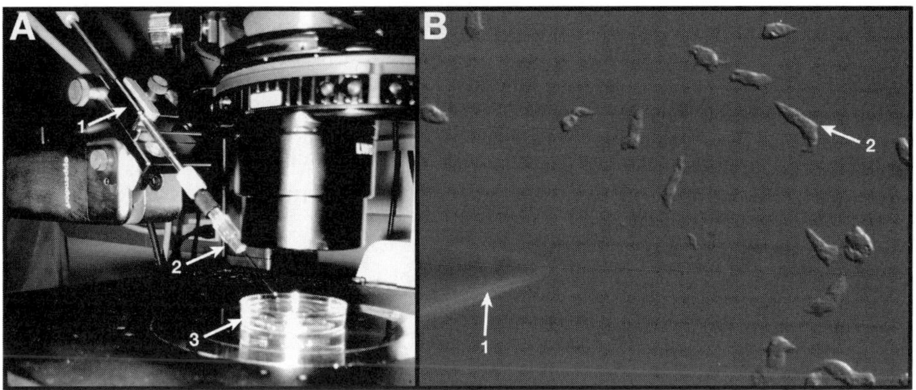

Fig. 2. Positioning a micropipet containing chemoattractant to stimulate *Dictyostelium*. Microscope and micropipet setup. **(A)** A Micromanipulator (1; Eppendorf-Netheler-Hinz GmbH, Hamburg, Germany) controls the positioning of the micropipet (2; fcmtotip, Eppendorf, New York) into the chambered coverslip (3) on the stage of an inverted microscope. **(B)** The needle (1) is positioned at the opposite side of the field as the *Dictyostelium* cells being analyzed (2).

7. Put the input tubing into the 7.5 μ*M* cAMP solution.
8. Calibrate the flow rate of the pump to achieve a 40-μL dispensing volume in 2 s by monitoring the flow output through the 1-mL pipets (*see* **Notes 7** and **8**).
9. Pulse cells with 30 n*M* cAMP at 6-min intervals, for 5 h, by delivering 40-μL pulses of 7.5 μ*M* cAMP into the 10-mL cultures, while shaking at 150 rpm (*see* **Notes 9** and **10**).
10. Fill a chambered coverslip with 12 m*M* Na/K phosphate buffer A and add 75 μL of the pulsed cells (~4 × 10⁶ cells).
11. Allow cells to adhere for 10 min.
12. Place the chamber on the inverted microscope and use the micromanipulator to position a micropipette containing 150 μ*M* cAMP to create a chemoattractant gradient to stimulate the cells (*see* **Notes 11** and **12** and **Fig. 2**).
13. Capture a 40× (63× or 100× oil image if analyzing fluorescent fusion protein localization during chemotaxis up a cAMP gradient) image of the cells every 6 s for 30 min (300 images) using imaging software and a CCD camera (*see* **Notes 4** and **13**).
14. Use ImageJ or Metamorph software to convert the stacked TIFF files into a QuickTime (Apple Computer) movie for easier viewing.

3.2.2. Analyzing Chemotaxis Toward cAMP: Method 2, Developed by the Devreotes Laboratory (8)

This protocol differs from that of Method 1 (*see* **Subheading 3.2.1.**) in five ways. The starvation buffer has slightly less phosphate than that of the Firtel group and includes the addition of $MgSO_4$ and $CaCl_2$. The cells are also starved at a 4X higher concentration and are shaken at a lower rpm during starvation. Starvation begins with 1 h without pulsing, followed by 4 h with pulsing (rather than five straight hours of pulsing). Finally, the cells are pulsed with a higher cAMP concentration and chemotaxis is analyzed with a cAMP gradient created with a 15-fold lower concentration of cAMP, which correlates to a 10-fold decrease in average receptor occupancy.

1. Grow 5×10^7 cells in exponential growth phase in nutrient medium (*see* **Note 1**).
2. Count cell density using a hematocytometer.
3. Harvest 5×10^7 cells by centrifuging at 400*g* for 5 min (*see* **Note 5**).
4. Wash cells twice by centrifuging at 400*g* for 5 min in 40 mL of 10 m*M* Na/K phosphate buffer B (*see* **Note 6**).
5. Resuspend the cells at 2×10^7 cells/mL in 10 mL of 10 m*M* Na/K phosphate buffer B, in a 125-mL Erlenmeyer flask.
6. Shake cells for 1 h at 120 rpm.
7. Make 50 mL of 25 µ*M* cAMP stock by diluting 30 m*M* cAMP into 10 m*M* Na/K phosphate buffer B.
8. Put the input tubing into the 25 µ*M* cAMP solution.
9. Calibrate the flow rate of the pump to achieve a 40-µL dispensing volume in 2 s by monitoring the flow output through the 1-mL pipets (*see* **Note 8**).
10. Pulse cells with 100 n*M* cAMP at 6-min intervals for 4 h by delivering 40-µL pulses of 25 µ*M* cAMP into the 10-mL cultures, while shaking at 120 rpm (*see* **Notes 9** and **10**).
11. Dilute cells to 2×10^5 cells/mL.
12. Spot 20 µL of 2×10^5 cells/mL (4×10^3 cells) onto a chambered coverslip.
13. Allow cells to adhere for 10 min.
14. Cover cells with 10 m*M* Na/K phosphate buffer B to prevent them from drying.
15. Place the chamber on the inverted microscope and use the micromanipulator to position a micropipet containing 10 µ*M* cAMP to stimulate the cells (*see* **Notes 11** and **12** and **Fig. 2**).
16. Capture a 40× (63× or 100× oil image if analyzing fluorescent fusion protein localization during chemotaxis up a cAMP gradient) image of the cells every 6 s for 30 min (300 images) using imaging software and a CCD camera (*see* **Notes 4** and **13**).
17. Use ImageJ or Metamorph software to convert the stacked TIFF files into a QuickTime (Apple Computer) movie for viewing and further analysis.

3.3. Analyzing Chemotaxis Toward Folic Acid: Developed by the Weeks Laboratory (12)

1. Grow 1×10^6 cells in exponential growth phase in nutrient medium (*see* **Note 1**).
2. Count cell density using a hematocytometer.
3. Seed approx 4×10^5 cells/cm^2 onto a chambered coverslip containing either Na/K phosphate starvation buffer A or B.

4. Allow cells to adhere for 10 min.
5. Wash cells once in the Na/K phosphate starvation buffer used in **step 3** by aspirating the medium/buffer and carefully pipetting buffer onto an inside corner of the dish. Be careful not to disturb the attached cells.
6. Add 3–4 mL of fresh Na/K phosphate starvation buffer (the same buffer used in **steps 3** and **5**) containing 20% nutrient medium to cover the cells.
7. Place the chamber on the inverted microscope and use the micromanipulator to position a micropipet containing 25 m*M* folic acid to stimulate the cells (*see* **Note 11** and **Fig. 2**).
8. Capture 40× (63× or 100× oil image if analyzing fluorescent fusion protein localization during chemotaxis up a folic acid gradient) images of the cells every 12–24 s for 1–2 h (300–600 images) using imaging software and a CCD camera (*see* **Note 4**).
9. Use ImageJ or Metamorph software to convert the stacked TIFF files into a QuickTime (Apple Computer) movie for viewing and further analysis.

3.4. Analyzing Cellular Responses to Global Chemoattractant Stimulation

3.4.1. Analyzing Cellular Responses to Global cAMP Stimulation

1. Prepare cells by following **Subheading 3.2.1.**, **steps 1–11** or **Subheading 3.2.2.**, **steps 1–14**.
2. Place the chamber on the inverted microscope, focus on a cell of interest, and begin capturing 63× or 100× images every second using imaging software and a CCD camera (*see* **Notes 4** and **14**).
3. Stimulate the cells with 30 μ*M* cAMP by rapidly pipetting 200 μL of 150 μ*M* cAMP per mL of buffer while taking images (*see* **Note 15**).
4. Use ImageJ or Metamorph software to convert the stacked TIFF files into a QuickTime (Apple Computer) movie for viewing and further analysis.

3.4.2. Analyzing Cellular Responses to Global Folate Stimulation

1. Grow 1×10^6 cells in exponential growth phase in nutrient medium (*see* **Note 1**).
2. Count cell density using a hematocytometer.
3. Seed cells at low density (5×10^4 cells/mL or 2×10^4 cells/cm^2 ≈ 1.5×10^5 cells in a 35 mm × 10 mm, 8 cm^2 chemotaxis chamber) in nutrient medium onto a chambered coverslip (*see* **Note 2**).
4. Allow cells to adhere for 10 min.
5. Wash cells three times in either Na/K phosphate starvation buffer A or B by aspirating the medium/buffer and carefully pipetting buffer onto an inside corner of the dish. Be careful not to disturb the attached cells.
6. Add 3–4 mL of fresh Na/K phosphate starvation buffer (the same buffer used in **step 5**) to cover the cells and let sit for 1 h (*see* **Note 3**).
7. Place the chamber on the inverted microscope, focus on a cell of interest, and begin capturing 63× or 100× images every second using imaging software and a CCD camera (*see* **Notes 4** and **14**).

8. Stimulate the cells with 50 μ*M* folic acid by rapidly pipetting 100 μL of 500 μ*M* folic acid per mL of buffer while taking images (*see* **Note 15**).

9. Use ImageJ or Metamorph software to convert the stacked TIFF files into a QuickTime (Apple Computer) movie for viewing and further analysis.

4. Notes

1. When growing cells, it is important to take into account that wild-type cells double every 8 h, but mutant strains can take longer. For example, shaking 1×10^7 cells (about one confluent 10-cm plate) in 20 mL of nutrient medium overnight (16 h or two doubling periods) usually yields about 4×10^7 cells at 2×10^6 cells/mL the next morning. Furthermore, some mutant strains, such as strains with inactive myosin II, exhibit conditional cytokinesis defects when grown in shaking cultures *(17,18)*. These cells become very large and multinucleate as a result of a lack of cell division in the absence of a solid substratum. For chemotaxis analyses, such strains should always be expanded on sterile tissue culture dishes rather than in shaking cultures.

2. The low cell density facilitates tracking individual cell movements within the field of view over the course of the assay.

3. Random motility is assayed in Na/K phosphate buffer, rather than nutrient medium, because nutrient medium contains folic acid, a chemoattractant. However, after >2 h in Na/K phosphate starvation buffer, *Dictyostelium* secretes pulses of cAMP, another chemoattractant. Thus, less than 2 h should elapse from the point of addition of starvation buffer to the end of analysis. To ensure that no autonomous chemoattractant signaling is occurring, seed the cells on and monitor their motility in a perfusion chamber, such as a Dvorak-Stotler chamber (Nicholson Precision Instruments, Gaithersburg, MD) or a Sykes-Moore chamber (Bellco Glass, Vineland, NJ) *(3)*. Connect the inlet tube of the chamber to a reservoir containing Na/K starvation buffer and the outlet tube to a peristaltic pump in order to replace the chamber fluid with fresh buffer every 8 s *(19)*.

4. If using fluorescent probes to monitor protein localization or activity during chemotaxis, exposure times and/or exposure frequency may need to be reduced to avoid photobleaching the cells. Fluorescent proteins are sensitive to ultraviolet (UV) light, and the microscope system and camera sensitivity are important. For example, using a Nikon model TE300 microscope and Coolsnap-HQ camera, images with 50- to 150-ms exposure times can be taken every second using a shutter system. For faster (<1 s) images, such as those taken during global stimulation, the shutter will need to remain permanently open. Although this exposes the cells to more UV light, the total recording time is shorter for these applications (5 min for global stimulation vs 30 min for chemotaxis analysis).

5. If sequentially analyzing multiple strains, it is best to re-count, wash, and begin pulsing the different strains at 1-h intervals to ensure that each strain is starved for the same amount of time before analysis. Analyzing each strain for chemotaxis toward cAMP entails recording the cells' movement for 30 min, and two recordings may be desired for internal controls and downstream analysis. Some-

times, inadvertently starving the cells for a longer time before analysis can result in more polarized cells that move faster and with better directionality towards cAMP.

6. It is important to ensure that the *Dictyostelium* cells are washed free of nutrients found in the nutrient medium before setting up the starvation program. Be sure to sufficiently dilute the cells in the Na/K starvation buffer when washing.

7. The key is to be able to deliver 30 nM pulses of cAMP to the cells without drastically altering the volume (and thus the concentration) of the cells. Pulsing 10 mL of cells at 5×10^6 cells/mL with 40-µL pulses adds a total volume of 2 mL after 5 h and dilutes the cells to approx 4×10^6 cell/mL. The conditions can be adjusted to deliver 20-µL pulses of 15 µM cAMP to add only 1 mL total volume over the course of the experiment. However, pulsing less than a 20-µL volume can cause problems with the pulse that can lead to the cAMP drop not having enough weight to counter the surface tension and falling into the shaking flask. Pulsing larger than a 40-µL volume will dilute the cells too much. In the absence of a pump, cAMP can be manually added to the cells every 6 min or, if it is known that the strains are capable of secreting and responding to cAMP, they can be incubated in the chambered coverslip for 6–8 h and allowed to develop on their own.

8. It is critical to check the tubing and pipets for bubbles and leaks. If pulsing more than one strain at a time, each input/output tubing combination needs to be calibrated to ensure that there is ≤10% difference in the amounts of cAMP delivered to the individual strains.

9. Wild-type cells should be elongated (polarized) after proper cAMP pulsing.

10. Some protocols call for a 30-min treatment with caffeine after pulsing and before the actual chemotaxis analysis, in order to inhibit adenylyl cyclase and reduce any basal chemotaxis signaling. However, this method is problematic and unadvisable because caffeine likely specifically and nonspecifically inhibits many other cellular activities. For example, it has recently been discovered that caffeine inhibits TOR Complex 2, which among many other substrates, phosphorylates and activates Akt and is required for chemoattractant-mediated activation of adenylyl cyclase (Lee and Firtel, unpublished data) *(20,21)*. Akt is a key signaling protein required for actin polymerization and pseudopod protrusion and its inhibition would severely affect the chemotaxis phenotype *(6)*.

11. In order to provide the maximum distance for the cells to travel before reaching the micropipet tip, position the micropipet at the edge of the field of view opposite to that of the cells to be analyzed.

12. The cAMP gradient generated with 1–10 µM cAMP has been estimated to create a 2–10% change in receptor occupancy across the cell *(8)*, the minimum gradient detected by eukaryotic cells. In the methods described here, cAMP diffuses from the tip of the micropipet. Other methods use positive pressure. In these cases, a lower concentration of cAMP is used in the micropipette.

13. Mutant strains that do not polarize after pulsing and/or do not chemotax toward the micropipet must be checked for expression of the cAMP receptors (cAR1 and cAR3) and other signaling proteins normally induced by cAMP pulsing and

necessary for adenylyl cyclase activation, such as PKA, Erk2, and Myb *(9,10)*. Either Northern blot analysis and/or immunoblotting for expression of such genes/proteins will determine whether the mutation blocks a chemoattractant-driven event or the developmental transition into chemotaxis competence.

14. Some signaling events occur faster than others and the frequency of recorded images should be adjusted accordingly. For example, GFP-RBD translocation is faster than that of GFP-PH, and images must be taken every 0.2 s to differentiate the two responses *(15)*.

15. A flow chamber with inlet and outlet tubes can also be used to globally stimulate cells with a uniform concentration. Both input and output tubes are filled with Na/K buffer to the same level. Fresh buffer containing the chemoattractant of interest is then added to the inlet tube, causing the chamber to fill with the chemoattractant by laminar flow without creating a gradient, in the course of approx 1 s *(14,22)*.

References

1. Parent, C. A. (2004) Making all the right moves: chemotaxis in neutrophils and *Dictyostelium. Curr. Opin. Cell Biol.* **16**, 4–13.
2. Friedl, P., Borgmann, S., and Brocker, E. B. (2001) Amoeboid leukocyte crawling through extracellular matrix: lessons from the *Dictyostelium* paradigm of cell movement. *J. Leukoc. Biol.* **70**, 491–509.
3. Soll, D. R., Wessels, D., Heid, P. J., and Zhang, H. (2002) A contextual framework for characterizing motility and chemotaxis mutants in *Dictyostelium discoideum. J. Muscle Res. Cell. Motil.* **23**, 659–672.
4. Funamoto, S., Milan, K., Meili, R., and Firtel, R. A. (2001) Role of phosphatidylinositol 3' kinase and a downstream pleckstrin homology domain-containing protein in controlling chemotaxis in *Dictyostelium. J. Cell Biol.* **153**, 795–809.
5. Funamoto, S., Meili, R., Lee, S., Parry, L., and Firtel, R. A. (2002) Spatial and temporal regulation of 3-phosphoinositides by PI 3-kinase and PTEN mediates chemotaxis. *Cell* **109**, 611–623.
6. Meili, R., Ellsworth, C., Lee, S., Reddy, T. B., Ma, H., and Firtel, R. A. (1999) Chemoattractant-mediated transient activation and membrane localization of Akt/PKB is required for efficient chemotaxis to cAMP in *Dictyostelium. EMBO J.* **18**, 2092–2105.
7. Iijima, M. and Devreotes, P. (2002) Tumor suppressor PTEN mediates sensing of chemoattractant gradients. *Cell* **109**, 599–610.
8. Parent, C. A., Blacklock, B. J., Froehlich, W. M., Murphy, D. B., and Devreotes, P. N. (1998) G protein signaling events are activated at the leading edge of chemotactic cells. *Cell* **95**, 81–91.
9. Aubry, L. and Firtel, R. (1999) Integration of signaling networks that regulate *Dictyostelium* differentiation. *Annu. Rev. Cell. Dev. Biol.* **15**, 469–517.
10. Manahan, C. L., Iglesias, P. A., Long, Y., and Devreotes, P. N. (2004) Chemoattractant signaling in *Dictyostelium discoideum. Annu. Rev. Cell. Dev. Biol.* **20**, 223–253.

11. Insall, R. H., Soede, R. D., Schaap, P., and Devreotes, P. N. (1994) Two cAMP receptors activate common signaling pathways in *Dictyostelium. Mol. Biol. Cell* **5,** 703–711.

12. Lim, C. J., Zawadzki, K. A., Khosla, M., Secko, D. M., Spiegelman, G. B., and Weeks, G. (2005) Loss of the *Dictyostelium* RasC protein alters vegetative cell size, motility and endocytosis. *Exp. Cell Res.* **306,** 47–55.

13. Wessels, D. and Soll, D. R. (1998) Computer-assisted characterization of the behavioral defects of cytoskeletal mutants of *Dictyostelium discoideum*, in *Motion Analysis of Living Cells* (Soll, D. R. and Wessels, D., eds.), Wiley-Liss, New York: pp. 101–140.

14. Postma, M., Roelofs, J., Goedhart, J., Gadella, T. W., Visser, A. J., and Van Haastert, P. J. (2003) Uniform cAMP stimulation of *Dictyostelium* cells induces localized patches of signal transduction and pseudopodia. *Mol. Biol. Cell* **14,** 5019–5027.

15. Sasaki, A. T., Chun, C., Takeda, K., and Firtel, R. A. (2004) Localized Ras signaling at the leading edge regulates PI3K, cell polarity, and directional cell movement. *J. Cell. Biol.* **167,** 505–518.

16. Jin, T., Zhang, N., Long, Y., Parent, C. A., and Devreotes, P. N. (2000) Localization of the G protein betagamma complex in living cells during chemotaxis. *Science* **287,** 1034–1036.

17. Robinson, D. N., Girard, K. D., Octtaviani, E., and Reichl, E. M. (2002) *Dictyostelium* cytokinesis: from molecules to mechanics. *J. Muscle Res. Cell. Motil.* **23,** 719–727.

18. de la Roche, M. A., Smith, J. L., Betapudi, V., Egelhoff, T. T., and Cote, G. P. (2002) Signaling pathways regulating *Dictyostelium* myosin II. *J. Muscle Res. Cell. Motil.* **23,** 703–718.

19. Wessels, D., Schroeder, N. A., Voss, E., Hall, A. L., Condeelis, J., and Soll, D. R. (1989) cAMP-mediated inhibition of intracellular particle movement and actin reorganization in *Dictyostelium. J. Cell Biol.* **109,** 2841–2851.

20. Sarbassov, D. D., Guertin, D. A., Ali, S. M., and Sabatini, D. M. (2005) Phosphorylation and regulation of Akt/PKB by the rictor-mTOR complex. *Science* **307,** 1098–1101.

21. Lee, S., Comer, F. I., Sasaki, A., McLeod, I. X., Duong, Y., Okumura, K., Yates Iii, J. R., Parent, C. A., and Firtel, R. A. (2005) TOR complex 2 integrates cell movement during chemotaxis and signal relay in *Dictyostelium. Mol. Biol. Cell* **16,** 4572–4583.

22. Potma, E., de Boeij, W. P., van Haastert, P. J., and Wiersma, D. A. (2001) Real-time visualization of intracellular hydrodynamics in single living cells. *Proc. Natl. Acad. Sci. USA* **98,** 1577–1582.

24

Characterization of Cross-Linked Actin Filament Gels and Bundles Using Birefringence and Polarized Light Scattering

Ruth Furukawa and Marcus Fechheimer

Summary

Fundamental processes in the life of *Dictyostelium*, such as locomotion, endocytosis, cytokinesis, and morphogenesis, are mediated by the actin cytoskeleton, which is composed of actin, myosins, and numerous actin-binding proteins. An understanding of these processes at the molecular level will require characterization of the structure, function, and dynamics of the actin and actin-binding proteins both in vivo and in vitro. *Dictyostelium* has more than a dozen actin cross-linking proteins that can mediate the formation of isotropic actin gels and/or anisotropic actin bundles. We describe the use of transmitted polarized light and polarized light scattering in studies of actin and actin-binding proteins during formation of nematic, gelled, or bundled structures. These methods have allowed quantitative studies of the effects of actin filament length, the concentration of actin, and the concentration of the cross-linking protein on the formation of cross-linked actin structures. Such methods hold great promise for characterization of novel cross-linking proteins, and for interpretation of phenotypes from strains lacking or expressing altered forms of these proteins. These methods are also applicable to studies of other systems such as the interactions of microtubules and microtubule associated proteins.

Key Words: Actin gels; actin bundles; birefringence; polarized light scattering.

1. Introduction

The actin cytoskeleton plays a major role in the life of *Dictyostelium* in diverse processes including endocytosis *(1)*, cytokinesis *(2)*, chemotaxis *(3)*, and morphogenesis *(4)*. The facility with which biochemistry, cell biology, genetics, and reverse genetics can be utilized in *Dictyostelium*, and the central role of cell motility in virtually all stages of the life cycle, have made

From: *Methods in Molecular Biology, vol. 346:* Dictyostelium discoideum *Protocols*
Edited by: L. Eichinger and F. Rivero © Humana Press Inc., Totowa, NJ

Dictyostelium a key model organism for studies of the cytoskeleton for the past 25 yr. The actin cytoskeleton depends on the behavior of actin, actin-binding proteins, and a large family of myosin motor proteins that have been intensively studied in *Dictyostelium (5–7)*. In addition, completion of the genome of *Dictyostelium* has uncovered a host of novel cytoskeletal proteins *(8,9)* in addition to those that had been discovered by conventional methods.

Examination of the functions of multiple cytoskeletal proteins will require production of cells lacking, overexpressing, or producing altered forms of specific gene products. New methods for light and electron microscopy employing rapid imaging of green fluorescent protein (GFP)-tagged cytoskeletal proteins and tomography developed using *Dictyostelium* will be extremely valuable for characterization of the dynamics and organization of the actin cytoskeleton in vivo *(10–12)*. Characterization of the association of cytoskeletal proteins with a detergent-insoluble fraction can be assessed using well developed methods for production of such preparations in *Dictyostelium (13,14)*.

Our laboratory has had a long-standing interest in proteins that cross-link actin filaments to form isotropic gels and bundles. *Dictyostelium* has more than a dozen proteins that fall into this class of actin-binding proteins, many of which are expressed simultaneously in both vegetative and developing cell types *(9,15)*. Understanding the similarities and differences between multiple cytoskeletal proteins with overlapping functions and interpretation of phenotypes is dependent on fundamental knowledge of biochemical activity and basic studies of the effects of actin-binding proteins on the assembly of actin and on the formation of supramolecular arrays of actin filaments. An excellent review of the methods used to study actin assembly details the primary approaches of viscometry, spectrofluorometry, electron microscopy, and sedimentation *(16)*. Similarly, methods for study of the formation of actin networks, including sedimentation and a variety of rheometers for measurements of viscoelastic properties, were described in an excellent review *(17)*.

Many actin cross-linking proteins can induce the formation of isotropic gelled or anisotropic bundled actin structures *(15)*. Although these morphologies can be qualitatively assessed using electron microscopy, more quantitative assays for study in solution add significantly to our ability to detect subtle differences in the biochemical activity of actin cross-linking proteins. Furthermore, formation of isotropic cross-linked gels or parallel bundles have distinct effects on the rheology of polymer solutions, and independent methods are required to characterize the state of the mixture. We have used polarized light and polarized light scattering in studies of the spontaneous orientation of actin into a nematic liquid crystal domain *(18)* and in studies of the formation of gelled or bundled structures in mixtures of actin with actin-binding proteins *(19,20)*. These methods have allowed quantitative studies of the effects of actin

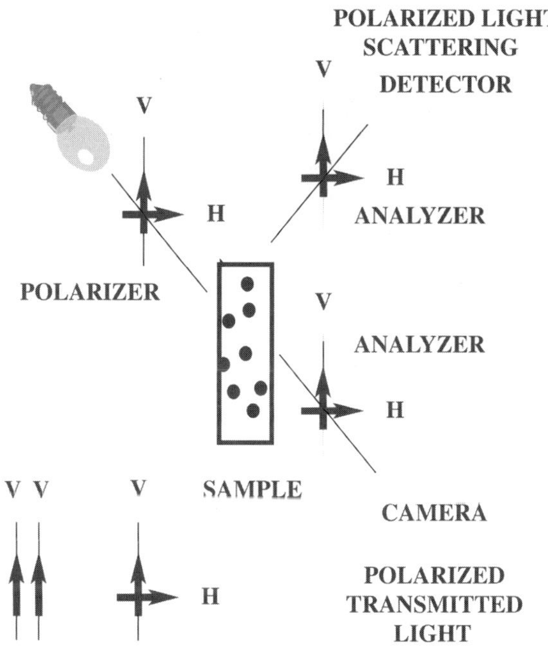

Fig. 1. Polarized light scattering. Schematic diagram of polarized light-scattering experiment. The position of the polarizer and analyzer in the light path and the definition of the isotropic (Vv) and anisotropic (Hv) components of the light scattering are shown.

filament length and the concentration of actin and of the cross-linking protein on the formation of cross-linked actin structures. Such methods hold great promise for characterization of other cross-linking proteins, for characterization of differences in the structure and function of multiple members of a family, and for interpretation of phenotypes from strains lacking or expressing altered forms of these proteins.

1.1. Theory of Polarized Light Scattering

Physicists have described anisotropic light scattering from liquid crystalline solutions and polymeric films and described the use of this approach for assessing the orientational correlation among the molecules *(21–23)*. Polarizing filters placed in the light path between the light source and the sample, and between the sample and the detector are termed polarizer and analyzer, respectively (*see* **Fig. 1**). The conditions in which the two filters are aligned vertically (Vv) or are crossed (Hv) are shown. The isotropic component of scattering (Vv) increases as the concentration or size of scattering particles increases no

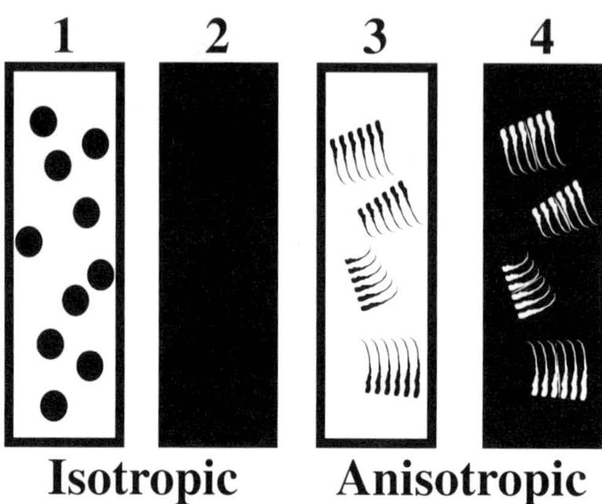

Fig. 2. Polarized light scattering. Solutions of isotropic (1) or anisotropic (3) particles are expected to appear black (2) or birefringent (4) when viewed in the Hv orientation through crossed polarizers.

matter what the orientation of the molecules in solution. The intensity of the anisotropic light scattering component (Hv) increases dramatically if the orientational correlation among the molecules is increased (e.g., by formation of a nematic liquid crystalline phase or a cross-linked filament bundle) *(24,25)*. Both Vv and Hv components of scattering must be measured to characterize the orientation of the molecules in the solution. Orientation in the samples is detected by an increase in the ratio of light scattered from samples with filters in the incident and scattered beams in crossed (Hv) vs parallel (Vv) orientations. The increase in anisotropic light scattering can also be observed as the transmission of light through crossed polarizers observed by the eye or by polarization microscopy. Isotropic solutions appear black when viewed through crossed polarizers, whereas anisotropic solutions reveal bright regions that correspond to aligned regions/bundles in the solution (*see* **Fig. 2**). Direct measurements of the Hv/Vv 90° light scattering components are used for quantitative measurements of these phenomena.

Because this method has been widely used in studies of the orientational correlation of other polymeric macromolecules, its application to analyses of the bundling of actin filaments was logical. We have demonstrated the ability of this assay to determine the concentration and length dependence of actin filaments required for the formation of liquid crystals *(18)*, to discriminate between forms of a cross-linking protein that induce formation of either a gel or a bundle *(19)*, and to determine the actin filament length and concentration dependence of an actin bundling protein *(20)*. In this chapter, we describe methods for observation of birefringence, and for quantitative studies of polarized light scattering to assess the orientation of filaments in mixtures of actin and actin-binding proteins. Although the following methods are enumerated for actin solutions, they are equally applicable to microtubule solutions. Tubulin solutions have been previously shown to exhibit a phase transition from a disordered to highly ordered nematic liquid crystalline phase *(26)*. Appropriate changes for the buffers and polymerization conditions are required.

2. Materials

2.1. Actin and Other Reagents

1. Buffer G: 2 m*M* Tris-HCl, pH 8.0, 0.2 m*M* ATP, 0.2 m*M* CaCl$_2$, 0.2 m*M* dithiothreitol (DTT), 0.02% NaN$_3$.
2. KCl: prepare a 4 *M* stock.
3. MgCl$_2$: prepare a 1 *M* stock.
4. PIPES or Tris: prepare a 1 *M* stock at the desired pH of the actin polymerization.
5. ATP: prepare a 1 *M* stock at pH 7.0.
6. DTT: prepare a 1 *M* stock at pH 7.0.
7. Rabbit skeletal muscle actin is prepared as described previously *(27–29)*. Further removal of capZ protein by another cycle of polymerization/depolymerization followed by gel filtration is recommended to ensure that the actin filament length is controlled *(28,30)*. *Dictyostelium* actin can be prepared by the method of Fechheimer and Furukawa *(31)*. The concentration of actin should be at least 72 µ*M* but less than 100 µ*M* to ensure that the concentration is below the critical concentration of G-actin in buffer G. The resultant G-actin can be held in dialysis in buffer G at 4°C for 1 wk with buffer changes every 24 h.
8. Gelsolin is purified from bovine serum as described previously *(18,32)*. Prepare a 60 n*M* stock.
9. Actin-binding proteins are purified according to specific protocols. It is beneficial to prepare at least a 10 µ*M* stock in its specific buffer (ABP buffer). Just prior to use, filter the buffer or protein solutions through a 0.22-µm filter (*see* **Note 1**) into a dust-free container and cap.

2.2. Instrumentation and Optics

1. Cuvets should be of high optical quality—preferably quartz or glass—because the lower-quality cuvets can be birefringent. Cuvets for light scattering should be fluorometer cuvets, that is, optically clear on four sides. Cuvets for optical birefringence can be clear on two sides, such as those used for absorbance. The cuvet capacity should be small so as to conserve on the amount of protein used. We use type 3–5 quartz cells (Starna) that are 5-mm square and that have a capacity of 700 µL for light scattering. For optical birefringence measurements, type 3–3.30 (Starna) cuvets are available that have a path length of 3 mm, width of 5.5 mm, and height of 30 mm with a capacity of 225 µL. Reduced-size cuvets require an adapter holder for use in fluorometers (available through either the fluorometer manufacturer or Starna).

2. Beads are the isotropic scattering particles used for instrument calibration. High-quality beads that are not aggregated must be used. We have used calibration-standard beads supplied from Polysciences, Inc., although in principle, other high-quality beads designed for electron microscopy or flow cytometry calibration can be used. Check to ensure that beads are monomeric. If a high concentration of aggregated beads is used, artifacts may arise from anisotropic scattering.

3. Spectrofluorometer. Specialized equipment using high-power lasers, sensitive detectors, and measurement at a variety of scattering angles can be utilized for light-scattering measurements. However, a simple spectrofluorometer equipped with polarization optics is sufficient for measurements of light scattering at a fixed angle of 90°. We have used a Perkin Elmer LS-5 equipped with polarization optics for the majority of our studies. The optics consist of polarizing filters (discussed later) in both the excitation and emission light paths held in a filter holder with two fixed stops at a 90° angle. Excitation and emission monochromators are set at the same wavelength. Using calibration beads as a standard, maximum sensitivity with this instrumentation was empirically found to occur at a wavelength of 450 nm.

4. Filters. Dichroic sheet polarizers (cat. no. 47 36 00, Carl Zeiss, Inc.) are used to record images of transmission of light through crossed polarizers. Polarizing filters of this type are used as polarizer and analyzer for differential interference optics in light microscopy, and can be utilized to observe filament alignment in macroscopic samples.

2.3. Preparation of Dust-Free Materials

All materials must be free of dust, which scatters light in both the isotropic and anisotropic orientations. This scattering can obscure measurements from the proteins of interest. Dust may result in nonreproducible readings that appear to fluctuate rapidly (*see* **Note 2**).

1. Create a dust-free beaker by filtering water obtained from a purification system (distilled or filtration system) through a 0.22-μm filter in a laminar flow hood into a 150-mL beaker. Discard this water. Repeat five to six times, ensuring that all inner surfaces of the beaker are rinsed. A peristaltic pump equipped with a filter is useful to generate the large volumes of water needed.

2. Create a large capacity dust-free container (2 L bottle). Pour dust-free filtered water from the 150-mL beaker into the bottle. Rinse all surfaces of the container, including the cap, several times with dust-free water by capping the bottle and rotating/shaking the water around. Discard the water between rinses.

3. Fill the container with dust-free water by filtering the water directly into the bottle. Cap.

4. Using dust-free water, rinse the containers, including the cuvets that will have buffer or protein solutions in them, by rinsing five to six times. Discard the water between rinses.

Table 1
**Sample Calculations for Experiments Varying the Concentration
of Actin-Binding Protein (ABP) with a Fixed Actin Concentration**

[ABP] final	µL actin (72 µM stock) + µL gelsolin (60 nM stock) + µL buffer G	µL ABP (10 µM stock)	µL ABP buffer	µL KCl + µL MgCl$_2$ + µL TRIS + µL ATP + µL DTT + µL H$_2$O
0	666.67 + 0.8 + 0.53 = 668	0	300	21.25 + 2 + 5.66 + 0.866 + 0.806 + 1.42 = 32
0.5 µM	666.67 + 0.8 + 0.53 = 668	50	250	21.25 + 2 + 5.66 + 0.866 + 0.806 + 1.42 = 32
3 µM	666.67 + 0.8 + 0.53 = 668	300	0	21.25 + 2 + 5.66 + 0.866 + 0.806 + 1.42 = 32

Volumes of each solution are calculated to give a final volume of 1 mL. Actin and gelsolin are combined with buffer G as one solution and the salts, Tris, ATP, dithiothreitol (DTT), and H$_2$O as another. ABP buffer contains 2 mM Tris, 50 mM KCl, and 0.2 mM DTT. Final solution concentrations are: actin, 48 µM; gelsolin:G-actin ratio, 1:1000; 10 mM Tris, 1 mM ATP, 1 mM DTT, 100 mM KCl, and 2 mM MgCl$_2$.

3. Methods

3.1. Preparing Solutions

The simplest determination of whether an actin-binding protein induces the gelation of F-actin vs bundle formation is to use optical birefringence. Use an actin concentration of 2 mg/mL (48 µM), because nematic liquid crystals do not form at this concentration at any F-actin length previously studied. Vary the concentration of the actin-binding protein from 0.5 µM to 3–4 µM (more if possible). Actin-binding proteins such as α-actinin can induce both F-actin gelation and bundle formation *(33–38)*.

Calculate the amounts of proteins and salts required to perform the experiment. In order to simplify calculations of actin, actin-binding protein, and salts required for polymerization volumes and sample preparation, it is necessary to keep the volume of the actin solution, actin-binding protein, and salts required for polymerization constant between the samples (for sample calculations holding either the actin or actin-binding protein concentration constant, *see* **Note 3** and **Tables 1** and **2**).

1. For protein solutions, rinse the outside of the pipette tip with dust-free water just prior to use, rotating carefully to rinse the entire outside surface. Rinse the inner surface of the tip by pipetting dust-free water into the tip and discarding it in another beaker.
2. To simplify pipetting the samples, the actin, gelsolin, and actin buffer are combined as one solution and the salts, buffer, ATP, DTT, and H$_2$O as another in

Table 2
**Sample Calculations for Experiments Varying Actin Concentration
with a Constant Concentration of Actin-Binding Protein (ABP)**

[actin] final	µL actin (72 µM stock) + µL gelsolin (60 nM stock)	µL buffer G	µL ABP + mL KCl + µL MgCl$_2$ + µL Tris + µL ATP + µL DTT + µL H$_2$O
12 µM	166.67 + 0.2 = 166.87	501.1	300 + 21.25 + 2 + 5.66 + 0.866 + 0.806 + 1.42 = 332
24 µM	333.33 + 0.4 = 333.73	334.27	300 + 21.25 + 2 + 5.66 + 0.866 + 0.806 + 1.42 = 332
48 µM	666.67 + 0.8 = 667.47	0.53	300 + 21.25 + 2 + 5.66 + 0.866 + 0.806 + 1.42 = 332

Volumes of each solution are calculated to give a final volume of 1 mL. Actin and gelsolin are combined with buffer G as one solution and the salts, Tris, ATP, dithiothreitol (DTT), and H$_2$O as another. ABP buffer contains 2 mM Tris, 50 mM KCl, and 0.2 mM DTT. Final solution concentrations are: ABP, 3 µM; gelsolin:G-actin ratio, 1:1000; 10 mM Tris, 1 mM ATP, 1 mM DTT, 100 mM KCl, and 2 mM MgCl$_2$.

dust-free tubes. Multiply the volumes of the actin and salt solutions by the number of the samples and allow a small excess for waste.

3. Pipet the actin, gelsolin, buffer G solution into all the cuvets. Work with each sample individually. Pipet the actin-binding protein and ABP buffer into the cuvet. Mix by inversion with dust-free parafilm. Polymerization may begin if the ABP buffer contains enough KCl or MgCl$_2$.

4. Pipet the salt, buffer, ATP, DTT, and H$_2$O solution. Mix by inversion and cap with parafilm. Prepare the next sample until all the samples are prepared.

5. Degas by placing cuvets in a small vacuum dessicator for a couple of minutes, and then cap. Removal of air will prevent formation of bubbles in the solution that can obscure all light-scattering measurements. Let incubate 20–24 h prior to measurement to achieve steady state polymerization, cross-linking, and alignment. Once polymerization is initiated, mix, degas, and cap promptly. Do not pipet, mix, or transfer the solutions after this time, as these procedures will produce shear-induced alignment of the filaments.

6. For bead solutions, do not filter, because the beads will be larger than 0.22 µm and will be trapped in the filter. Dilute the beads 1:1000 in dust-free buffer. A series of trial concentrations made from this stock should be made. Depending on the value of the most intense light-scattering sample and the diameter of the beads used, one concentration of beads will be chosen to serve as the calibration standard (*see* **Note 4**).

7. Place the sample in the fluorometer for light scattering and set the intensity to the largest number that does not fluctuate very much. This will represent the highest value to be measured and for which the dynamic range will be determined. Measure the intensity of each bead solution. Some of the concentrations will be off-

scale or will measure the limiting value of the instrument. The remainder of the samples will be on-scale. Note which concentration yields the highest value without large fluctuations in intensity. Choose that concentration of beads to be the calibration standard for the fluorometer.

3.2. Macroscopic Images of Birefringence

1. Illuminate a standard solution of polystyrene beads or a solution of bovine serum albumin (BSA) in a cuvet. Place the polarizer between the light and sample. Rotate the polarizer to a minimum light intensity passing through the sample as determined by eye. This fixes the horizontal setting. Adjust the holder set screw to fix the polarizer in the holder. Rotate the polarizer 90° to the vertical position (*see* **Note 5**).
2. With the illuminating light at vertical, place the analyzer after the sample. Rotate the analyzer until the light intensity passing through the sample is at a minimum. The two filters are now crossed in the Hv orientation.
3. Replace the bead or BSA sample with the protein solution to be studied without disturbing the polarizer or analyzer. Record the image by photography using a macro lens (50 mm). Tri-X film gives excellent results, but high-resolution digital cameras should also suffice if they have the required sensitivity and resolution. Repeat until all samples are viewed.
4. To illustrate the method, images of birefringent actin samples taken at a series of angles of polarizer and analyzer are shown (*see* **Fig. 3**). The birefringence is most prominent when polarizer and analyzer are crossed at an angle of 90°.
5. Images of actin solutions recorded at a series of actin concentrations reveal the concentration dependence for formation of aligned domains (*see* **Fig. 4A**). The phase transition to formation of a nematic liquid crystalline occurs in a narrow concentration range (*see* **Fig. 4A**), and depends on the actin filament length. Filament length is readily controlled by varying the ratio of gelsolin to actin *(18)*.

3.3. Polarized Light Scattering

1. Place the cuvet containing the buffer solution into the sample holder. Set the zero value.
2. Exchange the buffer sample with the bead solution. Remove the analyzer (polarizer after the sample). Loosen the set screw of the polarizer. Rotate the polarizer and read the intensity until it is a minimum (Horizontally polarized light [H]) (*see* **Note 6**). Tighten the set screw and rotate the polarizer 90° (Vertically polarized light [V]). The polarizer will remain in this position for measurement of the Hv and Vv components.
3. Insert the analyzer into its holder. Loosen the set screw, and rotate the analyzer and read the intensity until it is a minimum (Horizontally polarized light [H]). Tighten the set screw. This is now the Hv orientation of the component of scattered light.
4. Reset the zero value for buffer; now, the polarizer and analyzer have been adjusted precisely to the Hv orientation. Rotate the analyzer 90° to the vertical orientation. Record the value of the solvent light scattering in the Vv orientation.

A   **B**   **C**   **D**   **E**   **F**

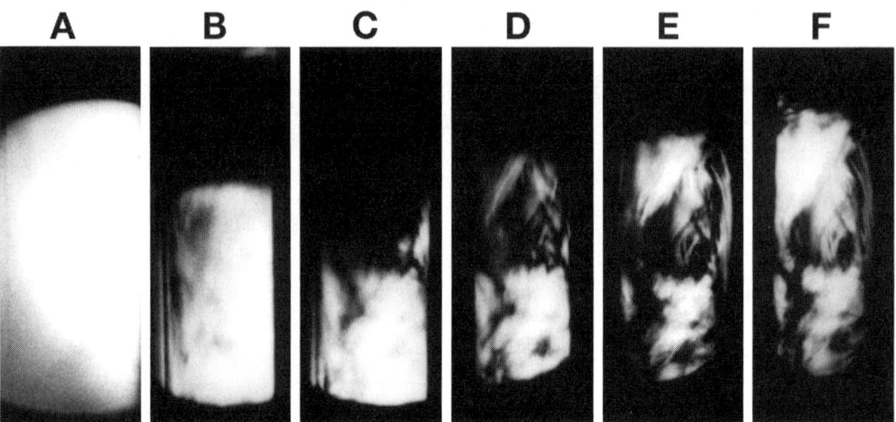

Fig. 3. Transmitted light image of a birefringent actin solution. Transmitted light images though 2.25 mg/mL actin at a gelsolin:G-actin ratio of 1:1111 as a function of the observed polarization angle with vertically polarized incident light. The angle of the analyzer with respect to the polarizer was **(A)** 0°; **(B)** 15°; **(C)** 45°; **(D)** 60°; **(E)** 70°; and **(F)** 90°. The round edge of the polarizer is seen as a circular boundary at the top and/or bottom of the cuvet in some images. The aligned actin is clearly visible through the crossed polarizers, and is most prominent as the angle between polarizer and analyzer approaches 90°.

5. Replace the buffer solution with the bead solution. Set the maximum value of the bead solution to the maximum value already determined for the Vv light scattering (*see* **Note 7**).

6. Replace the bead solution with the solutions to be measured. Record the values with the analyzer adjusted to both the horizontal (H) and vertical (V) settings for each sample. Subtract the value of the solvent from the Vv light scattering intensities. Calculate the ratio by dividing the Hv by the Vv measurement.

7. Make all readings in triplicate on independently prepared samples. Plot Hv light-scattering intensity (Hv/Vv) vs the experimental variable such as actin concentration, cross-linker concentration, or actin filament length.

8. Examples. An example showing the effect of actin concentration on formation of aligned domains is shown in **Fig. 4B**. Note that Hv/Vv light scattering increases over a very narrow range of actin concentration. A second example showing the difference between formation of actin filament gels and bundles is shown in **Fig. 5**. Both the 34 kDa protein and a 27 kDa derivative were able to cross-link actin filaments as shown by viscometry. Polarized light scattering was used to show an increase in the orientational correlation among the filaments in samples containing the 34-kDa protein, but not the 27-kDa fragment (*see* **Fig. 5**). These results were confirmed by electron microscopy *(19)*.

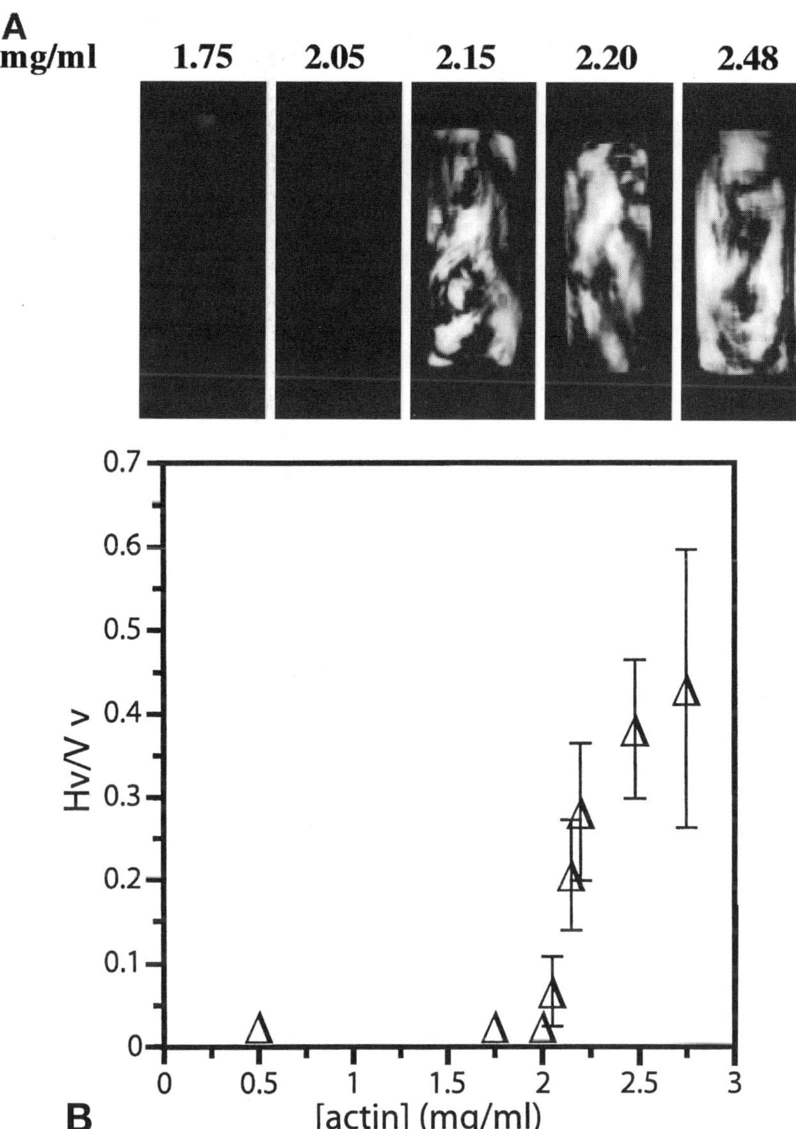

Fig. 4. Formation of oriented domains of actin filaments depends on concentration of actin filaments. (**A**) Polarized transmitted light images of actin solutions above and below the phase transition at fixed length. The gelsolin to actin ratio was 1:1778. (**B**) Hv/Vv light scattering intensity (Hv/Vv) as a function of actin concentration at a ratio of gelsolin to actin of 1:1778. Quantitative measurements of the Hv/Vv ratio of scattered light intensities provide a quantitative measure of the alignment that is visible in the transmitted light images.

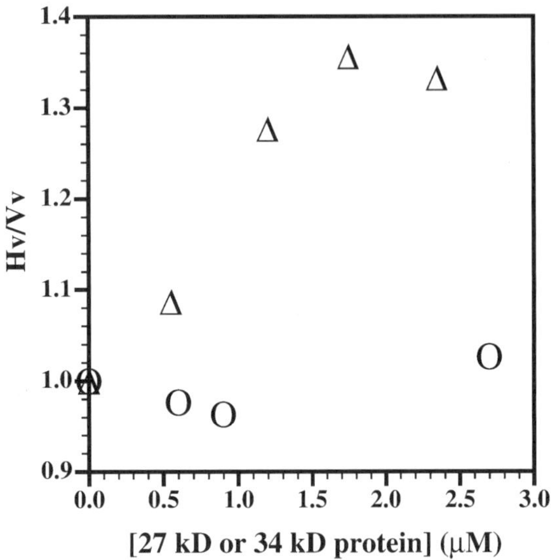

Fig. 5. Polarized light scattering can discriminate isotropic actin gels and anisotropic actin filament bundles. The polarized light scattering intensity (Hv/Vv) as a function of the concentration of the 34 kDa actin cross-linking protein (Δ) or its 27 kDa fragment (O). Both proteins show calcium regulated actin filament binding and cross-linking. Formation of filament bundles by the 34 kDa protein is reflected in the Hv/Vv light scattering. Mixtures of actin with the 27 kDa fragment form isotropic gels that do not reveal any increase in polarized light scattering intensity *(19)*.

4. Notes

1. Use only low protein binding and hydrophilic filters. Millipore Durapore (PVDF) has ultralow protein binding. This is important, as actin and actin-binding proteins will generally be very low in molar mass.
2. The presence of dust is the single greatest cause of noise and irreproducible results. Dust should be removed by rinsing all glassware, solutions, and pipet tips with dust-free water.
3. There are alternative ways to calculate the solution conditions. The volumes that are held constant are arbitrarily chosen. However, the total volumes of actin and actin buffer, as well as those of the actin-binding protein and its buffer, should be held constant in order to simplify the calculation of the volume of salts added to initiate polymerization, because the composition of the ABP buffer may contribute to one of the salts. The exact components of the buffer vary depending on whether the activity of the actin-binding protein is pH-dependent or calcium-dependent and whether the ionic conditions of actin polymerization are an experimental variable. A minimum of 10 mM of either Tris or PIPES is required to ensure that the pH of the solution is maintained. Polymerization of actin is initiated by salts in the form of KCl or MgCl$_2$. ATP and DTT should be added to a

final concentration of 1 mM. EGTA can also be included to lower the concentration of calcium for those proteins that cross-link actin at low calcium. If gelsolin is used to control the length of the filaments, calcium must be present initially to allow gelsolin to form nuclei with actin monomers. First, briefly mix the desired gelsolin and actin with calcium, and then add the salt and EGTA after an initial incubation period. Degas and cap the solutions and wait for steady state.

4. The concentration of beads used for the calibration is empirically determined depending on the bead size. Too large a number, or too high of a concentration, of beads will cause difficulties in measurements. The problems are due to multiple light scattering and can cause the value of the intensity measurement to fluctuate greatly. Problems can also arise as a result of settling of the beads in solution.

5. Although it is desirable that the polarizers are in adjustable holders that rotate precisely 90°, it is not absolutely necessary to have this equipment in order to visualize birefringence. A simple, yet effective, method is to illuminate a cuvet containing an isotropic solution of either BSA or spheres and adjust the vertical polarizer (between the light and the cuvet) until the light appears to be at a maximum intensity. Place the analyzer in front of the cuvet and rotate it until the light intensity is minimized. Replace the isotropic solution with the solutions to be analyzed.
The polarizer and analyzer must be adjusted so that they are never moved before the solutions to be analyzed are examined. This will ensure that the polarizers remain in the "crossed position."

6. Background noise of light passing through the polarizers is due to incorrect extinguishing of the polarizers. Thus, the polarizer should be set first and then the analyzer. The optics are rotated once to set the calibration standard and will minimize leakage of light which is not extinguished in the Hv mode.

7. The Vv and Hv components of the scattered light and Hv/Vv ratio have been measured for 38-nm-diameter polystyrene spheres, BSA, and F-actin. The ratio of Hv/Vv for beads and BSA is 0.023 ± 0.002 and 0.025 ± 0.006, respectively, over a 40-fold range of bead concentration, and a 100-fold range in concentration of BSA from 0.8 to 80 mg/mL The Hv/Vv ratio for 1.2 mg/mL F-actin is 0.029 ± 0.01. These measurements on samples known to lack anisotropically oriented structures serve to validate the method using our experimental technique and apparatus.

References

1. Maniak, M. (2002) Conserved features of endocytosis in *Dictyostelium*. *Int. Rev. Cytol.* **221,** 257–287.
2. Robinson, D. N. and Spudich, J. A. (2004) Mechanics and regulation of cytokinesis. *Curr. Opin. Cell Biol.* **16,** 182–188.
3. Van Haastert, P. J. and Devreotes, P. N. (2004) Chemotaxis: Signaling the way forward. *Nature Rev. Mol. Cell Biol.* **5,** 626–634.
4. Gomer, R., Gao, T., Knecht, D., and Titus, M. A. (2002) Cell motility mediates tissue size regulation in *Dictyostelium. J. Muscle. Res. Cell Motil.* **23,** 809–815.

5. Noegel, A. A., and Luna, E. J. (1995) The *Dictyostelium* cytoskeleton. *Experientia* **51,** 1135–1143.

6. De la Roche, M. A., Smith, J. L., Betapudi, V., Egelhoff, T. T., and Côté, G. P. (2002) Signaling pathways regulating *Dictyostelium* myosin II. *J. Muscle Res. Cell Motil.* **23,** 703–718.

7. Soldati, T. (2003) Unconventional myosins, actin dynamics, and endocytosis: a ménage à trois? *Traffic* **4,** 358–366.

8. Eichinger, L., Pachebat, J. A., Glockner, G., et al. (2005) The genome of the social amoeba *Dictyostelium discoideum. Nature* **435,** 43–57.

9. Rivero, F. and Eichinger, L. (2005) The microfilament system of *Dictyostelium discoideum*, in *Dictyostelium Genomics* (Loomis, W. F. and Kuspa, A. eds.), Horzon Bioscience Cromwell, Norfolk, UK: pp. 125–171.

10. Gerisch, G. and Müller-Taubenberger, A. (2003) GFP-fusion proteins as fluorescent reporters to study organelle and cytoskeleton dynamics in chemotaxis and phagocytosis. *Methods Enzymol.* **361,** 320–327.

11. Medalia, O., Weber, I., Frangakis, A. S., Nicastro, D., Gerisch, G., and Baumeister, W. (2002) Macromolecular architecture in eukaryotic cells visualized by cryoelectron tomography. *Science* **298,** 1209–1213.

12. Diez, S., Gerisch, G., Anderson, K., Müller-Taubenberger, A., and Bretschneider, T. (2005) Subsecond reorganization of the actin network in cell motility and chemotaxis. *Proc. Natl. Acad. Sci. USA* **102,** 7601–7606.

13. Spudich, A. (1987) Isolation of the actin cytoskeleton from amoeboid cells of *Dictyostelium. Methods Cell Biol.* **28,** 209–214.

14. Aguado-Velasco, M., Aguado-Velasco, C., and Kuczmarski, E. R. (1993) Isolation of myosin from *Dictyostelium* cytoskeletons. *Prot. Expr. Purif.* **4,** 328–332.

15. Furukawa, R. and Fechheimer, M. (1997) The structure, function, and assembly of actin filament bundles. *Int. Rev. Cytol.* **175,** 29–90.

16. Cooper, J. A. and Pollard, T. D. (1982) Methods to measure actin polymerization. *Methods Enzymol.* **85,** 182–210.

17. Pollard, T. D. and Cooper, J. A. (1982) Methods to measure actin filament networks. *Methods Enzymol.* **85,** 211–233.

18. Furukawa, R., Kundra, R., and Fechheimer, M. (1993) The formation of liquid crystals from actin filaments. *Biochemistry* **32,** 12,346–12,352.

19. Fechheimer, M., and Furukawa, R. (1993) A 27,000 dalton core of the *Dictyostelium* 34,000 dalton protein retains Ca^{+2}-regulated actin cross-linking but lacks bundling activity. *J. Cell Biol.* **120,** 1169–1176.

20. Furukawa, R. and Fechheimer, M. (1996) Role of the *Dictyostelium* 30 kDa protein in actin bundle formation. *Biochemistry* **35,** 7224–7232.

21. de Gennes, P.-G. (1974) *The Physics of Liquid Crystals.* Clarendon, Oxford, UK: p. 333.

22. Flygare, W. and Gierke, T. D. (1974) Light scattering in noncrystalline solids and liquid crystals. *Ann. Rev. Mat. Sci.* **4,** 255–285.

23. Chandrasekhar, S. (1977) *Liquid Crystals.* Cambridge University Press, Cambridge, UK: p. 342.

24. Chatelain, P. (1948) Sur la diffusion, par les cristaux liquides du type nematique, de la lumière polarisée. *Acta Crystal.* **1,** 315–323.

25. Chatelain, P. (1951) Etude theorique de la diffusion de la lumière par un fluide presentant un seul axe d'isotropie: Application aux cristaux liquides du type nematique. *Acta Crystal.* **4,** 453–457.

26. Hitt, A. L., Cross, A. R., and Williams, R. C. (1990) Microtubule solutions display nematic liquid crystalline structure. *J. Biol. Chem.* **265,** 1639–1647.

27. Spudich, J. A., and Watt, S. (1971) The regulation of rabbit skeletal muscle contraction. *J. Biol. Chem.* **246,** 4866–4871.

28. MacLean-Fletcher, S. D. and Pollard, T. D. (1980) Identification of a factor in conventional muscle actin preparations which inhibits actin filament self-association. *Biochem. Biophys. Res. Commun.* **96,** 18–27.

29. Pardee, J. D. and Spudich, J. A. (1982) Purification of muscle actin. *Methods Enzymol.* **85,** 164–181.

30. Casella, J. F., Barron-Casella, E. A., and Torres, M. A. (1995) Quantitation of CapZ in conventional actin preparations and methods for further purification of actin. *Cell Motil. Cytoskel.* **30,** 164–170.

31. Fechheimer, M. and Furukawa, R. (1991) Purification of the 30,000 dalton actin binding protein from *Dictyostelium discoideum*. *Methods Enzymol.* **196,** 84–91.

32. Cooper, J. A., Bryan, J., Schwab, B., Frieden, C., and Loftus, D. J. (1987) Microinjection of gelsolin into living cells. *J. Cell Biol.* **104,** 491–501.

33. Honda, M., Takiguchi, K., Ishikawa, S., and Hotani, H. (1999) Morphogenesis of liposomes encapsulating actin depends on the type of actin cross-linking. *J. Mol. Biol.* **287,** 293–300.

34. Limozin, L. and Sackmann, E. (2002) Polymorphism of cross-linked actin networks in giant vesicles. *Phys. Rev. Lett.* **89,** 168,103–168,106.

35. Maciver, S. K., Wachsstock, D. H., Schwarz, W. H., and Pollard, T. D. (1991) The actin filament severing protein actophorin promotes the formation of rigid bundles of actin filaments crosslinked with alpha-actinin. *J. Cell Biol.* **115,** 1621–1628.

36. Wachsstock, D. H., Schwatz, W. H., and Pollard, T. D. (1993) Affinity of α-actinin for actin determines the structure and mechanical properties of actin filament gels. *Biophys. J.* **65,** 205–214.

37. Meyer, R. K. and Aebi, U. (1990) Bundling of actin filaments by alpha-actinin depends on its molecular length. *J. Cell Biol.* **110,** 2013–2024.

38. Grazi, E., Trombetta, G., Magri, E., Cuneo, P., and Schwienbacher, C. (1994) Alpha-actinin from chicken gizzard: at low temperature, the onset of actin gelling activity correlates with actin bundling. *Biochem. J.* **298,** 129–133.

25

Quantitative and Microscopic Methods for Studying the Endocytic Pathway

Francisco Rivero and Markus Maniak

Summary

Endocytosis is a process that is essential to the life of all eukaryotic cells. Laboratory strains of *Dictyostelium* are extremely efficient in the uptake of both particles and fluid. Many different cellular processes feed into the endocytic pathway, and many organelle-associated and cytoplasmic proteins, including the ones from the cytoskeleton, contribute to the efficiency of transit. Therefore mutants, especially in genes of unknown function, must be characterized regarding their endocytic performance. We describe the most common tools and protocols to visualize and quantify all of the individual steps in endocytic transit.

Key Words: Phagocytosis; pinocytosis; exocytosis; fluorescencent probes; endolysosomal pH.

1. Introduction

1.1. The Endocytic Pathway of Amoebae

Endocytosis is a process that is essential to the life of all eukaryotic cells. Therefore, its basic components have remained unchanged for millions of years. This enables us to extend discoveries made in "lower" eukaryotes to cells of "higher" organisms such as mammals. On the other hand, there are a few physiological features that distinguish the endocytic pathways of mold and man, which can be traced back to the difference of multicellular vs unicellular lifestyle. One example is the necessity to operate an osmoregulatory system in every cell unless you are blessed with a blood circulation and kidney function. Because the relationship between the contractile vacuole and the endocytic pathway is not described in this book, the reader is referred to the discussions in **ref. *1***.

From: *Methods in Molecular Biology, vol. 346:* Dictyostelium discoideum *Protocols*
Edited by: L. Eichinger and F. Rivero © Humana Press Inc., Totowa, NJ

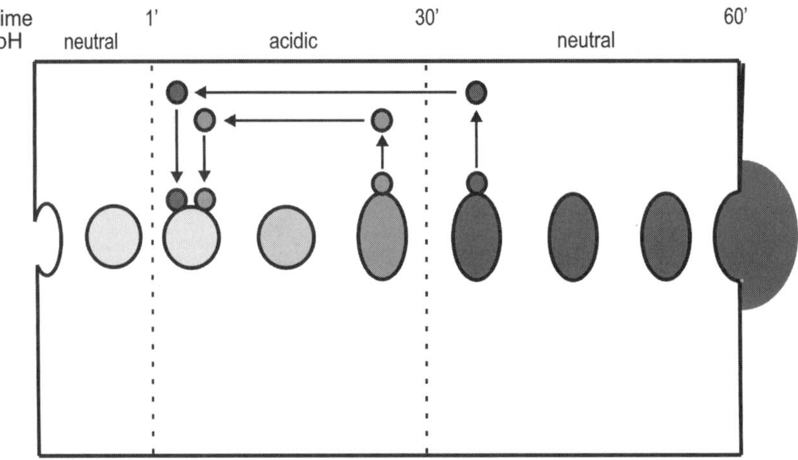

Fig. 1. Representation of a cell (white box) in liquid medium (gray). Between endocytosis (left) and exocytosis (right), the medium is concentrated (light to dark gray). Recycling pathways are connected by arrows. The pH changes inside the endosome and the corresponding times are indicated at the top of the figure.

In essence, the endocytic pathway of *Dictyostelium* can be defined as a transit of cargo through the cell with concomitant digestion. From the viewpoint of cargo, transit starts by uptake at the plasma membrane, packaging into endosomes that sequentially encounter acidic and neutral pH values together with different enzymatic activities, and release of indigestible remnants by exocytosis. In the course of this transit, many components are delivered to the maturing endosome, retrieved from it, and recycled, often by the means of small carrier vesicles (reviewed in **ref. 2**) (*see* **Fig. 1**). Many different cellular processes feed into the endocytic pathway, and many organelle-associated and cytoplasmic proteins, including the ones from the cytoskeleton, contribute to the efficiency of transit (reviewed in **ref. 1**). Therefore mutants, especially in genes of unknown function, must be characterized regarding their endocytic performance. In the following, we will describe the most common tools and protocols to visualize and quantify all of the individual steps in endocytic transit.

1.2. Tools for Studying the Endocytic Pathway

Laboratory strains of *Dictyostelium*, unless they bear dramatic mutations, are extremely efficient in the uptake of both particles and fluid, because this is the predominant way to bring into the cell nutrients needed for growth and multiplication. Accordingly, measuring the growth rate in a bacterial suspension or in axenic medium is a first approximation to whether a strain is capable of normal endocytosis or not. The major drawback here is that the rate of cell multiplication also depends on the efficiency of cell division, which may be controlled independently of endocytic performance. For a precise analysis, it is therefore instrumental to quantify the uptake of cargo directly.

In the study of phagocytosis, it is extremely convenient that *Dictyostelium* cells will internalize almost any sort of particles, edible or not. Investigators have used erythrocytes, yeast, iron beads, latex particles, and various bacteria for different purposes, such as the magnetic or biochemical fractionation of phagosomes (*see* Chapter 26) and studies of pathogenicity (*see* Chapter 31). For the quantification of uptake, two properties of particles are especially helpful. First, the particle should be indigestible, so that it remains intact inside the cell until exocytosis occurs. This is true for latex beads and yeast particles, but we do not recommend assays measuring intracellular survival of bacteria such as the gentamycin-protection protocol *(3)*. Second, a fluorescent label on or within the particle provides a means for easy detection in the microscope or in the spectrofluorimeter. The choice of fluorescent label should be made taking into account that different fluorophores will display different sensitivity to environmental pH. The emission intensity of the most common green fluorophore used for labeling, fluorescein isothiocyanate (FITC), decreases dramatically when encountering the acidic pH value of an early endosome, whereas the signal from tetramethylrhodamine isothiocyanate (TRITC) remains almost unchanged. Below, we describe phagocytosis assays using FITC-labeled bacteria and latex beads as well as TRITC-labeled yeast particles.

To study the uptake of liquid medium, many of the arguments made for the choice of particles hold true. Fluorescent labels will provide convenience of use in microscopy and cuvet-based assays over radioactively labeled tracers. Bovine serum albumin (BSA) as a marker protein will be degraded and metabolized, whereas dextran will remain stable until exocytosis. In principle, one could think that it does not matter which size of molecule the fluorophore is coupled to, but the different kinetics observed with large tracers (TRITC-dextran) or small molecules (Lucifer yellow) indicate that they access different endocytic pathways *(4)* or are recycled with different kinetics *(1)*. Here, we will describe two different assays that make use of FITC- or TRITC-labeled dextrans to study the uptake of fluid (measuring mainly macropinocytosis). We will further indicate how the properties of the probes can be exploited to visualize and measure changes in endosomal pH during transit and outline a protocol to quantify exocytosis.

Inevitably, plasma membrane is internalized concomitantly with the uptake of particles and fluid. Therefore, one could think that it might be even more convenient to design protocols measuring the kinetics of a membrane probe, e.g., in the presence or absence of particles. Here, we will not describe the procedures in detail, but rather refer the reader to original work. Surface labeling of membrane proteins allows us to follow their internalization and recycling back to the membrane in a pulse-chase setup by means of Western blotting of cell homogenates, but these techniques are less suitable for microscopy and quantification *(5)*. On the other hand, styryl dyes that are incorporated into the hydrophobic phase of the membrane are more convenient to use, but less well characterized. The flow of FM 1-43 has been used to suggest a highly efficient pathway of membrane internalization that is distinct from macropinocytosis but otherwise unknown *(6)*. If added to the medium, FM 4-64 immediately labels the plasma membrane, but then quickly accumulates in the contractile vacuole and next in mitochondria by unknown transfer mechanisms *(7)*.

Measuring the activity of lysosomal enzymes provides yet another means to quantify the relative amounts of exocytosis vs recycling and may therefore be used to generate complementary results to those obtained by determining endocytic transit of a small fluorescent probe (like Lucifer yellow) as compared with a large tracer like TRITC-dextran. Because some lysosomal enzymes enter endosomes early in the pathway to profit from the acidic luminal pH value, they may be efficiently recycled and reused. Other enzymes that are delivered to endosomes later, when their pH has returned to neutral, tend to accumulate in the medium. We routinely assay the distribution of acid phosphatase, an enzyme that is efficiently retained in the cell, and α-mannosidase, a protein that is released in large amounts into the medium by assays originally developed by Dimond and co-workers *(8)*.

2. Materials

2.1. Culture of Dictyostelium *Cells*

Usually, *Dictyostelium* cells are cultured in nutrient medium in suspension on a rotary shaker (160 rpm) at 21°C. In some cases, it may be necessary to perform the analyses on cells growing axenically on culture plates or on cells growing on agar plates with bacteria. In these cases, cells can be resuspended and, if needed, washed free of bacteria prior to the measurements. In general, cells are harvested at room temperature by centrifugation at 500*g* for 3 min. Wherever possible, one should avoid centrifuging and resuspending the cells excessively before the subsequent assay. Unless otherwise indicated, the assays described here are performed at 21°C on a rotary shaker operating at 160 rpm. A batch of medium should be kept at 21°C for washing and resuspending cells where required.

It is critical to ensure identical growth conditions of the strains tested. Because the batch of the medium and cultivation in the presence of antibiotics have been found to affect the assay, we grow cells with the minimal required concentration of selective antibiotics, and dilute all cultures 24 h before assay in the same batch of medium lacking antibiotics, so that the desired cell density is reached when the assay is set up. It is not only recommended that cells be cultivated under the same conditions for some days preceding the assay, but also that the respective parent strain be measured every time. In most assays, it is convenient to use the results obtained from the parent wild-type strain to adjust the data gathered from multiple independent experiments for the presentation in a single graph.

2.2. Buffers and Other Reagents

1. Phosphate-buffered saline (PBS): 8 g NaCl, 0.2 g KCl, 1.44 g Na_2HPO_4, and 0.24 g KH_2PO_4 in 1 L dH_2O, pH 7.4. A 10X concentrated solution can be prepared for storage.
2. Soerensen phosphate buffer: 2 mM Na_2HPO_4, 14.6 mM KH_2PO_4, pH 6.0.
3. Labeling buffer: 50 mM Na_2HPO_4, pH 9.2. Add TRITC (T5646, Sigma) or FITC (F7628, Sigma) (freshly dissolved in dimethylsulfoxide [DMSO]) to 0.1 mg/mL final concentration and use immediately.
4. Lysis buffer: 50 mM Na_2HPO_4, pH 9.3, 0.2% Triton X-100.
5. FITC-labeled, carboxylate-modified polystyrene microspheres, 1.0 µm in diameter, can be obtained from Polysciences. Prepare a suspension of 4×10^{10} particles/mL. Latex beads with a wide array of fluorophores are commercially available and can be used in cases in which FITC cannot be used.
6. Polyethylene glycol (PEG) 8,000 (Fluka 81272). Prepare a 20% aequous solution.
7. FITC-dextran 70,000 M_r (Sigma FD-70S), TRITC-dextran 76,000 M_r (Sigma T1162), Lucifer yellow CH (Fluka 62649). Make a 100 mg/mL stock solution in dH_2O, aliquot, and store at –20°C protected from the light.
8. MES buffer: 50 mM 2-morpholinoethanesulfonic acid (MES), pH 6.5. Prepare a 20X concentrated stock.
9. Assay solutions for lysosomal enzyme determinations. For acid phosphatase determination: 37 mM sodium acetate, pH 5.0, 16.7 mM p-nitrophenyl phosphate (pNPP). For α-mannosidase determination: 5 mM sodium acetate, pH 5.0, 5 mM p-nitrophenol-α-mannopyranoside.
10. Stop solution for lysosomal enzyme determinations: 1 M Na_2CO_3.

2.3. Preparation of Fluorescent Bacteria

1. Inoculate 300 mL LB medium with *Escherichia coli* B/r and grow overnight at 21°C. Harvest bacteria by centrifugation at 5000g for 10 min. Wash the bacterial pellet twice with Soerensen buffer. Adjust the suspension to 1×10^{10} bacteria/mL. This can be achieved by measuring the optical density (540 nm) of a diluted aliquot of the bacterial suspension: 1 OD = 1×10^9/mL. It is important to resuspend the bacteria well before measuring. The bacterial suspension can be kept for up to 1 wk in a cold-room with slow shaking.

2. For labeling, resuspend the pellet of 50 mL cells in 50 mL of labeling buffer containing FITC and incubate 2 h at 37°C on a rotary shaker. Wrap with aluminium foil to avoid bleaching.

3. Wash bacterial pellet four times with Soerensen buffer. Adjust the suspension to 2×10^{10} bacteria/mL and store at −20°C in 1- or 2-mL aliquots protected from the light (*see* **Note 1**).

2.4. Preparation of Fluorescent Yeast Cells

1. Suspend 5 g of yeast (YSC-2, Sigma) in 50 mL PBS in a 100-mL Erlenmeyer flask. Maintain for 30 min in a boiling water bath and stirred.

2. Harvest yeast cells by centrifugation at 3000*g* for 5 min. Wash yeast pellet five times with PBS and twice with Soerensen buffer. Adjust the suspension to 1×10^9 particles/mL. Unlabeled yeast can be stored at −20°C in aliquots.

3. For labeling, resuspend the pellet of 2×10^{10} cells in 20 mL of labeling buffer containing TRITC and incubate 30 min at 37°C on a rotary shaker.

4. Wash yeast pellet twice with labeling buffer lacking TRITC and four times with Soerensen buffer. Adjust the suspension to 1×10^9 particles/mL and store at −20°C in 1- or 2-mL aliquots protected from the light (*see* **Note 1**).

2.5. Preparation of Trypan Blue Solution

1. Dissolve trypan blue (Merck 1.11732) at a final concentration of 2 mg/mL in 20 m*M* citrate containing 150 m*M* NaCl. Adjust pH to 4.5 with HCl.

2. Stir for 30 min, filter through filter paper and then through a 0.45-μm pore-size filter to remove particulate component. Store at 4°C.

3. Methods

3.1. Fluorimetric Analysis of Phagocytosis

We describe protocols for quantitation of phagocytosis using three different particles: latex microspheres, bacteria, and yeast. A critical step in these assays is the separation of cells containing ingested material from the bulk of uningested material. For latex microspheres and bacteria, this problem has been overcome by centrifuging the cell suspension through a cushion of a highly viscous solution of PEG or sucrose *(9)*. Using this method, particles with a diameter below 2 μm remain on top of the cushion, whereas amoebae are collected in the pellet. By contrast, free yeasts and those that stick to the cell surface cannot be separated from *Dictyostelium* cells by centrifugation. In this case, a quenching step with Trypan Blue is added *(10,11)* (*see* **Note 2**).

3.1.1. Fluorimetric Analysis of Phagocytosis
Using FITC-Labeled Latex Beads

1. Harvest cells if necessary and resuspend at 2×10^6/mL in fresh nutrient medium.

2. Dispense 100 μL of cell suspension in an Eppendorf tube, spin (10 sc in a table-

top centrifuge at maximum speed), remove the supernatant, and freeze the pellet at –20°C. This sample is used to determine protein content.

3. Dispense 10 mL of cell suspension in a 30-mL Erlenmeyer flask and incubate for 15 min on a rotary shaker to allow cells to recuperate.

4. Add 100 µL of FITC-labeled latex beads suspension (previously vortexed vigorously to dissociate aggregates) and incubate on a rotary shaker. The amount of beads added corresponds to a 200:1 ratio relative to *Dictyostelium* cells.

5. Starting immediately (time point 0), withdraw 1-mL samples and add to conical tubes containing 2 mL of ice-cold Soerensen buffer to rapidly stop phagocytosis. Continue at selected time points (15, 30, 45, 60 min). Samples can be collected on ice until the end of the experiment or can be processed until the lysis step.

6. Centrifuge (800*g*, 10 min, 4°C preferably in a swing-out rotor) through an ice-cold 10-mL cushion of PEG 8,000 in a 15-mL conical tube. Wash cell pellet twice with cold Soerensen buffer.

7. Carefully resuspend pellet in 1 mL of lysis buffer and measure in a fluorimeter (excitation 470 nm; emission 515 nm) (*see* **Note 3**).

8. Plot relative fluorescence against time after subtraction of fluorescence at time zero and correction for differences in protein content (*see* **Note 4**).

3.1.2. Fluorimetric Analysis of Phagocytosis Using FITC-Labeled Bacteria

The assay is performed exactly as the assay using fluorescent latex beads (*see* **Subheading 3.1.1.**), with the difference that 0.5 mL of FITC-labeled bacteria is added to the cell suspension, which corresponds to an approx 500:1 ratio relative to *Dictyostelium* cells.

3.1.3. Fluorimetric Analysis of Phagocytosis Using TRITC-Labeled Yeast Cells

1. Proceed as in **Subheading 3.1.1.**, **steps 1–3**.

2. Thaw an aliquot of fluorescent yeasts (*see* **Subheading 2.4.**), sonicate briefly, and vortex vigorously to dissociate aggregates. Add 120 µL of fluorescent yeast cells and incubate on a rotary shaker. The amount of yeasts added corresponds to a sixfold excess relative to *Dictyostelium* cells.

3. Starting immediately (time point 0), withdraw 1-mL samples and add to Eppendorf tubes containing 100 µL of Trypan Blue solution. Continue at selected time points (15, 30, 45, 60, 90, 120 min).

4. Incubate samples for 3 min with agitation (preferably on a slowly running, vortex-like instrument), pellet cells for 2 min at 800*g* in a tabletop centrifuge, and remove supernatant carefully but completely (*see* **Note 5**).

5. Carefully resuspend pellet in 1 mL of Soerensen buffer and measure immediately in a fluorimeter (excitation 544 nm; emission 574 nm) (*see* **Note 3**).

6. Plot relative fluorescence against time after subtraction of fluorescence at time 0 and correction for differences in protein content (*see* **Note 2**).

3.2. Fluorimetric Analysis of Fluid-Phase Endocytosis and Exocytosis

The protocols described as follows make use of fluorescently labeled dextran at various concentrations. It has been shown that in a range between 0.5 and 10 mg/mL, uptake of labeled dextran is directly proportional to its concentration in the medium *(9)* and proceeds linearly with time for at least 1 h at 21°C. After this time, the rate of uptake slows *(12–14)*. The most commonly used dextran is of 70.000 M_r, but it can be substituted with 150.000 M_r without any obvious alterations. FITC-dextran is the fluorescent marker most commonly used to assay pinocytosis (*see* **Notes 5** and **6**). However, the assay of TRITC-dextran pinocytosis is useful for analyzing strains that express a green fluorescent protein (GFP) fusion protein that would interfere during the measurement as a result of overlapping excitation and emission spectra.

3.2.1. Pinocytosis Assay Using FITC-Dextran

1. Harvest cells if necessary and resuspend at 5×10^6/mL in fresh nutrient medium.
2. Dispense 100 µL of cell suspension in an Eppendorf tube, spin (10 s in a tabletop centrifuge at maximum speed), remove the supernatant, and freeze the pellet at –20°C. This sample is used to determine protein content.
3. Dispense 5 mL of cell suspension in a 30-mL Erlenmeyer flask and incubate for 15 min on a rotary shaker to allow cells to recuperate.
4. Add 100 µL of FITC-dextran solution from the 100 mg/mL stock and incubate on a rotary shaker.
5. At selected time points (0, 15, 30, 45, 60, 90, 120, 180 min), withdraw 500-µL samples and add to 2 mL microfuge tubes containing 1.5 mL of ice-cold Soerensen buffer. Samples can be collected on ice until the end of the experiment or can be processed immediately until the lysis step (**step 7**).
6. Pellet cells (2 min at 800*g* in a tabletop centrifuge), wash once with ice-cold Soerensen buffer, and pellet again.
7. Resuspend pellet in 1 mL of lysis buffer and measure in a fluorimeter (excitation 470 nm; emission 515 nm) (*see* **Note 3**).
8. Plot relative fluorescence against time after subtraction of fluorescence at time 0 and correction for differences in protein content (*see* **Note 4**).

3.2.2. Pinocytosis Assay Using TRITC-Dextran

1. Proceed as in **Subheading 3.2.1.**, **steps 1–3**.
2. Add 100 µL of TRITC-dextran solution from the 100 mg/mL stock and incubate on a rotary shaker.
3. At selected time points (0, 15, 30, 45, 60, 90, 120, 180 min), withdraw 500-µL samples and add to Eppendorf tubes containing 50 µL of Trypan Blue solution.
4. Turn tube over once and immediately pellet cells (2 min at 800*g* in a tabletop centrifuge) and remove supernatant carefully but completely (*see* **Note 5**).

5. Wash cells once in 1 mL of Soerensen buffer, centrifuging as described previously.
6. Carefully resuspend pellet in 1 mL of Soerensen buffer and measure immediately in a fluorimeter (excitation 544 nm; emission 574 nm) (*see* **Note 3**).
7. Plot relative fluorescence against time after subtraction of fluorescence at time 0 and correction for differences in protein content (*see* **Note 4**).

3.2.3. Exocytosis Assay

The assay described as follows is set up to quantitate exocytosis only. The assay can be performed as continuation of a pinocytosis assay. In this case, prepare double amount of cell suspension and allow for a total of 180 min loading with the dye before the exocytosis assay begins. Then proceed with **step 4** below.

1. Proceed as in **Subheading 3.2.1.**, **steps 1–4** or **Subheading 3.2.2.**, **steps 1** and **2**, depending on the marker used.
2. Immediately after addition of the fluorescent dextran, withdraw a 500-µL sample. This sample will be used to determine the background fluorescence and is processed as indicated in **step 6** below.
3. Incubate the cell culture, wrapped in aluminum foil, on a rotary shaker for 180 min.
4. Collect the cell suspension in a 15-mL conical tube, centrifuge at 500*g* for 3 min, wash the cell pellet twice with nutrient medium, and resuspend cells in 5 mL fresh nutrient medium.
5. Dispense the cell suspension in a 30-mL Erlenmeyer flask and incubate on a rotary shaker.
6. Immediately (time point 0) withdraw 500-µL samples and process and measure as for the pinocytosis assay (*see* **Subheading 3.2.1.**, **steps 5–7**, or **Subheading 3.2.2.**, **steps 3–6**, omitting **step 5**, depending on the marker used). Continue at selected time points (15, 30, 45, 60, 90, 120, 180 min).
7. Plot relative fluorescence against time after subtraction of background fluorescence. For each sample the 0 time point is assigned 100% relative fluorescence and the rest of the time points are calculated relative to the corresponding 0 time point. No correction for cell mass variation is needed.

3.3. Presentation of Results

In general, results are presented as relative fluorescent units, taking the parental wild-type strain as reference. This allows adjustment of the data gathered from multiple independent experiments for the presentation in a single graph (*see* **Fig. 2**). It is also possible to derive the amount of ingested particles or ingested volume by comparison with a standard curve. For this, a dilution series of nutrient medium with fluid-phase marker or fluorescent particles is dispensed in tubes containing 1 mL of lysis buffer (for FTIC-dextran and FITC-labeled latex beads or bacteria) or Soerensen buffer (for TRITC-dextran and

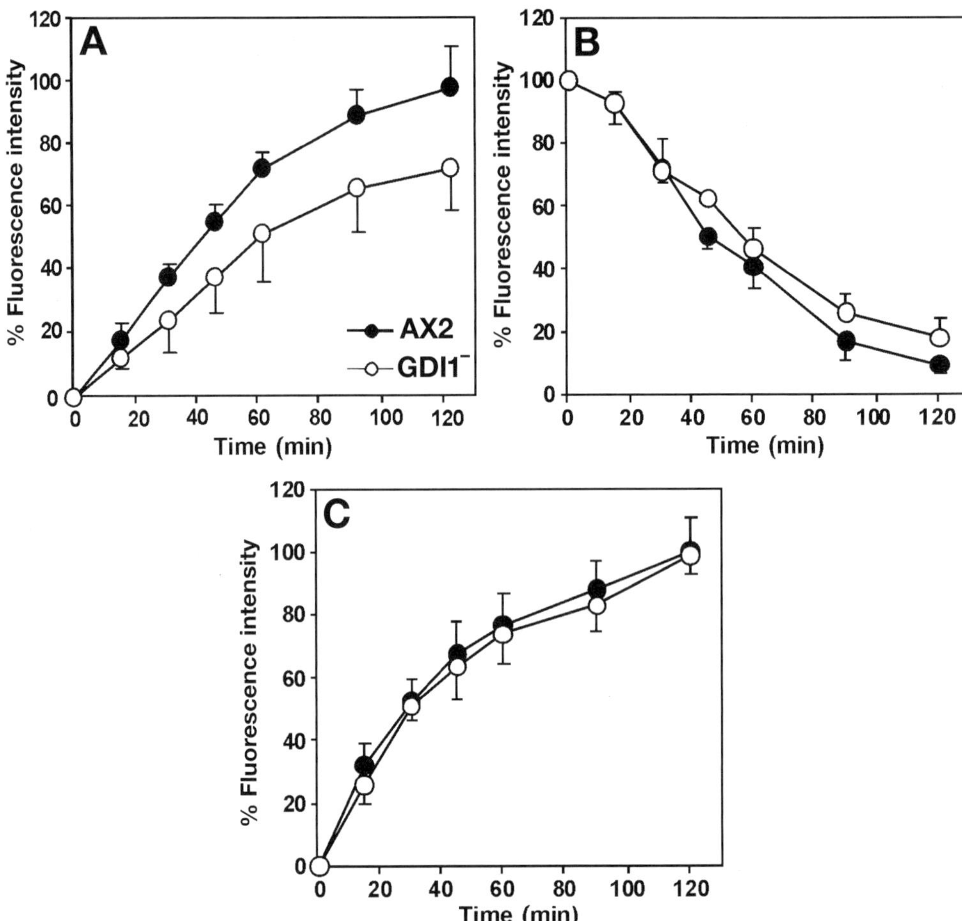

Fig. 2. Quantitative endocytosis and exocytosis. AX2 cells and cells deficient in RhoGDI (GDI1⁻) were compared. (**A**) Pinocytosis of fluorescein isothiocyanate (FITC)-dextran. Cells were resuspended in fresh axenic medium in the presence of 2 mg/mL FITC-dextran and fluorescence from the internalized marker was measured at the indicated time points. (**B**) Fluid-phase exocytosis of FITC-dextran. Cells were pulsed with FITC-dextran (2 mg/mL) for 3 h, washed, and resuspended in fresh axenic medium. Fluorescence from the marker remaining in the cell was measured. (**C**) Phagocytosis of tetramethylrhodamine isothiocyanate (TRITC)-labeled yeast cells. *Dictyostelium* cells were resuspended in fresh axenic medium and challenged with a fivefold excess of fluorescent yeast cells. Fluorescence from internalized yeasts was measured at the designated time points. Data are presented as relative fluorescence, AX2 being considered 100%, and all values are the average ± standard deviation of at least three independent experiments, with error bars depicted only in one direction. Reproduced from **ref. 23**, with permission.

TRITC-labeled yeasts). Fluorescence is measured as indicated for the corresponding assay using the same settings of the fluorimeter as for the cell samples. Results can then be plotted as number of ingested particles or volume of internalized fluid per milligram of protein against time.

3.4. Measurement of Endo-Lysosomal pH

3.4.1. Spectral Shift Setup

The pH-dependent behavior of FITC can be exploited for the determination of the pH of endo-lysosomal compartments. For this, a dual excitation ratio method using FITC-dextran as a pH probe is commonly used *(13–15)*. The protocol described as follows can be used to monitor the effects of drugs and stress conditions on acidification of the endo-lysosomal compartment over time *(16)*.

1. Proceeed as in **Subheading 3.2.1.**, **steps 1–4**. Before adding the fluorescent dye, remove a 500-µL sample. This sample will be used to determine the background fluorescence and is processed as indicated in **step 3** below.
2. Incubate on a rotary shaker for 3 h. This time is sufficient for the complete loading of all the endo-lysosomal compartments. At this point, drugs can be added or treatments can be started. Processing of samples for measurement is then performed as follows.
3. Withdraw 500-µL samples and add to 2-mL microfuge tubes containing 1.5 mL of ice-cold MES buffer. Pellet cells (2 min at 800g in a tabletop centrifuge), wash twice with ice-cold MES buffer, and pellet again.
4. Carefully resuspend pellet in 1 mL of 20X MES buffer and measure in a fluorimeter (excitation 450 nm and 495 nm; emission 515 nm) (*see* **Note 3**).
5. The fluorescence excitation ratio (I_{495}/I_{450}) is calculated after subtraction of the background fluorescence. The endosomal pH is determined from a standard curve generated by adjusting 10 µg/mL FITC-dextran in 20X MES buffer to various pH values between 4.0 and 8.0 and measuring the fluorescence excitation ratio.

3.4.2. Dual Fluorophore Setup

In contrast to the method outlined under **Subheading 3.4.1.**, the following protocol determines the change of endosomal pH value over time and exocytosis concomitantly. The pH-independent dye TRITC is taken as a measure of the volume remaining in the cell, whereas the decrease of FITC fluorescence indicates the acidity of the compartments through which the marker passes during transit.

1. For each culture to be measured, mix 40 mg of TRITC-dextran with 4 mg of FITC-dextran in 10 mL of nutrient medium.
2. Harvest 10^8 cells and immediately resuspend in 10 mL of the fluorescent medium by gently pipetting up and down through a wide orifice, and transfer into a 30-mL Erlenmeyer flask.
3. Allow the cells to internalize medium for 10 min on a rotary shaker.

4. Collect the cell suspension in a 15-mL conical tube, centrifuge at 500*g* for 3 min. The fluorescent supernatant can be kept for further use (*see* **Note 7**).
5. Resuspend cells in 10 mL of nonfluorescent nutrient medium. Dispense the cell suspension in a 30-mL Erlenmeyer flask and incubate on a rotary shaker.
6. Immediately withdraw a 1-mL sample (time point 0) and add to an Eppendorf tube containing 100 μL of Trypan Blue solution. Mix once by turning the Eppendorf tube over. Pellet cells (2 min at 800*g* in a tabletop centrifuge), aspirate the supernatant carefully and completely.
7. Carefully resuspend the pellet in 1 mL of Soerensen phosphate buffer by gently pipetting up and down a few times. Immediately measure the fluorescence of TRITC (excitation 544 nm, emission 574 nm) and FITC (excitation 470 nm, emission 515 nm) (*see* **Notes 3** and **5**).
8. Repeat **steps 6** and **7** at 15 min intervals until 120 min. Calculate the FITC/TRITC emission ratio and relate this value to a standard curve obtained from measuring the fluorophore mix at pH values between 4.0 and 8.0.

3.5. Recycling of Markers and Lysosomal Enzymes

3.5.1. Fluorescent Probes

In this setup, we exploit the rationale of intracellular sorting of small vs large dyes as a result of differential access to small compartments *(17)*. A small dye is preferentially found in recycling vesicles, whereas a large dye will remain within large endosomes *(1)*. Therefore, the release of the two markers into the medium will occur with different kinetics. The experiment is conducted as outlined in **Subheading 3.4.2.**, except that FITC-dextran (> 70,000 M_r) is replaced by 4 mg Lucifer yellow (~500 M_r), and wavelength for excitation and emission recording is 455 nm and 535 nm, respectively.

3.5.2 Lysosomal Enzymes

Enzymatic activities are measured in both medium and cell lysate using specific substrates that are converted to *p*-nitrophenol. The product of the reaction is quantitated colorimetrically.

1. For convenience, we grow cells on a Petri dish, starting with an inoculum of 5×10^5 cells. On d 3, aspirate and save culture medium. Adherent cells are lysed in 1 mL of water containing 0.5% Triton X-100 yielding 2.5×10^7 cell-equivalents per mL. Growing cells for shorter times will give a time course in which the accumulation of enzymatic activity in the medium over time can be assayed.
2. For each sample, prepare 480 μL of assay solution containing pNPP (for acid phosphatase determination) or 300 μL of assay solution containing *p*-nitrophenyl-α-mannopyranoside (for α-mannosidase determination). Add 20 μL (for acid phosphatase) or 200 μL (for α-mannosidase) of cell extract (or the same ammount of medium) and incubate at 37°C for 30 min. Also prepare a blank with Triton solution or fresh medium.

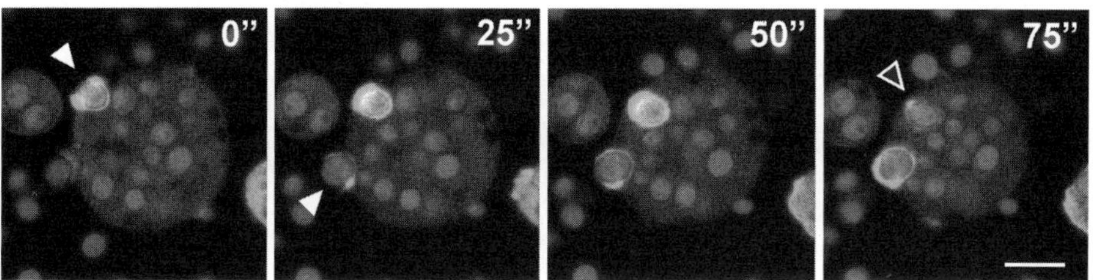

Fig. 3. Redistribution of green fluorescent protein (GFP)-actin during phagocytosis monitored in vivo by uptake of tetramethylrhodamine isothiocyanate (TRITC)-labeled yeast cells. Images were acquired every 5 s with a Leica TCS SP confocal laser scanning microscope equipped with an Ar laser (excitation at 488 nm) and a HeNe laser (excitation at 543 nm). Signals from both fluorophores were acquired simultaneously and merged. In the grayscale picture shown yeast cells can be easily distinguished. GFP-actin assembles at the sites of phagocytic cup formation (close arrow heads) and is released when the uptake is nearly completed (open arrow head). Bar = 10 µm.

3. Stop the reaction by adding 500 µL of a 1 M Na_2CO_3 solution. Measure absorbance at 420 nm. Subtract background of the corresponding blank sample.

To make results comparable, protein concentrations must be determined in the cell extracts according to standard procedures.

3.6. Microscopic Analysis of Phagocytosis and Fluid-Phase Endocytosis

The following brief technical description is the basis for imaging endocytic steps in living cells and is suitable for tracing the fate of any of the particles or fluid phase markers described above. Specifically using a mix of FITC- and TRITC-dextrans produces a spectacular real-time representation of macropinocytosis followed by endosomal acidification *(18)*. Microscopic analysis is particularly useful and has been widely employed for in vivo monitoring of GFP-tagged proteins during fluid or particle uptake *(11)* (*see* **Fig. 3**). The setup described here implies that you have access to an inverted fluorescence microscope with a cooled charge-coupled device (CCD) camera or preferably a confocal laser scanning system, equipped with a 63× or 100× objective (*see* **Note 8**). Because of the vastly different microscope properties, we cannot specify any instrument settings.

1. To construct a chamber for in vivo imaging, a plastic ring of 45 mm diameter (or a grid of similar dimensions) 4 mm high is immersed in molten paraffin and placed on a custom-made, 50-mm-square coverslip (e.g., from Karl Hecht, Germany). Alternatively, the ring can be fixed with a thin layer of silicon grease.
2. Spread 10^6 cells over the area and let adhere for 30 min while the sample is in a moist chamber. Before placing the chamber on the microscope stage, be

sure to wipe off any humidity that has condensed on the lower surface of the coverslip.

3. Now, the liquid can be aspirated and replaced by medium or Soerensen buffer containing any of the particles or fluid-phase tracers described previously. To study later stages, pulse-chase setups can be easily arranged by aspirating and replacing media *(19)*.

4. After addition of the fluorescent probe, it may be necessary to adjust the focal plane of the sample. Start image capture (*see* **Notes 9** and **10**). Nice representations of endocytic processes are obtained with one image captured every 5–10 s, whereas exocytosis is much faster as judged form particle imaging *(20,21)*, but has not yet been frequently demonstrated with fluid phase markers *(22)*.

5. Images are processed and arranged using the software that accompanies the image acquisition equipment.

4. Notes

1. Bacteria and yeast cells may release minor amounts of soluble fluorophore when thawed, but only negligible amounts are taken up by pinocytosis by *Dictyostelium* cells.

2. There are some advantages of using yeast over latex beads. Both assays must distinguish adherent vs internalized particles. With yeast, this is done by quenching the fluorescence of extracellular particles. Latex beads are not quenchable and must be washed away from the surface. Also, the time resolution of the yeast assay should be more precise, because if particle uptake occurs after quenching, it does not contribute to the signal. With latex beads, the value is fixed by cooling the samples on ice and measuring after cell lysis. This also precludes microscopic analysis during the assay, which is a possibility of the yeast-based assay.

3. The settings for fluorescence measurements are dependent on the equipment available; therefore, no generalization can be made. Offset and gain parameters should be optimized to obtain a good signal-to-noise ratio. This is particularly important for assays with TRITC-labeled yeasts, in which measurements cannot be postponed to the end of the experiment. Because efficiency of labeling might vary for every new batch of TRITC-labeled yeasts, before performing an assay with a new batch, it is advisable to optimize measurement parameters by taking readings of a wild-type culture after 120 min incubation with labeled yeasts. We generally assume that phagocytosis assays require lowest sensitivity, whereas pinocytosis and transit assays work best at intermediate to high settings, respectively.

4. The reliability of the assay strongly depends on appropriate corrections for cell mass differences among samples. Without correction, adjusting samples to the same cell density is unreliable because of sampling errors and, more importantly, because mutant strains may have different cell sizes. One appropriate correction method is the determination of the protein content in an aliquot of each sample before adding the fluorescent probe. Any differences in protein amount per cell must be traced back to changes in cell volume and/or nuclear number.

5. It is not advisable to collect samples over a time course in the staining step and process all samples at the end. Instead, one should proceed through the entire protocol for each time point (quenching with Trypan Blue is optimal at 3 min). Even if multiple samples are processed in parallel, there is usually enough time to obtain the fluorescence readout before collecting the next time point. Therefore, once started, the assay requires almost full-time dedication until finished. The reason for this is that Trypan Blue seems to enter *Dictyostelium* cells by pinocytosis and meet previously internalized particles or fluid phase marker if left for enough time.

6. One disadvantage of TRITC-dextrans is their low degree of labeling. When buying from commercial sources be sure that the coupling efficiency equals or exceeds 0.003 mol TRITC/mol glucose, otherwise increase the concentration of fluorescent probe in the assays accordingly.

7. The medium containing labeled dextrans is only briefly in contact with the cells. Therefore, it can be reused three times, provided that it is centrifuged to be cell-free, filter-sterilized, and stored at 4°C.

8. The main advantage of the confocal microscope is that damage caused by illumination of the sample is reduced, which can, in principle, also be achieved by a shutter system on a conventional fluorescence microscope. Another advantage of confocal microscopy is the characteristic plane discrimination, which allows the imaging of nonfluorescent objects in a bath of fluorescent medium.

9. Especially in the case of commercially available labeled particles or yeasts labeled according to **Subheading 2.4.**, the fluorescence signal may be so intense that it is also inadvertently detected in other channels of the microscope. Producing a batch of weakly labeled yeast may help to resolve this issue.

10. For long-term imaging, evaporation of the medium is a common problem. Here, it is advisable to cover the chamber with a coverslip of the appropriate size.

References

1. Maniak, M. (2002) Conserved features of endocytosis in *Dictyostelium*. *Int. Rev. Cytol.* **221**, 257–287.
2. Maniak, M. (2003) Fusion and fission events in the endocytic pathway of *Dictyostelium*. *Traffic* **4**, 1–5.
3. Zhang, H., Gomez-Garcia, M. R., Brown, M. R., and Kornberg, A. (2005) Inorganic polyphosphate in *Dictyostelium discoideum*: influence on development, sporulation, and predation. *Proc. Natl. Acad. Sci. USA* **102**, 2731–2735.
4. Neuhaus, E. M., Almers, W., and Soldati, T. (2002) Morphology and dynamics of the endocytic pathway in *Dictyostelium discoideum*. *Mol. Biol. Cell* **13**, 1390–1407.
5. Neuhaus, E. M. and Soldati, T. (2000) A myosin I is involved in membrane recycling from early endosomes. *J. Cell. Biol.* **150**, 1013–1026.
6. Aguado-Velasco, C. and Bretscher, M. S. (1999) Circulation of the plasma membrane in *Dictyostelium*. *Mol. Biol. Cell.* **10**, 4419–4427.
7. Heuser, J., Zhu, Q., and Clarke, M. (1993) Proton pumps populate the contractile vacuoles of *Dictyostelium* amoebae. *J. Cell Biol.* **121**, 1311–1327.

8. Dimond, R. L., Burns, R. A., and Jordan, K. B. (1981) Secretion of lysosomal enzymes in the cellular slime mold, *Dictyostelium discoideum. J. Biol. Chem.* **256,** 6565–6572.

9. Vogel, G., Thilo, L., Schwarz, H., and Steinhart, R. (1980) Mechanism of phagocytosis is mediated by different recognition sites as disclosed by mutants with altered phagocytic properties. *J. Cell Biol.* **86,** 456–465.

10. Hed, J., Hallden, G., Johansson, S., and Larsson, P. (1987) The use of fluorescence quenching in flow cytofluorometry to measure the attachment and ingestion phases in phagocytosis in peripheral blood without prior cell separation. *J. Immunol. Methods* **101,** 119–125.

11. Maniak, M., Rauchenberger, R., Albrecht, R., Murphy, J., and Gerisch, G. (1995) Coronin involved in phagocytosis: dynamics of particle-induced relocalization visualized by a green fluorescent protein tag. *Cell* **83,** 915–924.

12. Thilo, L. and Vogel, G. (1980) Kinetics of membrane internalization and recycling during pinocytosis in *Dictyostelium discoideum. Proc. Natl. Acad. Sci. USA* **77,** 1015–1019.

13. Aubry, L., Klein, G., Martiel, J. L., and Satre, M. (1993) Kinetics of endosomal pH evolution in *Dictyostelium discoideum* amoebae. Study by fluorescence spectroscopy. *J. Cell Sci.* **105,** 861–866.

14. Padh, H., Ha, J., Lavasa, M., and Steck, T. L. (1993) A post-lysosomal compartment in *Dictyostelium discoideum. J. Biol. Chem.* **268,** 6742–6747.

15. Cardelli, J. A., Richardson, J., and Miears, D. (1989) Role of acidic intracellular compartments in the biosynthesis of lysosomal enzymes. The weak bases ammonium chloride and chloroquine differentially affect proteolytic processing and sorting. *J. Biol. Chem.* **264,** 3454–3463.

16. Temesvari, L. A., Rodríguez-Paris, J. M., Bush, J. B., Zhang, L., and Cardelli, J. A. (1996) Involvement of the vacuolar proton-translocating ATPase in multiple steps of the endo-lysosomal system and in the contractile vacuole system of *Dictyostelium discoideum. J. Cell Sci.* **109,** 1479–1495.

17. Berthiaume, E. P., Medina, C., and Swanson, J. A. (1995) Molecular size-fractionation during endocytosis in macrophages. *J. Cell Biol.* **129,** 989–998.

18. Maniak, M. (2001) Fluid-phase uptake and transit in axenic *Dictyostelium* cells. *Biochim. Biophys. Acta.* **1525,** 197–204.

19. Jenne, N., Rauchenberger, R., Hacker, U., Kast, T., and Maniak, M. (1998) Targeted gene disruption reveals a role for vacuolin B in the late endocytic pathway and exocytosis. *J. Cell Sci.* **111,** 61–70.

20. Maniak, M. (1999) Endocytic transit in *Dictyostelium. Protoplasma* **210,** 25–30.

21. Clarke, M., Kohler, J., Arana, Q., Liu, T., Heuser, J., and Gerisch, G. (2002) Dynamics of the vacuolar H(+)-ATPase in the contractile vacuole complex and the endosomal pathway of *Dictyostelium* cells. *J. Cell Sci.* **115,** 2893–2905.

22. Lee, E. and Knecht, D. A. **(2002)** Visualisation of actin dynamics during macropinocytosis and exocytosis. *Traffic* **3,** 186–192.

23. Rivero, F., Illenberger, D., Somesh, B. P., Dislich, H., Adam, N., and Meyer, A.-K. (2002) Defects in cytokinesis, actin reorganization and the contractile vacuole system in cells deficient in RhoGDI. *EMBO J.* **21,** 4539–4549.

26

Preparation of Intact, Highly Purified Phagosomes from *Dictyostelium*

Daniel Gotthardt, Régis Dieckmann, Vincent Blancheteau, Claudia Kistler, Frank Reichardt, and Thierry Soldati

Summary

Phagocytosis plays a fundamental role in the immune system for the defense against invading microorganisms and the clearing of apoptotic and cancerous cells. The common amoeba *Dictyostelium discoideum* is a recognized model for professional immune phagocytes and is now commonly used to study host–pathogen interactions. *Dictyostelium* is genetically and biochemically tractable and is a most versatile experimental system. The classical protocol for purifying phagosomes formed by ingestion of latex beads particles has been adapted to *Dictyostelium*. It was improved in yield, purity, and synchronicity, allowing isolation of milligram amounts of phagosomal proteins and lipids. This method has been used successfully to highlight membrane trafficking and phagosome maturation. Here, we present a step-by-step protocol including detailed notes necessary for ensuring access to a large number of highly synchronized phagosomes of high purity and integrity.

Key Words: Phagosomes, organelle purification, phagocytosis, *Dictyostelium*.

1. Introduction

Phagocytosis is a fundamental biological process that is morphologically and mechanistically conserved during evolution, from simple protozoa to specialized animal cells. The overall process is complex and consists of the recognition and engulfment of the particle via the formation of a phagocytic cup, followed by a multifaceted maturation process that ensures killing and degradation (or egestion) of the particle. The social amoeba *Dictyostelium discoideum* is a recognized model for professional immune phagocytes and is commonly used to study host–pathogen interactions with *Mycobacteria (1)* and

From: *Methods in Molecular Biology, vol. 346:* Dictyostelium discoideum *Protocols*
Edited by: L. Eichinger and F. Rivero © Humana Press Inc., Totowa, NJ

Legionella (2). These pathogenic microorganisms manipulate the phagocytic pathway to evade killing. Our studies are aimed at the understanding of the physiological conditions and changes accompanying phagosome maturation, as well as the manipulations exerted by pathogens. Such studies require the purification of phagosomes of high purity and intactness. Ingestion of low-density latex beads by phagocytic cells allows one to purify the bead-containing phagosomes via their flotation in sucrose-step gradients. The method was originally introduced by Wetzel and Korn for *Acanthamoeba (3)* and adapted successfully to macrophages by Desjardins and collaborators *(4)*. Because *Dictyostelium* is a very effective phagocyte with rates of phagocytosis 10- to 20-fold higher than than those of macrophages, the protocol was adapted to this model organism and further refined. Compared with macrophages, *Dictyostelium* can easily be grown in shaking culture up to an optimal density of 10^7 cells/mL. In a routine experiment, we use 100 mL of cell suspension (10^9 cells total) and an initial ratio of about 180 latex beads/cell that optimally leads to the formation of about 40 phagosomes/cell, or 4×10^{10} phagosomes in total. This yield is between 10- and 100-fold higher than a typical preparation from macrophages *(4)*.

The following protocol has been optimized for fast access to a large number of pure, intact, and synchronized phagosomes. Synchronicity was ensured by a preincubation in the cold of cells and beads at high concentration. It is possible to concomitantly biotinylate the cell surface to allow monitoring of the uptake and recycling of plasma membrane proteins during maturation. Phagosome purity was significantly improved by a brief incubation of the cell lysate with ATP *(5)*, which releases the rigor mortis actin/myosin interaction, thereby avoiding piggy-backing of enmeshed organelles. We also describe here a pulse-chase protocol that gives access to all stages of phagosome maturation *(5,6)*.

2. Materials

2.1. Buffers and Equipment

1. Soerensen buffer (SB): 15 mM KH$_2$PO$_4$, 2 mM Na$_2$HPO$_4$, pH 6.0.
2. Soerensen/Sorbitol buffer (SSB): Soerensen buffer containing 120 mM sorbitol.
3. HESES: 20 mM HEPES-KOH, pH 7.2, 0.25 M sucrose.
4. Homogenization buffer (HB): HESES, 2X Complete ethylenediamine tetraacetic acid (EDTA)-free (protease inhibitor cocktail; Roche, Hertfordshire, UK).
5. Membrane buffer: 20 mM HEPES-KOH, pH 7.2, 20 mM KCl, 2.5 mM MgCl$_2$, 1 mM dithiothreitol (DTT), 20 mM NaCl.
6. Storage buffer: 25 mM HEPES-KOH, pH 7.2, 1.5 mM Mg-acetate, 1 mM NaHCO$_3$, 1 µM CaCl$_2$, 25 mM KCl, 1 mM ATP, 1 mM DTT, 1X Complete EDTA-free, 100 mM sucrose.

7. Ball homogenizer (Isobiotec, Heidelberg, Germany, barrel 8.000 mm, ball diameter 7.990 mm, resulting in a void clearance of 5 μm).

2.2. Cell Culture

Dictyostelium discoideum cells of wild-type strain Ax2 are grown axenically in HL5c medium *(7)* (ForMedium Ltd, Norwich, UK) in shaking culture (at 180 rpm) at 22°C to a density of 5×10^6 cells/mL.

2.3. Preparation of Latex Beads

Always prepare fresh. 0.5 mL of 0.8-μm latex beads suspension (approx 3.57×10^{11} beads/mL, Sigma, St. Louis, USA) are spun down in an Eppendorf tube (10,000g, 5 min). The beads are washed twice in HL5c to remove the detergent and sodium azide contained as preservatives in the suspension supplied by the manufacturer, and finally resuspended in 0.5 mL of SB and kept on ice. Extensive washing of the latex beads appears critical, as insufficient washing blocks phagocytosis. Note that, for routine production of phagosomes, the beads are coated with HL5c medium, which prevents bead clumping. Alternatively, if surface biotinylation is intended, the beads are only washed with H_2O, but the tendency to clump is higher. Before use, the bead suspension is sonicated for 5 min in a bath sonicator.

2.4. Preparation of Sucrose Step Gradients

Prepare sucrose step gradients in disposable centrifuge tubes (Ultra-clear tubes 25×89 mm, Beckman) by layering the following sucrose solutions on top of each other: 6 mL of 60%, 12 mL of 35%, 12 mL of 25% sucrose in 20 mM HEPES-KOH, pH 7.2. At this step, the gradients can be stored overnight without vibration. The last layer composed of 3 mL of 10% sucrose in 20 mM HEPES-KOH, pH 7.2 is added after the phagosome samples have been loaded, to obtain an undisturbed 25–10% interface and to balance the tubes exactly.

3. Methods

3.1. Phagosome Purification Protocol

1. Cells from an overnight culture are counted with a hemocytometer, and 10^9 cells are centrifuged at 500g for 5 min. Subsequent steps are performed at 4°C or on ice when possible, except indicated otherwise.
2. Gently resuspend cell pellet in 5 mL of ice cold preincubation buffer (SSB pH 8.0). For concomitant biotinylation, *see* **Note 1**.
3. Add 0.5 mL of the bead suspension to the cells (approx 1.8 10^{11} beads, ratio beads:cells about 180:1). Mix by inverting the tube.
4. Incubate for 15 min on ice.
5. Pour the bead/cell mixture into 100 mL of medium kept at 22°C in a 250-mL flask (*see* **Note 2**).

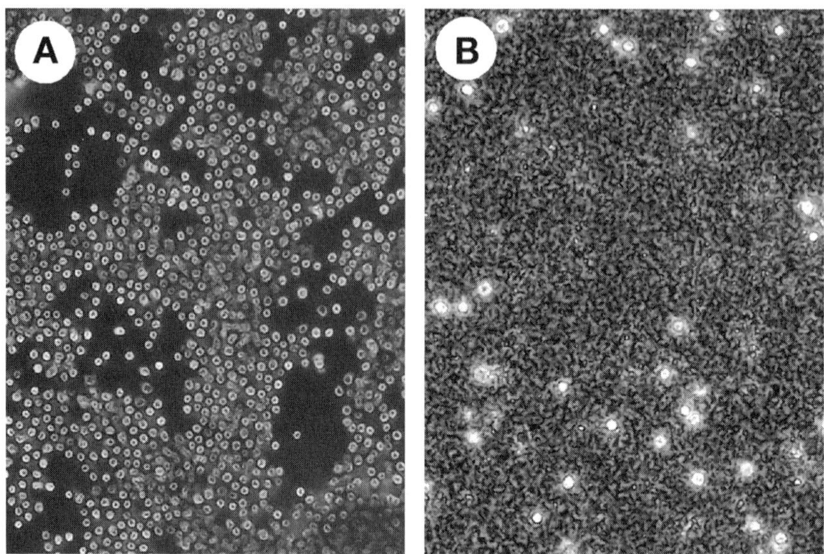

Fig. 1. Optimal cell breakage. Phase-contrast pictures of cells before (**A**) and after (**B**) 11 passages through a ball homogenizer. The cell suspension was dropped on a slide and overlaid with a coverslip. Pictures were taken with The Scope (ScalarScopes, http://www.scalarscopes.com) and a 10× objective. Optimal breakage is obtained when 95% of the cells have been lysed and have lost their refractile appearance, which usually happens at between 8 and 12 passages. At this stage, the few intact cells left literally "swim" in an ocean of debris and organelles.

6. Incubate at 22°C while shaking at 120 rpm for the appropriate period of time (*see* **Subheading 3.3.**).
7. At the chosen time point, in order to stop phagocytosis, pour the 100-mL sample into 330 mL of ice-cold SSB prealiquoted in centrifugation bottles (500 mL, 69 × 160 mm Beckman rotor JA-10) kept on ice (for **steps 7–10**, *see* **Note 3**).
8. Centrifuge cells for 6 min at 800g.
9. Resuspend the cell pellet in 50 mL ice-cold HESES and centrifuge again for 4 min at 500g (clinical centrifuge).
10. Repeat this washing step once to thoroughly wash away uningested latex beads.
11. Resuspend cell pellet (approx 2 mL) in 2 mL of HB containing a protease-inhibitor cocktail (Complete EDTA-free, Roche) at 2X concentration, resulting in a 1X final concentration.
12. Homogenize cells by eight passages through a ball homogenizer (*see* **Fig. 1** and **Note 4**).
13. Adjust the final concentrations of ATP to 10 mM, of MgCl$_2$ to 10 mM, and of sucrose to about 45–50% from freshly made stocks (0.1 M ATP in 40% sucrose,

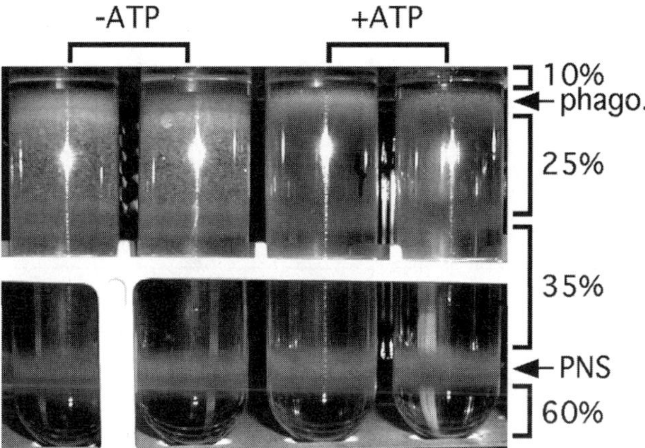

Fig. 2. Effect of ATP on the flotation of latex-beads phagosomes. Pictures of sucrose step gradients (percentages indicated on the right) showing the separation of a post nuclear supernatant (PNS) loaded at the bottom of the gradient. After incubation with 10 m*M* ATP, the phagosomes float quantitatively up to the 10–25% interface, whereas in absence of ATP incubation, most of the material is dispersed throughout the 25% layer.

20 m*M* HEPES-KOH buffered to pH 7.2; 1 *M* MgCl$_2$; 71.4% sucrose in 20 m*M* HEPES-KOH pH 7.2) (for **steps 13** and **14**, *see* **Note 5**).

14. Mix gently for 15 min using an overhead-tumbler or a wheel.

15. Load the density-adjusted homogenate between the 60% and 35% layer of the sucrose step gradients using a syringe and a needle (100 mm ×1.5 mm), and overlay with 3 mL of 10% sucrose solution.

16. Centrifuge gradients in a Beckman rotor SW 28 at 28,000 rpm (100,000*g* average), 4°C for 3 h (*see* **Fig. 2** and **Note 6**).

17. Collect the interphase between 10% and 25% and dilute with membrane buffer to a final volume of 15 mL, mix by inverting the tube.

18. Take a 50-µL aliquot to measure scattering at 600 nm to calculate the number of phagosomes in the sample (*see* procedure under **Subheading 3.2.** and **Fig. 3**).

19. Dilute this suspension further with membrane buffer to a final volume of 37 mL (to **steps 19–21**, *see* **Note 7**).

20. Pellet phagosomes in a Beckman rotor SW 28 at 28,000 rpm (100,000*g* average) for 45 min

21. Resuspend the pellet in storage buffer (adjust the amount of buffer according to the scattering measurements so as to reach the same concentration of phagosomes in all samples) (*see* procedure under **Subheading 3.2.**).

22. Snap-freeze aliquots in liquid nitrogen and store at –80°C until further use (*see* **Fig. 4**).

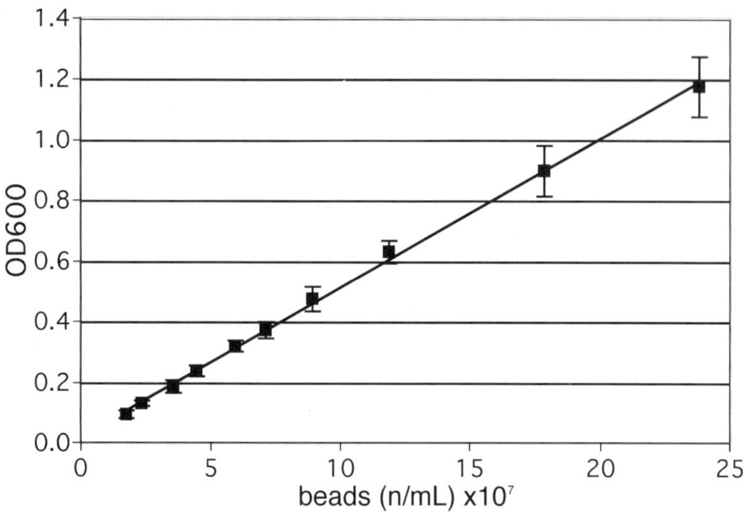

Fig. 3. Measurement of phagosome concentration. A calibration curve was obtained by measuring the light scattering at 600 nm of latex bead-containing solutions of the indicated concentrations. A regression line illustrates the linearity of the measurements between 2×10^7 and 2×10^8 beads/mL. Such a curve is used to determine the concentration of latex beads phagosomes after sucrose gradient purification.

3.2. Measurement of the Phagosome Concentration and Yield

Light scattering at 600 nm is used to measure the concentration of latex-bead phagosomes in each sample. A standard curve can be made with serial dilutions of the beads, generating a linear correlation between scattering and bead concentration in a range between 2×10^7 and 2×10^8 beads/mL (*see* **Fig. 3**). We determined that scattering was caused only by the latex beads and not contaminating particles and/or organelles by treating the latex beads-phagosome sample with sodium dodecyl sulfate (SDS) and pelleting. The resulting "clean" latex beads yielded the same scattering values as when residing inside phagosomes (not shown).

3.3. Pulse-Chase Feeding

The protocol described under **Subheading 3.1.** was adapted slightly to purify phagosomes at different stages of maturation (*see* **Note 8**).

1. Our typical setup is based on six time points:
   - 5 min pulse
   - 15 min pulse
   - 15 min pulse/15 min chase
   - 15 min pulse/45 min chase

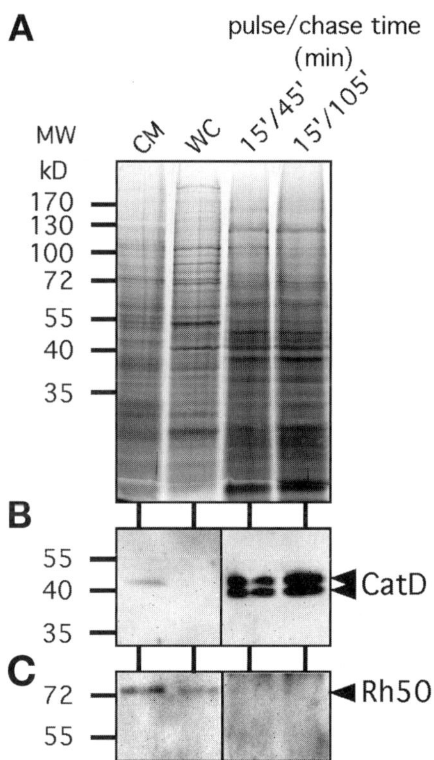

Fig. 4. Characterization of the phagosomal fractions. A silver-stained 10% sodium dodecyl sulfate (SDS)-polyacrylamide gel electrophoresis (PAGE) gel **(A)** of the same fractions as used for western blotting experiments **(B–C)**. As quantified by Bradford assay, lanes were loaded with 2 µg of proteins from the following fractions: a whole cell lysate from Ax2 wild-type cells (WC), a crude membrane fraction (CM), and two phagosomal fractions from the 15-min pulse and 45-min chase (15'/45') or 1 h 45-min chase (15'/105') time points, respectively. Western blot analysis illustrates the very significant enrichment of cathepsin D **(B)** and the depletion of Rhesus 50 **(C)** in the two phagosomal fractions. The different fractions were separated on a 10% SDS-PAGE gel in nonboiling and nonreducing conditions and then transferred and treated as described *(3)*. Rabbit polyclonal anti-cathepsin D (dilution 1:500) is a kind gift from J. Garin and rabbit polyclonal anti-Rhesus 50 (1:1500) is a kind gift of P. Cosson. The secondary antibody is a horseradish peroxidase (HRP)-coupled goat anti-rabbit immunoglobulin (Ig)G (BioRad, Hercules, CA) used at 1:10,000 dilution.

- 15 min pulse/1 h 45 min chase
- 15 min pulse/2 h 45 min chase

2. Use the above-mentioned protocol (*see* **Subheading 3.1.**, **steps 1–8**) with 10^9 cells for each time point.

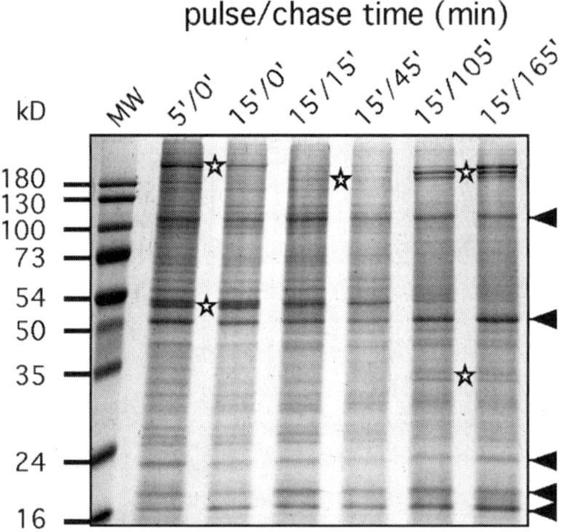

Fig. 5. Pulse-chase experiment. A Coomassie Blue-stained 10% sodium dodecyl sulfate (SDS)-polyacrylamide gel electrophoresis (PAGE) gel of phagosomal fractions obtained after a pulse-chase feeding experiment with latex beads. It illustrates that, while some bands are visible with little fluctuation throughout the gel (arrowheads), the appearance and disappearance of different bands (asterisks) is a signature of the maturation process.

3. The samples of the 5- and 15-min pulse are then treated as in **Subheading 3.1.**, **step 9** and subsequent steps, whereas for the chased samples, the cell pellets of **step 8**. are first resuspended in 5 mL of ice-cold HL5c and further phagosome maturation initiated by flash dilution in 100 mL of HL5c at 22°C.
4. Stop phagosome maturation in the chased samples as described in **Subheading 3.1.**, **step 7** on, and purify phagosomes (*see* **Figs. 4** and **5**).

4. Notes

1. The preincubation step is performed in order to maximize the number of latex beads adsorbed onto the cells. If desired, it is also possible to concomitantly biotinylate plasma membrane proteins, using 5 mg of Imuno-Pure NHS-LC-Biotin from Pierce (Rockford, IL). The high pH is thus necessary for efficient biotinylation, but the cells should not be kept for extended periods of time at this pH.
2. To start phagocytosis, a 20-fold excess (100 mL) of HL5c medium at 22°C is added to the sample, instantaneously raising the temperature and thus generating a sharp synchronous wave of uptake.
3. To stop the phagocytic process, the 100-mL cell suspension is poured into 3.3 volumes of ice-cold SSB and immediately centrifuged. The use of 120 m*M* sorbitol is crucial in order to increase the bead flotation and yet not perturb cell pelleting. These two cycles of resuspension/centrifugation are sufficient to eliminate a maximum of uningested beads.

4. Homogenization using a ball homogenizer results in homogeneous cell breakage together with preservation of organelle integrity. The osmolarity of the buffer, the void clearance, and the number of passages is optimized to yield about 95% cell breakage (*see* **Fig. 1**). Higher ratio of cell breakage only results in increased nuclear lysis and contamination of cytoplasm with chromatin. Subsequent treatments carefully avoid vortexing, fast pipetting through narrow openings, or other harsh procedures that damage phagosomes.

5. Use of a physiological concentration of ATP for a few minutes in the cold avoids artefactual formation of a rigor mortis meshwork of actin and myosins that was shown to entrap contaminating organelles. Omission of ATP has visible consequences (*see* **Fig. 2**), as it prevents fast and clean flotation of the phagosomes (this can be compensated by longer centrifugation). It also results in the co-fractionation of vast amounts of an actin-myosin II meshwork (as judged by Coomassie staining of the resulting fractions after SDS-polyacrylamide gel electrophoresis [PAGE]), which was shown to entrap contaminating organelles *(5)*.

6. If centrifugation time is reduced to under 3 h, the fractionation on the gradients was not as sharp, and the phagosome yield is about 20–30% lower. The rotor must stop without brake to avoid vibrations.

7. The 14.95 mL of phagosome suspension are further diluted with membrane buffer up to 37 mL and pelleted in the SW 28 again for 1 h at 28,000 rpm (100,000*g* average). This dilution is necessary in order to decrease the sucrose concentration and thus decrease the buoyancy of the latex-beads phagosomes. The pellet is resuspended in storage buffer using adjusted amounts of this buffer according to the scattering measurements to normalize the concentration of phagosomes (*see* **Subheading 3.2.** and **Fig. 3**).

8. Because of the extreme dynamics of the first minutes, time points of 5 and 15 min cover the first signaling, cytoskeleton, and membrane-trafficking phases. The 5-min sample is processed and homogenized before the 15-min chase is ended, whereas the other four pellets are resuspended in 5 mL ice-cold medium and the chase period is started by pouring into 100 mL of 22°C medium. Only then is the 15-min pulse sample processed and homogenized. The chase is stopped after 15 min, 45 min, 1 h 45 min, and 2 h 45 min to cover most maturation stages until exocytosis of the particles. SDS-PAGE analysis of equal volumes of phagosome samples results in a gel in which each lane is loaded with an identical number of phagosomes, resulting in a relatively homogeneous band pattern (*see* **Fig. 5**), indicating equal loading. Nevertheless, some differences are clearly visible and highlight the progression of the maturation process and the unique identity of phagosomes at each stage.

Acknowledgments

This work was mainly carried out at the Max-Planck Institute for Medical Research in Heidelberg and was supported by a grant from the Deutsche Forschungsgemeinschaft. We also acknowledge support from The Wellcome Trust, the UK Biotechnology and Biological Sciences Research Council (BBSRC), and the Swiss National Science Foundation. A big "thank you" goes to all the lab members who have participated in the establishment of the protocols and have suggested amendments and improvements.

References

1. Solomon, J. M., Leung, G. S., and Isberg, R. R. (2003) Intracellular replication of *Mycobacterium marinum* within *Dictyostelium discoideum*: efficient replication in the absence of host coronin. *Infect. Immun.* **71,** 3578–3586.
2. Hagele, S., Kohler, R., Merkert, H., Schleicher, M., Hacker, J., and Steinert, M. (2000) *Dictyostelium discoideum*: a new host model system for intracellular pathogens of the genus *Legionella*. *Cell. Microbiol.* **2,** 165–171.
3. Wetzel, M. G. and Korn, E. D. (1969) Phagocytosis of latex beads by *Acanthamoeba castellanii* (Neff). 3. Isolation of the phagocytic vesicles and their membranes *J. Cell Biol.* **43,** 90–104.
4. Desjardins, M., Huber, L. A., Parton, R. G., and Griffiths, G. (1994) Biogenesis of phagolysosomes proceeds through a sequential series of interactions with the endocytic apparatus. *J. Cell Biol.* **124,** 677–688.
5. Gotthardt, D., Warnatz, H. J., Henschel, O., Bruckert, F., Schleicher, M., and Soldati, T. (2002) High-resolution dissection of phagosome maturation reveals distinct membrane trafficking phases. *Mol. Biol. Cell* **13,** 3508–3520.
6. Lefkir, Y., Malbouyres, M., Gotthardt, D., et al. (2004) Involvement of the AP-1 adaptor complex in early steps of phagocytosis and macropinocytosis. *Mol. Biol. Cell* **15,** 861–869.
7. Sussman, M. (1987) Cultivation and synchronous morphogenesis of *Dictyostelium* under controlled experimental conditions. *Methods Cell Biol.* **28,** 9–29.

27

Assaying Cell–Cell Adhesion

Salvatore Bozzaro

Summary

A major feature in *Dictyostelium* development is the transition from the unicellular to the multicellular stage, a process brought about by chemotaxis and cell–cell adhesion. Growth-phase cells are weakly cohesive, whereas aggregation-competent cells adhere strongly to each other. In addition, aggregating cells display an ethylenediamine tetraacetic acid (EDTA)-resistant form of adhesion, which is developmentally regulated. Measuring cell–cell adhesion can thus be a simple and convenient method by which to assess the developmental progression of cells, to characterize mutants, and to discriminate between development and functional defects in cell adhesion molecules or membrane–cytoskeletal interactions. A quantitative cell adhesion assay is obviously crucial for identifying novel cell adhesion factors.

Key Words: CsA glycoprotein; contact sites A and B; EDTA-sensitive cell adhesion; EDTA-resistant cell adhesion; cell aggregation; cell agglutination; agglutinometer; light scattering.

1. Introduction

Development of *Dictyostelium* cells is accompanied by discrete changes in cell adhesiveness. Growth-phase cells are barely cohesive; on agar or glass, they adhere to each other only transiently, whereas in shaken suspensions, they form loose aggregates, which are easily dissociated with ethylenediamine tetraacetic acid (EDTA). During the first hours of development, cells become gradually more cohesive, and shortly before aggregation, i.e., after 4 to 6 h of starvation, depending on the strain, they display EDTA-resistant cell contacts.

Appearance of EDTA-resistant cell adhesion is due to the expression on the cell surface of the 80-kDa glycoprotein contact site A (csA), which mediates a homophilic form of adhesion. *CsA* gene disruption completely abolishes EDTA-resistant adhesion during the aggregation stage, whereas its constitutive

From: *Methods in Molecular Biology, vol. 346:* Dictyostelium discoideum *Protocols*
Edited by: L. Eichinger and F. Rivero © Humana Press Inc., Totowa, NJ

expression induces EDTA-resistant adhesion in growing cells (*1,2*; for reviews, *see* **refs.** *3,4*). During normal development, the *csA* gene is induced by starvation and strongly enhanced by pulsatile cAMP signaling, making it one of the best markers of acquired aggregation competence *(5)*. Transcription is repressed after tight aggregate formation, and the csA glycoprotein disappears, although slowly, from the cell surface of most, but not all, cells. EDTA-stable adhesion, however, persists as the csA activity is complemented, and eventually taken over, by gp150, which is the product of the *lagC* gene. Gp150 is expressed after tight aggregate formation and mediates a heterophilic form of cell adhesion *(6)*.

EDTA-sensitive adhesion is present throughout development, possibly mediated by more proteins. Several lines of evidences indicate that DdCAD-1, which shares some similarities with cadherins, plays a major role in EDTA-sensitive cell adhesion throughout development *(4,7–9)*. A second glycoprotein, Gp126/130, which was supposed to be involved in cell adhesion and phagocytosis, may have an intriguing role as negative regulator of cell adhesion *(10)*.

For assaying cell–cell adhesion, two methods have been mainly used in *Dictyostelium*. The first one exploits changes in light scattering and makes use of the agglutinometer developed by Beug and Gerisch *(11)*. The agglutinometer allows one to simultaneously measure cell aggregation in small volumes of 24 samples under identical conditions. The method is very reliable, highly reproducible, quantitative, and sensitive enough to detect subtle effects altering cell adhesion. The drawback is that it requires equipment, which is not commercially available. If a lab plans to engage in cell adhesion studies, it is worthwhile to build this equipment (*see* discussion below).

The second method, the single-cell assay, is less sensitive as well as cumbersome for in-depth analysis; however, it does not require any special equipment, and thus it is convenient for labs that are not routinely involved in assaying cell adhesion. Both methods will be described in the following.

2. Materials

2.1. Buffers and Solutions

1. Soerensen phosphate buffer: 0.017 *M* Soerensen phosphate buffer, pH 6.05 ± 0.05, using KH_2PO_4 and Na_2HPO_4, as acid and base, respectively. A 50-fold concentrated stock can be conveniently prepared by dissolving 99.86 g of KH_2PO_4 and 14.2 g of Na_2HPO_4 (or 17.8 g of $Na_2HPO_4 \cdot 2H_2O$) in a total volume of 1 L of deionized and distilled water. The stock is diluted 1:50 in water, resulting in a pH of 6.05 ± 0.05 (*see* **Note 1**). The Soerensen phosphate buffer is sterilized by autoclaving at 120°C for 20 min, and can be used for several weeks, if opened under sterile conditions.

2. EDTA solution: 20 or 40 mM EDTA solution in 0.017 M Soerensen phosphate buffer. Adjust the pH to 6.1 with a few drops of 1 N NaOH.
3. 1 mM cAMP solution in Soerensen phosphate buffer.
4. 10% (v/v) glutaraldehyde or trichloroacetic acid (TCA) solution in water.
5. Pepsin/HCl: 5 mg pepsin in 5% HCl solution.

2.2. Cell Cultures

Wild-type or mutant *Dictyostelium* cells are grown in axenic medium or on bacteria up to a final concentration of 4–6×10^6 cells per mL, depending on the strain, as described by Urushihara (*see* Chapter 7). To induce development, cells are washed free of medium or bacteria by centrifuging at 200g for 3 min and washing twice in half volume of ice-cold Soerensen phosphate buffer. After the last centrifugation, the cell pellet is resuspended in 0.4 the initial volume of Soerensen phosphate buffer and vortexed strongly, and the cells counted in a haemocytometer in order to adjust the concentration to 1×10^7 cells per mL. The cell suspension is then incubated in an Erlenmayer flask and shaken on a gyratory shaker at 150 rpm and $22 \pm 1°C$ for the desired time. For optimal development, the cell volume should amount to about one-third of the nominal volume of the flask. Under these conditions, cells of the wild-type strains V12M2 and NC4 or axenic strain AX2 will become aggregation-competent after 4 to 5 h starvation.

3. Methods
3.1. Assaying Cell–Cell Adhesion by Measuring Light Scattering with the Agglutinometer of Beug and Gerisch
3.1.1. The Agglutinometer

The agglutinometer of Beug and Gerisch exploits light scattering as a principle to measure aggregate formation: particles, such as single cells, absorb light proportionally to their number. At constant cell concentration, unscattered light or extinction (E) is inversely proportional to the size of cell aggregates: with typical cells of 10–12 µm diameter, the extinction curve will drop linearly for aggregates of up to 10–15 cells, and slowly plateau thereafter *(11)*.

Cells are kept in suspension by a rotating device that allows cells to adhere to one another, depending on their cohesiveness. The mean aggregate size will depend on the shear forces, which are generated by the speed of rotation and counteract the cohesive forces of the cells. Cell size and shape, as well as the optical density (OD) of the solution, will affect light scattering; however, comparing the OD of each sample with a reference cuvet, in which completely dissociated but otherwise identical cells are incubated, eliminates such variations.

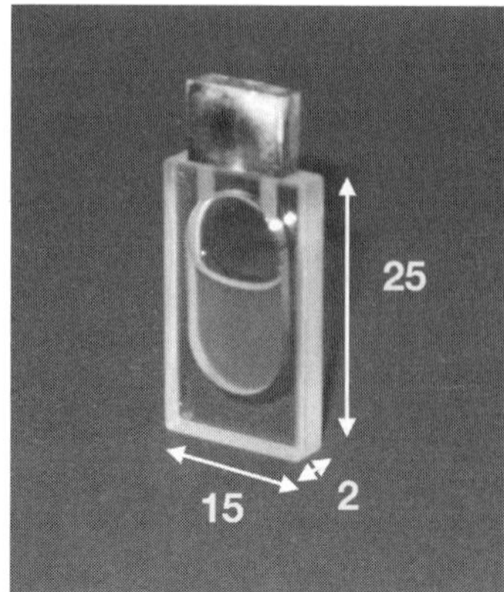

Fig. 1. The agglutinometer of Beug and Gerisch. The original agglutinometer, shown on the left, consists of (**A**) a rotating device, which can be loaded with cuvets, and is driven by (**B**) a motor, regulated by (**C**) a control unit. The cuvets (right) are made of glass and are filled with 0.2 mL cell suspension (dimensions in mm). The apparatus is mounted on an Eppendorf photometer for measuring light scattering in the cell suspension.

I will briefly describe the agglutinometer of Beug and Gerisch before explaining in detail how the assay is done. Although the agglutinometer is available only in two labs worldwide (Gerisch's lab at the Max-Planck-Institute for Biochemistry, Germany, and in my lab), the description may help in better understanding the advantages and limitations of the assay, in eventually designing one's own agglutinometer, or establishing collaborative experiments.

The equipment is shown in **Fig. 1**, and consists of cuvets, a rotating carrier for the cuvets, an automatic changing device, and the optical system.

1. The cuvets are made of glass, with an oblong cavity, and are filled with 0.2 mL cell suspension through canals, which, during rotation, are sealed by tightly fitting polyvinyl chloride (PVC) plugs (**Fig. 1**, right). The air volume in the cuvet provides for sufficient oxygenation during the experiment. The inner surface of the cuvet is continuously wetted by rotation, such that adhesion of cells to the cuvet walls is prevented. The shape of the cavity and the constant rotation provide the fluid motion that generates the shear forces interfering with cell association (*see* **Note 2**).

2. The rotating carrier can be loaded with 24 cuvets and is rotated at an adjustable constant speed by a synchronous motor, a gear presetting 80, 40, 16, and 8 rpm (**Fig. 1**, left). In an improved, electronically controlled version, designed by R. Merkl *(12)*, the speed can be varied continuously between 16 and 80 rpm. For measurement, the cuvet carrier is arrested at a position shown in **Fig. 2**. The cuvet-changing device moves the carrier one-twenty-fourth of a full revolution,

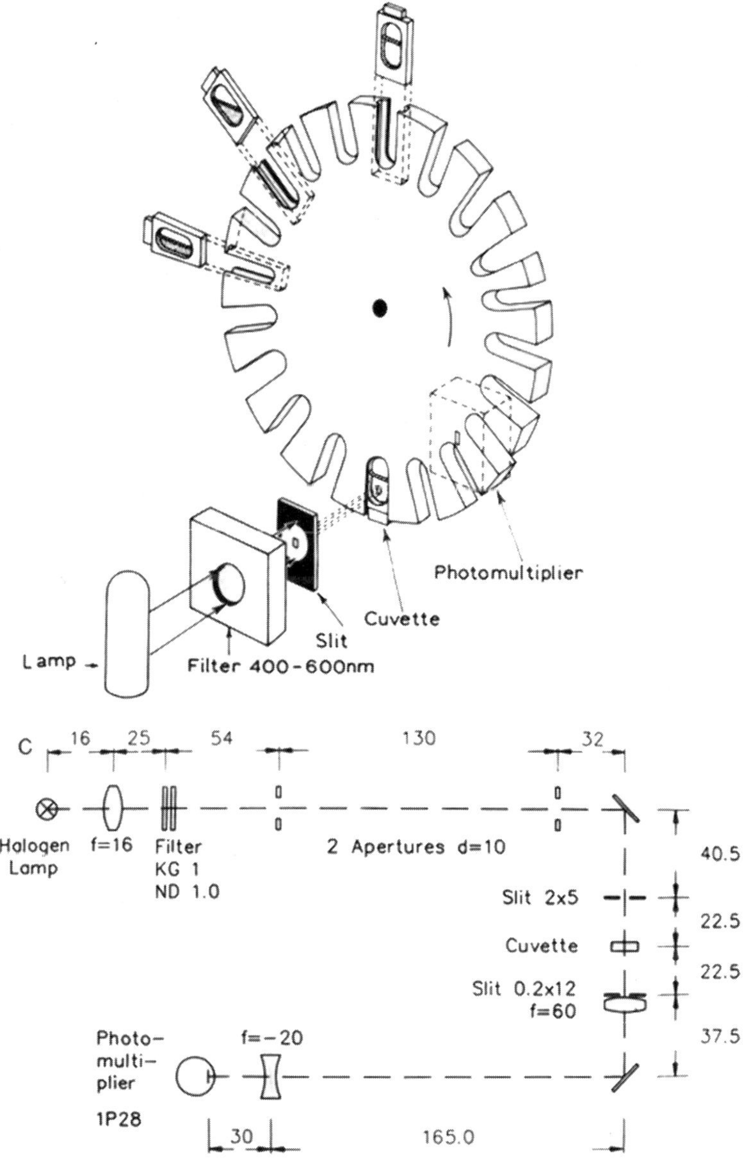

Fig. 2. Quantification of cell adhesion with the agglutinometer. Unscattered light (*E, Extinction*) is a function of the aggregate size and is measured in 23 cuvets loaded with suspended cells and a blank reference cuvet filled with Soerensen phosphate buffer. The cuvets, inserted in the rotating carrier, are subjected to constant shear forces, generated by the fluid motion and depending on the speed of the motor (*see* **Fig. 1**). At fixed time points during the rotation, light extinction is measured in each cuvet by the optical system that is specified in the lower panel. Dimensions are in millimeters. Modified from **ref. *12***.

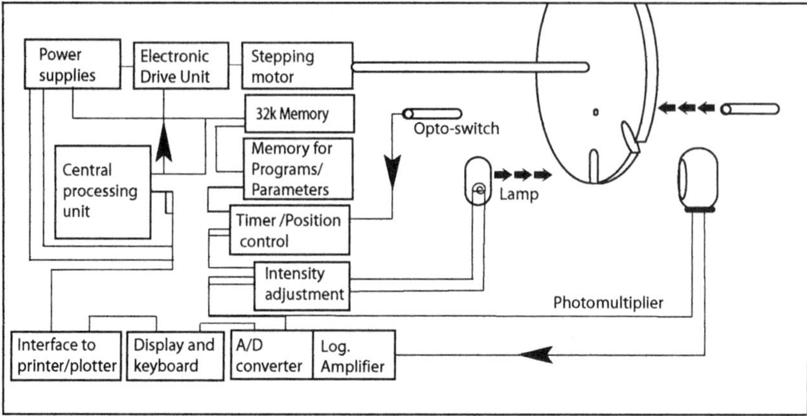

Fig. 3. Block diagram of the control unit of the electronically controlled version of the agglutinometer. A fully automated version of the agglutinometer has been designed by R. Merkl *(12)*. An electronic control allows the system to work automatically and to carry out the light-scattering measurements without the motion being stopped. The *central processing unit* controls the system, peripheral devices and the data flow. All functions are programmed using a keyboard controlled by the *display and keyboard* circuits. The *timer* generates pulses that are sent to the *electronic drive unit,* which controls the motion of the stepping motor. The *opto switch* triggers measurement of the reference cuvet. By counting the number of pulses sent to the stepping motor after the opto switch signal, the *position control* determines the time lapse for each cuvet to be ready for measurement. The signal of the *photomultiplier* is fed through a *logarithmic amplifier* and digitized by the *A/D converter*. A malfunction of the essential parts of the system is indicated by a warning lamp (not shown). Modified from **ref. *12*.**

      such that measuring 24 cuvets takes about 30 s. In the electronic version, the device is controlled by a microprocessor, such that at fixed time intervals during the run measurements are automatically taken without interrupting the rotation, the data are stored, and the aggregation kinetics plotted at the end of the experiment.

3. The optical system consists of a halogen lamp, a filter with a 400- to 600-nm pass band, a slit in front of the cuvet, and a second slit in front of the photomultiplier tube to screen the scattered light (**Fig. 2**). In the original agglutinometer, motor and cuvet carrier were mounted on an Eppendorf photometer (**Fig. 1**). In the electronic version, an *ad hoc* optical system as outlined in **Fig. 3** was designed.

The agglutinometer measures the rates of aggregation of single cells as well as dissociation of suspended aggregates as a function of applied shear forces. Under normal conditions, equilibrium is reached between 30 and 50 min of rotation, after which time the sizes of the aggregates are practically constant.

The most important application of the agglutinometer has been in the identification and characterization of cell surface molecules involved in intercellular adhesion *(13,14)*. The apparatus has also been used to discriminate between EDTA-sensitive and EDTA-stable adhesiveness *(11,15)*, or to quantitate lectin agglutination of *Dictyostelium* cells or erythrocytes *(11,16)*. Because acquisition of EDTA-resistant adhesion is developmentally regulated, the agglutinometer has also been routinely applied over the years to the characterization of mutants, in order to distinguish between developmental mutants or mutants specifically impaired in adhesion either because of a defect in adhesion molecules or in cytoskeletal components that stabilize adhesion.

3.1.2. Assaying EDTA-Sensitive and EDTA-Resistant Cell Adhesion During Development

The most common use of the agglutinometer is assessing the appearance of EDTA-resistant adhesion during development. Basically, all other assays are variations on this theme; thus, I will describe in detail this assay and then indicate briefly, under **Subheading 3.3.**, how it can be adapted for specific purposes. All of the assays are run at 23°C.

1. Switch on the photometer and set the light scattering *(E, extinction)* in the range of 0 to 1. Place the cuvets required for the experiment in a holder, taking care to keep clean the walls that will be exposed to the light path.
2. Pipet 0.2 mL Soerensen phosphate buffer in cuvet 1 to be used as blank.
3. Pipet 0.1 mL Soerensen phosphate buffer in cuvets 2 and 3 (duplicate).
4. Pipet 0.1 mL of 20 mM EDTA solution in cuvets 4 and 5 (duplicate).
5. At the beginning of starvation, remove 2 mL of cell suspension (1×10^7cells/mL) from the shaking culture, wash once in half volume of ice-cold Soerensen phosphate buffer by pelleting at $200g$ for 3 min, resuspend in 1 mL of the same buffer (final concentration 2×10^7 cells/mL), vortex vigorously, and immediately pipet 0.1 mL of cell suspension each in cuvets 2 to 5. Insert the plugs and place the cuvets in the cuvet carrier anticlockwise, starting with cuvet 1. Switch on the control unit, with the rotating speed set at 40 rpm.
6. At times 5, 20, 30, 40, and 50 min, read the E values of each cuvet, adjusting the 0 point with the blank cuvet. Once adjusted, the 0 value changes minimally (*see* **Note 3**).
7. After the last measurement, put the cuvets on a glass slide and check at the microscope condition and morphology of the cell samples to avoid misinterpretation of the light-scattering data. The dissociating effect of EDTA on wild-type cells at the beginning of starvation will be evident, as cells are completely single (*see* **Notes 4** and **5**).
8. Repeat **steps 1–7** with cells starved for 1–6 h.
9. Plotting the data. For each sample, average the E values obtained between 30 and 50 min. The mean value of duplicate samples is plotted over time. To account for possible variations in cell size or shape, the data are normalized by calculating

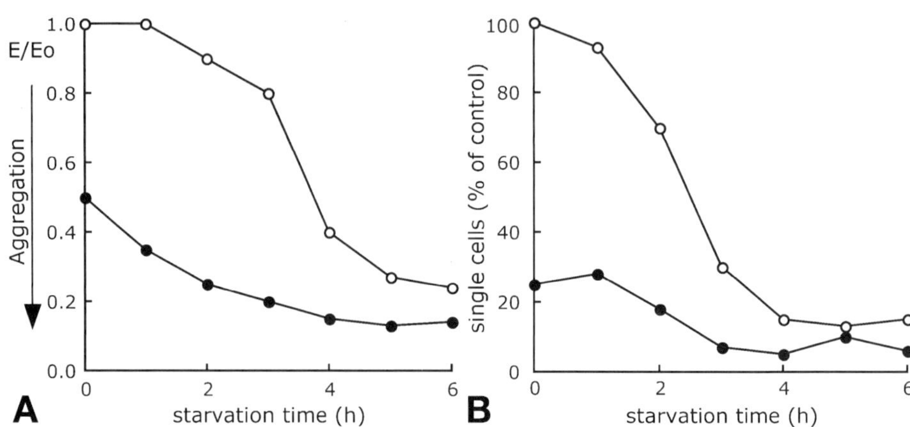

Fig. 4. Assessing ethylenediamine tetraacetic acid (EDTA)-sensitive and EDTA-resistant adhesion during development by using the light scattering (with agglutinometer) or the single-cell assay. **(A)** At the starvation time indicated in the abscissa, AX2 cells were washed, resuspended in Soerensen phosphate buffer with (open symbols) or without (closed symbols) EDTA, and incubated in the agglutinometer as described in the text. Unscattered light (E) was measured after 30 to 50 min of incubation and the values averaged and normalized for E_0 (EDTA-treated cells at time 0). Mean values of duplicate samples are shown. At the beginning of development, cells form aggregates that are totally dissociated with EDTA. Starting at 2 h of development, cell aggregates become increasingly resistant to EDTA dissociation. The finding that in the absence of EDTA the rate of aggregation is higher than in the presence of EDTA, even at t_{5-6}, indicates that an EDTA-sensitive form of cell adhesion persists also during the aggregation stage. **(B)** At the starvation time indicated in the abscissa, AX2 cells were washed, resuspended in Soerensen phosphate buffer with (open symbols) or without (closed symbols) 10 mM EDTA, and incubated in vials on a gyratory shaker as described in the text. After 30 min of incubation, cell aggregates were fixed with glutaraldehyde, diluted 10-fold, and the single cells left in the suspension were counted in a haemocytometer. Mean values of duplicate samples are shown. *Control:* single cells in the presence of EDTA at time 0. Acquisition of EDTA-resistant adhesion during development is also evident with the single cell assay. *See* text for comparison between the single cell and the light scattering assay.

the ratio of the E value of the samples to the E_0 value, i.e., the extinction value of totally dissociated, but otherwise similarly treated, cells. Typical examples of EDTA-sensitive and EDTA-resistant adhesion curves over time are shown in **Fig. 4A** (*see* **Notes 6–8**).

10. Clean the cuvets immediately after the experiment to avoid cell lysis and sticking of debris to the glass walls. The cleaning procedure is described in **Note 9**.

3.2. Assaying EDTA-Sensitive and EDTA-Resistant Cell Adhesion During Development Using the Single-Cell Assay

Single-cell measurement has been widely used in adhesion studies with mammalian cells and also applied to *Dictyostelium* cells. The rationale is that aggregate formation results in gradual disappearance of single cells from the suspension, and thus it is possible to measure aggregation indirectly by determining the ratio of single cells left in the suspension at the end of the incubation to total cells at the beginning. Also in this case, equilibrium between single cells and aggregates will be reached, which depends on the cohesive forces of the cells and the shear forces applied to the liquid medium. Because the method actually measures disappearance of single cells, it does not discriminate with regard to whether smaller or larger aggregates are formed, unless the assumption is made that the number of single cells left at the end of the experiment is directly proportional to the average size of the aggregates, which is not necessarily true. It is important, therefore, to realize that the single-cell assay does not allow one to compare the average size of aggregates formed, e.g., by different mutant strains, which is instead possible with the light-scattering assay.

The single-cell assay requires a gyration shaker or a rotating drum. Gyration shakers are available in all labs in which *Dictyostelium* cells are grown or developed in suspension, and is used in the assay described as follows.

Glass scintillation vials or beakers with a diameter of 2.5 cm are best suited for the assay.

For measuring acquisition of EDTA-resistant cell–cell adhesion during development, the assay is as follows:

1. Pipet 0.4 mL of ice-cold Soerensen phosphate buffer in vials 1 and 2 (duplicate samples).
2. Pipet 0.4 mL of 20 m*M* EDTA solution in vials 3 and 4 (duplicate).
3. At the beginning of starvation, remove 4 mL of cell suspension (1×10^7 cells/mL) from the shaking culture, wash once in half volume of ice-cold Soerensen phosphate buffer, resuspend in 2 mL of the same buffer (final concentration 2×10^7 cells/mL), vortex vigorously, and immediately pipet 0.4 mL of cell suspension in each of the vials. Transfer the vials to the shaker und incubate at 120 rpm for 30 min.
4. At the end of the incubation, rapidly add 0.2 mL of 10% glutaraldehyde (or TCA) solution to each sample, if possible without stopping cell shaking, and incubate for additional 10 min (*see* **Note 10**).
5. Add 7 mL Soerensen phosphate buffer to each vial.
6. Cover each vial with parafilm, gently invert the vial for resuspending the aggregates, and pipet a drop into a haemocytometer, without further dilution. Count single cells only (*see* **Note 11**).
7. Repeat **steps 1–6** at times 1–6 h of starvation.

8. Plotting the data. The mean value of single-cell number in duplicate samples, expressed as percentage of total cells, is plotted over time. A typical curve is shown in **Fig. 4B**.

By comparing the curves in **Fig. 4A,B**, it is apparent that both adhesion assays detect developmental changes in EDTA-stable adhesion. In the single-cell assay, however, the difference between EDTA-resistant and EDTA-sensitive adhesion is minimal, sometimes absent after 4 h of development. In addition, changes in EDTA-resistant adhesion leading to formation of smaller aggregates are not detected by this assay. It is also important to realize that although the single-cell assay can detect inhibitory effects of substances to be tested, potential agglutinating effects cannot be easily inferred from single-cell counting. A qualitative assessment is only possible by microscopic examination of the cells, unless an electronic cell counter or cell sorter is available to determine size distribution of the aggregates. Despite these limitations, the single-cell assay is a good alternative to the agglutinometer assay for studying cell adhesion (*see* **Notes 12** and **13**).

3.3. Applications and Modifications of the Cell Adhesion Assay

3.3.1. Cell Adhesion Assay and the Analysis of Mutants

Unless one is interested in investigating the molecular basis of cell adhesion, the most common use of the cell adhesion assay is in the identification and characterization of mutants.

Cell–cell adhesion may be affected by: (1) functional mutations in cell adhesion molecules (structural mutations, altered posttranslational modifications, impaired transport to the cell surface); (2) developmental mutations, which affect the expression of cell adhesion molecules; or (3) cytoskeletal mutations affecting cell shape. The cell adhesion assay, combined with careful observation of cell morphology and with other assays, may lead to rapid identification of mutants belonging to one of these three categories.

Failure to form EDTA-resistant adhesion at aggregation stage can only be due to impaired expression on the cell surface of the csA glycoprotein. Disrupting the *csA* gene or blocking transport of the csA glycoprotein to the cell surface is sufficient to completely inhibit EDTA-stable adhesion at aggregation stage *(1,3)*. *CsA*-null mutants are nevertheless able to aggregate by chemotaxis and to form fruiting bodies when incubated on agar *(1)*. Only under stringent conditions, e.g., when tested on soil, they fail to aggregate *(20)*. For detecting functional mutations in the csA glycoprotein, the cell adhesion assay in the presence of EDTA is therefore crucial. Conversely, a mutant defective in EDTA-stable adhesion, but displaying rather normal aggregation on agar, is, with very high probability, defective in functional csA.

Developmental mutants may fail to express csA, and therefore EDTA-stable adhesion. The cell adhesion assay in the presence of EDTA allows one to determine the severity of the developmental inhibition, but more importantly, it gives some clues as to the defective developmental pathway, as will be discussed under **Subheading 3.3.2.**

Finally, defects in cytoskeletal proteins may impair adhesion by affecting cell shape. In this case, both EDTA-sensitive and EDTA-resistant adhesion will be affected in the cell adhesion assay, and this is a good indication for defects in cytoskeletal components, rather than adhesion molecules. As will be shown under **Subheading 3.3.6.**, with some ingenuity it is possible to measure EDTA-stable contacts even in totally round cells.

3.3.2. Assaying EDTA-Resistant Adhesion as a Rapid Test for Characterizing Mutants Defective in cAMP Signaling and Relay

The gene encoding the csA glycoprotein is one of the best markers of aggregation-specific genes, because its expression is induced at a very low level by starvation and is strongly stimulated by endogenous cAMP signaling. Mutants defective in adenylyl cyclase activation, and therefore in cAMP signal relay, can be induced to develop by exogenously supplied cAMP pulses. cAMP pulsing bypasses the requirement of adenylyl cyclase activity, therefore cell responses to cAMP enable to characterize mutations in receptor and G protein-linked pathways *(5,18)*.

Assaying EDTA-resistant contacts in aggregateless mutants, which are treated with cAMP pulses, is a convenient and straightforward way to identify mutants defective in cAMP relay, and to distinguish, under certain circumstances, mutations downstream of the receptor or the G protein, before engaging in more complex biochemical assays.

Acquisition of EDTA-resistant adhesion is assayed as outlined above (*see* **Subheading 3.1.2.**), but the test cells are first treated with cAMP pulses. For each strain to be tested, the procedure is as follows:

1. Wash cells free of medium or bacteria as described previously and resuspend at a concentration of 1×10^7 cells/mL in Soerensen phosphate buffer. Transfer 30 mL each in two 100-mL beakers (one for control and the second for cAMP treatment) and place them on the lab shaker.
2. Fill one of two 10-mL disposable syringes with 5 mL of 0.04 mM cAMP solution and the other with Soerensen phosphate buffer. The syringe is equipped with a needle connected to a piece of capillary tubing (e.g., PE or ETFE capillary tubing of 1.8/0.8-mm outer/inner diameter). The tubing must be cut long enough to be positioned above the beaker, such that the solution will be dispensed drop-wise directly in the beaker. The diameter of the tubing should be chosen to allow drops of 15 ± 2 µL volume to be formed (as it is the case with the one suggested),

corresponding to a final concentration of 20 ± 2 nM cAMP in the beaker (*see* **Note 14**).

3. Place the syringes on a multisample perfusor, and regulate the syringe driving such that the solution is administered to the cell suspension at a speed of one drop every 6 ± 0.5 min. Control cells will be pulsed with Soerensen phosphate buffer and the cAMP-treated cell sample with cAMP.

4. Switch on the agglutinometer and prepare two cuvets for each control and test sample (duplicates), in addition to the reference cuvet 1, containing 0.2 mL Soerensen phosphate buffer.

5. At t_0 of starvation and at every hour thereafter up to 6 h of starvation, wash 1 mL of cell suspension in Soerensen phosphate buffer, resuspend in half volume, and transfer 0.1 mL of control or cAMP-treated cells in cuvets 2–3 or 4–5, respectively, each containing 0.1 mL of 20 mM EDTA solution.

6. Measure EDTA-resistant adhesion in the agglutinometer and plot the data as described above in **Subheading 3.1.2.**, **steps 6–8**.

A typical experiment showing the effects of cAMP pulsing on EDTA-resistant adhesion of wild-type AX2 or mutant HSB1 that is defective in cAMP relay *(19)*, is displayed in **Fig. 5**.

As is evident from **Fig. 5**, cAMP pulsing accelerates the appearance of EDTA-resistant contacts in wild-type cells. Mutants specifically defective in cAMP relay are strongly impaired in EDTA-resistant adhesion, but they can be fully rescued, insofar as adhesion is concerned, by cAMP pulsing. This is confirmed also by Northern blot analysis of *csA* gene expression (**Fig. 5**, top). Interestingly, the assay allows one to distinguish mutations upstream of or at the G protein from mutations downstream of the heterotrimeric G protein and upstream of the adenylyl cyclase, such as mutations involving adaptor proteins CRAC, PIA, or the RasGEF aimless, as well as the adenylyl cyclase A (*acaA*). In all these latter cases, rates of EDTA-resistant adhesion similar to the one shown in **Fig. 5** are found. In case of mutations upstream of or at the heterotrimeric G protein, EDTA-resistant adhesion is almost lacking, but more importantly, cAMP pulses elicit delayed and quantitatively reduced EDTA-stable adhesion or have no effect at all, depending on the mutation. Thus, coupling cAMP pulsing to the cell adhesion assay is a powerful tool in characterizing mutants defective in signaling.

3.3.3. Assaying the Effect of Antibodies, Lectins, or Drugs on Cell Adhesion by Light Scattering with the Agglutinometer: Problems and Solutions

The small volume in which adhesion can be assayed is one of the major advantages of the agglutinometer, particularly when expensive or valuable substances must be tested. In addition, a given substance can be tested at different concentrations in the same experiment, as 23 sample cuvets are available. To

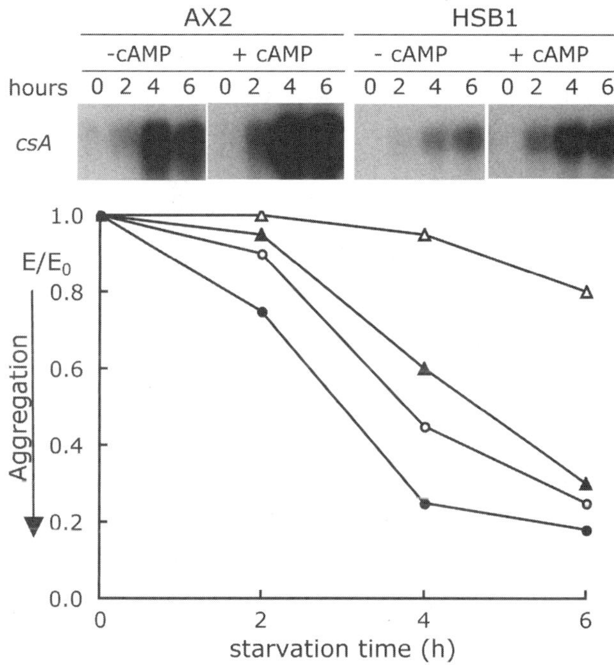

Fig. 5. Acquisition of ethylenediamine tetraacetic acid (EDTA)-resistant adhesion following cAMP pulsing. Control AX2 (circles) or mutant HSB1 (triangles) cells were treated with buffer alone (open symbols) or 20 n*M* cAMP pulses (closed symbols) every 6 min as described in text. At the time indicated on the abscissa, cells were washed, resuspended in Soerensen phosphate buffer, and incubated with 10 m*M* EDTA. Adhesion was measured by the agglutinometer assay (*see* **Fig. 4**). In parallel, total RNA was extracted from the same cell suspensions at the time indicated, subjected to Northern blot, and hybridized with probes specific for the *csA* gene (upper panel). See text for comments.

distinguish between effects on EDTA-sensitive or EDTA-resistant adhesion, it is convenient to use growth-phase cells, which form only EDTA-sensitive contacts, or aggregation-competent cells, respectively. In the first case, the assay is done in the absence of EDTA, using cells just washed free of medium or bacteria (starvation time: t_0). In the second case, 4- to 6-h starved cells are used, depending on the strain, and the assay is done in the presence of 10 m*M* EDTA. The protocol is the same as outlined above (*see* **Subheading 3.1.2.**), except that 40 m*M* EDTA solution is used. This allows for a volume of up to 0.05 mL of the tested substance to be filled in the cuvet. If necessary, the available volume can be brought up to 0.1 mL, by concentrating the cell suspension fourfold, thus reducing the cell volume to be added to the cuvet to 0.05 mL.

3.3.3.1. Cell Adhesion-Blocking Antibodies

Antibodies against cell adhesion molecules or other surface antigens, being divalent, can agglutinate cells. This is most often the case with polyspecific rabbit antibodies, as they are directed against different epitopes. Monoclonal antibodies may agglutinate cells, particularly when they are raised against carbohydrate epitopes shared by several membrane glycoproteins. The agglutination effect is independent of the antibodies being directed against cell adhesion molecules, and it is no evidence at all for their target antigens being involved in cell adhesion. Agglutination, on the other hand, may obscure the potential inhibitory effect on cell adhesion of antibodies reacting with adhesion molecules. For cell adhesion assays, rabbit polyclonal antibodies are best tested as univalent antibody fragments (Fab), which are obtained by papain treatment. In the case of mouse monoclonal antibodies, however, the higher sensitivity to papain should be taken into account. Even when papain treatment does not lead to evident antibody degradation, the preparation may fail to block adhesion, as a result of low affinity or low titer of the Fab preparation. It is beyond the purpose of this chapter to describe how to successfully raise polyspecific and high-titer adhesion-blocking antibodies or how to purify their target antigens. Some indications can be found in **ref. *12***.

The adhesion assay with the agglutinometer is very sensitive for testing the adhesion-blocking activity of Fab. A major advantage is that once the inhibitory activity of a given preparation of Fab (or more rarely of intact monoclonal antibodies) has been established, a kinetic curve can be elaborated and used for testing the neutralizing effect of putative target antigens, either crude or purified. A schematic representation of the experimental strategy is shown in **Fig. 6**. The assays are as described above (*see* **Subheading 3.1.2.**), except that concentrated solution of both EDTA and cells are used to allow for addition of substances to be tested (Fab and antigens), as mentioned previously.

3.3.3.2. Lectins

Lectins, such as concanavalin A (ConA), wheat germ agglutinin (WGA), or the *D. discoideum* lectin discoidin, agglutinate *Dictyostelium* cells in a concentration-dependent manner. When tested in their monomeric form, neither ConA nor WGA inhibit EDTA-resistant adhesion, although the csA glycoprotein contains ConA- and WGA-positive carbohydrate chains. The agglutinometer is sensitive enough to measure agglutination by lectins in a concentration-dependent manner. This is a further advantage of the assay when studying drugs interfering with cell adhesion, as the agglutinometer is sensitive to both inhibitory and stimulatory effects on cell adhesion.

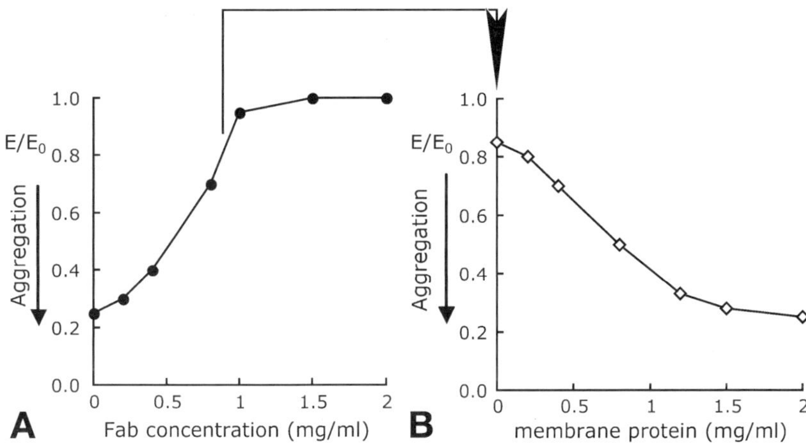

Fig. 6. Strategy for identification of cell adhesion molecules using antibodies against cell surface antigens. The immunological approach, which has been successfully applied for identification of cell adhesion molecules in *Dictyostelium discoideum* and *Polysphondylium pallidum*, is outlined. Rabbit monovalent antibodies (Fab) raised against crude membranes from aggregation-competent cells are first exhaustively absorbed with growth-phase cells to make them specific for aggregation-competent cells. The Fab preparation is tested for its adhesion-blocking activity in cell adhesion assay in the presence of ethylenediamine tetraacetic acid (EDTA) with the agglutinometer **(A)** Fab concentration-dependent inhibition of cell adhesion is obtained. **(B)** Adhesion-blocking Fab at subsaturating concentration is used to partially block adhesion in the absence or presence of increasing amounts of crude membrane fractions that neutralize the Fab activity. The assay allows one to determine both the adhesion-blocking activity of the Fab preparation and the presence in crude membranes of antigens acting as putative adhesion molecules. Further enrichment and purification of the relevant antigen(s), using the assay on **(B)** at each purification step, allows one to identify the putative adhesion molecules. The same experimental strategy can obviously be used to assay the activity of antibodies raised against purified antigens. Although the approach has been successfully applied also to other systems besides *Dictyostelium* and *Polypshondylium*, leading to the identification of N-CAM and cadherin, among others, in mammals, it must be kept in mind that evidence for such antigens to work as adhesion molecules is operational, and must be corroborated by genetic studies.

3.3.3.3. Drugs and Chemicals Affecting Cell Shape

Optimal cell–cell adhesion depends on the plasticity of the cell surface and on membrane projections, such as filopodia, that allow close interdigitations between cells, therefore extending the contact area of apposing membranes. Any substance interfering with the cytoskeleton, and resulting in cells rounding up, may affect cell adhesion. It is important, therefore, to distinguish between specific or unspecific effects on cell adhesion, a specific effect being one that directly interferes with the binding of adhesion molecules to one another. An unspecific effect will impair both EDTA-sensitive and EDTA-resistant adhesion, resulting in increased unscattered light (E) in the adhesion assay under both conditions. When aggregation-competent cells are incubated in the agglutinometer in the presence of cytochalasin and EDTA, for example, unscattered light is higher than in untreated control cells. A similar result is obtained when EDTA-resistant adhesion is measured in the presence of drugs that uncouple oxidative phosphorylation from O_2 consumption, such as sodium azide or 2,4-dinitrophenol. Microscopic examination of the cells at the end of the experiment will show that increased light scattering is due to the cell rounding effect of all of these drugs, which leads to formation of loose aggregates, but does not hinder the cells' attachment to one another via EDTA-resistant contacts. Formation of EDTA-resistant adhesion can thus be detected in the agglutinometer even under these extreme conditions, allowing one to study incidental effects on cell adhesion of drugs that affect primarily cytoskeleton or ATP production, as well as mutations in genes involved in cell motility and cell shape control.

The versatility of the cell adhesion assay, particularly when coupled with the agglutinometer, makes it a powerful tool in *Dictyostelium* cell biology.

4. Notes

1. Under these conditions, the pH of the diluted buffer is between 6.0 and 6.1, and there is no need for pH adjustment.
2. Cuvets and plugs are available on demand by Hellma (Mühlheim, Germany) under the reference number 110.041.02).
3. The 5-min measurement is meaningless for the assay, but it is recommended as a checkpoint control, particularly when several samples are tested at once. The E values should be roughly similar in all cuvets; if this is not the case, some errors have been made in cell concentration or in pipetting. If drugs or other substances are being tested, cell damage may have occurred.
4. Lysed cells or dead bacteria sticking to the cell surface, in the case of cells grown on bacteria, may give rise to small clumps. EDTA at a concentration of 10 mM may round up the cells, particularly at the beginning of starvation, but it does not lyse them, at least during the time of the experiment. If lysis is observed, check the conditions of the cell culture and the solutions that have been used. If drugs or other chemicals to be tested have been added to the buffer, be sure that they do not damage the cells.
5. Although not routinely required, it is however recommended, in crucial experiments, to also document the condition of the cell suspension in the cuvets, using conventional imaging systems attached to the microscope. Low magnification (10 or 20× objective) should be chosen to give a panoramic picture of cells or

aggregates. Cell aggregates dissociate rapidly once they have attached to glass, particularly during the first 3–4 h of starvation and in the presence of EDTA. Pictures should be taken within 2–3 min upon placing the cuvets on the glass slide. If many cuvets must be checked, keep them in motion in the rotating device until pictures can be taken. For sharp pictures, focus close to the bottom of the cuvet and use transmitted light, no darkfield or contrast phase. The cell density in the cuvet is quite high, thus darkfield or contrast phase result in blurred pictures.

6. If monovalent antibodies (Fab), which totally block adhesion, are available, they can be used for establishing the E_0 value by adding a reference cuvet at each time point. Otherwise, the E value of EDTA-treated cells at time 0 h can be used for wild-type and most mutant cells. Cell size decreases during development, but size differences between t_0 and t_6 cells are minimal, thus the t_0 value in the presence of EDTA is a very good approximation of single cells also for later stages.

7. EDTA concentrations as low as 1 or 5 mM are sufficient to dissociate or keep single growth-phase AX2 or V12M2 wild-type strains, respectively. An excess of 10 mM concentration is usually used, which does not damage the cells and allows distinguishing between EDTA-sensitive and EDTA-resistant adhesion at the aggregation stage. Increasing the EDTA concentration to 20 mM at aggregation or postaggregative stages results in a dissociating effect that is only minimally higher than that for 10 mM EDTA, and is thus not recommended.

8. Cells at mound or slug stage can also be used as test cells; however, they are difficult to dissociate. A satisfactory dissociation of most, although not all, cells can be obtained by resuspending mounds or migrating slugs in cold Soerensen phosphate buffer and pipetting repeatedly the aggregates through a Pasteur pipet.

9. The cleaning procedure for the cuvets is as follows. Connect a Pasteur pipet to a vacuum or water pump and open the pump. Unplug the cuvet, insert the Pasteur pipet tip into one of the two canals of the cuvet, and suck out the cell suspension. Immerse the cuvet in distilled water, holding it with plastic tweezers (to avoid scratching the cuvet walls, wrap the tweezer tips with parafilm or envelop them in thin flexible tubing). Flush the inner cavity of the cuvet for 10–15 s by sucking water with the Pasteur pipet through one canal while holding the second one under water. Transfer the cuvet to a beaker containing a solution of pepsin/HCl, and continue aspirating with the Pasteur pipet, first washing and then filling the cuvet with pepsin/HCl solution (do not leave air bubbles inside). Keep the cuvets, filled with this solution, in the beaker overnight. By using the Pasteur pipet as above, again flush the cuvet first with distilled water, then with 70% ethanol, and finally with acetone. After the last step, dry up the cuvet cavity by sucking air in through the canal. Clean the external surface with a cosmetic tissue before storing the cuvets. Clean the cuvet plugs simply by washing with water. If the cuvets are to be used repeatedly the same day, the washing step in pepsin/HCl can be omitted.

10. Glutaraldehyde or TCA, at the indicated concentrations, act quite rapidly as fixatives, blocking further cell–cell adhesion as well as dissociation of the aggregates. Keeping the cells under shaking for a few minutes, after addition of the

fixative, facilitates rapid mixing of the fixative with the aggregates and reduces their cross-linking, in particular with glutaraldehyde.

11. Because a 10-fold dilution has been made in **step 5**, the total cell concentration will be 1×10^6 cell/mL, i.e., 100 cells should be counted, if cells were totally single. Doublets should not be counted as single cells, but this is a matter of choice.

12. A serious drawback of the single-cell assay is that potential cell lysis may be obscured by the fixation procedures as a result of the substances being tested. To identify lysed cells sticking to, or mixed with, aggregates requires much experience.

13. The shaking speed depends on the rotation diameter of the lab shaker and should be adjusted accordingly to avoid either cell sedimentation or centrifugation effects. The minimal sample volume that can be used depends on the diameter of vials (beakers or test tubes), and the latter is also linked to the rotation diameter of the shaker. In the assay described above, vials and sample volumes were optimized with a lab shaker having a rotating diameter of 50 mm.

14. The initial cAMP concentration must be adjusted according to the cell volume to be treated by calculating the dilution factor of the cAMP drop. The drop volume should be as small as possible compared with the volume of the cell suspension, particularly if the cell suspension is pulsed over many hours and several samples are withdrawn. With a period of 6 min, there are 50 pulses in 5 h, corresponding, under our conditions, to addition of 0.75 to 1.0 mL of cAMP to an initial volume of 30 mL cell culture. Take into consideration that too small a volume of the cell suspension at the beginning of the experiment may lead at later time points to the concentration of cells or cAMP pulses being significantly diluted or much higher, respectively.

References

1. Harloff, C., Gerisch, G., and Noegel, A. A. (1989) Selective elimination of the contact site A protein in *Dictyostelium discoideum* by gene disruption. *Genes Devel.* **3,** 2011–2019.

2. Faix, J., Gerisch, G., and Noegel, A. A. (1990) Constitutive overexpression of the contact site A glycoprotein enables growth-phase cells of *Dictyostelium discoideum* to aggregate. *EMBO J.* **9,** 2709–2716.

3. Ponte E. and Bozzaro, S. (1995) Cell adhesion in the life cycle of *Dictyostelium. Experientia* **51,** 1175–1188.

4. Siu, C. H., Harris, T. J. C., Wang, J., and Wong. E. (2004) Regulation of cell-cell adhesion during *Dictyostelium* development. *Sem. Cell Dev. Biol.* **15,** 633–641.

5. Gerisch, G. (1987) Cyclic AMP and other signals controlling cell development and differentiation in *Dictyostelium. Annu. Rev. Biochem.* **56,** 853–879.

6. Wang, J., Hou, L., Awrey, D., Loomis, W. F., Firtel, R.A., and Siu, C. H. (2000) The membrane glycoprotein gp150 is encoded by the *lagC* gene and mediates cell-cell adhesion by heterophilic binding during *Dictyostelium* development. *Devel. Biol.* **227,** 734–745.

7. Knecht, D. A., Fuller, D. L., and Loomis. W. F. (1987) Surface glycoprotein gp24 involved in early adhesion of *Dictyostelium discoideum. Devel. Biol.* **121,** 277–283.

8. Wong, E. F. S., Brar, S. K., Sesaki, H., Yang, C., and Siu, C. H. (1996) Molecular cloning an characterization of DdCAD-1, a Ca^{2+}-dependent cell-cell adhesion molecule in *Dictyostelium discoideum*. *J. Biol. Chem.* **271**, 16,399–16,408.

9. Wong, E., Yang, C., Wang, J., Fuller, D., Loomis, W. F., and Siu, C. H. (2002) Disruption of the gene encoding the cell adhesion molecule DdCAD-1 leads to aberrant cell sorting and cell-type proportioning during *Dictyostelium* development. *Development* **129**, 3839–3850.

10. Chia, C. P., Gomathinayagam, S., Schmaltz, R. J., and Smoyer, L. K. (2005) Glycoprotein gp 130 of *Dictyostelium discoideum* influences macropinocytosis and adhesion. *Mol. Biol. Cell* **16**, 2681–2693.

11. Beug, H. and Gerisch, G. (1972) A micromethod for routine measurement of cell agglutination and dissociation. *J. Immunol. Methods* **2**, 49–57.

12. Bozzaro, S., Merkl, R., and Gerisch, G. (1987) Cell adhesion: its quantification, assay of the molecules involved, and selection of defective mutants in *Dictyostelium* and *Polysphondylium*. *Methods Cell Biol.* **28**, 359–385.

13. Müller, K. and Gerisch, G. (1978) A specific glycoprotein as the target site of adhesion blocking Fab in aggregating *Dictyostelium* cells. *Nature* **274**, 445–449.

14. Bozzaro, S., Tsugita, A., Janku, M., Monok, G., Opatz, K., and Gerisch, G. (1981) Characterization of a cell surface glycoprotein as a contact site in *Polysphondylium pallidum*. *Experim. Cell Res.* **134**, 181–191.

15. Beug, H., Katz, F. E., and Gerisch, G. (1973) Dynamics in antigenic membrane sites relating to cell aggregation in *Dictyostelium discoideum*. *J. Cell Biol.* **56**, 647–658.

16. Bozzaro, S. and Gerisch, G. (1978) Contact sites in aggregating cells of *Polysphondylium pallidum*. *J. Mol. Biol.* **120**, 265–279.

17. Ponte, E., Bracco, E., Faix, J., and Bozzaro, S. (1998) Detection of subtle phenotypes: the case of the cell adhesion molecule csA in *Dictyostelium*. *Proc. Natl. Acad. Sci. USA* **95**, 9360–9365.

18. Parent, C. A. and Devreotes, P. N. (1996) Molecular genetics of signal transduction in *Dictyostelium*. *Annu. Rev. Biochem.* **65**, 411–440.

19. Pergolizzi, B., Peracino, B., Silverman, J., et al. (2002) Temperature-sensitive inhibition of development in *Dictyostelium* due to a point mutation in the *piaA* gene. *Devel. Biol.* **251**, 18–26.

28

Periodic Activation of ERK2 and Partial Involvement of G Protein in ERK2 Activation by cAMP in *Dictyostelium* Cells

Mineko Maeda

Summary

Intercellular signaling mediated by cAMP plays a pivotal role in coordinating cell movement into aggregates at the early stage of *Dictyostelium* development when the extracellular level of cAMP periodically changes at 6- to 7-min intervals. We have shown that MAP kinase ERK2 is activated via the cAMP receptor CAR1 in phase with this periodic change. This was revealed by assessing the level of ERK2 activation with immunoblots using two kinds of antibodies, commercially available anti-phospho-p44/p42 MAP-kinase antibody and anti-*Dictyostelium* ERK2 antibody. In this chapter, we describe the methods we have used to assess the level of activated ERK2 and partial involvement of G protein in cAMP-induced ERK2 activation.

Key Words: *Dictyostelium*; cAMP; MAP kinase; ERK2; immunoblot; phosphory-lated ERK2; *erkB⁻*; oscillation.

1. Introduction

MAP kinase ERK2 was identified after screening for aggregation-defective mutants *(1)*. The enzyme has a conserved amino acid sequence, TEYVATRWYRAP, in its subdomain VII. Its expression is developmentally regulated. At the start of development, a 1.8-kb mRNA is present and its level increases two- to fivefold during early development. The mutant cells (*erkB⁻*) lacking this enzyme respond chemotactically to gradients of cAMP in the nanomolar range, but they are unable to properly relay the cAMP signal following a pulse of exogenous cAMP. As a consequence, their phenotype is aggregation-minus. Exogenous cAMP induces activation of ERK2 and phosphorylation on its tyrosine residues, which is detectable with anti-phospho-p44/p42 MAP kinase antibody in wild-type cells

From: *Methods in Molecular Biology, vol. 346:* Dictyostelium discoideum *Protocols*
Edited by: L. Eichinger and F. Rivero © Humana Press Inc., Totowa, NJ

(2–4). After phosphorylation, the mobility of ERK2 is retarded on sodium dodecyl sulfate (SDS)-polyacrylamide gel electrophoresis (PAGE) as revealed with anti-*Dictyostelium* ERK2 antibody *(4)*. By using these antibodies, it becomes apparent that ERK2 is periodically activated in phase with periodic changes in the cAMP level in aggregation stage cells *(4)*. This result contributes to understanding the mechanism underlying the oscillations in extracellular cAMP level that plays a pivotal role in the transition from growth to differentiation. cAMP-induced ERK2 activation is mediated through the cell-surface cAMP-receptor CAR1. However, the involvement of CAR1-coupled G protein in the activation is controversial *(2,4)*. Here, by using anti-phospho-p44/p42 MAP kinase antibody, we show that CAR1-coupled G protein is involved in, but contributes only partially to, the activation of ERK2.

2. Materials

2.1. Strains and Culture Medium

1. Strains: *Dictyostelium discoideum* Ax2, *erkB⁻*, *carA⁻*, *acaA⁻*, *dagA⁻*, and *gpb⁻* (or *g* β⁻).
2. HL-5 axenic medium: 26 g Oxoid bacteriological peptone, 13 g Oxoid yeast extract, 18 g glucose, 0.87 g KH_2PO_4, 2.3 g $Na_2HPO_4 \cdot 12H_2O$, 90 µL of 0.1 µg/µL vitamin B_{12}, 180 µL of 2 µg/µL folic acid in 1.8 L deionized water *(5)* (*see* **Note 1**).
3. Phosphate buffer (PB): 1.55 g KH_2PO_4 and 0.8 g $Na_2HPO_4 \cdot 12H_2O$ in 1 L deionized water.

2.2. Immunoblotting

1. 10–20% polyacrylamide gradient gel (PAGEL Model AE-6000 NPG-1020 L, Atto Corporation) (*see* **Note 2**).
2. Running buffer: 3.03 g Trizma base, 14.42 g glycine, 50 mL of 20% SDS in 1 L of deionized water.
3. 2X sample buffer: 0.001% (w/v) bromophenol blue, 2% (w/v) SDS, 10% (v/v) glycerol, 100 m*M* dithiothreitol (DTT), 60 m*M* Tris-HCl (pH 6.8).
4. A semidry transferring apparatus such as Model AE5570, ATTO.
5. Polyvinyl difluoride membrane (Immobilon P, Millipore) and filter paper (3 MM CHR Whatman) that are cut to the same size as the gel being used.
6. Transferring buffer: 2.9 g Trizma base, 1.45 g glycine, 0.19 g SDS, 100 mL methanol in 400 mL deionized water.
7. TBS-T: 0.2% Tween 20, 3.03 g Trizma base, 8.0 g NaCl, 0.2 g KCl in 1 L deionized water. pH is adjusted with HCl to 8.0.
8. Blocking buffer: 5% (w/v) skim milk in TBS-T.
9. Antibodies: anti-phospho-p44/p42 MAP-kinase antibody (#9101, Cell Signaling Technology), anti-dERK2 antibody, and horseradish peroxidase (HRP)-conjugated anti-rabbit immunoglobulin (Ig) antibody (#7074, Cell Signaling Technology).

10. Detection: ECL reagent (Enhanced Chemiluminescence system, Amersham) and the LuminoImage Analyzer (LAS1000, FujiFilm).

2.3. Purification of Glutathione-S-Transferase (GST) and GST-Tagged ERK2 and Preparation of GST- and GST-ERK2C Columns

1. Terrific broth: 12 g Bactopeptone, 24 g Bactoyeast extract, 4 mL glycerol, 100 mL of 10X PB in 900 mL deionized water. 10X PB is composed of 11.57 g KH_2PO_4 and 82.15 g K_2HPO_4 in 500 mL deionized water.
2. Phosphate-buffered saline (PBS): 137 g NaCl, 8.1 g $Na_2HPO_4 \cdot 12H_2O$, 1.47 g KH_2PO_4 and 2.68 g KCl in 1 L deionized water (pH 6.4).
3. Lysis buffer: 10 mM ethylenediamine tetraacetic acid (EDTA) (pH 7.0), 3 mM DTT, 1 mM phenylmethyl sulfoxide (PMSF), 10 µg/µL pepstatin A, 2 µg/µL leupeptin, 10 µg/µL aprotinin.
4. 50% (v/v) glutathione bead slurry (Glutathione Sepharose, Amersham-Pharmacia) in PBS.
5. 0.2 M sodium borate (pH 9.0).
6. 0.2 M ethanolamine (pH 8.0).
7. Lysozyme (solid).
8. Dimethylpimelimidate (solid).
9. 1% (w/v) Merthiolate.

2.4. Affinity Purification of Anti-dERK2 Antibody

1. 10 mM Tris-HCl (pH 7.5).
2. 10 mM Tris-HCl (pH 8.8).
3. 100 mM Glycine (pH 2.5).
4. 0.5 M NaCl in 10 mM Tris-HCl (pH 7.5).
5. 1 M Tris-HCl (pH 8.0).
6. Sodium azide (solid).
7. Cellulose membrane filter (pore size 0.20 µm; Advantec 25CS020AS).

3. Methods

3.1. Large-Scale Preparation of GST and GST-ERK2C

1. *Escherichia coli* XL1-blue is transformed with plasmids pGEX-4T-1 (Amersham-Pharmacia), pGST-ERK2C or pGST-full length ERK2.
2. Inoculate a single colony of each clone into 50 mL of Terrific broth and incubate overnight at 37°C.
3. The next day, add 450 mL of Terrific broth to each culture. Then, culture for 10 h at 22°C. Add 0.5 mL of 100 mM isopropyl-β-D-thiogalactopyranoside (IPTG) and culture overnight at 22°C (*see* **Note 3**).
4. Harvest cells by centrifugation at 10,000g for 10 min at 4°C. Then, add 2.5 mL of ice-cold PBS containing 5 mM EDTA in a plastic 15-mL conical centrifugation tube. Mix and resuspend thoroughly by vortexing.

5. Add 5 mg of lysozyme to the cell suspension and place it on ice for 5–10 min to break the cell wall (*see* **Note 4**).
6. Add 7.5 mL of lysis buffer and 1.1 mL of 10% (v/v) Triton-X 100 to the suspension and mix well.
7. Cells are disrupted by sonication and then centrifuged at 20,000*g* for 30 min at 4°C to pellet the cell debris (*see* **Note 5**).
8. Add 2 mL of 50% glutathione bead slurry to the resulting supernatant and gently shake for 1 h at room temperature. Centrifuge for 2 min at 1000*g* to pellet the beads, which bind glutathione-*S*-transferase (GST) or GST-tagged proteins (*see* **Note 6**).

3.2. Preparation of GST and GST-ERK2C Columns

1. Wash the GST beads or GST-ERK2C beads twice with PBS and then twice with 10 volumes of 0.2 *M* sodium borate (pH 9.0) and pellet by centrifugation at 10,000*g* for 30 s.
2. To covalently link GST or GST-ERK2 to agarose beads, resuspend the beads in 10 bed-volumes of 0.2 *M* sodium borate (pH 9.0).
3. Add dimethylpimelimidate (solid) to each bead suspension to a final concentration of 20 m*M*. Gently mix for 30 min at room temperature on a rocker.
4. Wash the beads once in 0.2 *M* ethanolamine (pH 8.0) and then incubate for 2 h in 0.2 *M* ethanolamine (pH 8.0) with gentle rocking (*see* **Note 7**).
5. The beads are resuspended in PBS containing 0.01% merthiolate.
6. Fill the beads in a 5-mL syringe plugged with a small amount of cotton and wash with 10-bed volumes of 10 m*M* Tris-HCl (pH 7.5) for affinity purification (do not dry them out).

3.3. Affinity Purification of Anti-Dictyostelium ERK2 Antibody

Anti-*Dictyostelium* ERK2 (anti-dERK2) antibody is raised against a GST-ERK2C fusion protein in rabbits. ERK2C is the 27-amino acid peptide (IIKKKKEERKKQTNPTKPDTTAPTLST) of *Dictyostelium* ERK2 located at its C-terminal region. Anti-dERK2 antibody is affinity purified on immobilized GST-ERK2C as follows (*see* **Note 8**).

1. Dilute anti-GST-dERK2 antiserum with 10 volumes of PBS.
2. Filtrate the diluted antiserum through a cellulose membrane filter (pore size 0.20 µm) to remove debris.
3. Pass the filtrated anti-GST-ERK2C antiserum through the GST column, in order to eliminate anti-GST antibody from the serum. Save the flow-through fraction.
4. Repeat the previous step three more times. Use the flow-through fraction for this step. Before every passage, activate the GST column by sequentially washing with 10 bed-volumes of 10 m*M* Tris-HCl (pH 7.5), 10 bed-volumes of 100 m*M* glycine (pH 2.5), 10 bed-volumes of 10 m*M* Tris-HCl (pH 8.8), and 10 bed-volumes of 10 m*M* Tris-HCl (pH 7.5).
5. Apply the finally obtained flow through fraction onto the GST-ERK2C column (*see* **Note 9**).

6. Pass the flow-through fraction through the column three times.
7. Wash the column with 20 bed-volumes of 10 m*M* Tris-HCl (pH 7.5) and then with 20 bed-volumes of 500 m*M* NaCl in 10 m*M* Tris-HCl (pH 7.5).
8. Elute the bound antibodies by passing 10 bed-volumes of 100 m*M* glycine (pH 2.5) and collect the eluate in a tube containing 1 bed-volume of 1 *M* Tris-HCl (pH 8.0).
9. Dialyze overnight at 4°C against PBS containing 0.02% (w/v) sodium azide.

The specificity of the anti-dERK2 antibody is shown in **Fig. 1A**. The specificity of the anti-phospho p44/p42 antibody is also shown in **Fig. 1B**.

3.4. Immunoblotting

1. Protein samples or aliquots (10 µL) of cell lysates are electrophoresed for 2 h with 20 mA in a 10–20% gradient gel for SDS-PAGE according to the method of Laemmli *(6)*.
2. Immunoblotting is performed according to the method described by Harlow and Lane *(7)*. Proteins are electrophoretically transferred with a semi-dry transfer apparatus. A sheet of Immobilon P is immersed in deionized water for 1 min after a brief treatment with methanol. The gel is briefly washed with deionized water and overlaid on the membrane supported by five sheets of filter paper wetted with transfer buffer placed on the carbon anode. Next, five sheets of filter paper wetted with transfer buffer are overlaid (*see* **Note 10**).
3. Finally, the carbon cathode is set. Run electric current for 1 h at 1 mA/cm^2 gel from the gel to the membrane.
4. After transfer, briefly wash the membrane in deionized water and then shake it overnight at 4°C in blocking buffer.
5. After blocking, incubate for 1 h at 37°C in 2–3 mL of primary antibody on a rocker. As primary antibodies, use rabbit anti-phospho-p44/42 MAP kinase antibody (500-fold diluted in blocking buffer) and rabbit anti-dERK2 antibody (10-fold dilution in blocking buffer).
6. After washing four times (15 min each) in TBS-T, incubate the membrane for 1 h at room temperature in 10 mL of secondary antibody. Use HRP-conjugated anti-rabbit IgG antibody (5000-fold diluted in blocking buffer).
7. Wash the membrane four times (15 min each) in TBS-T. The signal can be detected with ECL reagent and analyzed with a LuminoImage Analyzer.

3.5. Analysis of Activated ERK2 After cAMP Stimulation by Immunoblotting

1. *Dictyostelium* cells are harvested from the exponential growth phase (3–5 × 10^6 cells/mL) and washed twice with PB.
2. Washed cells are resuspended at 5 × 10^6/mL in PB and then stimulated for 4 h with 100 n*M* cAMP every 6 min to induce cAMP receptors, Gα2, and other components of the aggregation response (*see* **Note 11**).
3. Wash the cells twice with PB to remove cAMP. Then, resuspend them in PB at 5 × 10^7/mL and shake at 150 rpm. After 10 min of incubation, cells are stimulated with various concentrations of cAMP.

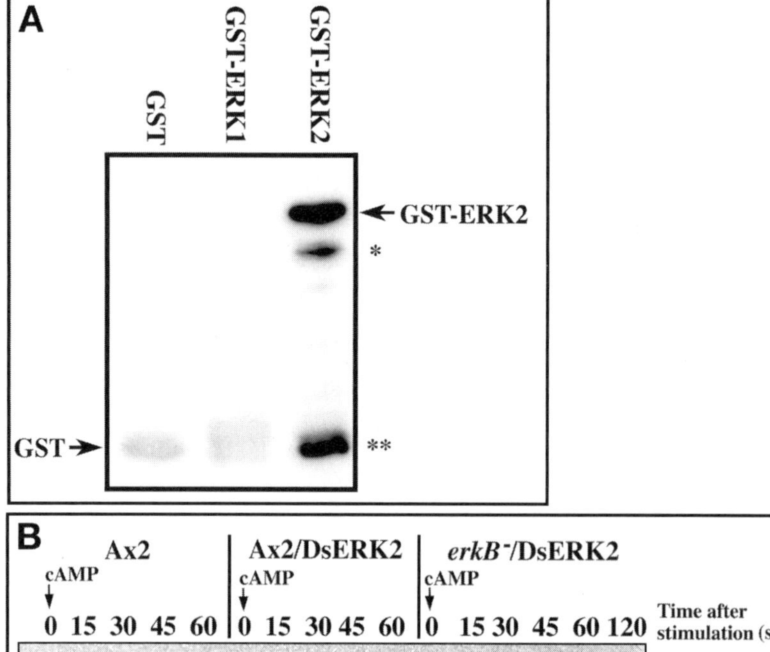

Fig. 1. Specificity of anti-dERK2 antibody to recombinant proteins. (A) In order to test the specificity of the anti-dERK2 antibody, glutathione-*S*-transferase (GST), GST-DictyERK1, and GST-DictyERK2 (full length) are analyzed by immunoblotting. The results clearly show that the anti-dERK2 antibody specifically binds to full-length DictyERK2 and its cleaved moieties (indicated by * and **). (B) Anti-phospho-p44/p42 MAP-kinase antibody specifically binds to ERK2 activated with cAMP. Activation is detected after cAMP stimulation in wild-type ERK2 and DsRed-tagged ERK2 (DsERK2). In the *erkB⁻* mutant, which expresses DsRed-tagged ERK2, only the activated form of DsERK2 is detected. This result indicates that the anti-phospho-p44/p42 MAP-kinase antibody used in this study specifically detects the activated form of *Dictyostelium* ERK2. *DsERK2 indicates activated DsRed-tagged ERK2. DsRed is a fluorescent protein which is fused to *Dictyostelium* ERK2 and expressed in *Dictyostelium* under the control of the *actin15* promoter.

4. An aliquot (30 µL) is transferred into a tube containing 30 µL of 2X sample buffer every 15 s immediately after cAMP stimulation for 2 min. Samples are then boiled for 3 min. These cell lysates are stored at –20°C for analysis.
5. Lysates (10-µL aliquots each) are analyzed by immunoblotting using anti-phospho-p44/p42 antibody and anti-dERK2 antibody.

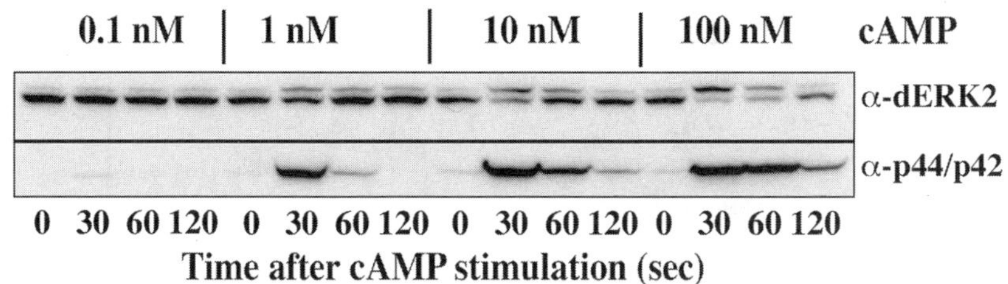

Fig. 2. Kinetics of cAMP-induced activation assessed by anti-dERK2 and anti-phospho-p44/p42 MAP-kinase antibodies. Activation of ERK2 with various concentrations of cAMP is analyzed. The same samples are analyzed by both anti-dERK2 and anti-phospho-p44/p42 MAP-kinase antibodies. Anti-dERK2 antibody shows clear increase in the amount of ERK2 with, slower mobility after stimulation with 1 n*M* cAMP. ERK2 with slower mobility is only detected with anti-phospho-p44/p42 MAP-kinase antibody, which provides clear strong signals after stimulation with 1n*M* or higher concentrations of cAMP.

The results clearly show that *Dictyostelium* ERK2 is activated by cAMP in a concentration-dependent manner to reach a peak after 30 s of stimulation (*see* **Fig. 2**). When activated, ERK2 is detected with anti-phospho-p44/p42 antibody and single, strong signals are observed. When anti-dERK2 antibody is used, two bands are observed, and the strength of the signal in the band with slower mobility transiently increases after stimulation with cAMP. The position of the band with slower mobility corresponds to that of the signal detected with anti-phospho-p44/p42 antibody. Furthermore, such signals are not observed in the mutant lacking ERK2 (*see* **Fig. 1B**) *(4)*, indicating that these signals result from the activation of ERK2 with cAMP. The kinetics of ERK2 activation observed with these antibodies are basically similar to those observed by in gel assay *(2)*. Also, analysis of several mutants with anti-phospho-p44/p42 antibody gives results consistent with those previously obtained by in-gel assay (*see* **Fig. 3**) *(2)*. In the mutant *g* β⁻, which lacks G protein β-subunit, ERK2 is also activated by cAMP. The level of activation is low, but this result is consistent with previous one assessed by in gel assay *(2)*. Thus, we conclude that G protein is involved in, but only partially responsible for, ERK2 activation with cAMP.

3.6. Periodic Change in the Level of Phosphorylated ERK2

1. *Dictyostelium* cells are harvested from the exponential growth phase ($3–5 \times 10^6$ cells/mL) and washed twice with PB.
2. Washed cells are resuspended at 5×10^6/mL in PB and then stimulated for 4 h with 100 n*M* cAMP every 6 min to induce cAMP receptors, Gα2, and other components of the aggregation response.

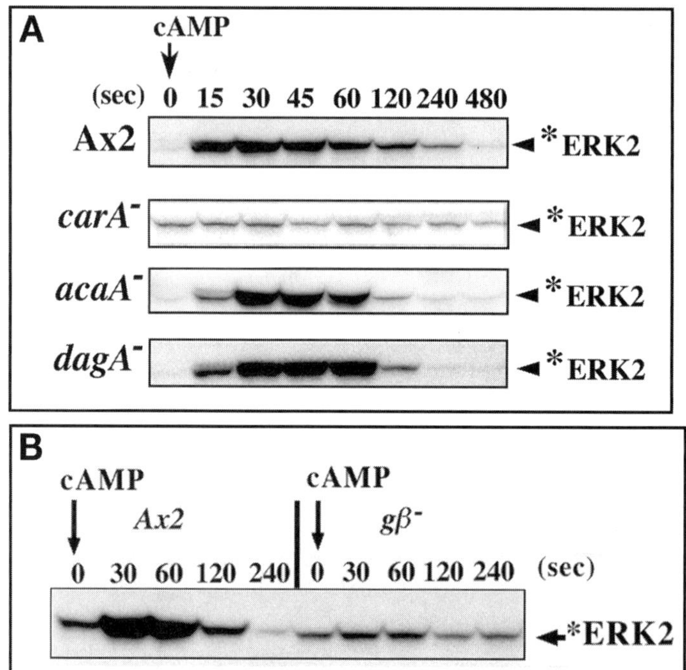

Fig. 3. Kinetics of ERK2 activation with cAMP in various mutants. (**A**) ERK2 is activated in *acaA⁻* lacking adenylyl cyclase A and *dagA⁻* lacking cytosolic regulator of adenylylcyclase (CRAC) as seen in Ax2. On the other hand, ERK2 activation in *carA⁻* cells lacking CAR1 is greatly reduced. These results are consistent with our previous results assessed by in-gel assay *(2)*. (**B**) Partial activation of ERK2 in a *g* β⁻ strain, which lacks the G protein β-subunit. ERK2 is activated in this mutant as previously shown by in-gel assay *(2)*, although its activation level is low. This result indicates G protein is partially involved in ERK2 activation by cAMP. *ERK2 indicates activated ERK2.

3. Thereafter, cells are washed twice, resuspended at 5×10^7/mL in PB, and shaken at 150 rpm. After 10 min of incubation, aliquots (100 µL) of the suspension are withdrawn and mixed with 100 µL of 2X sample buffer at 1-min intervals for 30 min.
4. Cells are fully lysed by boiling for 3 min.
5. The level of phosphorylated ERK2 is analyzed by immunoblotting with anti-phospho-p44/p42 MAP-kinase antibody. A 10-µL aliquot of each sample is used for this analysis. As shown in **Fig. 4**, the level of ERK2 changes at 6-min intervals.

4. Notes

1. Water used for preparation of all solutions is deionized with a MilliQ apparatus (Millipore).

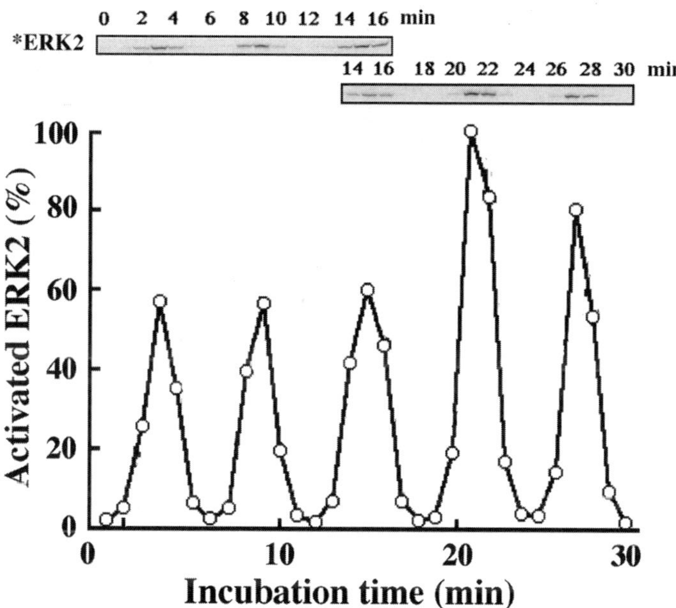

Fig. 4. Periodic activation of ERK2 in aggregation-competent cells. The activated level of ERK2 periodically changes with a phase of 6 min. The level of signals were estimated by National Institutes of Health image software on a MacIntosh G4 computer.

2. A 10–20% gradient polyacrylamide gel provides clearer signals than a 10% polyacrylamide gel, thus it is suitable to detect and identify an ERK2 band, which shifts slightly after stimulation with cAMP.

3. The formation of inclusion bodies by recombinant ERK2 during expression in *E. coli* can be reduced by slower growth at a lower temperature than 37°C. We recommend 20°C.

4. After 10 min, the cell suspension becomes very viscous. If this does not occur, add more lysozyme and incubate for another 5–10 min.

5. After sonication, the suspension becomes less turbid and clearer than before sonication. The color of the suspension changes from pale brown to dark brown.

6. Do not elute GST or GST-tagged recombinant proteins with glutathione solution. Keep the bead pellet in PBS and store at 4°C until use.

7. Treatment with ethanolamine is important in order to block formation of covalent bonds between nonspecific proteins and beads.

8. The preparation of full-length recombinant ERK2 protein is difficult. Actually, hexahistidine-tagged ERK2 was produced as inclusion bodies in bacteria, and it was thus barely solubilized. GST-tagged ERK2 protein in full length was also difficult to prepare in large scale. Several cleavage products of the protein are always observed, but the extent of cleavage seems to depend on the concentra-

tion of protease inhibitors in the lysis buffer. On the other hand, truncated ERK2 protein tagged with GST, e.g., GST-ERK2C, can be produced in large scale. Thus, GST-ERK2 in a piece of gel, which is excised and briefly stained with Coomasie Brilliant Blue solution, is used as the antigen for rabbit immunization.

9. The flow-through fraction still contains anti-dERK2 antibody even after three passages, but at a greatly reduced level.

10. Immobilon P membrane and 10 sheets of filter paper are cut to the same size as the gel used for the electrotransfer. This is important in order to avoid short-circuit current between the two carbon electrodes and thus to obtain good transfer of the proteins in a gel.

11. To effectively induce CAR1, Gα2, and ACA, 100 nM cAMP pulses are used. This concentration provides good induction, but cAMP pulses at 30 nM or less are insufficient.

Acknowledgments

I greatly thank T. Yamamoto, D. Ikeno, H. Kuwayama, and Y. Miyazaki for their technical assistance. I also thank Dr. P. Devreotes who provided various mutants used in this study.

References

1. Segall, J. E., Kuspa, A., Shaulsky, G., et al. (1995) A MAP kinase necessary for receptor-mediated activation of adenylyl cyclase in *Dictyostelium. J. Cell Biol.* **128,** 405–413.

2. Maeda, M., Aubry, L. Insall, R., Gaskins, C., Devreotes, P. N., and Firtel, R. A. (1996) Seven helix chemoattractant receptors transiently stimulate MAP kinase in *Dictyostelium*: role of heterotrimeric G-proteins. *J. Biol. Chem.* **271,** 3351–3354.

3. Knetsch, M. L. W., Epskamp, S. J. P., Schenk. P. W., Wang, Y. W., Segall, J. E., and Snaar-Jagalska, B. E. (1996) Dual role of cAMP and involvement of both G-proteins and ras in regulation of ERK2 in *Dictyostelium discoideum. EMBO J.* **15,** 3361–3368.

4. Maeda, M., Lu, S., Shaulsky, G., et al. (2004) Periodic signaling controlled by an oscillatory circuit that includes protein kinases ERK2 and PKA. *Science* **304,** 875–878.

5. Watts, D. J. and Ashworth, J. M. (1970) Growth of myxamoebae of the cellular slime mould *Dictyostelium discoideum* in axenic culture. *Biochem. J.* **119,** 171–174.

6. Laemmli, U. K. (1970) Cleavage of structural proteins during the assembly of the head of bacteriophage T4. *Nature* **227,** 680–685.

7. Harlow, E. and Lane, D. (1988) *Antibodies: a laboratory manual.* Cold Spring Harbor Laboratory, Cold Spring Harbor, NY.

29

An Improved Method for *Dictyostelium* Centrosome Isolation

Irene Schulz, Yvonne Reinders, Albert Sickmann, and Ralph Gräf

Summary

The *Dictyostelium dicoideum* centrosome consists of a box-shaped, layered core structure surrounded by dense nodules embedded in amorphous material, which make up the so-called corona. Thus, it differs markedly from centriole-containing centrosomes in animal cells or the plaque structure of yeast spindle pole bodies. For a long time, purification of *Dictyostelium* centrosomes was hampered by its extraordinarily tight linkage to the nucleus, which resisted all attempts to dissociate centrosomes and nuclei without destruction of the centrosome itself. Fortunately, we were able to solve this problem, and have already published a centrosome isolation protocol that is based on treatment of nucleus/centrosome complexes with sodium pyrophosphate and shear forces, followed by centrosome isolation through sedimentation and filtration techniques. However, isolated centrosomes prepared according to this protocol still contained too many impurities to allow mass spectrometrical analyses. Here, we present an improved protocol for the isolation of *Dictyostelium* centrosomes that contain considerably less contaminations with cytosolic and nuclear proteins.

Key Words: *Dictyostelium*; centrosome; microtubule-associated center; spindle pole body.

1. Introduction

The centrosome is a proteinaceous, nonmembranous organelle linked to the nucleus. Its main function lies in its role as a microtubule-organizing center in interphase cells and especially during mitosis, where it organizes the mitotic spindle. With a size of 0.5 to 1 µm and a composition of approx 100 different protein components, it represents the largest protein complex in a eukaryotic cell *(1)*. Although the function of the centrosome is quite similar in different organisms, its morphology varies amazingly. In contrast to the mammalian

From: *Methods in Molecular Biology, vol. 346:* Dictyostelium discoideum *Protocols*
Edited by: L. Eichinger and F. Rivero © Humana Press Inc., Totowa, NJ

centrosome, the *Dictyostelium* centrosome does not contain centrioles *(2)*. Instead, it consists of a three-layered, box-shaped core structure surrounded by a corona. The corona represents the functional equivalent of the pericentriolar matrix of animal centrosomes and is characterized by dense nodules embedded in amorphous material. The nodules contain γ-tubulin and, thus, are considered as sites of microtubule nucleation *(3)*. Many efforts to characterize several centrosomal components and to elucidate their function within this intriguing organelle have been undertaken, using methods such as electron microscopy and live-cell imaging *(1)*. Nevertheless, our understanding of the *Dictyostelium* centrosome's exact role in the regulation and initiation of microtubule growth and its replication remains rudimentary.

Thus, it is important to identify most if not all the proteins that build up this complex stucture. One strategy was to use isolated centrosomes to generate a palette of monoclonal antibodies as molecular tools for functional analysis of the corresponding centrosomal antigens *(4)*. Another strategy is to make use of advanced protein analytics, including sophisticated chromatography techniques, two-dimensional gel electrophoresis (*see* **Fig. 1**), and mass spectrometry, to identify the protein inventory of purified centrosomes, as it was successfully performed for yeast spindle pole bodies *(5)* and mammalian centrosomes *(6)*. Such a proteomic approach requires an isolation protocol yielding centrosomes of high purity.

For a long time, it was impossible to isolate *Dictyostelium* centrosomes, because the fibrous linkage structure that couples the centrosome to the nucleus resisted all attempts to separate the centrosome from the nucleus without destroying the centrosome itself *(7)*. However, in 1998, a breakthrough in *Dictyostelium* centrosome isolation was achieved when we found out that treatment of isolated nucleus/centrosome complexes with sodium pyrophosphate and shear forces caused disintegration of the nucleus without affecting centrosome integrity through an unknown mechanism *(8)*. This procedure eventually allowed an approx 1500-fold enrichment of centrosomes through the employment of filtration and differential centrifugation techniques as they were similarly used by others for the isolation of centrosomes from other species *(9–13)*. However, our preparations still contained considerable contaminations with DNA and DNA-associated proteins. Therefore, we have improved our centrosome isolation protocol and added two further steps to the previously published protocols *(8,14)*.

Our first improvement addresses the initially crude isolation procedure for nucleus/centrosome complexes, in which the centrosomes were simply collected as a pellet after cell lysis and low-speed centrifugation. Now, purification of nuclei after cell lysis through a sucrose cushion allows a more efficient separation of the centrosomes, which are still tightly coupled to the nucleus from many cytoplasmic contaminations. The subsequent steps were taken over from the published proto-

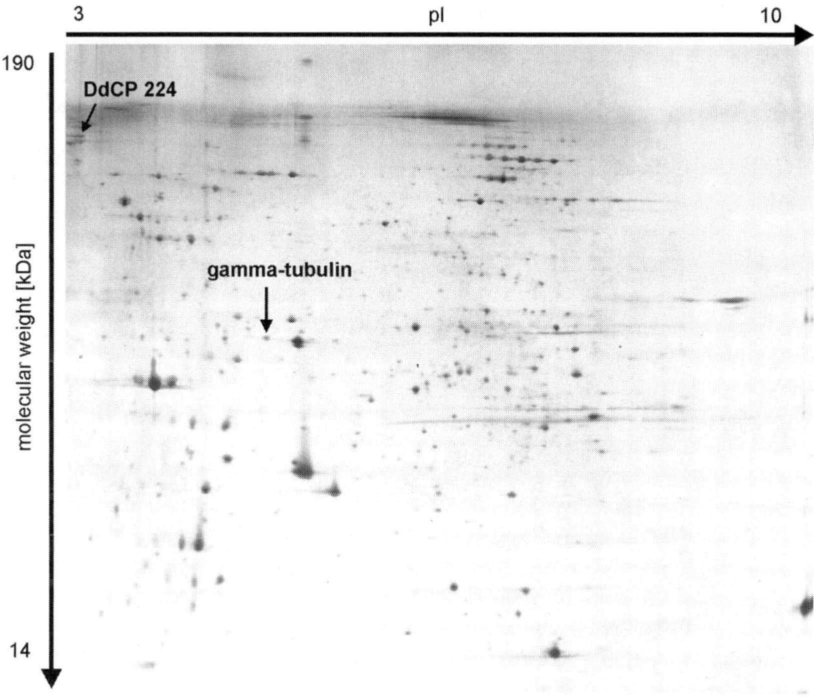

Fig. 1. Two-dimensional gel electrophoresis of denatured isolated centrosomes prepared according to the full protocol described in this work. Electrophoresis was performed as described by Sickmann et al. *(17)*. Silver staining was carried out according to Blum et al. *(18)*.

col. After disintegration of the nuclei by pyrophosphate treatment, centrosomes are purified by two consecutive sucrose step gradient centrifugations. Although already highly enriched, centrosomes are still contaminated with a lot of nuclear material, which can be visualized by 4',6-diamidino-2-phenylindole (DAPI) staining followed by fluorescence microscopy (*see* **Fig. 2**). To overcome this problem, a second major modification of the initial protocol was introduced, which is digestion of the centrosome fraction with protease-free (= RNase-free) DNase. This procedure does not only cleave DNA, but also helps to get rid of DNA-associated proteins such as histones or polymerases. Finally, sedimentation through a sucrose cushion separates the centrosomes from DNase and its cleavage products.

The improved protocol described below can be used with a starting amount of up to 8×10^9 *Dictyostelium* cells. The quality of the purified centrosomes is most easily controlled by immunofluorescence microscopy using centrosome-specific antibodies, and DAPI staining (*see* **Subheading 3.2.1.**).

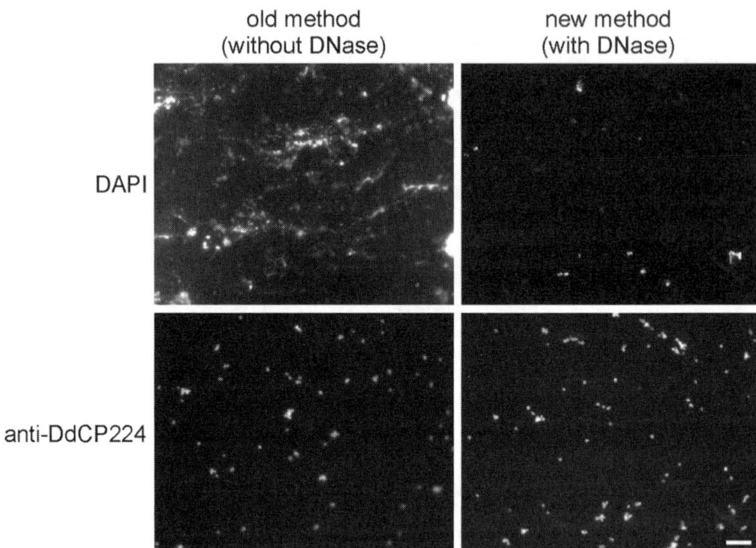

Fig. 2. Immunofluorescence microscopy of isolated centrosomes labeled with the monoclonal antibody 2/165 *(4)* against the centrosomal protein DdCP224 and staining of DNA with DAPI. Scale bar = 10 µm.

2. Materials

2.1. Isolation of Centrosomes

All solutions except AX medium should be rendered particle-free by filtration through a 0.45-µm filter.

2.1.1. Cell Preparation

1. AX medium: add 14.3 g peptone, 7.15 g yeast extract (both from Oxoid, Basingstoke, UK), 18 g glucose, 0.62 g Na_2HPO_4, and 0.49 g KH_2PO_4 to 1 L of deionized water and autoclave for 20 min.
2. Phosphate buffer: 14.6 mM KH_2PO_4, 2 mM Na_2HPO_4, pH 6.0.
3. 5 mM cytochalasin A stock solution (Sigma, Deisenhofen, Germany); dissolve in dimethyl sulfoxide and store at –20°C.

2.1.2. Cell Lysis and Purification of Nuclei

1. Triton X-100 25% (v/v) stock solution; store at 4°C for less than 1 mo.
2. Protease inhibitor cocktail (50-fold concentrated): 50 mM Pefabloc SC, 1.25 mg/mL leupeptin, 0.5 mg/mL tosyl-arginin-methylester, 0.5 mg/mL soybean trypsin inhibitor, 0.05 mg/mL aprotinine, 0.05 mg/mL pepstatine, 100 mM benzamidine, 50 mM Na-ATP (*see* **Note 2**), pH 7.0; store at –70°C.
3. Dithiothreitol (DTT): 1 M stock solution, store at –20°C.

4. Lysis buffer: 100 m*M* Na-PIPES, pH 6.9, 2 m*M* MgCl$_2$, 10% (w/v) sucrose; add prior to use: 0.3% Triton X-100, 1X protease inhibitor cocktail, 2 μ*M* cytochalasin A, 1 m*M* DTT.
5. Sucrose solution: 55% (w/v) sucrose, 100 m*M* Na-PIPES, pH 6.9, 2 m*M* MgCl$_2$; add prior to use: 0.1% Triton X-100, 1 m*M* DTT, 1X protease inhibitor cocktail.
6. 5-μm mesh polycarbonate filter (diameter 47 mm, Nuclepore, Whatman Inc., Clifton, NJ).
7. SW40 rotor and tubes (Beckmann), or equivalent.

2.1.3. Pyrophosphate Treatment

1. Pyrophosphate buffer: 100 m*M* Na-PIPES, 2 m*M* MgCl$_2$, 30% (w/v) sucrose; prior to use add 40 m*M* tetra-sodium diphosphate and adjust pH to 6.9 with HCl, then add 1 m*M* DTT, 1% Triton X-100 and 1X protease inhibitor cocktail (*see* **Note 3**).
2. 12 mg/mL heparin solution in pyrophosphate buffer; prepare fresh every time.

2.1.4. Sucrose Density Gradient Centrifugation

1. Gradient buffer: 10 m*M* Na-PIPES, 2 m*M* MgCl$_2$; add prior to use: 0.1% Triton X-100, 0.2 *M* DTT, 1X protease inhibitor cocktail.
2. Sucrose stock solution: 80% (w/v) sucrose, 10 m*M* Na-PIPES, pH 6.9, 2 m*M* MgCl$_2$; add prior to use 0.1% Triton X-100, 0.2 *M* DTT, 1X protease inhibitor cocktail.
3. SW40 rotor and tubes (Beckmann), or equivalent.
4. SW50.1 rotor and tubes (Beckmann), or equivalent.
5. Syringe with 27-gauge needle.

2.1.5. DNase Treatment and Sedimentation

1. DNase: 2.5 mg/mL stock solution (RNase free!); store at –20°C.
2. SW50.1 rotor and tubes (Beckmann), or equivalent.
3. Syringe with 27-gauge needle.

2.2. Analysis of Isolated Centrosomes

2.2.1. Immunofluorescence Microscopy and DAPI Staining

1. Phosphate-buffered saline (PBS) buffer, 10X: 15 m*M* KH$_2$PO$_4$, 79 m*M* Na$_2$HPO$_4$, 1.38 *M* NaCl, 27 m*M* KCl, pH 7.4.
2. Mounting medium: 120 mg/mL polyvinyl alcohol 4-88, 30% glycerol, 0.2 *M* Tris-HCl, pH 8.8 (*see* **Note 4**).
3. DAPI, 0.1 μg/mL solution in PBS.
4. Primary antibodies: Centrosome-specific antibodies such as anti-DdCP224 monoclonal antibody 2/165 *(4)* or anti-γ-tubulin polyclonal antibodies *(8)*.
5. Secondary antibodies: AlexaFluor or Cy3 conjugates (Molecular Probes).

2.2.2. Measurement of Protein Yield

1. Urea sample buffer: 150 m*M* Tris-HCl, pH 6.8, 6 *M* urea, 10% sodium dodecyl sulfate (SDS), 100 m*M* DTT, 0.1% bromophenol blue (optional), store at –20°C.

2. Bovine serum albumin (BSA) stock solution: 1 mg/mL in urea sample buffer, store at –20°C.
3. Wash solution: 90% (v/v) methanol, 10% (v/v) acetic acid.
4. Amido Black solution: 2.6 mg/mL amido black 10B (Merck) in wash solution, store for less than 1 mo.

3. Methods

3.1. Isolation of Centrosomes

In order to minimize proteolytic degradation, all solutions are used ice-cold and, if possible, all manipulations are performed on ice, except where indicated differently. The individual isolation steps are summarized in **Fig. 3**.

3.1.1. Cell Preparation

1. Grow *Dictyostelium discoideum* cells (axenic strain AX2) at 21°C in 2 L of AX-medium distributed in three 2-L Erlenmeyer flasks to a density of approx 4×10^6 cells/mL on a rotary shaker (150 rpm).
2. Sediment cells at 500g for 5 min at 4°C.
3. Resuspend the cell pellets in approximately 80 mL of chilled phosphate buffer and distribute into three 50-mL tubes. Centrifuge as above.
4. Wash cells two more times with 20 mL of phosphate buffer per tube. Add 2 μM cytochalasin A prior to the last washing step (*see* **Note 2**).

3.1.2. Cell Lysis and Purification of Nuclei

1. Prepare six Beckmann SW40 tubes each filled with 2 mL of 55% sucrose solution supplemented with protease inhibitor cocktail, DTT, and Triton X-100.
2. Resuspend cell pellets in a total volume of 50 mL of lysis buffer and vortex for 1 min (*see* **Note 5**).
3. Filtrate lysate through a 5-μm mesh polycarbonate filter using a syringe.
4. Immediately load onto the sucrose cushion in the Beckmann SW 40 tubes and centrifuge at 80,000g (32,000 rpm for a Beckmann SW 40 rotor) for 30 min at 4°C.
5. Remove supernatant with a pipette without disturbing the pellet.

3.1.3. Pyrophosphate Treatment

1. Resuspend the pellets containing the nucleus/centrosome complexes in a total volume of 40 mL of pyrophosphate buffer (*see* **Note 3**), distribute into two 50-mL tubes and vortex for 1 min.
2. Immediately centrifuge at 2500g for 10 min at 4°C. Collect supernatant.
3. Add 1.2 mL of heparin solution and incubate on ice for 5 min (*see* **Note 6**).

3.1.4. First Sucrose Density Gradient

1. Use the 80% sucrose stock solution and gradient buffer to prepare aliquots of supplemented sucrose gradient solutions. Altogether, you will need 6 mL of 80%,

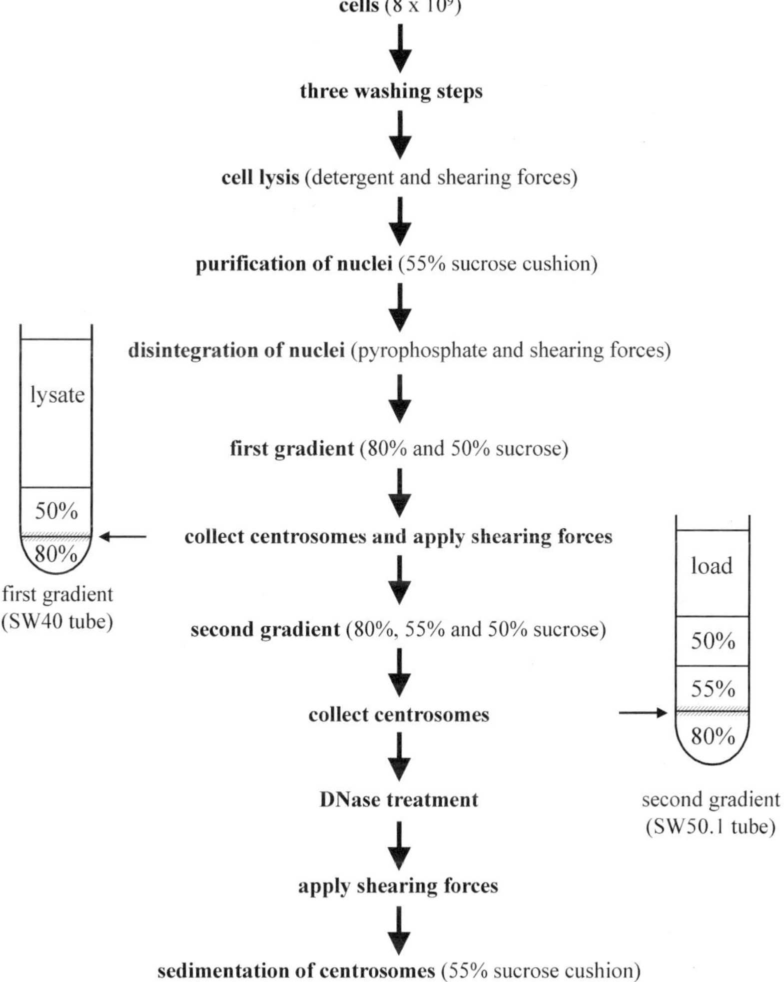

Fig. 3. Overview of *Dictyostelium* centrosome isolation steps.

12 mL of 55%, 8 mL of 50% sucrose solution, and 7 mL of gradient buffer. Supplement with protease inhibitor cocktail, DTT, and Triton X-100.

2. Prepare four Beckmann SW40 tubes with a step gradient with each 0.5 mL of 80% and 1.0 mL of 50% sucrose solution.

3. Filtrate supernatant described under **Subheading 3.1.2.** through a 5-μm mesh polycarbonate filter and carefully load onto the sucrose density gradient. Centrifuge at 55,000*g* (21,000 rpm for a Beckmann SW 40 rotor) for 1 h at 4°C.

4. After centrifugation, two weak bands should be visible, one at the border between the 50% and 80% sucrose fractions which contain the centrosomes and a second

one in the upper part of the 50% sucrose fraction (*see* **Fig. 3**). Collect 1 mL from the bottom, using a glass capillary and a peristaltic pump at a flow rate of 0.4 mL/min.

3.1.5. Second Sucrose Density Gradient

1. Prepare four Beckmann SW50.1 tubes with a step gradient each with 1 mL of 80%, 1 mL of 55%, and 1 mL of 50% sucrose solution.
2. Dilute the four collected 1-mL fractions (*see* **Subheading 3.1.4.**) with 3 mL of supplemented gradient buffer and thoroughly pass the suspension two times through a syringe with a 27-gauge needle.
3. Load onto the second sucrose density gradient and centrifuge at 55,000g (21,000 rpm for a Beckmann SW 40 rotor) for 1 h at 4°C.
4. Fractionate the gradient from the bottom with the aid of a glass capillary and a peristaltic pump at a flow rate of 0.4 mL/min. Discard the first 0.5 mL and then collect 1 mL.

3.1.6. DNase Treatment and Sedimentation

1. Mix the four 1-mL fractions with 4 mL of gradient buffer and add 12.5 µg/mL DNase. Incubate for 30 min at room temperature (*see* **Note 7**).
2. Apply shearing forces by passing the suspension two times through a syringe with a 27-gauge needle.
3. Load onto sucrose cushions (2 mL, 55% sucrose, supplemented with protease inhibitor cocktail, DTT, and Triton X-100) filled in four SW 50.1 tubes and sediment the centrosomes at 55,000g (21,000 rpm for a Beckmann SW 40 rotor) for 1 h at 4°C.
4. Carefully remove supernatant with a pipet. Resuspend the almost invisible pellet with gradient buffer (supplementation is not necessary) and pool suspension from all four gradients. Check the centrosomes by immunofluorescence microscopy (*see* **Subheading 3.2.1.**) if desired; this suspension can be used. For biochemical analysis, it will be necessary to centrifuge the centrosomes again. For this purpose, divide the suspension into a useful number of aliqots (*see* **Note 8**) and sediment the centrosomes at 17,500g for 20 min at 4°C.

3.2. Analysis of Isolated Centrosomes

3.2.1. Immunofluorescence Microscopy and DAPI Staining (see **Note 9**)

1. Insert a round, untreated coverslip (diameter 12 mm) into a suitable flat-bottom centrifuge tube (e.g., self-designed plastic plates that fit into a swing-out rotor, with 13 mm wells). Alternatively, Corex tubes equipped with self-designed plexiglas inset with one flat and one round end as described by Blomberg-Wirschell and Doxsey *(15)* may be used. In this case, the void volume between the inset and the glass surface should be as small as possible. Dilute one fiftieth of one gradient in 1 mL of phosphate buffer and centrifuge at 2500g for 20 min at 4°C. This step also removes the detergent, which interferes with adherence of centrosomes to the glass surface.

2. Remove supernatant and either fix with methanol for 5 min at –20°C or with 3.7% formaldehyde in 1X PBS-buffer at room temperature.
3. Remove supernatant and wash with 1X PBS buffer for 5 min.
4. Incubate with first (centrosome-specific) antibody (diluted in 1X PBS with 0.1% BSA) for 45 min.
5. Wash three times for 5 min with 1X PBS.
6. Incubate with secondary antibody for 45 min.
7. Wash with 1X PBS as in **step 5**, but include 0.1 µg/mL DAPI in the second washing step.
8. Mount coverslips on slides using a small droplet of mounting medium and remove excess liquid.

3.2.2. Measurement of Protein Yield (see **Notes 9 and 10**)

1. Dissolve one half gradient of the centrosome preparation in 100 µL of urea sample buffer mixture and heat for 5 min at 60°C. Split into two samples with equal volume to get duplicate values.
2. Use 4, 8, and 12 µg of BSA in sample buffer as a standard and fill with sample buffer to a total volume of 50 µL. Pipet samples in duplicate.
3. Add 0.6 mL of amido black solution to every sample and incubate for 5 min at room temperature. Centrifuge in a microfuge for 4 min at top speed.
4. Carefully aspirate supernatant and wash two times by vortexing with wash solution. Centrifuge as above after each washing step.
5. Dissolve pellet with 0.7 mL of 0.1 M NaOH and read absorbance at 615 nm in a spectrophotometer.

4. Notes

1. It is not necessary or even useful to carry out this long preparation protocol for all purposes. In some cases, the only reason for using isolated centrosomes is the total removal of microtubules in order to analyze whether a certain protein is a genuine centrosomal protein and, thus, requires no microtubules for its centrosomal localization. Because of the inefficiency of microtubule-depolymerizing drugs in *Dictyostelium*, this cannot be achieved by treatment with, for example, nocodazole. Instead, isolated, microtubule-free centrosomes are analyzed by fluorescence microscopy employing antibodies against the protein of interest or green fluorescent protein (GFP) fluorescence, in cases where the protein of interest is expressed as a GFP fusion protein. For this kind of experiment, it is sufficient to stop the preparation procedure after the first sucrose gradient (*see* **Subheading 3.1.4.**, **step 4**), although performance of the second gradient might even improve quality of the fluorescence microscopy images. For such purposes, it is also not necessary to use a sucrose cushion to purify the nuclei (*see* **Subheading 3.1.2.**). It is sufficient to sediment them by centrifugation in 50-mL tubes for 10 min at 3000g and 4°C. If strains expressing a GFP fusion protein are used, all buffers should be prepared without DTT, because reducing reagens lead to strongly decreased GFP fluorescence.

2. ATP and cytochalasin A are used to reduce copurification of actomyosin complexes, which readily form upon cell lysis with Triton X-100. Whereas ATP prevents rigor binding of myosin to actin, cytochalasin A depolymerizes actin fibers. Because the latter requires some time to fulfill its purpose, it is already added at the last washing step.
3. The pyrophosphate needs about 15 min to dissolve completely; therefore, remember to prepare the buffer in time.
4. The polyvinyl alcohol 4-88 is first dissolved in water by stirring and heating to 70°C for 30 min. Afterward, the glycerol and the Tris-HCl (pH 8.8), using a 1 *M* stock solution, is added. To remove particles, the mixture is centrifuged for 10 min at 3000*g*. Aliquots are stored at –20°C.
5. The lysis of the cells is a critical step in this preparation. For this, you should use 50-mL tubes filled with a maximum of 25 mL. Make sure that the liquid is swirled up and down along the whole tube while vortexing.
6. Heparin is a polyanion, which is used to bind histones in order to facilitate solubilization of chromatin and, thus, to reduce DNA contaminations.
7. After the previous purification steps, the centrosomes are now stable enough to be incubated at room temperature. However, it is very important to use the highest purity of DNase, which is "RNase-free," because this is likely also to be free of proteases.
8. For a standard one-dimensional gel for staining or Western blot analysis, it is sufficient to load one-sixteenth of a preparation started with 8×10^9 cells and using four tubes for each gradient centrifugation (approx 10 µg). For two-dimensional gels, centrosomes collected from at least two tubes are required (*see* **Fig. 1**).
9. Because the centrosomes isolated according to this procedure adhere less well to coverslips than the less pure centrosomes prepared using the previous protocol, it is not possible to determine the total number of the isolated centrosomes from an immunofluorescence microscopy experiment. The yield of protein can be measured with an amido black assay (*see* **Subheading 3.2.2.**) and should range between 120 and 150 µg per preparation starting with 8×10^9 cells.
10. Accurate measurement of the protein content requires that the isolated centrosomes be dissociated and that all centrosomal proteins be completely dissolved. This is most easily achieved by treatment of isolated centrosomes with urea sample buffer, which contains a high amount of urea and SDS. Unlike most other protein assays, the Amido Black assay *(16)* is not sensitive to high contents of urea or SDS.

References

1. Gräf, R., Daunderer, C., and Schulz, I. (2004) Molecular and functional analysis of the *dictyostelium* centrosome. *Int. Rev. Cytol.* **241,** 155–202.
2. Moens, P. B. (1976) Spindle and kinetochore morphology of *Dictyostelium discoideum. J. Cell Biol.* **68,** 113–122.
3. Euteneuer, U., Gräf, R., Kube-Granderath, E., and Schliwa, M. (1998) *Dictyostelium* gamma-tubulin: molecular characterization and ultrastructural localization. *J. Cell Sci.* **111,** 405–412.

4. Gräf, R., Daunderer, C., and Schliwa, M. (1999) Cell cycle-dependent localization of monoclonal antibodies raised against isolated *Dictyostelium* centrosomes. *Biol. Cell* **91,** 471–477.

5. Wigge, P. A., Jensen, O. N., Holmes, S., Soues, S., Mann, M., and Kilmartin, J. V. (1998) Analysis of the *Saccharomyces* spindle pole by matrix-assisted laser desorption/ionization (MALDI) mass spectrometry. *J. Cell Biol.* **141,** 967–977.

6. Andersen, J. S., Wilkinson, C. J., Mayor, T., Mortensen, P., Nigg, E. A., and Mann, M. (2003) Proteomic characterization of the human centrosome by protein correlation profiling. *Nature* **426,** 570–574.

7. Omura, F. and Fukui, Y. (1985) *Dictyostelium* MTOC: Structure and linkage to the nucleus. *Protoplasma* **127,** 212–221.

8. Gräf, R., Euteneuer, U., Ueda, M., and Schliwa, M. (1998) Isolation of nucleation-competent centrosomes from *Dictyostelium* discoideum. *Eur. J. Cell Biol.* **76,** 167–175.

9. Bornens, M., Paintrand, M., Berges, J., Marty, M. C., and Karsenti, E. (1987) Structural and chemical characterization of isolated centrosomes. *Cell Motil. Cytoskeleton* **8,** 238–249.

10. Rout, M. P. and Kilmartin, J. V. (1990) Components of the yeast spindle and spindle pole body. *J. Cell Biol.* **111,** 1913–1927.

11. Mitchison, T. J. and Kirschner, M. W. (1986) Isolation of mammalian centrosomes. *Methods Enzymol.* **134,** 261–268.

12. Moritz, M., Braunfeld, M. B., Fung, J. C., Sedat, J. W., Alberts, B. M., and Agard, D. A. (1995) Three-dimensional structural characterization of centrosomes from early *Drosophila* embryos. *J. Cell Biol.* **130,** 1149–1159.

13. Vogel, J. M., Stearns, T., Rieder, C. L., and Palazzo, R. E. (1997) Centrosomes isolated from *Spisula solidissima* oocytes contain rings and an unusual stoichiometric ratio of alpha/beta tubulin. *J. Cell Biol.* **137,** 193–202.

14. Gräf, R. (2001) Isolation of centrosomes from *Dictyostelium*. *Methods Cell Biol.* **67,** 337–357.

15. Blomberg-Wirschell, M. and Doxsey, S. J. (1998) Rapid isolation of centrosomes. *Methods Enzymol.* **298,** 228–238.

16. Popov, N., Schmitt, M., Schulzeck, S., and Matthies, H. (1975) Reliable micromethod for determination of the protein content in tissue homogenates. *Acta Biol. Med. Ger.* **34,** 1441–1446.

17. Sickmann, A., Dormeyer, W., Wortelkamp, S., Woitalla, D., Kuhn, W., and Meyer, H. E. (2002) Towards a high resolution separation of human cerebrospinal fluid. *J. Chromatogr. B. Analyt. Technol. Biomed. Life Sci.* **771,** 167–196.

18. Blum, H., Beier, H., and Gross, H. J. (1987) Improved silver staining of plant proteins, RNA and DNA in polyacrylamide gels. *Electrophoresis* **8,** 93–99.

30

Epigenetics in *Dictyostelium*

Markus Kaller, Wolfgang Nellen, and Jonathan R. Chubb

Summary

Dictyostelium has a good potential to serve as a model for the study of chromatin function and epigenetic gene regulation. The organism bridges the complexity of higher eukaryotic systems and the simplicity of yeast in that it harbors pathways that are similar to the former and is accessible to genetic manipulation like the latter. The findings that, in contrast to previous assumptions, *Dictyostelium* DNA contains methylated cytosine and that the RNA interference machinery may be involved in chromatin remodeling, open up new avenues to investigate epigenetic aspects in one of the most simple developing organisms. The protocols in this chapter describe the more recent techniques established for other organisms, with an emphasis on special precautions for application in *Dictyostelium*.

Key Words: Chromatin; histone modification; DNA methylation; epigenetics; ChIP.

1. Introduction

Chromatin remodeling has become an important area of research and is closely linked to epigenetic gene regulation. In recent years, non-Mendelian inheritance and information not encoded in the DNA has attracted substantial attention and provided new insights into developmental processes that, when impaired, may cause diseases in humans that were difficult to trace by standard genetics *(1)*. Efforts are now focused on the understanding of the mechanisms through which large sections of the genome may be reversibly inactivated by modification of DNA and/or chromatin. Model systems range from highly complex mammals to yeasts, but there is a gap in the complexity scale that could be conveniently filled by *Dictyostelium*. Unlike the yeasts, *Dictyostelium* cells differentiate and show distinct cell specialization, allowing analysis of the role of chromatin modifiers in developmental gene regulation. In fact, the developmental cycle is accompanied by differential chromatin modification *(12)*. Simi-

From: *Methods in Molecular Biology, vol. 346:* Dictyostelium discoideum *Protocols*
Edited by: L. Eichinger and F. Rivero © Humana Press Inc., Totowa, NJ

lar to higher eukaryotes, retrotransposons appear to be silenced by different epigenetic mechanisms that may be linked to the developmental cycle. This includes DNA methylation (Kuhlmann et al., manuscript in preparation), previously thought to absent in the organism *(2)*. As shown in **Table 1**, most of the components of the chromatin remodeling systems, such as histone modification enzymes, chromatin proteins, DNA methyltransferase and others, are found in the genome, some of them in various isoforms that suggest a large combinatorial network of regulation. Furthermore, a complete RNA interference machinery *(3)* (*see also* Chapter 13) is present. As in other organisms, this may contribute to RNA-mediated DNA methylation *(4)* and/or RNA-mediated generation of heterochromatin *(5)*. Although *Dictyostelium* chromosomes are small, high-resolution fluorescence microscopy allows for a refined description of nuclear organization and chromatin features that may be combined with fluorescence *in situ* hybridization (FISH).

2. Materials

2.1. Genes and Tools

In this section, we list the key players in chromatin remodeling that have been identified in *Dictyostelium* genome databases. Some of these have only been annotated as a result of amino acid sequence similarity with respective proteins in other eukaryotes. For others, further investigations have been carried out and various additional tools are available.

Homologs of all of the core nucleosomal histones can be identified, and variant histones (e.g., H3.3 and H2AZ) and a linker histone are also present. It seems that the three major histone H3 genes are all H3.3-like *(6)*, based on the amino acid substitutions in their histone fold and N-terminal domains. Histone deacetylases and acetyl transferases, ATP-dependent chromatin-remodeling enzymes, chromodomain proteins (including three HP1 isoforms), bromodomain proteins, histone methyltransferases, including a su(var)3-9 homolog, a DNA methyltransferase, and a putative histone demethylase have been found. In addition, a complete set of the known proteins required for RNA interference has been identified (*see also* Chapter 13). The genes that have been experimentally confirmed or that display convincing similarity to homologues in other organisms are listed in **Table 1**. It should be noted that several components identified in *Dictyostelium* and higher eukaryotes are not found in the yeast *Saccharomyces cerevisiae* and some are not found in *Schizosaccharomyces pombe*, which have more simple chromatin systems. The *Dictyostelium* genome was screened with standard search tools and the list is not likely to be complete, particularly because chromatin domains are often poorly conserved at the sequence level. It is, however, obvious that several of these proteins

Table 1
Chromatin Proteins in *Dictyostelium*

Chromatin protein	Class	Dicty gene	Examples	Homologs	*Dictyostelium* reagents
Histone lysine methyltransferase (HMTase)	H3-K4	*setA*	DDB0188336	Set1p, hSet1(Hs)	KO, meH3-K4 antibodies, GFP-tagged, catalytic deads
	H3-K9	*suvA*	DDB0190352	Su(var)3-9 (Dm)	me2H3-K9 antibodies (Upstate and Abcam)
	H3-K36		DDB0189799	Set2p?	KO, meH3-K36 antibodies
	H3-K79		DDB0216874	Dot1p	
Histone arginine methyltransferase			DDB0217760	CARM1	
			DDB0183976	PMRT1	
Histone demethylase	H3-K4me		DDB0188339	LSD1	
Histone Acetyl-transferase (HAT)	GCN5	*GCN5*	DDB0220684	GCN5	
	MYST		DDB0167422	Sas2p	
	TAF$_{II}$250		DDB0220687	TAF$_{II}$250	
	HAT1		DDB0167416	Hat1p	
Histone demethylase) (HDAC	Rpd3		DDB0189724	Rpd3p	
	HdaI		DDB0206434	Hda1p	
	Sir2		DDB0219946	Sir2p	
chromodomain		*hcpA*	DDB0220646	HP1 (Hs)	KO, GFP-tagged,
		hcpB	DDB0220645		
		hcpC	DDB0220647		myc-tagged
			DDB0220640	Chd1p	
bromodomain			DDB0220685	15+ more	
Histone H3		*HstH3*	DDB0191157		
			DDB0216291		
Histone H4			DDB0216310	H4	
Histone H1		*hstA*	DDB0191459	H1	
Histone H2A			DDB0216271	H2A	
			DDB0216302	H2AZ	
Histone H2B			DDB0216310	H2B domain	
DNA methyl transferase	Dnmt2	*dnmA*	DDB0231095	Dnmt2	KO, GFP-tagged, myc-tagged
ATP-dependent chromatin-remodeling enzymes			DDB0220518	SNF2	
			DDB0231763	ISWI	

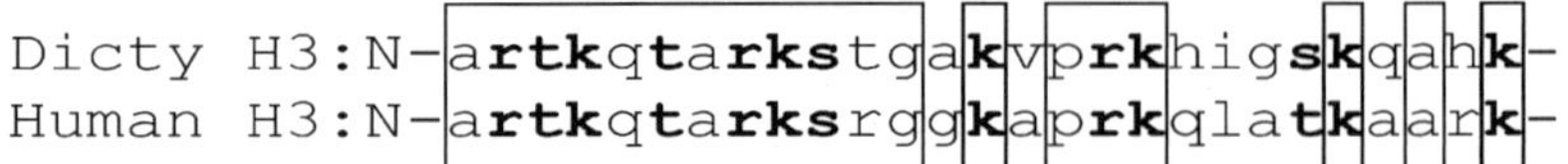

Fig. 1. Lysine, arginine and serine/threonine residues are largely conserved in the *Dictyostelium* histone H3 tail.

occur in large families. For example, there are nine members in the family of putative histone deacetylases, but it should be noted that the substrates of these proteins may not be exclusively histones. Other domains that are typical of chromatin proteins have been identified and include, for example, PHD zinc fingers, jumonji domains, and Tudor domains.

The majority of the amino acid residues that are known to be posttranslationally modified in histones (lysine, arginine, and serine/threonine) are present in the *Dictyostelium* histone homologs (*see* **Fig. 1**). Several posttranslational modifications at these residues can be detected in *Dictyostelium* cells using commercially available antisera. Strong nuclear-wide staining is observed with antibodies (Abcam, Upstate, and others) against methylated lysine 4 of histone H3 (H3-K4). Staining is lost in a Set1 knockout mutant, confirming the identity of the gene product as a H3-K4 methyltransferase. Antisera against di-methyl H3-K9 reveal a distinct nuclear distribution, and different antibodies against acetylated lysines give rise to speckled nuclear staining patterns.

Table 1 also contains the currently available tools for chromatin evaluation. These include antibodies that can be obtained commercially or through the *Dictyostelium* community, knockout strains, tagged overexpression strains, recombinant proteins, and so on.

2.2. Chromatin Immunoprecipitation

1. RLB: 0.32 M sucrose, 10 mM Tris-HCl pH 7.5, 5 mM MgCl$_2$, 1% Triton X-100.
2. GPA: 10 mM Tris-HCl pH 7.5, 10 mM ethylenediamine tetraacetic acid (EDTA).
3. GPB: 10 mM Tris-HCl pH 7.5, 0.7 % sodium dodecyl sulfate (SDS).
4. TBS: 10 mM Tris pH 7.5, 150 mM NaCl.
5. IP buffer: 0.5% Triton X-100 in TBS, 1 mM EDTA.
6. W buffer: 10 mM Tris-HCl pH 8.0, 500 mM LiCl, 1% deoxycholate, 1% Nonidet P40 (NP40), 1 mM EDTA.
7. E buffer: 50 mM Tris-HCl pH 7.5, 1% SDS, 1 mM EDTA.
8. Proteinase K: 20 mg/mL stock (Invitrogen 25530-049).
9. Protein A-sepharose: 4 Fast Flow (Amersham 17-5280-01).
10. Qiagen polymerase chain reaction (PCR) columns (kit 28104).
11. Phosphate buffer: 20 mM potassium phosphate pH 6.2.

12. Protease inhibitors: Roche EDTA-free cocktail (Roche 11061400).
13. Sonicator: Fisher Sonic Dismembrator-need to optimize for other models (*see* **Note 1**).

2.3. DNA Methylation Analysis

2.3.1. DNA Preparation

1. Lysis buffer: 50 mM HEPES pH 7.5, 40 mM MgCl$_2$, 20 mM KCl, 5% sucrose, 1% NP40.
2. SDS lysis buffer: 0.7% SDS in TE buffer.
3. RNaseA solution: 1 mg/mL RNaseA in water; boil for 5 min to inactivate DNase.

2.3.2. Bisulfite Treatment

1. NaOH: prepare 3 M and 10 M solutions.
2. Hydroquinone solution: 10 mM hydroquinone (55 mg in 50 mL of water). Prepare fresh.
3. Bisulfite solution: in a 50 mL tube, dissolve 4.1 g of metabisulfite (Na$_2$S$_2$O$_5$) in 8 mL of H$_2$O (final concentration: 3.1 M). Water should be degassed under vacuum. Add 250 µL of hydroquinone solution and 300 µL of 10 M NaOH. The final pH should be 5.0.

2.4. Pull-Down of Ectopically Overexpressed Proteins

It is important to avoid the use of EDTA or other chelating agents in the lysis buffer, because they can result in loss of the His-tagged protein from the beads.

1. Lysis buffer: 20 mM Tris-HCl pH 7.5, 10 mM imidazole, 300 mM NaCl, 0.5% NP40, 0.5 mM phenylmethylsulfonyl fluoride (PMSF). Keep PMSF as a stock solution in 100% ethanol or isopropanol and add directly before use, as PMSF half-life in aqueous solutions is approx 30 min.
2. Washing buffer: 20 mM Tris-HCl pH 7.5, 50 mM imidazole, 300 mM NaCl.
3. Elution buffer: 20 mM Tris-HCl pH 7.5, 250 mM imidazole, 30 mM NaCl.
4. Sonicator: Dr. Hielscher Sonicator UP 200S with microtip S3 (must be optimized for other models).
5. Ni Sepharose™ 6 Fast Flow (Amersham Biosciences).
6. Poly-Prep® Chromatography Columns (Bio Rad).

2.5. Electrophoretic Mobility Shift Assay

1. For DNA end labeling: T4 Polynucleotide Kinase (10U/µL) and 10X forward reaction buffer (MBI Fermentas GmbH).
3. Storage buffer: 10 mM Tris-HCl pH 8.0, 300 mM NaCl, 0.1 mM EDTA.
4. 1X assay buffer: 10 mM Tris-HCl pH 8.0, 150 mM NaCl, 5 mM MgCl$_2$.
5. 10X TBE buffer: 1 M Tris, 0.9 M boric acid, 20 mM EDTA. Dilute 1:10 with water.

2.6. Immunoprecipitation

1. Cell lysis buffer: 20 m*M* HEPES pH 7.5, 40 m*M* MgCl$_2$, 20 m*M* KCl, 5% glucose (w/v).
2. RIPA buffer: 50 m*M* Tris-HCl, pH 7.4, 1% NP40, 0,25% sodium deoxycholate, 150 m*M* NaCl, 1 m*M* EDTA, 1 m*M* PMSF.
3. 1X phosphate-buffered saline (PBS): 10 m*M* potassium phosphate pH 7.4 (8.02 mL of 1 *M* K$_2$HPO$_4$ and 1.98 mL of KH$_2$PO$_4$ in 1 L water), 150 m*M* NaCl.
4. Sonicator: Dr. Hielscher Sonicator UP 200S with microtip S3 (must be optimized for other models).

3. Methods

3.1. Chromatin Immunoprecipitation

Chromatin Immunoprecipitation (ChIP) is designed to investigate the association of specific chromatin structures, i.e., specific chromatin proteins, with a particular DNA sequence. Formaldehyde-cross-linked chromatin is sheared by sonication, then immunoprecipitated with an antibody against a chromatin protein or a protein modification. The DNA is eluted from the precipitated chromatin and the abundance of a particular sequence is assessed by semi-quantitative or, preferably, real-time PCR. PCR fragments of 150–200 bp are best, although larger fragments may be necessary if AT-rich sequences complicate the design of primers with adequate annealing temperatures. When repetitive sequences are analyzed, Southern blots may also be used.

It is important to ensure efficient re-suspension in RLB and not to cut corners on the washes, because even during the later washes, one can still detect DNA coming off of the beads. Concentrated DNA solutions are usually recovered, so eluting the DNA can be done in as much as 200 µL of buffer (or water). Real-time PCR reactions are easily perturbed by high DNA concentrations, so dilute DNA is not a problem here.

3.1.1. Cross-Linking of Proteins

1. Resuspend 6×10^7 cells harvested from growing cultures or from development plates or filters in 50 mL of phosphate buffer (cell number may be varied).
2. Add formaldehyde to a final concentration of 1.2% and gently agitate for 15 min.
3. Add glycine to 360 m*M* (1.34 g/50 mL) and agitate for 5 min.
4. Spin down cells by centrifugation at approx 560*g* for 5 min. Cell pellets may be stored frozen at –80°C.

3.1.2. Cell Lysis and Sonication

1. Resuspend cells in 5 mL of RLB cell lysis buffer and centrifuge at 3200*g*.
2. Repeat **step 1** with the pellet and discard supernatant.

3. Resuspend the pellet in 80 μL of RLB, add 0.5 mL of GPA buffer, and then add 0.5 mL of GPB buffer when the pellet is fully resuspended. Leave on ice for 10 min.
4. Sonicate the suspension six times with 10-s pulses at 10 μ*M* in a 2-mL Eppendorf tube. Cool the suspension on ice in between the sonications and avoid to generate froth (*see* **Note 1**).
5. Centrifuge at 20,000*g* for 5 min at 4°C in a microcentrifuge and collect the supernatant. Aliquots of 50 μL are frozen at −80°C or directly used in the experiment. Keep in mind that one aliquot is required to determine the input material!

3.1.3. Precipitation

1. To preclear the chromatin preparation from proteins that bind nonspecifically to the matrix, add 50 μL of proteinA-sepharose beads and incubate on a rotating wheel for 30 min at 4°C.
2. Centrifuge at 2700*g* for 2 min at 4°C, collect the cleared supernatant, and divide between samples (tests, no antibody, control antibody). Fill with IP buffer (with protease inhibitors) to 1 mL (e.g., 250 μL of cleared extract plus 750 μL of IP buffer).
3. Add antibody (2–3 μL) and add 50 μL of protein A-sepharose beads to each sample. Incubate on a rotating wheel overnight at 4°C. One sample is incubated with a control antibody (e.g., histone H3, *see* **Note 2**) and another with beads alone (*see* **Note 3**).
4. Centrifuge for 2 min at 2700*g* and store the supernatant.
5. Wash the pelleted beads with 1 mL of IP buffer, allow to stand for 3 min, and centrifuge. Repeat this washing step once.
6. Wash the beads with 1 mL of W buffer, allow to stand for 3 min, and centrifuge. Repeat this washing step three times.

3.1.4. Elution

1. Resuspend the beads in 150 μL of E buffer, vortex, and incubate at 37°C for 15 min.
2. Vortex the beads again and centrifuge for 3 min at 2700*g*. Collect the supernatant, and re-extract the beads with 100 μL E buffer. Pool the collected supernatants.
3. To the supernatant, add proteinase K to 200 μg/mL. Incubate this mixture at 56°C for 4 h to overnight. Do the same with the input fraction (which is topped off with E buffer to the same volume as the samples).
4. Purify the DNA on a Qiagen PCR column.

3.2. Considerations on the Use of DNA Microarrays

A standard approach for identifying genes regulated by a particular chromatin protein or modification is to hybridize RNA from chromatin protein mutant cells to DNA microarrays. This method can also reveal the genomic organization of mis-expressed genes, and can detect clusters of co-regulated genes. Although targets can be identified this way, these are typically diluted out by indirect effects of the mutation. For example, a gene regulated by a specific histone modification is a regulator of another gene that is not directly influenced by the histone modification. To remove indirect effects, a good approach is the ChIP on CHIP method. Here, chromatin immunoprecipitated by an antibody against the histone modification is also hybridized to the array in order to detect genomic sites of the modification. These can then be compared to the sites of mis-expression in the mutant, and indirect effects are removed. However, some genes that might experience modification may not be perturbed even in the absence of the modification, perhaps as a result of redundancy in the histone code. In this case, the loci should be studied using ChIP to assess the other modifications. The role of these modifications, in combination with the modification of interest, can then be assessed. Obviously, these types of studies are best done in a model organism in which genetics can be combined with the ability to obtain large quantities of homogeneous cell populations. Obviously, *Dictyostelium* fits these criteria better than any metazoan model.

3.3. C-Methylation in DNA

Bisulfite sequencing is used to specifically analyze the cytosine methylation of a defined stretch of DNA. Chemically, the method is based on the conversion of cytosine to cytosine sulfonate by metabisulfate. After deamination, the conversion to uracil sulfonate takes place. This is then desulfonated with NaOH to uracil. Because 5-methyl-cytosine is unaffected by the chemical modification, methylated C-residues are not converted to U. Bisulfite-treated DNA is amplified by PCR using primers directed against the modified sequence, assuming that within the primer target sequence, no methylation occurred. The amplified DNA fragment can be cloned and individual clones may be sequenced. Alternatively, the PCR reaction may be directly sequenced and the methylation status of every single cytosine can be determined.

3.3.1. Preparation of Genomic DNA

1. 1×10^8 axenically grown *Dictyostelium* cells are harvested and washed in ice-cold phoshate-buffer at 540g for 10 min at 4°C.
2. The cell pellet is resuspended in 50 mL of lysis buffer (*see* **Subheading 2.3.1.**) and the cells lysed by vigorous shaking until the suspension is clearing (cells lysed, nuclei still intact).
3. The nuclei are collected at 4000g and lysed in 5 mL of SDS-lysis buffer with 100 µL of proteinase K solution (25 mg/mL of water) for 45 min at 65°C.
4. The lysate is extracted twice with 1 volume phenol/chloroform (1:1mix) by vortexing at medium speed. Phases are separated by centrifugation for 20 min at 10,000g.
5. Genomic DNA is precipitated with one-tenth volume of sodium acetate and 2 volumes of ethanol at room temperature.

6. DNA is collected by centrifugation at 10,000*g* for 30 min at 4°C and the supernatant is carefully discarded.
7. DNA is washed with 70% ethanol, air-dried, and dissolved in an appropriate volume of ultrapure water.
8. DNA is treated with RNaseA solution (1 µg/mL final concentration) for 30 min at 37°C. RNase treatment is important, because RNA may interfere with the bisulfite reaction!

It may be necessary to enrich for the DNA fragment of interest in order to reduce the PCR background. For this, at least 50–100 µg of genomic DNA is digested with restriction enzymes that generate a fragment of defined size. The digested DNA is separated on an agarose gel with a wide pocket and the appropriate size range is cut out. The enriched fragment is eluted using standard gel elution kits.

3.3.2. Bisulfite Treatment

1. 5 µg of genomic, RNase-treated DNA in a volume of 50 µL are denatured by addition of 6 µL of 3 *M* NaOH for 20 min at room temperature. It may be necessary to complete denaturation by an additional incubation for 3 min at 90°C (*see* **Note 4**).
2. 600 µL of freshly prepared bisulfite solution are added to the sample and the surface is sealed with PCR wax or a layer of mineral oil to prevent oxidation.
3. For the conversion reaction, the sample is incubated for 3 h at 55°C, then briefly heated to 95°C and further incubated for 12–16 h at 55°C.
4. DNA is extracted with Geneclean or glasmilk, precipitated, and dissolved in 100 µL of distilled water.
5. For desulfonation, 11 µL of 3 *M* NaOH are added and the sample is incubated at 37°C for 20 min.
6. DNA is again precipitated with 35 µL of 4 *M* ammonium acetate and 3 volumes of 100% ethanol, washed with 70% ethanol and redissolved in 20 µL of H_2O.
7. 2–4 µL are used for the PCR reaction.

3.3.3. Design of Primers

Especially for asymmetric methylation sites as they are usually found in *Dictyostelium*, both strands should be analyzed separately, because the methylation pattern on one strand does not correspond to that on the other strand. The analyzed strand (top or bottom) is determined by the choice of primers. Primers are designed assuming complete conversion. For the forward primer, instead of a G that should hybridize to a C on the analyzed DNA strand, an adenosine residue is incorporated into the primer. At the 3' end, the primer should contain an adenosine that corresponds to a C residue that has been converted to an uracil. This ensures that only converted DNA molecules are amplified. The forward primer should contain as many G residues as possible to achieve a sufficiently high annealing temperature. Because the methylation status of the DNA is not known *a priori*, primers may not fit if a C residue complementary to the forward primer is methylated. Therefore, the number of cytosines in the forward priming site should be as small as possible, but they should end with an adenine (corresponding to a converted cytosine).

For the reverse primer, every cytosine is exchanged for a thymidine and the 3' end should be a thymidine corresponding to an adenosine that in turn corresponds to a converted cytosine. Annealing temperatures of the forward and reverse primer should be similar. Forward and reverse primers should span a DNA fragment of 250 to 400 bp; larger fragments are difficult to analyze. PCR reactions are carried out as usual (*see* **Note 5**)

3.4. Considerations on the Use of Epitope-Tagged and Overexpressed Proteins

It is a standard rule that epitope-tagging of proteins must be employed with caution, as tags can interfere with folding, enzymatic activity, and protein–protein interactions. Chromatin proteins are no exception, and considering that many operate in megadalton-sized complexes, it is critical to be aware of the potential effects on the performance of such a complex with one of its proteins carrying an adjunct. Caution is advised: do not trust a tagged protein unless it can rescue a phenotype, use small tags and wherever possible, and use an antibody against the native protein.

Considering RFP-tagged proteins, earlier versions of RFP *(7)*, as a result of their tendency to form oligomers, altered protein localization significantly. For example, expression of RFP-tagged HcpA, one of the *Dictyostelium* HP1 isoforms, resulted in unspecific clustering at the nuclear membrane, which was not observed for the corresponding GFP fusion. The use of momomeric RFP (mRFP) should circumvent these problems *(8)*.

3.5. Pull-Down of Ectopically Overexpressed Proteins in Dictyostelium

The pull-down method allows assay of functionality of (tagged) proteins that are ectopically overexpressed in *Dictyostelium*. It has been evaluated to verify the functionality of GFP-tagged HP1 isoforms (Kaller et al., submitted), but may also be suitable for other proteins that are not involved in chromatin organization. Similar protocols have been used to study interactions between the fission yeast HP1 homologue Swi6 and DNA polymerase α *(9–11)*.

The general idea is to immobilize a known 6xHis-tagged interaction partner of the protein of interest on Ni sepharose beads. The preincubated beads are then challenged with *Dictyostelium* cell or nuclear extracts. After washing, the His-tagged protein on the beads is eluted with imidazole, and co-eluted pro-

teins are analyzed by SDS-polyacrylamide gel electrophoresis (PAGE) and/or
Western blotting.

In the best case, one may be able to reconstitute a multiprotein complex on
the column, although this might be sterically inhibited as a result of the immobilized protein on the beads. As controls, this experiment requires a beads-only
fraction without His-tagged protein, and a fraction with wild-type (in case of
ectopically overexpressed proteins) or knockout (in case one assays for endogenous proteins) cell extracts

3.5.1. Preparation of Protein-Covered Ni Sepharose Beads

The conditions required for expression of the His-tagged recombinant protein in *Escherichia coli* may vary for each protein. For HP1 proteins, we use
the *E. coli* BL21 (DE3) pLysS strain (Invitrogen). Protein expression is induced
with 1 m*M* isopropyl-β-D-thiogalactopyranoside (IPTG) for 3 h at 30°C. Induced cell cultures are centrifuged for 5 min at 3220*g*. The cell pellet is either
stored at –20°C or used directly. The following steps are all done on ice or at
4°C to avoid protein degradation.

1. Resuspend the cell pellet of induced *E. coli* culture in 10 mL of lysis buffer (*see*
   Subheading 2.4.). Lyse the cells by sonification on ice for three times 30 s with
   30-s intervals. The settings of the sonifier are 100% intensity with 50% duty
   cycle.
2. Centrifuge the cell lysate at 10,400*g* for 10 min at 4°C. Collect the supernatant,
   and take one aliquot as the input fraction
3. Equilibrate Ni sepharose columns (column volume 300–500 µL) with several
   column volumes of lysis buffer.
4. Load the supernatant on the Ni sepharose column.
5. Wash the columns with at least 20 column volumes of washing buffer. The optimal imidazole concentration in the washing buffer that is needed in order to eliminate unspecific binding of proteins to the column without washing out the protein
   of interest must be determined empirically for each protein.
6. The His-tagged protein can then be eluted, dialyzed and reloaded on fresh Ni
   sepharose beads, or the preincubated beads can directly be challenged with
   Dictyostelium extracts.

3.5.2. Preparation of Dictyostelium Cell Extracts

1. Centrifuge 8 × 10⁷ cells for 3 min at 290*g* and wash twice with phosphate buffer.
2. Resuspend the cell pellet in 10 mL of lysis buffer and lyse cells by sonification
   on ice four times for 15 s with 30-s intervals. The settings of the sonicator are the
   same as described in **Subheading 3.5.1., step 1**.
3. Centrifuge the cell lysate at 10,400*g* for 10 min at 4°C. Collect the supernatant,
   and keep one aliquot as the input fraction.

3.5.3. Pull-Down of Dictyostelium Proteins

1. Divide the supernatant into two parts and transfer together either with the preincubated or with empty beads to fresh 15-mL plastic tubes (*see* **Note 3**). Rotate the mixture on a rotating wheel for 1–2 h or overnight at 4°C.
2. Transfer the beads-cell lysate mixture into empty columns. Allow the beads to sediment. Collect the flowthrough, and keep an aliquot as the flowthrough fraction.
3. Wash the beads with at least 20 column volumes of washing buffer
4. Elute the bound proteins with elution buffer stepwise in a 400-μL volume for each step. The first fraction may contain no or only very few protein. Keep aliquots of each eluted fraction for further analysis (e.g., SDS-PAGE, Western blot).

3.6. Electrophoretic Mobility Shift Assay

Some chromatin-associated proteins, such as HP1 proteins, display direct binding to DNA (and RNA), which contributes to subnuclear targeting. A standard procedure to detect nucleic acid binding activity of a protein is the electrophoretic mobility shift assay (EMSA).

A radioactively labeled DNA substrate is incubated with increasing amounts of purified recombinant protein in the absence or presence of unlabeled competitor DNA. The mixture is then subjected to electrophoresis. Formation of a DNA–protein complex leads to changes in charge and size, which cause a retardation (shift) of the associated DNA in the gel.

1. End-labeling of DNA. Prepare the following reaction mixture:

    Dephosphorylated DNA (*see* **Note 6**), 1-20 pmol of 5×-termini
    10X forward reaction buffer, 2 μL
    $[\alpha^{32}\text{-ATP}]$, 20 pmol
    Water to 19 μL
    T4 Polynucleotide Kinase (10 U), 1 μL
    Incubate at 37°C for 45 min. Inactivate the enzyme for 10 min at 60°C.

2. Purify the DNA on a Sepharose-G50-column by centrifugation at $300g$ for 10 min.
3. For further purification, the DNA can be precipitated with 100% ethanol. After precipitation, dry the DNA and resuspend it in 1X assay buffer.
4. Mix increasing amounts of protein (kept in storage buffer) with 3 μL of labeled DNA. Fill this mixture with 1X assay buffer to 20 μL.
5. Keep the reaction mixtures on ice for 1 h or for 30 min at room temperature.
6. Add glycerol to a final concentration of 10%.
7. Load the samples on a 5% polyacrylamide gel in 1X TBE. The gel should have been prerun for at least 30–60 min.
8. Run the gel at 120 V for 3– h. Expose on radiosensitive phosphorimager screens to visualize the radiolabeled bands.

3.7. *Immunoprecipitation*

Cell or nuclear extracts are incubated with antibodies directed against a specific protein. Protein–antibody complexes are captured with Sepharose beads covered with protein A, which binds immunoglobulin (Ig)G. Thereby, the protein against which the antibody is directed, but also proteins associated with the protein of interest, can be precipitated. The precipitated proteins are then analyzed by SDS-PAGE and/or Western blotting. Further analysis, such as mass spectrometry, can be used to identify the co-precipitated proteins. For immunoprecipitation of chromatin-associated proteins, purification of nuclei prior to immunoprecipitation may help to minimize unspecific binding of proteins that are of cytoplasmic origin.

3.7.1. Purification of Nuclei

1. Centrifuge 2×10^8 cells for 3 min at 290g at 4°C.
2. Wash the cell pellet with 30 mL of cold phosphate buffer.
3. Resuspend the cells in 27 mL of cell lysis buffer (*see* **Subheading 2.6.**, **item 1**), then add 3 mL of 10% NP40. Invert the suspension carefully until it becomes clear, which indicates lysis of the cells. Under these conditions, the nuclei remain intact.
4. Centrifuge the nuclei at 1600g for 15 min at 4°C. Use a swinging bucket rotor, which allows the nuclear pellet to be collected at the bottom of the tube.
5. Discard the supernatant, resuspend the nuclei in 1.5 mL of RIPA buffer, and transfer the suspension into a 2 mL tube.
6. For complete lysis of the nuclei, sonify twice for 5 s (90% duty cycle, 40% intensity)
7. Centrifuge at 10,400g for 10 min at 4°C. Transfer the supernatant into a fresh tube.

3.7.2. Immunoprecipitation

1. Add 50 µL of a 50% proteinA-Sepharose slurry in 1X PBS, and rotate the mixture for 30 min to 1 h at 4°C.
2. Spin down beads for 5 s at 20,800g. Transfer the supernatant into a fresh tube. This preclearing step is required to minimize unspecific binding of proteins to the beads.
3. Divide the supernatant (from **Subheading 3.7.1.**, **step 7**) into three parts. Add either (a) antibody against the protein of interest, (b) an unspecific antibody, or (c) no antibody. The amount of antibody that quantitatively precipitates the protein of interest must be determined empirically.
4. Rotate on a rotating wheel overnight at 4°C. In cases in which antibodies that hardly bind to protein A, such as mouse IgG$_1$, are being used, incubate with a rabbit anti-mouse bridging antibody for another 1.5 h.
5. Add 50 µL of proteinA-Sepharose slurry and incubate for further 1–2 h.
6. Spin down the beads at 20,800g for 5 s. Keep an aliquot of the supernatant.
7. Wash the beads with 800 µL of cold RIPA buffer. Repeat this washing step three times.

8. To the beads, add 60 µL of 2X Laemmli buffer.
9. Boil for 5 min at 95°C. This boiling step releases the protein–antibody complex from the beads. Briefly spin down the beads at 20,800*g* and transfer the supernatant, which is the immunoprecipitated fraction, into a fresh tube.

4. Notes

1. Sonication for ChIP. It is important not to let the mixture overheat, as this reverses the cross-links. Sonication should yield a smear of DNA fragments that run on agarose gels at a size between 400 and 1000 bp. Longer sonication does not appear to give smaller fragments. For a particular sonicator, one should run tests to determine the minimal number and lowest intensity of pulses required to give this smear.
2. Controls for ChIP. As controls for the ChIP procedure, the input fraction must to be analyzed in the PCR step. A "no antibody" control is processed through the entire protocol to determine whether nonspecific binding to the beads occurs, which may still be observed even after the clearing step. A "control antibody," such as histone H3, should be used as an alternate reference to the input; however, the distribution can vary across genes, so it is best to use both if possible. A knockout cell line is the best negative control. For controlling the PCRs, use a water control, in addition to the input, no antibody, and control antibody.
3. Controls for the pull-down: even after stringent washing conditions, the eluted fractions may contain a high background of unspecifically bound protein. For this reason, it is important to run a beads-only control, in order to confirm that the protein–protein interaction to be observed is not due to unspecific binding to the beads.
4. Bisulfite sequencing. Complete denaturation is required to achieve complete conversion. This is especially important when examining repetitive sequences.
5. Bisulfite sequencing. Do not use proofreading polymerases—they do not amplify converted DNA! Conventional Taq polymerase is recommended.
6. End-labeling of DNA. Chemically synthesized primers usually do not contain a 5× phosphate, and thus produce 5×-dephosphorylated DNA. For end-labeling of PCR products, use forward reaction buffer.

Acknowledgments

This work was supported by a grant of the Deutsche Forschungsgemeinschaft (Ne 285/8) and of the European Union (FOSRAK 005120) to W.N.

References

1. Bickmore, W. A. and van der Maarel, S. M. (2003) Perturbations of chromatin structure in human genetic disease: recent advances. *Hum Mol Genet.* **12**, R207–R213.
2. Smith, S. S. and Ratner, D. I. (1991) Lack of 5-methylcytosine in *Dictyostelium discoideum* DNA. *Biochem J.* **277**, 273–275.
3. Martens, H., Novotny, J., Oberstrass, J., Steck, T. L., Postlethwait, P., and Nellen, W. (2002) RNAi in *Dictyostelium*: the role of RNA-directed RNA polymerases and double-stranded RNase. *Mol Biol Cell.* **13**, 445–453.

4. Mette, M. F., Aufsatz, W., van der Winden, J., Matzke, M. A., and Matzke, A. J. (2000) Transcriptional silencing and promoter methylation triggered by double-stranded RNA. *EMBO J.* **19,** 5194–5201.

5. Verdel, A., Jia, S., Gerber, S., et al. (2004) RNAi-mediated targeting of heterochromatin by the RITS complex. *Science* **303,** 672–676.

6. Ahmad, K. and Henikoff, S. (2002) The histone variant H3.3 marks active chromatin by replication-independent nucleosome assembly. *Mol. Cell.* **9,** 1191–1200.

7. Knop, M., Barr, F., Riedel, C. G., Heckel, T., and Reichel, C. (2002) Improved version of the red fluorescent protein (drFP583/DsRed/RFP). *Biotechniques* **33,** 592–598.

8. Fischer, M., Haase, I., Simmeth, E., Gerisch, G., and Müller-Taubenberger, A. (2004) A brilliant monomeric red fluorescent protein to visualize cytoskeleton dynamics in *Dictyostelium. FEBS Lett.* **577,** 227–232.

9. Nakayama, J., Allshire, R. C., Klar, A. J., and Grewal, S. I. (2001) A role for DNA polymerase alpha in epigenetic control of transcriptional silencing in fission yeast. *EMBO J.* **20,** 2857–2866.

10. Ahmed, S., Saini, S., Arora, S., and Singh, J. (2001) Chromodomain protein Swi6-mediated role of DNA polymerase alpha in establishment of silencing in fission Yeast. *J. Biol. Chem.* **276,** 47,814–47,821.

11. Nakayama, J., Rice, J. C., Strahl, B. D., Allis, C. D., and Grewal, S. I. (2001) Role of histone H3 lysine 9 methylation in epigenetic control of heterochromatin assembly. *Science* **292,** 110–113.

12. Chubb, J. R., Bloomfield, G., Xu, Q., et al. (2006) Developmental timing in *Dictyostelium* is regulated by the Set1 histone methyltransferase. *Dev. Biol.* **292,** 519–532.

31

Dictyostelium discoideum as a Model to Study Host–Pathogen Interactions

Can Ünal and Michael Steinert

Summary

Dictyostelium discoideum is a molecularly amenable host model system for several human pathogens, including *Legionella pneumophila*, *Mycobacterium avium*, *Mycobacterium marinum*, *Pseudomonas aeruginosa*, and *Cryptococcus neoformans*. *Dictyostelium* wild-type cells have proven useful in screening and identifying numerous bacterial und fungal virulence factors. Moreover, *Dictyostelium* mutant cells can be used to identify genetic host determinants of susceptibility and resistance to infections. Marker genes such as the green fluorescence protein (GFP) gene allow the in vivo monitoring of infection-relevant host factors. Here, we present methods that have already contributed to the deciphering of important aspects of the *Dictyostelium–Legionella* interaction. Moreover, the described phagocytosis assay, infection assay, and the confocal in vivo monitoring of GFP-tagged host factors can easily be adapted to other host–pathogen interactions.

Key Words: *Legionella*; phagocytosis; infection assay; intracellular replication; virulence; confocal in vivo monitoring.

1. Introduction

Infectious diseases represent one of the most severe health problems worldwide. In order to identify new targets for anti-infective therapies, it is important to analyze the cellular and molecular mechanisms that are leading to diseases. The cross-talk between a pathogen and its host can include processes such as adherence, invasion, disruption of membrane trafficking events, intracellular growth, host defense evasion, and, eventually, the exit from the host. Many different infection models like animals, organ cultures, or cell lines have been developed for the study of various aspects of infections. The limitations of many experimental approaches, however, are the lack of molecular genetic amenability of the respective host model as well as human and animal rights

From: *Methods in Molecular Biology, vol. 346:* Dictyostelium discoideum *Protocols*
Edited by: L. Eichinger and F. Rivero © Humana Press Inc., Totowa, NJ

considerations. The cellular similarity of *Dictyostelium discoideum* cells to macrophages, the complete genome sequence, and the genetic tractability of the haploid social amoeba generate many interesting opportunities to study infection relevant processes *(1–3)*.

Dictyostelium wild-type cells can be used as screening system for mutagenized pathogens. By agar plating assays, it was shown that the extracellular pathogen *Pseudomonas aeruginosa* utilizes quorum sensing-mediated virulence pathways to infect *D. discoideum (4,5)*. When the amoebae were plated together with different *P. aeruginosa* wild-type and mutant strains, only virulent bacteria produced an intact lawn. In contrast, certain avirulent mutant strains were eliminated by the amoebae, which resulted in plaques within the bacterial lawn.

Phagocytosis assays with mutated *Dictyostelium* host cells or mutated intracellular pathogens such as *Legionella pneumophila*, *Mycobacterium* spp., and *Cryptococcus neoformans* can reveal important aspects of the first steps of infection *(6–9)*. Moreover, the use of specific cellular inhibitors can help to identify the involved signal transduction pathways. In the case of *L. pneumophila*, the causative agent of Legionnaires' disease, it could be demonstrated that the uptake by *Dictyostelium* cells occurs by conventional phagocytosis, which includes heterotrimeric G proteins and the phospholipase C (PLC) pathway. Additionally, these experiments revealed that cytoplasmic calcium levels, cytoskeleton-associated proteins (coronin, actinin/filamin, villidin), and the calcium binding proteins of the endoplasmic reticulum (ER), calreticulin and calnexin, specifically influence the uptake of *Legionella (10,11)*.

Infection assays in which the intracellular growth of a pathogen is analyzed can illuminate the outcome of a host–pathogen interaction. For the infection processes of *Legionella*, a roadmap of the involved host cell factors is being developed. The comparison of infected macrophages with *Dictyostelium* cells by transmission electron microscopy demonstrated that both cell types harbor engulfed legionellae within organelle-studded vacuoles that are associated with rough endoplasmic reticulum (ER) *(12–14)*. The determination of intracellular bacterial growth also shows that many virulence factors like the type IV secretion systems (Icm/Dot) are similarly relevant in macrophages and *Dictyostelium* cells *(15)*. These approaches demonstrated that certain host responses have been conserved throughout evolution *(10,12)*.

For some infection-relevant host factors, further analyses were performed by green fluorescent protein (GFP) fusion proteins. Confocal microscopic time series with GFP-tagged calnexin and calreticulin demonstrated the accumulation of both host proteins in the phagocytic cup of *L. pneumophila*-infected *Dictyostelium* cells. These calcium-binding proteins also decorated the replicative vacuole of *L. pneumophila* at later stages of infection *(10)*. Co-localization studies with GFP-tagged bacteria and antibodies directed against specific lysosomal markers (DdLIMP) of *Dictyostelium* corroborated that *Legionella* inhibits the endosomal maturation pathway *(6)*. In the following sections, we describe the basic setup for studying the complex interactions of *L. pneumophila* and *D. discoideum*.

2. Materials
2.1. Bacterial and Cell Culture

1. Buffered charcoal-yeast (BCYE) agar for *Legionella*: 5 g N-(2-acetamido)-2-amino-ethanesulfonic acid (ACES; Gerbu), 10 g yeast extract (Oxoid) are dissolved in 900 mL double distilled water (ddH_2O), and the pH is adjusted to 6.9 with 10 N KOH. Then, 2 g activated charcoal (Fluka) and 15 g Agar (BD Difco) are added and the volume is completed to 980 mL by adding ddH_2O (*see* **Note 1**). After the agar has been autoclaved and cooled down to 50°C, it is supplemented with 0.4 g L-cysteine in 10 mL ddH_2O and 0.25 g $Fe(NO_3)_3 \cdot 9H_2O$ in 10 mL ddH_2O (*see* **Note 2**).
2. HL5 medium for *Dictyostelium*: 7.15 g yeast extract (Oxoid), 14.3 g Bacto™ proteose peptone Nr.2 (BD Difco), 1.28 g Na_2HPO_4, 0.49 g KH_2PO_4 are dissolved in 900 mL ddH_2O and the pH is adjusted to 7.5 with 10 N KOH. After autoclaving and cooling, 15.4 g glucose monohydrate dissolved in 100 mL ddH_2O is added (*see* **Note 3**).
3. 50X stock Soerensen buffer: 99.86 g Na_2HPO_4, 17.8 g KH_2PO_4 are dissolved in 1 L ddH_2O and the pH is adjusted to 6.0 with 10 N KOH. After autoclaving, the stock solution can be kept at room temperature for extended time periods.

2.2. Phagocytosis Assay

1. Infection medium: this is a 1:1 mixture of HL5 medium and 1X Soerensen buffer (dilute one part 50X Soerensen buffer [*see* **Subheading 2.1.**, **item 3**] with 49 parts ddH_2O and autoclave) (*see* **Note 4**).
2. 10 mg gentamicin is dissolved in 1 mL ddH_2O and sterilized by filtration. Aliquots are stored at –20°C.
3. 1X Soerensen buffer (*see* **Subheading 2.2.**, **item 1**).
4. Autoclaved ddH_2O.
5. BCYE agar plates (*see* **Subheading, 2.1.**, **item 1**).

2.3. Infection Assay

1. Infection medium (*see* **Subheading 2.2.**, **item 1**).
2. Autoclaved ddH_2O.
3. BCYE agar plates (*see* **Subheading 2.1.**, **item 1**).

2.4. Confocal In Vivo Monitoring of GFP-Tagged Host Factors During Invasion by Rhodamine-Labeled Legionella

1. 10X phosphate-buffered saline (PBS) stock solution: 80 g NaCl, 2 g KCl, 12.5 g Na_2HPO_4, 2 g KH_2PO_4 (adjust to pH 7.4 with HCl). The stock solution should be

autoclaved before storing at room temperature. Prepare 1X PBS by diluting one part stock solution with nine parts ddH_2O.

2. For labeling of *Legionella*, adjust the pH of 1X PBS to 7.8 with HCl.
3. 30 m*M* stock solution of 5,6-carboxymethylrhodamine, succinimidylester [5(6) TAMRA, SE] (Molecular Probes) in dimethylsulfoxide (DMSO). Store this solution at –20 °C.
4. 50 m*M* Tris-HCl (pH 7.5).
5. 1X Soerensen buffer (*see* **Subheading 2.2.**, **item 1**).
6. Zeiss LSM 510 laser scanning microscope.
7. Coverslips or chamber slides (Falcon™).

3. Methods

The infection by a pathogen enforces a host reaction. To examine host functions, it is important to characterize and quantify the infection, which is a product of pathogen uptake and intracellular replication. This can be accomplished by a phagocytosis assay and an infection assay, respectively. The assays described can be used for many purposes, including the analysis of mutants, cellular inhibitors, signal transduction, protein expression profiles, and microscopic studies. Especially, confocal microscopy is one of the most straightforward methods, because it allows the in vivo imaging of cells and their components.

Because *L. pneumophila* is infectious to humans, experimental work with this pathogen must be performed under S2 conditions (*see* **Note 5**).

3.1. Cultivation of L. pneumophila and Dictyostelium discoideum

1. *L. pneumophila* is grown on BCYE agar at 37°C in 5% CO_2 atmosphere for 3 d (*see* **Note 6**).
2. The *D. discoideum* wild-type strain (AX2) and the mutant strains are grown in 30 mL HL5 medium at 24.5°C, either as shaking culture in a 100-mL flask or in 75-cm^2 cell-culture flasks (*see* **Note 7**).

3.2. Phagocytosis Assay

1. *D. discoideum* cells of a 3-d-old culture are harvested ($200g$, 7 min, at room temperature) and resuspended in the same volume of infection medium. If mutants are grown in the presence of antibiotics, a washing step ($200g$, 7 min, at room temperature) with infection medium is required.
2. 25×10^5 cells are seeded into 25-cm^2 cell-culture flasks and the volume is adjusted to 5 mL with freshly mixed infection medium. The final cell densitiy is 5×10^5 cells/mL. Before bacterial inoculation, the cells should have settled down for 30 min at 25.5 °C (*see* **Note 8**).
3. A 3-d-old plate culture of *L. pneumophila* is suspended in 1 mL ddH_2O and the cell density is adjusted to 10^9 cells/mL (*see* **Note 9**).

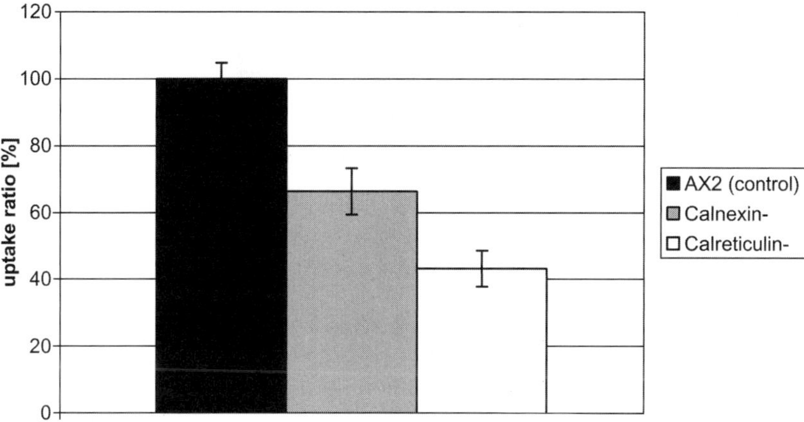

Fig. 1. Effects of calnexin- and calreticulin-host cell mutations in *Dictyostelium* on uptake of *Legionella pneumophila*. The values shown represent uptake in percentage of wild-type. The standard deviation is indicated by an error bar.

4. 25 µL of the prepared bacterial suspension are added to each cell culture flask (mutiplicity of infection [MOI] 10). To determine the exact inoculation doses, serial dilutions of a 100-µL sample from each culture flask are plated on BCYE agar.
5. Following an invasion period of 2 h, the remaining extracellular bacteria are killed by a gentamicin treatment (100 µg/mL). 50 µL of a gentamicin stock solution are added to every cell culture flask. After 1 h incubation at 25.5°C, the *Dictyostelium* cells are washed three times with 5 mL Soerensen buffer (*see* **Note 10**).
6. After washing, the cells are resuspended in 5 mL ddH$_2$O by knocking the flask on the edge of a table. 1 mL of the suspension is transferred into a 1.5-mL reaction cap. Cells are lysed by centrifugation (13,600g, room temperature, 6–8 min) and vigorous shaking. In order to determine the number of intracellular bacteria, serial dilutions of this lysate are plated on BCYE agar. An example result is shown in **Fig. 1**.

3.3. Infection Assay

The preparation of the infection assay resembles the phagocytosis assay (*see* **Subheading 3.2.**, **steps 1–3**), but with the following changes.

1. The bacterial suspension is diluted to 10^6 cells/mL, and 50 µL of this dilution are added to every cell culture flask (MOI 0.02) (*see* **Note 11**).
2. 100 µL from each cell culture flask are transferred into a 1.5-mL reaction tube containing 900 µL ddH$_2$O. In order to determine the cfu/mL of the inoculum, 100 µL of this suspension and further dilutions are plated on BCYE agar.

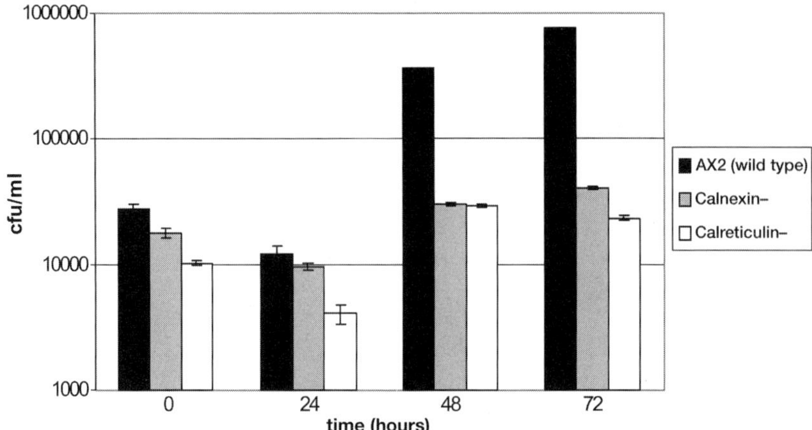

Fig. 2. Intracellular growth of *Legionella pneumophila* in *Dictyostelium discoideum* AX2 wild-type, calnexin- and calreticulin-minus cells. The values shown are the mean numbers of colony forming units grown on buffered charcoal-yeast agar plates. The standard deviation is indicated by an error bar for each time point after infection.

3. After different time intervals (24, 48, 72, and 96 h), serial dilutions of cell lysates (*see* **Subheading 3.2.**, **step 6**) are plated on BCYE agar (*see* **Note 12**). An example result is shown in **Fig. 2**.

3.4. Confocal In Vivo Monitoring of GFP-Tagged Host Factors During Invasion by Rhodamine-Labeled Bacteria

1. A culture of *D. discoideum* cells expressing a GFP-tagged protein is prepared as described in **Subheading 3.1.**, **step 2**. The cells are harvested and washed twice with Soerensen buffer to remove antibiotics.
2. 1 mL with 10^5 cells is brought onto a glass coverslip or, alternatively, into a chamber slide. The cells should adhere to the surface for 5 mins at 25.5°C. Use a humidified chamber to prevent drying of the sample.
3. A 3-d-old, plate culture of *L. pneumophila* is suspended in 1 mL 1X PBS (pH 7.8). After the cell density is adjusted to 10^9 bacteria/mL, the cells are washed three times (2000*g*, 5 min, at room temperature) and finally resuspended in 1 mL PBS (pH 7.8).
4. 10 µL of the rhodamine stock solution (0.3 m*M* final concentration) are added to the bacteria and the cells are rotated slowly in the dark at room temperature for 30 min.
5. After labeling, bacteria are washed four times (*see* **Note 13**) with 50 m*M* Tris/ HCl (pH 7.5) and finally resuspended in 1X Soerensen buffer.
6. Bacteria are added to the *Dictyostelium* cells at a MOI of 10 and the cells are incubated at 25.5°C for 0–24 h.

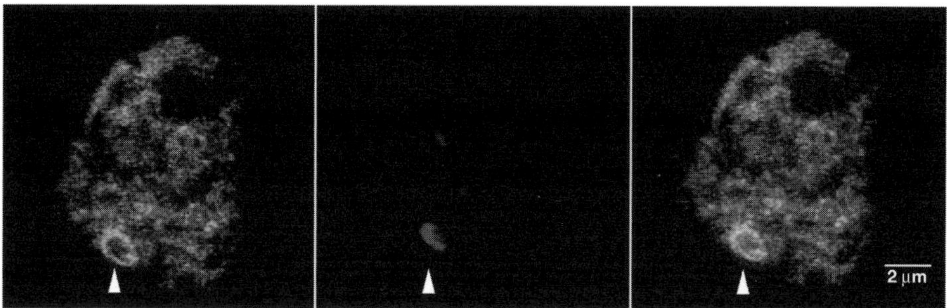

Fig. 3. Confocal in vivo monitoring of the *Dictyostelium discoideum* protein calreticulin after uptake of *Legionella pneumophila*. Left panel shows the green fluorescent protein-tagged host protein, and the middle panel shows the rhodamine-labeled bacteria. The double exposure (right) shows the association of bacteria with calreticulin. Size bar as indicated.

7. The observation is performed with a Zeiss LSM 510 laser-scanning microscope equipped with a 63×/1.4 Plan-Neofluar objective. GFP-tagged host factors are visualized with the help of an argon laser emitting light at 488 nm. Rhodamine-labeled *Legionella* are visualized at 543 nm with a helium-neon laser. An example result is shown in **Fig. 3**.

4. Notes

1. Activated charcoal tends to aggregate. Hence, prepared agar should be mixed thoroughly before and after autoclaving.
2. $Fe(NO_3)_3 \cdot 9H_2O$ and L-cysteine precipitate if autoclaved. Therefore, both solutions should be freshly prepared. Store sterile filtered $Fe(NO_3)_3 \cdot 9H_2O$ solution (25 g/L) at 4°C.
3. The glucose solution should be autoclaved separately, because glucose tends to caramelize in the presence of divalent cations.
4. The infection medium has to fulfill two criteria: (a) it should not promote the growth of *Dictyostelium* or extracellular bacteria; (b) it should contain a residual amount of nutrients, so that differentiation and morphogenesis of *Dictyostelium* is prevented. An alternative infection medium is LoFlo (www.dictybase.org).
5. The inhalation of aerosolized legionellae can result in a severe pneumonia. Therefore, aerosol formation must be avoided. Furthermore, contaminated material has to be autoclaved and desinfected carefully.
6. Passaging of the bacteria should not be performed more than three times, because *Legionella* tend to loose their virulence during this procedure. The stock culture is prepared by suspending a 2- to 3-d-old plate culture in 1 mL ddH_2O and adding 1 mL sterile 86% glycerine.
7. *Dictyostelium* precultures are prepared in 25-cm^2 cell-culture falsks with 10 mL of HL5 medium inocculated with spores or cells from liquid nitrogen. It takes

2–3 d for the majority of *Dictyostelium* spores to germinate. Cell cultures of *D. discoideum* are prepared by inoculating 30–50 mL HL5 medium with 1–2 mL of a fresh preculture.

8. Human infections occur at 37°C, but *Dictyostelium* does not survive temperatures above 27°C. The highest possible infection temperature in the *Dictyostelium* model is 25.5°C.

9. A 1:5 dilution of bacterial suspension is transferred into a photometer cuvet and the absorption is measured at 550 nm. An absorption of 1.31 corresponds to 10^9 cells/mL.

10. Washing in culture flasks requires careful removing of the supernatant and careful adding of Soerensen buffer. Avoid pipetting the liquid directly on the cell lawn.

11. The low MOI of 0.02 allows one to perform the assay without gentamicin treatment.

12. *Legionella* is not able to replicate extracellularly in the infection medium. For other bacterial species, extracellular replication must be considered.

13. Washing can be stopped as soon as the supernatant is colorless.

References

1. Steinert, M. and Heuner, K. (2005) *Dictyostelium* as host for pathogenesis. *Cell. Microbiol.* **7,** 307–314.

2. Steinert, M., Leippe, M., and Roeder, T. (2003) Surrogate hosts: invertebrates as models for studying pathogen-host interactions. *Int. J. Med. Microbiol.* **293,** 1–12.

3. Rupper, A. and Cardelli, J. (2001) Regulation of phagocytosis and endophagosomal trafficking pathways in *Dictyostelium discoideum. Biochem. Biophys. Acta* **1525,** 205–216.

4. Pukatzki, S., Kessin, R. H., and Mekalanos, J. J. (2002) The human pathogen *Pseudomonas aeruginosa* utilizes conserved virulence pathways to infect the social amoeba *Dictyostelium discoideum. Proc. Natl. Acad. Sci. USA* **99,** 3159–3164.

5. Cosson, P., Zulianello, L., Join-Lambert, O., et al. (2002) *Pseudomonas aeruginosa* virulence analysed in a *Dictyostelium discoideum* host system. *J. Bacteriol.* **184,** 3027–3033.

6. Hägele, S., Köhler, R., Merkert, H., Schleicher, M., Hacker, J., and Steinert, M. (2000) *Dictyostelium discoideum*: a new host model system for intracellular pathogens of the genus *Legionella. Cell. Microbiol.* **2,** 165–171.

7. Skriwan, C., Fajardo, M., Hägele, S., et al. (2002) Various bacterial pathogens and symbionts infect the amoeba *Dictyostelium discoideum. Int. J. Med. Micobiol.* **291,** 615–624.

8. Solomon, J. M., Leung, G. S., and Isberg, R. R. (2003) Intracellular replication of *Mycobacterium marinum* within *Dictyostelium discoideum*: efficient replication in the absence of host coronin. *Infect. Immun.* **71,** 3578–3586.

9. Steenbergen, J. N., Nosanchuk, J. D., Malliaris, S. D., and Casadevall, A. (2003) *Cryptococcus neoformans* virulence is enhanced after growth in the genetically malleable host *Dictyostelium discoideum. Infect. Immun.* **71,** 4862–4872.

10. Fajardo, M., Schleicher, M., Noegel, A., et al. (2004) Calnexin, calreticulin and cytoskeleton associated proteins modulate uptake and growth of *Legionella pneumophila* in *Dictyostelium discoideum. Microbiol.* **150,** 2825–2835.

11. Schreiner, T., Mohrs, M. R., Blau-Wasser, R., et al. (2002) Loss of the F-actin binding and vesicle-associated protein comitin leads to a phagocytosis defect. *Euk. Cell.* **1,** 906–914.
12. Solomon, J. M. and Isberg, R. R. (2000) Growth of *Legionella pneumophila* in *Dictyostelium discoideum*: a novel system for genetic analysis of host-pathogen interactions. *Trends Microbiol.* **10,** 478–480.
13. Solomon, J. M., Rupper, A., Cardelli, J. A., and Isberg, R. R. (2000) Intracellular growth of *Legionella pneumophila* in *Dictyostelium discoideum*, a system for genetic analysis of host-pathogen interactions. *Infect. Immun.* **68,** 2939–2947.
14. Otto G. P., Wu, M. Y., Clarke, M., et al. (2004) Macroautophagy is dispensable for intracellular replication of *Legionella pneumophila* in *Dictyostelium discoideum*. *Mol. Microbiol.* **51,** 63–72.
15. Hilbi, H., Segal, G., and Shuman, H. (2001) Icm/Dot-dependent upregulation of phagocytosis by *Legionella pneumophila*. *Mol. Microbiol.* **42,** 603–617.

32

Pharmacogenetics

*Defining the Genetic Basis of Drug Action
and Inositol Trisphosphate Analysis*

Kathryn E. Adley, Melanie Keim, and Robin S. B. Williams

Summary

Medicinal drugs do not always have clearly understood mechanisms of action, especially as regards psychiatric treatment. Identification of genes involved in drug resistance is an important step toward elucidating the genetic basis of disease and the molecular mechanism of drug action. However, this approach is impractical in higher animals, as ablation and screening of every gene in an animal is not currently possible. *Dictyostelium* has proven a good model system for molecular pharmacological research as a result of its genetic tractability, ease of gene knockout, and creation of isogenic lines. In this system, we have identified genes that confer resistance to bipolar disorder drugs. This work has implicated inositol (1,4,5) trisphosphate (InsP$_3$) signaling as a common mechanism of action for these drugs in patients.

Key Words: Pharmacogenetics; drug screening; inositol trisphosphate; lithium; valproic acid.

1. Introduction

If the way in which a drug works in an organism or cell is to be identified, the drug must have an effect that can be monitored. If a pharmaceutical agent has a recognizable effect, then we can screen mutants in an organism and identify those genes which confer sensitivity or resistance to the drugs (*1*). This approach can be labeled "pharmacogenetics," or the study of the effect of genetic factors on reactions to drugs, and is a growing field of research concerning individuals' responses to pharmaceutical agents. Although the genetic basis of any drug with a phenotypic effect can be examined using this approach,

From: *Methods in Molecular Biology, vol. 346:* Dictyostelium discoideum *Protocols*
Edited by: L. Eichinger and F. Rivero © Humana Press Inc., Totowa, NJ

we have previously used it to identify genes controlling resistance to the two commonly used manic depression drugs, valproic acid (VPA) and lithium *(2,3)*.

Both lithium and VPA were accidentally discovered as manic depression treatments; thus, the way that they inhibit mood swings in this disorder remains unknown. One theory concerning this is the "inositol depletion" theory, suggested by Berridge and co-workers *(4)*, in which these drugs are described as working by "dampening down" an overactive inositol phosphate signaling cascade. Lithium exerts its effect on this cascade by reducing inositol recycling *(5,6)*, and VPA's effect is likely to be through the reduction in inositol synthesis *(7,8)*. Both drugs inhibit development in *Dictyostelium*, providing a phenotype for pharmacogenetic screening.

We present here a method for defining the genetic target of a drug using *Dictyostelium*. The experimental procedures outlined here start following the identification of a phenotypic effect of a drug. We describe the screening of a restriction enzyme-mediated integration (REMI) library, the identification of the genes causing drug resistance/sensitivity by inverse polymerase chain reaction (PCR) rescue *(9)*, the preparation of a construct to recapitulate this mutant using transposon technology to disrupt the identified gene *(10)*, the screening for putative recapitulation of this gene disruption by PCR, Southern analysis to confirm gene ablation, and the analysis of $InsP_3$ levels.

2. Materials
2.1. Cell Culturing and Storage

1. All *Dictyostelium* cells (*see* **Note 1**) are grown in liquid culture using axenic medium: 14.3 g/L peptone, 7.2 g/L yeast extract, 15.4 g/L D-glucose, 0.51 g/L Na_2HPO_4, 0.41 g/L KH_2PO_4, vitamin B_{12} 0.1 mg/L, biotin 0.02 mg/L, riboflavin 0.2 mg/L supplemented with streptomycin sulfate 100 µg/mL (Gibco 11860-038) with the addition of blasticidin-HCl 10 µg/mL (ICN 150477) for REMI library-derived or transformed cells.
2. Cells in axenic medium can be grown in still plates (10 mL medium) or, for larger quantities and for log-phase growth ($0.5–2.5 \times 10^6$ cells/mL), in bunged conical flasks, shaking (150 rpm, 22ºC).
3. *Dictyostelium* strains are grown on solid Sussman's medium (SM) plates: 10 g/L glucose, 1 g/L $MgSO_4 \cdot 7H_2O$, 2.2 g/L KH_2PO_4, 1 g/L Na_2HPO_4, 10 g/L peptone (DIFCO), 1 g/L yeast extract (OXOID), 18g/L Bacto agar (DIFCO), following spreading with *Klebsiella pneumonia* (*see* **Note 2**) (use 3 mL culture grown at 22°C for overnight in LB, and store at 4°C for up to 2 wk).
4. Flat-blade spreader (SLS SLS2100).
5. Bioassay dishes, 245 mm^2 (Nunc 240835).
6. Cells are developed after washing in phosphate buffer (16.5 mM KH_2PO_4, 3.8 mM K_2HPO_4, pH 6.2) on Millipore nitrocellulose filters (VWR 007 1689-06) and

Millipore absorbent pads (VWR 007 1687-28) at 1×10^7 cells per 3 cm filter in the presence or absence of drug.

7. 3-MM filter paper (VWR 234091301).
8. Cells are stored under liquid nitrogen in freezing medium (horse serum 95% [Sigma H1270], dimethylsulfoxide 5%) in cryogenic vials (Nunc 377267) using a Nalgene cryo 1°C freezing container. Ensure lids are loosely fastened to allow liquid nitrogen evaporation on warming.

2.2. Identifying the Ablated Gene and Recapitulation of the Mutant

1. Use restriction, ligation, and polymerase enzymes from commercial suppliers.
2. Use custom PCR primers from commercial suppliers, with Tm $\geq$ 52°C.
3. TE: 10 mM Tris-HCl and 1 mM ethylenediamine tetraacetic acid (EDTA) adjusted to pH 8.0.
4. Glycogen 10 mg/mL (Roche, cat no. 901 393) in water is filter-sterilized.
5. TBE: 44.5 mM Tris borate pH 8.3, 1 mM Na$_2$EDTA.
6. DNA loading buffer (Bioline BIO 37045).
7. Microspin S400 spin HR columns (Amersham 275140-01).
8. TOPO® cloning (Invitrogen).
9. Qiagen plasmid maxi (12163) and mini (12123) columns.
10. EZ::TN transposon kit (Epicentre TNP92110).
11. Electrocompetent DH5αE cells (Invitrogen 11319-019) and 0.1 mm electroporation cuvets (Biorad 165-2083).
12. *Dictyostelium* cells are transformed in electroporation buffer (phosphate buffer (above) with 50 mM sucrose, filter-sterilized, and stored at 4°C) using 4 mm electroporation cuvet (Equibio ECU104) with the subsequent addition of CaMg (0.1 M CaCl$_2$ and 0.1 M MgCl$_2$).
13. Multi-channel pipet and sterile multi-channel reservoir (Anachem F-37877-000).
14. *Dictyostelium* lysis buffer: 50 mM KCl, 10 mM Tris-HCl pH 8.3, 2.5 mM MgCl$_2$, 0.45% NP40 substitute (Sigma 74385), 0.45% Tween 20, Proteinase K (1 µL of 20 µg/mL stock for every 25 µL of buffer, Roche 31115879).

2.3. Southern Analysis

1. DNAzol solution (Invitrogen 10503-027). The supplier is essential here, as other suppliers products are not equivalent.
3. Megaprime DNA labeling system (RPN 160415, Amersham Life Science).
4. Depurination solution: 0.25 M HCl.
5. Denaturation solution: 1.5 M NaCl, 0.5 M NaOH.
6. Neutralization solution: 1.5 M NaCl, 0.5 M Trizma® Base, pH 7.5 with HCl.
7. Hybond N+ nylon membrane (Amersham Pharmacia Biotech RPN 203B).
8. SSC, 20X: 3 M NaCl, 0.3 M sodium citrate.
9. Hybridization buffer (Modified from Church and Gilbert, **ref. *11***): 0.5 M phosphate buffer pH 7.2, 7% (w/v) SDS.
10. Wash solution 1 (2X SSC, 0.1% SDS).

11. Wash solution 2 (0.5X SSC, 0.1% SDS).
12. X-ray film (Kodak X-OMAT AR, 165 1454).
13. Bag sealer and plastic tubing (Jencons 295-003).

2.4. Inositol Trisphosphate Analysis

A modified method described for the TRK 1000 (Amersham Pharmacia Biotech) InsP$_3$ analysis kit is used.

1. Protein determination employs a Coomassie brilliant blue R-250 staining solutions kit (Biorad 161-0435).
2. Perchloric acid solution: 20% v/v in H$_2$O.
3. Neutralizing indicator solution: 1.5 M KOH, 60 mM HEPES, universal indicator. Combine 4.2 g KOH to 3 mL 1 M HEPES and approx 200 µL of universal indicator (pH 3-10; Sigma 36803) in 50 mL.
4. Indicator strips pH 5-10 (BDH 31 506).
5. Large orifice 200G tips (Fisher Scientific, PQP-749-010J).
6. Aspiration apparatus plumbed into a radioactive disposal sink.
7. Gel loading tips (Gelsaver Tip 1–200 µL, BIOplastics 171932).
8. Measure tritium levels in samples following resuspension in Ecolite scintillation fluid (ICN 882475) with scintillation vials (Bibby 28985).

3. Methods
3.1. Defining Drug Effects

The first step in a screen for drug resistance is to determine the phenotypic effects of the drug on wild-type cells. These may include effects on both growth (the amount of drug to arrest growth or kill cells) and developmental stages of the life cycle (the amount of drug to affect development speed or to alter fruiting body morphology). These effects must be quantified at various drug concentrations, ideally around the therapeutic concentration at which the drug is used. Once this has been established, screening of a REMI library may begin. Screening should ideally be under both liquid growth (axenic medium) and solid development (plating on a *Klebsiella* lawn on solid media) conditions in the presence of the drug. Resistant isolates should then be analyzed for developmental resistance by filter assay.

3.2. Selection of Drug-Resistant Clones
3.2.1. Developmental Selection

1. Combine cells from a *Dictyostelium* REMI library (*see* Chapter 12), using five times the original titre, in disposable 30 mL tubes. Vortex and pour on SM agar in 245 mm^2 bioassay dishes. Gently spread the *Dictyostelium/Klebsiella* mixture to cover the dish, using the flat blade of a spreader to loosely spread cell suspension and then applying the thin edge of spreader to thoroughly distribute cells

over plate (both top to bottom and side to side, repeatedly at right angles, using the narrow, blunt end of the spreader). Use the drug at an appropriate concentration to yield approx 500–1000 colonies per dish. Spread one plate of wild-type cells under the same conditions as a control.

2. Isolate resistant colonies (showing resistance to effects on growth or development or morphology) after 10–14 d and transfer to axenic medium containing blasticidin.

3. Confirm developmental resistance by filter assay.

4. Freeze around 2×10^8 cells in 1 mL of freezing medium at −80°C using a Nalgene cryo 1°C freezing container and transfer to liquid nitrogen.

3.2.2. Growth Selection

1. Combine cells from a *Dictyostelium* REMI library (*see* Chapter 12), using five times the original titre, in 9 cm tissue culture dishes containing 10 mL axenic medium with blasticidin. Supplement media with increasing concentrations of drug. Use wild-type cells as a control.

2. Replace media/drug every 3 d until control cells are dead or cells have been treated for 3 wk.

3. Plate resistant pools of cells onto SM/*Klebsiella* plates following serial dilution to get single colonies. Re-inoculate single isolates into axenic medium with drug (compared to wild-type cells) to confirm resistance. Also, examine resistance on solid media with increasing drug concentrations and by filter assay.

4. Freeze around 2×10^8 cells in 1 mL of freezing medium at −80°C using a Nalgene cryo 1°C freezing container and transfer to liquid nitrogen.

3.2.3. Developmental Filter Assay

1. Grow wild-type cells (0.5–2×10^6 per mL) in shaking axenic medium for 3 d.

2. Wash 1×10^7 cells in phosphate buffer.

3. Pre-wet nitrocellulose filters in phosphate buffer and soak filter pads in phosphate buffer with increasing levels of drug and transfer to a Petri dish (one for each drug concentration).

4. Wet 30 cm² of 3 MM filter paper with phosphate buffer and roll flat using a pipet. Place the filter on the paper and transfer 1 mL of cell suspension evenly to filter, allowing phosphate buffer to be slowly absorbed (*see* **Note 3**).

5. Transfer the filter to an absorbent pad saturated with phosphate buffer with or without drug.

6. Add 250 µL of phosphate buffer or phosphate buffer/drug solution to each whole pad.

7. Incubate at 22°C in a closed Petri dish, and observe regularly using a dissection microscope (*see* **Note 4**).

3.3. Identification of the Genetic Basis for Drug Resistance

Once drug-resistant mutants have been selected, the next step is identification of the disrupted gene. Inverse PCR is a simple and rapid technique for the isolation of genomic DNA sequences flanking a REMI insertion site *(12)*. The restriction enzyme *Alu*I cuts at the 4-bp recognition sequence (AGCT) frequently in plasmid DNA, but less often in *Dictyostelium* genomic DNA as a result of its strong A/T bias, giving a mean fragment length of 1050 bp. REMI insertional plasmids commonly include a blasticidin-resistance cassette (Bsr). *Alu*I cuts within the coding region of the Bsr but not in the intervening sequence between the insertion point and the coding region. Digestion of genomic DNA from REMI mutants, therefore, creates approx 1-kb fragments, with the disrupted gene fragment containing a known 5' sequence from the Bsr cassette. The DNA is then ligated, and the disrupted gene fragment isolated using specific PCR primers.

3.3.1. Inverse PCR Rescue

Inverse PCR rescue, outlined in **Fig. 1**, is based on a previously published procedure *(9)*:

1. Digest 20 µg of genomic DNA with 50 U of *Alu*I in a total volume of 100 µL at 37°C overnight (*see* **Note 5**). Check digestion by gel electrophoresis.
2. Dilute digested DNA to 400 µL with TE. Extract DNA once with phenol:chloroform (1:1) and once with chloroform *(14)*. Precipitate the DNA with 1X volume of isopropanol, and wash with 400 µL of 75% ethanol.
3. After air-drying the pellet, resuspend in 20 µL of TE, pH 8.0.
4. Combine 5 µL of digested DNA in a 400 µL total volume with 40 µL of 10X T4 DNA ligase buffer. Add 2 µL of T4 DNA ligase to 200 µL DNA/buffer mix (the remaining 200 µL being a control) and incubate overnight at 16°C.
5. Heat the ligation reactions for 10 min at 56°C to inactivate the ligase and precipitate with 2 µL of glycogen (10 mg/mL), 20 µL of 3 *M* sodium acetate, pH 5.5, and 800 µL of ice-cold ethanol. Incubate on ice for 30 min, pellet the DNA at 12,000*g* 30 min at 4°C, and wash the pellet with 300 µL of 75% ethanol. Dissolve the air-dried pellet in 50 µL of TE, pH 8.0.
6. PCR-amplify the cloned genomic DNA with unligated, digested DNA as controls (*see* **Note 6** for program). In a 50 µL reaction:

1 µL	Ligated/control DNA
5 µL	2 m*M* dNTPs
5 µL	10X Taq reaction buffer
2 µL	50 m*M* MgCl$_2$
0.2 µL	5 U/µL of Taq polymerase
2X 0.5 µL	100 pmol/µL of P$_A$ and P$_B$ (*see* **Note 7**)

   Add dH$_2$O to 50 µL
7. Amplified products are analyzed using gel electrophoresis (run 5 µL, with 1 µL loading buffer on a 1% agarose gel); the remainder is purified using an S400 spin column and cloned using any available method (e.g., TOPO® cloning) (*see* **Note 8**).

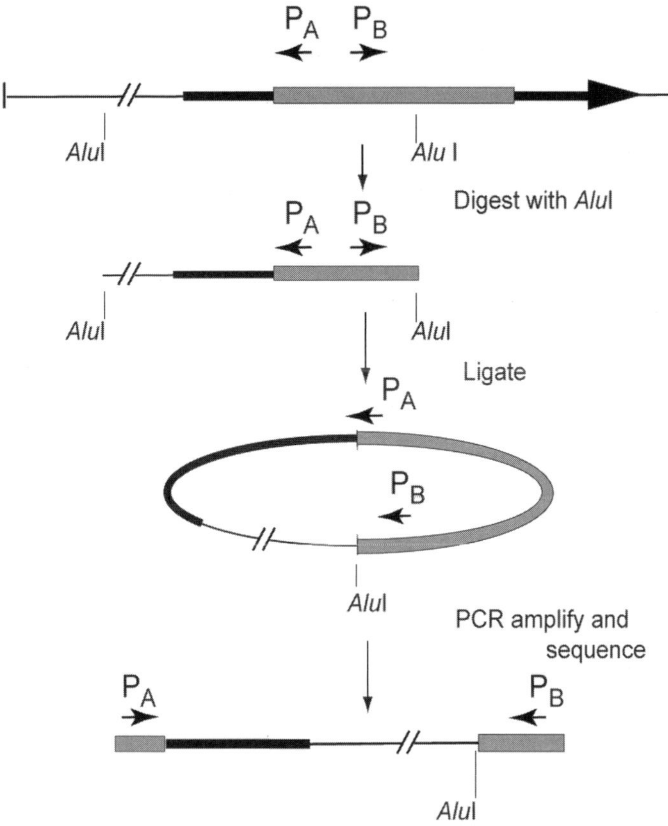

Fig. 1. Schematic diagram of inverse PCR rescue. This method relies on the Blasticidin construct (gray) containing an *Alu*I site. (**A**) Primers are designed at one flanking side of the insertion cassette used to make the library (made with *Dpn*II cut pDIV5[13]) and a primer internal to the Blasticidin gene (P$_A$ and P$_B$; *see* **Note 7**). (**B**) Cleavage of the mutant DNA with *Alu*I and (**C**) ligation allows (**D**) PCR amplification of the genomic region flanking the insertion site.

8. Determine rescued genomic DNA sequence using anchored primers within the cloning vector (*see* **Note 9**) and use the Basic Local Alignment Search Tool (BLAST) server at the online resource, http://dictybase.org, to identify the ablated gene (*see* **Note 10**).

3.4. Recapitulation of the Gene Knockout

Once the gene conferring drug resistance has been identified, the mutation must be recapitulated in a wild-type background to confirm that ablation of the

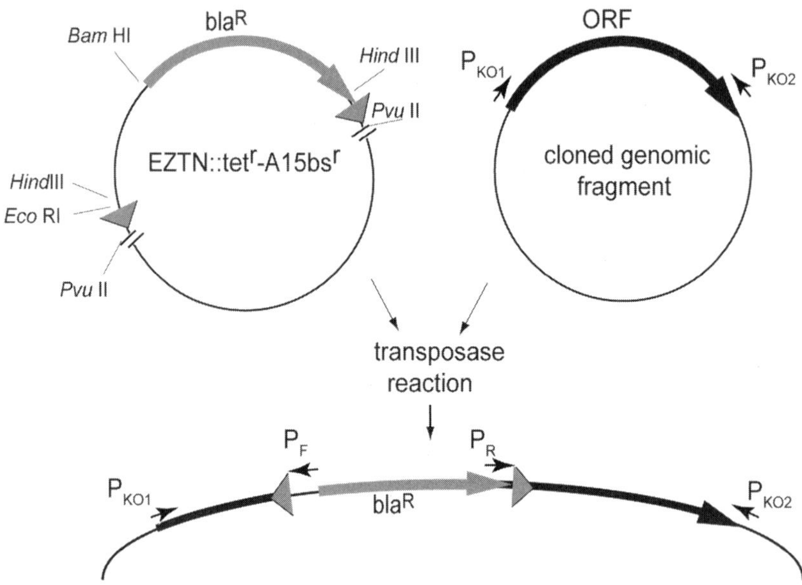

Fig. 2. Schematic drawing showing the construction of a knockout cassette using transposon technology. (**A**) *Pvu*II-digested transposon vector containing the Blasticidin resistance (bla^R) gene is combined with the target vector containing the cloned gene of interest (ORF). (**B**) The position and orientation of the transposon can be determined by PCR analysis using primers binding within the genomic clone and the transposon.

identified locus causes drug resistance/sensitivity. To do this, the gene is amplified from genomic DNA and transposon-hopping technology is used to insert an antibiotic (blasticidin) resistance gene into the cloned coding sequence (*see* **Fig. 2** and Chapter 12), enabling the rapid production of a knockout cassette. The resultant cassette is electroporated into wild-type cells, homologous integrants are isolated by a quick PCR screening approach (*see* **Fig. 3**), and confirmed in a rapid Southern analysis protocol (*see* **Fig. 4**).

3.4.1. Amplification of Gene-Controlling Drug Action

1. Design primers to amplify a 2–3-kb region containing the coding region identified (*see* **Subheading 3.3.** and **Note 11**).
2. Amplify the genomic region using 1 μg of genomic DNA, clone using commercially available kits (e.g., TOPO cloning kit, *see* **Note 8**), and confirm amplified region by sequencing from anchored cloning vector primers (*see* **Notes 9** and **12**).
3. Purify the cloned genomic region using standard protocols (Qiagen columns as described in suppliers notes).

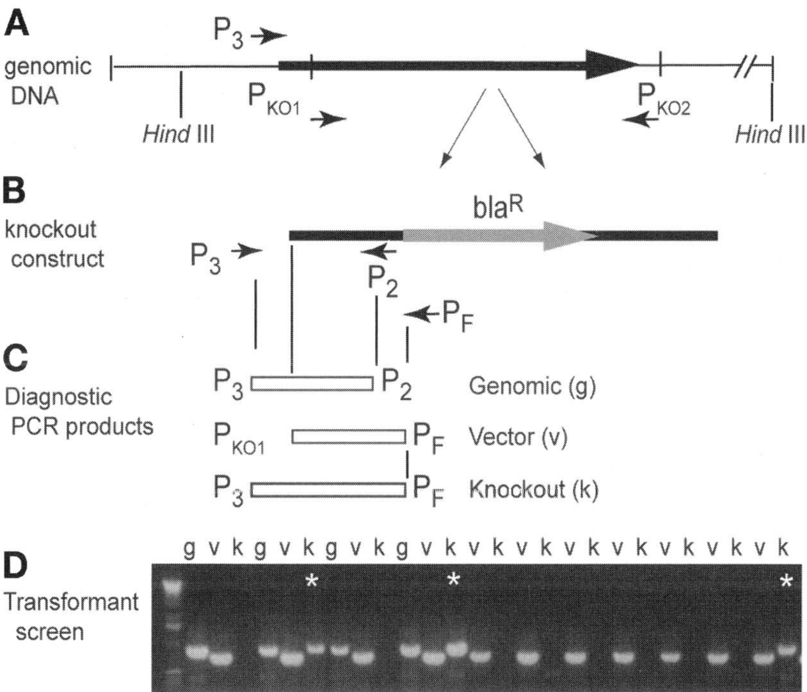

Fig. 3. Schematic drawing showing the PCR screen for potential homologous integrants transformed with a knockout cassette. Homologous crossover between (**A**) the wild-type gene and (**B**) the knockout construct containing the Blasticidin resistance (bla^R) gene can be detected using three primer combinations. (**C**) A diagnostic PCR product amplified from a primer annealing to the gene outside the knockout construct (P3) and within the construct (P2) provides a genomic control (g) for both the P3 primer and the target DNA. A diagnostic PCR product amplified from a primer used to generate the knockout construct (PKO1) and annealing to the transposon (PF) provides a control for both the vector (v) and the PF primer. Amplification of a product from a primer annealing to the flanking region of the knockout construct (P3) and a primer annealing to within the transposon (PF) indicates homologous integration of the cassette (k). (**D**) Clones producing this product (asterisk) are recapitulated gene knockouts.

3.4.2. Transposon-Targeted Gene Disruption

This method is based on transposon hopping (EZ::TN; Epicenter) (*10*).

1. Digest 10 µg of EZTN:tetR-A15-bsr plasmid (*see* **Note 13**) with *Pvu*II (5 µL enzyme, 80 µL volume, 3 h, providing enough for 20 transposon insertions). Purify the cleaved DNA using an S400 spin column.

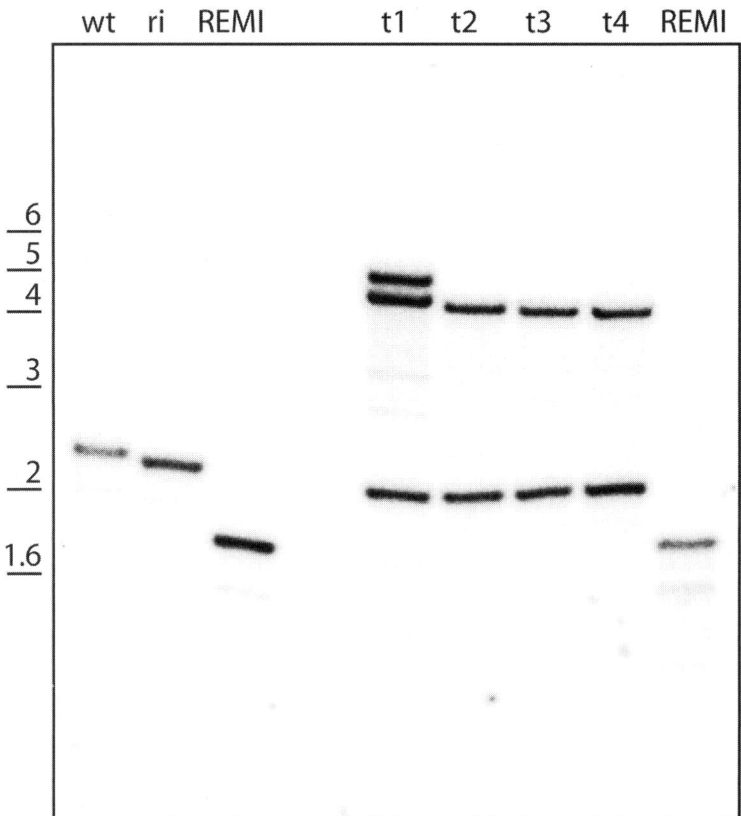

Fig. 4. Typical Southern analysis of multiple *Dictyostelium* Blasticidin resistant transformants. DNA was isolated, digested with *Bcl*I, separated by gel electrophoresis, blotted to Hybond N+, and probed with the region from a gene to be knocked out (*see* **Subheading 3.5.**). Hybridization patterns for wild-type (wt) and a random integrant (ri) show the endogenous *Bcl*I fragment size and the original mutant (REMI) to be recapitulated shows the altered fragment size. Four transformants (t1–t4) are analyzed here, all showing the endogenous 2.4-kb *Bcl*I fragment, and a second or third hybridizing band indicating that the knockout vector has integrated at a nonhomologous site, leaving the endogenous gene unaltered. Size markers are in kb.

2. Use a 100-μL PCR tube to combine 0.1 μg of cloned genomic DNA, 3 μL of *Pvu*II-digested EZTN:tetR-A15-bsr transposon, 0.5 μL of transposase, and 0.5 μL of transposase reaction buffer in a 5-μL total volume.

3. Incubate the transposase reaction at 37°C in a PCR machine for 2 h, stop the reaction with 0.5 μL of stop solution (supplied with transposase), and heat to 70°C for 10 min. Precipitate DNA (add 35 μL dH₂O, 1 μL of glycogen 10 mg/mL,

66 µL of isopropanol), microcentrifuge (table top centrifuge at top speed, 20 min, 4°C), wash the pellet (50 µL of 75% ethanol), and resuspend in 10 µL of dH$_2$O.

4. Add 10 µL of transposase reaction to 40 µL of electrocompetent DH5α cells and transfer to sterile, 0.1 mm electroporation cuvets on ice, electroporate (2.5 kV, 25 µF, 100 Ω), immediately add 1 mL of room-temperature SOC, transfer to 15 mL Falcon tube, and shake for 1 h at 37°C.
5. Plate 200-µL and 500 µL aliquots of the transformed cells onto LB plates containing kanamycin (50 µg/mL), ampicillin (50 µg/mL), and tetracycline (10 µg/mL). Colonies will appear after 24–36 h.
6. Transfer 30 colonies to 5 mL LB plus 50 mg/mL ampicillin only, in 20 mL disposable tubes, and purify plasmids using Qiagen miniprep columns as indicated.

3.4.3. PCR Analysis of Transposase Integrants

1. PCR screen each plasmid (*see* **Note 14**) in a 50 µL reaction:

0.5 µL	Plasmid DNA
5 µL	2 m*M* dNTPs
5 µL	10X Taq reaction buffer
2 µL	50 m*M* MgCl$_2$
0.2 µL	5 U/µL of Taq polymerase
2X 0.5 µL	100 pmol/µL of the following primer pairs:

   a. PKO1 and PF
   b. PKO1 and PR
   c. PKO2 and PF
   d. PKO2 and PR (*see* **Note 15** and **Fig. 2**)

   Add dH$_2$O to 50 µL
2. Analyze each product by gel electrophoresis (*see* **Fig. 3D**).
3. Select a plasmid which shows an integration at least 100 bp away from either end of the sequence to allow homologous cross-over during recombination.

3.4.4. Isolation of Single Transformants

1. Grow wild-type cells (0.5–2 × 10^6 per mL) in shaking axenic medium for 3 d.
2. Linearize 45 µg of knockout vector by restriction enzyme digest (for TOPO-pCR2.1 based vector, use 5 µL of *Bst*XI (*see* **Note 16**), 10 µL of enzyme buffer in 100 µL volume, incubate at 37°C overnight), and purify using a S400 spin column.
3. Wash 1 × 10^7 cells per electroporation in 10 mL of ice-cold electroporation buffer and resuspend in 800 µL of electroporation buffer.
4. Add 15 µg of digested plasmid per electroporation, invert three times, and incubate on ice for 10 min.
5. Place the cells in a chilled 4 mm electroporation cuvet, electroporate twice (infinite Ω, 3 µF and 1.6 kV, *see* **Note 17**), and incubate on ice for 10 min.
6. Add 8 µL of CaMg to each cuvet, mix gently, and incubate at room temperature for 15 min.

7. Transfer the cells from the cuvet (*see* **Note 18**) to a 50 mL Falcon tube containing 40 mL of axenic medium, mix by gentle inversion, and pour into a pipetting reservoir. Using an eight-channel multi-channel pipet, transfer 100 µL aliquots into four labeled 96-well plates. Incubate at 22°C overnight (*see* **Note 19**).
8. After 24 h, add 100 µL of axenic medium per well containing 20 µg/mL (2X) blasticidin and incubate for 10 d.
9. Examine wells using an inverse microscope to identify blasticidin-resistant clones.

3.4.5. Rapid DNA Extraction for PCR Analysis

This method is based on that published by Charette and Cosson *(12)*.

1. For each blasticidin-resistant clone, transfer 200 µL of cell suspension from the 96-well dish (*see* **Note 20**) to a 200-µL PCR tube and spin down (2000*g*, 1 min, room temperature).
2. Remove 95% of the supernatant, taking care not to lose the pellet.
3. Add 100 µL of *Dictyostelium* lysis buffer and vortex well.
4. Incubate at 95°C for 1 min in a hot block or PCR machine.
5. Use the crude suspension for PCR analysis (store at –20°C).

3.4.6. PCR Screen for Homologous Integrants

1. For each clone, carry out three amplification reactions as indicated in **Fig. 3** (*see* **Note 21** for program). In a 50 µL reaction, combine:

2 µL	Crude DNA
5 µL	2 m*M* dNTPs
5 µL	10X Taq reaction buffer
2 µL	50 m*M* MgCl$_2$
0.2 µL	5 U/µL of Taq polymerase
2X 0.5 µL	100 pmol/µL of the following primer pairs:

   a. P3 (outside construct) and P2 (in construct)
   b. PKO1 (in construct) and PF (blasticidin-resistance gene)
   c. P3 and PF

   Add dH$_2$O to a final volume of 50 µL
2. Analyze each product by gel electrophoresis (10 µL per well, 1% gel). *See* **Fig. 3D**.
3. To ensure that a single clone is isolated, prepare a dilution series of cells from wells indicating a homologous integrant and plate onto SM medium with *Klebsiella*. Transfer single colonies to axenic medium with blasticidin and repeat the PCR screen (*see* **Note 22**).

3.5. Confirmation of Gene Ablation by Southern Analysis

Following identification of a clonal isolate containing the ablated gene, it is often necessary to show that the isolate does not contain multiple insertions. Thus, we describe here a rapid method for isolating high-quality DNA for

Southern analysis, and a protocol for digesting, separating, blotting, and probing a Southern blot (*see* **Fig. 4**). Although we do not describe radiolabeled probe production, we would recommend the Megaprime DNA labeling system (Amersham Life Science).

3.5.1. Rapid DNA Extraction for Southern Analysis

1. Pellet 10^7 cells either from log-phase shaking or low-density still plate axenic medium at 2000*g* for 2 min.
2. Gently resuspend the pellet in 500 µL of DNAzol solution and transfer to an Eppendorf tube. Leave at room temperature for 5 min.
3. Spin down at 13,000*g* for 10 min (room temperature) and discard pellet.
4. Precipitate DNA from supernatant using 250 µL of 100% ethanol, stand at room temperature for 5 min, and pellet the DNA at 13,000*g* for 5 min.
5. Wash the pellet with 500 µL of 70% ethanol and centrifuge for 2.5 min at 13,000*g*.
6. Air-dry the pellet for 10–15 min and resuspend DNA in 20 µL of 8 m*M* NaOH.
7. Digest 10 µL of DNA over night in a total volume of 50 µL with 50–100 U of appropriate restriction enzyme in 1X manufacturer's buffer. Add 5 µL DNA loading dye.
8. Run total volume of the digested DNA on a TBE-agarose ethidium bromide gel at 80–130 V until the bromophenol blue band is 1 cm from the bottom. Visualize the gel under ultraviolet (UV)light to confirm digestion of the DNA and photograph with a ruler placed on either side of the gel.
9. Remove markers and any unnecessary gel parts before blotting.

3.5.2. Preparing a Southern Blot

1. Soak gel in depurination solution for 10 min while shaking gently on rocking platform (or until Bromophenol blue turns yellow).
2. Rinse gel in dH₂O and soak for 30 min in denaturation solution while shaking gently on rocking platform
3. Rinse gel in distilled water and soak for 30 min in neutralization solution while shaking gently on rocking platform.
4. Transfer DNA onto charged nylon membrane by capillary transfer using 10X SSC as transfer solution *(14)*.
5. Mark sides of filter with a pencil and cut off a corner on one side before fixing the DNA to the charged nylon membrane by UV cross-linking. Store at –20ºC.

3.5.3. Hybridization and Autoradiography

1. Place the membrane in a rotary hybridization oven tube and prehybridize with 30 mL of hybridization buffer for 30 min at 63°C (2–10 rpm).
2. Discard prehybridization solution and replace with 10 mL of hybridization buffer containing the freshly boiled radiolabeled probe. Hybridize overnight at 63°C in rotary oven.
3. Wash membrane twice for 15 min at 63°C in wash solution 1 in a rotary oven.

4. Wash membrane twice for 15 min at 63°C in wash solution 2 in a rotary oven.
5. Seal membrane into plastic whilst still damp and expose to X-ray film with screen amplifiers at –80°C until ready for development (usually overnight).

3.6. Inositol Trisphosphate Analysis

This protocol is adapted from that provided by the manufacturer (TRK 1000; Amersham Pharmacia Biotech).

3.6.1. Drug Treatment of Cells

1. Maintain cells in axenic medium in shaking suspension at 0.5–3×10^6 cells per mL for 3 d, wash cells in phosphate buffer, and resuspend at 5×10^7 cells per mL in phosphate buffer.
2. For basal (growing cell) InsP$_3$ levels, place 1 mL of cell suspension in a 15-mL Falcon tube and aerate (*see* **Note 23**) for 10 min (22°C).
3. For analysis of cells in development, place cell suspension in a 50 mL Falcon tube (minimum 3 mL) and aerate for 5 h. To determine if drug treatment alters InsP$_3$ levels during development, this step can include duplicate cell samples containing the drug.
4. Transfer 200 µL aliquots of aerated cells to ice cold Eppendorf tubes containing 40 µL of 20% perchloric acid. Take triple samples and keep on ice for at least 10 min.
5. Transfer 200 µL aliquots of aerated cells to ice cold Eppendorf tubes and freeze in duplicate for protein determination.
6. For protein determination, sonicate cell suspension twice (2 s pulse) and use Coomassie protein determination kit, as instructed (Biorad).

3.6.2. InsP$_3$ Extract Preparation

1. Spin acidified cell extracts ($12,000g$, 4°C, 10 min), and transfer 200 µL of supernatant to freshly labeled, chilled Eppendorf tubes. (Extracts can be frozen here for later purification.)
2. For each sample, add neutralising indicator solution (~30 µL) until color indicates pH of approx 7.0 (clear or very pale pink color). Then, carefully add 0.5 µL quantities of neutralizing indicator solution, vortex, and test 1 µL aliquots with pH indicator strips until pH is 7.5, making sure to keep samples cold. Once all samples are balanced, go back and test all samples again using 1 µL aliquots on pH indicator strips to ensure an equivalent pH over the whole sample set (*see* **Note 24**). If sample pH exceeds 9.0, the samples may be ruined, so be careful.
3. Spin at $12,000g$ at 4°C for 10 min, and transfer 30 µL of supernatant into ice-cold, labeled tubes in duplicate or triplicate (samples may be stored following this step at –20°C).

3.6.3. InsP$_3$ Assay

1. In a disposable 30 mL screw-top tube, prepare enough assay master mix for each standard/control in duplicate and sample in triplicate, containing 30 µL of assay buffer and 30 µL of radiolabeled InsP$_3$ diluted in dH$_2$O.

2. Prepare labeled Eppendorf tubes on ice for unknowns, standards, and controls.
3. On ice, aliquot duplicate 30 μL standards (*see* **Note 25**), dH$_2$O, and stock InsP$_3$ (NSB) for controls into tubes. Thaw Eppendorf tubes containing 30 μL of unknown samples on ice.
4. Transfer 60 μL of ice-cold master mix to the bottom of ice-cold Eppendorf tubes for sample/standard/controls.
5. Divide Eppendorf tubes into batches that can be centrifuged at one time at 4°C.
6. Thaw crude InsP$_3$-binding protein preparation on ice (*see* **Note 26**).
7. Vortex binding protein preparation prior to and after taking each third aliquot. Using a large-orifice, 200-gauge tip, transfer 30 μL aliquots of binding protein into tubes containing sample/assay mix. Vortex tubes immediately and keep on ice. Repeat until the batch is complete. Incubate 5 min on ice.
8. Microfuge samples (12,000g, 4°C, 2 min) and immediately aspirate off supernatant using suction apparatus with gel loading tip. Ensure you remove all the liquid and no pellet. Be careful, as liquid is radioactive.
9. Samples can then be left at room temperature. Add 200 μL of water to each tube and let stand (at least 30 min).
10. Repeat until all sample, standard, and control tubes have been tested.
11. Transfer tube contents to scintillation vial. Add 2 mL of Ecolite scintillation fluid. Vortex thoroughly.
12. Count samples in scintillation counter for 5 min per sample (*see* **Note 27**).
13. Calculate InsP$_3$ sample levels as detailed in kit instructions. This can then be expressed per unit cell number or per unit protein.

4. Notes

1. *Dictyostelium* cells should be kept on ice prior to experimentation or while handling.
2. *Klebsiella pneumonia* is available from the *Dictyostelium* Bacterial Strain Catalogue at http://dictybase.org/.
3. Filters (once cells have settled) and filter pads can be cut into halves or quarters to maximize drug conditions tested.
4. Drug treatment may slow or quicken development, so filters should be examined regularly. Changes in both the rate of development and the morphology of the fruiting body should be noted. Keep dishes moist using wet tissue paper at bottom of the storage box.
5. Depending on the blasticidin cassette employed, other restriction sites can be used (e.g., *Sac*I).
6. To amplify the genomic region containing the gene to be ablated, the genomic DNA was subjected to:

   94°C for 2 min, then 29 cycles of
   94°C for 30 s
   58°C for 30 s
   68°C for 4 min
   Final extension step at 68°C for 10 min.

7. For REMI libraries constructed using pDIV5 *(13)*, as described previously *(9)*, use

    P_A: TATCTAGGTAATACGACTCACTATAGGG
    P_B: ATAGGACGAGTAACTGTTTGTGCAG.

    Other blasticidin cassettes may require specific primers.
8. TOPO cloning works best with fresh PCR products.
9. For TOPO pCR2.1 vector, use primers:

    PR 5'-GAGCCAATATGCGAGAACACCCGAGAA-3'
    PF 5'-GCCAACGACTACGCACTAGCCAAC-3'

10. Determined sequence will contain region both directly adjacent to insert site and sequence from the distal part of the *Alu*I restriction fragment.
11. Primers should contain two or three G/C residues in the five most-3' region, not hybridize adjacent to a poly A/T rich region, and have a melting temperature above 55°C. Choose a region with two adjacent parallel primer sites on each side of the amplified region (e.g., *see* **Fig. 3**, P3 and PKO1). The inner pair of flanking primers is used to make the construct. The outer pair is used to screen for homologous integrants.
12. It is only necessary to sequence the flanking regions, as point mutations are not important.
13. The EZTN:tetRA15-bsr plasmid can be obtained from dictyBase (*see* http://www.dictybase.org/StockCenter/PlasmidList.html, cat no. 10).
14. To screen bacterial colonies for insert position, use the following program:

    94°C for 2 min, then 29 cycles of
    94°C for 30 s
    2°C below lowest primer annealing temperature for 30 s
    68°C for 2 min

    Final extension step at 68°C for 10 min.
15. All four primer combinations are used to identify multiple insertions.
16. Genomic DNA must not have this site.
17. The time constant should be between 0.3 and 0.5 ms.
18. Ensure that all cells are removed using a gel-loading tip to reach the bottom of the cuvet.
19. Incubate in a closed box with a damp tissue lining to reduce evaporation.
20. Pipet the medium up and down to dislodge cells from the bottom of the well.
21. To amplify genomic DNA from crude 96-well DNA preparations, use the following program:

    94°C for 2 min, then 29 cycles of
    94°C for 30 s
    2°C below lowest primer annealing temperature for 30 s
    68°C for 4 min

    Final extension step at 68°C for 10 min.
22. It is advisable to also sequence across the blasticidin cassette in the single transformant. If both wild-type band and diagnostic knockout bands are found,

     repeat wild-type cell transformation using different electroporation conditions.

23. Use a standard fish tank air bubbler, piping, and valves for multiple samples attached to disposable glass tubing (Blaubrand, cat. no. 7087-18) to provide a fine point for air outlet.

24. It is essential that all samples have the same pH. Thus, it is critical to ensure that all samples, once brought to around pH 7.5, are tested again.

25. Standards can be thawed up to three times.

26. When first defrosting the binding protein, divide into aliquots and store in glass at –80°C.

27. Occasionally, run wet tissue over bench and add to scintillation fluid to check for contamination.

References

1. Williams, R. S. (2005) Pharmacogenetics in model systems: defining a common mechanism of action for mood stabilisers. *Prog. Neuropsychopharmacol. Biol. Psychiatry* **29,** 1029–1037.

2. Williams, R. S., Eames, M., Ryves, W. J., Viggars, J., and Harwood, A. J. (1999) Loss of a prolyl oligopeptidase confers resistance to lithium by elevation of inositol (1,4,5) trisphosphate. *EMBO J.* **18,** 2734–2745.

3. Williams, R. S., Cheng, L., Mudge, A. W., and Harwood, A. J. (2002) A common mechanism of action for three mood-stabilizing drugs. *Nature* **417,** 292–295.

4. Berridge, M. J., Downes, C. P., and Hanley, M. R. (1989) Neural and developmental actions of lithium: a unifying hypothesis. *Cell* **59,** 411–419.

5. Leech, A. P., Baker, G. R., Shute, J. K., Cohen, M. A., and Gani, D. (1993) Chemical and kinetic mechanism of the inositol monophosphatase reaction and its inhibition by Li$^+$. *Eur. J. Biochem.* **212,** 693–704.

6. York, J. D., Ponder, J. W., and Majerus, P. W. (1995) Definition of a metal-dependent/Li(+)-inhibited phosphomonoesterase protein family based upon a conserved three-dimensional core structure. *Proc. Natl. Acad. Sci. USA* **92,** 5149–5153.

7. Eickholt, B. J, Towers, G. J., Ryves, W. J., et al. (2005) Effects of valproic acid derivatives on inositol trisphosphate depletion, teratogenicity, GSK-3β inhibition and viral replication—a screening approach for new bipolar disorder drugs based on the valproic acid core structure. *Mol. Pharmacol.* **67,** 1–8.

8. Shaltiel, G., Shamir, A., Shapiro, J., et al. (2004) Valproate decreases inositol biosynthesis. *Biol. Psychiatry* **56,** 868–874.

9. Keim, M., Williams, R. S., and Harwood, A. J. (2004) An inverse PCR technique to rapidly isolate the flanking DNA of *Dictyostelium* insertion mutants. *Mol. Biotechnol.* **26,** 221–224.

10. Abe, T., Langenick, J., and Williams, J. G. (2003) Rapid generation of gene disruption constructs by in vitro transposition and identification of a *Dictyostelium* protein kinase that regulates its rate of growth and development. *Nucleic Acids Res.* **31,** e107.

11. Church, G. M. and Gilbert, W. (1984) Genomic sequencing. *Proc. Natl. Acad. Sci. USA* **81,** 1991–1995.

12. Charette, S. J. and Cosson, P. (2004) Preparation of genomic DNA from *Dictyostelium discoideum* for PCR analysis. *Biotechniques* **36,** 574–575.
13. Harwood, A. J., Plyte, S. E., Woodgett, J., Strutt, H. and Kay, R. R. (1995) Glycogen synthase kinase 3 regulates cell fate in *Dictyostelium. Cell* **80,** 139–148.
14. Sambrook, J., Fritsch, E., and Maniatis, T. (2001) *Molecular Cloning: a Laboratory Manual.* Cold Spring Harbor Laboratory, Cold Spring Harbor, NY.

33

How to Assess and Study Cell Death in *Dictyostelium discoideum*

Artemis Kosta*, Catherine Laporte*, David Lam, Emilie Tresse, Marie-Françoise Luciani, and Pierre Golstein

Summary

In this chapter, we describe how to conveniently demonstrate, assess, and study cell death in *Dictyostelium* through simple cell culture, clonogenic tests, and photonic (with the help of staining techniques) and electronic microscopy. Cell death can be conveniently generated using minor modifications of the monolayer technique of Rob Kay et al., and either wild-type HMX44A *Dictyostelium* cells or the corresponding *atg1⁻* autophagy gene mutant cells. Methods to follow cell death qualitatively and quantitatively facilitate detailed studies of vacuolar death in wild-type cells and of nonvacuolar, "condensed" death in *atg1⁻* mutant cells.

Key Words: *Dictyostelium*; cell death; methods.

1. Introduction

Dictyostelium starvation-induced development ultimately leads to a sorocarp which comprises spores and stalk cells. Stalk cells are highly vacuolated *(1–6)* and dead, as demonstrated by their inability to regrow in culture medium *(7)*. *Dictyostelium* stalk cells may thus be considered one of the earliest known occurrences of developmental programmed cell death in eukaryote evolution.

Why study cell death in *Dictyostelium*? One of the main types of cell death, caspase-dependent apoptosis, has been intensively investigated in a limited number of biological models (essentially in two invertebrates, *Caenorhabditis elegans* and *Drosophila melanogaster*, and two vertebrates, man and mouse). The main caspase-independent cell death types (autophagic or necrotic cell

---

*The first two authors contributed equally to this work.

From: *Methods in Molecular Biology, vol. 346:* Dictyostelium discoideum *Protocols*
Edited by: L. Eichinger and F. Rivero © Humana Press Inc., Totowa, NJ

death), which are now raising considerable interest, were not immediately prominent in these models and, perhaps accordingly, have been far less investigated. In other, simpler biological models, these caspase-independent cell death types may be more experimentally accessible and may show less molecular redundancy, while being, it is hoped, phylogenetically conserved. *Dictyostelium* shows cell death, is (relatively) simple, and shows marked genetic tractability, and thus constitutes a tempting model organism with which to study cell death.

How to study cell death in *Dictyostelium*? Programmed cell death in *Dictyostelium* can be studied in vitro using conditions mimicking developmental circumstances that are more amenable to microscopic observation and to further genetic manipulations than in vivo stalk cell death. We took advantage of a protocol for differentiation in monolayers *(8)*, and of a *Dictyostelium* mutant strain called HM44 *(9)* derived from the original *Dictyostelium* strain V12M2, adapted for axenic growth as HMX44 (J.G. Williams, University of Dundee) and subcloned in our laboratory as HMX44A. Differentiation to stalk cells (i.e., to vacuolated, dead cells) seems to result from the sequential action of at least two main factors under starvation conditions: first, cyclic AMP (cAMP), and second, the dichlorinated hexanone differentiation-inducing factor DIF-1, which acts on starved cAMP-subjected cells and promotes their differentiation into stalk cells *(10–13)*. HM44 produces very little DIF but is sensitive to exogenous DIF *(9)*. Upon starvation and addition of exogenous DIF, HM44 differentiates into stalk cells, however without morphogenizing into a sorocarp.

In brief, *Dictyostelium* HMX44A cells were subjected to the usual *(8,14,15)* sequence of incubation in starvation medium in the presence of cAMP, which does not lead in itself to cell death, followed by another period of incubation together with DIF, which triggers a cascade of events leading to cell death. This cascade included differentiation into paddle cells, which rounded up and then vacuolized, with cytoplasmic condensation, focal chromatin condensation, and very late membrane lesions *(14,15)*. No early DNA fragmentation could be detected by standard or pulsed field gel electrophoresis *(14)* or by terminal deoxyribonucleotide transferase (TdT)-mediated dUTP nick end labeling (TUNEL) assays (unpublished). In addition, no phosphatidylserine externalization could be detected (unpublished). This cell death is caspase- and paracaspase-independent *(16,17)* and vacuolization, but not cell death itself, is dependent on an intact *atg1* autophagy gene *(18)*. Although in this chapter we usually refer to cell death as studied in our laboratory, other groups have also studied cell death in *Dictyostelium*, often in nondevelopmental contexts *(19–24)*.

In this chapter, a *Dictyostelium* cell will be considered dead when its membrane is altered as reflected by an abnormally increased permeability to certain dyes, and/or when it has lost its ability to multiply as reflected by loss of clonogenicity. Some of the methods used have been described in detail in a previous report *(25)*, together with genetic approaches to the molecular mechanism of cell death that will not be touched upon here. In this chapter, we will consider how to induce an equivalent of developmental cell death in culture, how to qualitatively study cell death by photonic and electronic microscopy, and how to quantitatively assess cell death through clonogenicity tests and flow cytometry.

2. Materials

1. HL-5 medium for culture of axenic strain cells: 14.3 g/L bacteriological peptone (OXOID LTD, Basingstoke, Hampshire, UK, L37) (*see* **Note 1**), 7.15 g/L yeast extract (EZMix yeast extract, Sigma, Y1626), 9 g/L maltose (Sigma, M5885), 0.93 g/L $Na_2HPO_4 \cdot 7H_2O$ (3.6 m*M*), 0.49 g/L KH_2PO_4 (3.6 m*M*), water. Rather than column-demineralized water, we use source water to reduce variations in the quality of water (*see* **Note 2**). Sterilize by autoclaving for 30 min. After cooling, filter through a 0.22-μm filter and store at 4°C. This medium is stable for several months at 4°C (*see* **Note 3**).
2. cAMP (cyclic AMP; 3',5' adenosine cyclic monophosphate, sodium salt) (Sigma A6885). Make stock solution 60 m*M* in water, filter-sterilize, keep at –20°C in 1-mL aliquots.
3. DIF-1 [1-(3,5-dichloro-2,6-dihydroxy-4-methoxyphenyl)-hexan-1-one] (Affiniti Research Products). Make stock solution 10 m*M* in absolute ethanol. Working stocks are diluted to 0.1 m*M* (in absolute ethanol) in 1-mL aliquots. Store at –20°C.
4. Soerensen buffer (SB) 50X stock solution: 100 m*M* Na_2HPO_4, 735 m*M* KH_2PO_4, pH 6.0; filter-sterilize and store at 4°C. SB 1X buffer is obtained by adding 10 mL of 50X solution to 490 mL of autoclaved source water.
5. Lab-Tek: Nalge Nunc, Lab-tek chambered Coverglass @1 German borosilicate sterile, two-well (in fact, two plastic chambers set on a slide-size borosilicate coverslip).
6. Propidium iodide (PI, Sigma P4170) stock solution: 80 μ*M* (53 μg/mL) in sterile water; keep at 4°C protected from light. Gloves should be worn, because PI is carcinogenic.
7. Fluorescein diacetate (Sigma F7378) stock solution: 10 mg/mL in acetone; store at 4°C away from light.
8. Calcofluor (Sigma F6259; now sold under cat no. F3397; also named "fluorescent brightener 28," "calcofluor white M2R," "C.I. [Colour Index] 40622," or "tinopal LPW"): stock solution 1% (w/v) in H_2O; keep at 4°C protected from light.
9. Trypan Blue (Sigma T8154) is sold as a 0.4% stock solution. Dilute four times with SB and store at room temperature under sterile conditions (may be frozen; 0.1% sodium azide can be added to prevent contamination). Gloves should be worn when manipulating Trypan Blue, which is teratogenic and may also include carcinogenic compounds.

10. Glutaraldehyde Grade I (Sigma, G5882): 1–2% glutaraldehyde in 200 mM HEPES buffer, pH 7.4 from a 25% aqueous solution. Always use fresh fixative solution. Keep the stock solution at –20°C. Once a vial has been thawed, aliquot and keep it at –20°C.

11. Osmium tetroxide (EMS, 19150): 1% OsO_4 in phosphate-buffered saline (PBS) buffer, pH 7.4 (Gibco cat. no. 10010015) from a 4% aqueous solution. OsO_4 works also fine as 1% aqueous solution. Keep the unopened 4% solution at 4°C. Once opened, store in an extremely clean vial with a good stopper. Put parafilm around the lid and keep the vial in a bigger vial, again with a good lid and parafilm around.

12. Uranyl acetate (EMS, 22400): 1–2% uranyl acetate aqueous solution. Prepare in a dark bottle and store at 4°C. Filter before use.

13. Low Melting Point Agarose (Sigma, A9414): 2% agarose in PBS buffer, pH 7.4. Bring to boiling point in a microwave oven and store at 4°C. When you need it, bring to boiling point and put in a water bath at 40°C.

14. EPON kit (Polysciences, 08792): Contents of the kit: 1X 500 g Poly/Bed 812, 1X 450 g dodecenyl succinic anhydride (DDSA), 1X 450 g nadic methyl anhydride (NMA), 1X 100 g DMP-30, complete instructions included. Prepare following the instructions in the kit. Most commercially available resins give you the weight per epoxy equivalent (WPE) of the EPON component. The "WPE" is listed on the bottle and it can differ from batch to batch.

15. Lead citrate solution, prepared as follows: In a 50-mL, clean flask, add: 1.33 g of lead nitrate (EMS, 17800), 1.76 g of sodium citrate (EMS, 21140), 30 mL of boiled, cooled, CO_2-free distilled water. Cover the flask. Shake vigorously for at least 1 min. Add 8 mL of 1 N CO_2-free NaOH (EMS, 21170). Mix until clear. The pH should be 12.0 ± 0.1. If the pH is low, add more NaOH to the clear solution. If the pH is above 12.1, start over, this time adding a smaller amount of NaOH. Add CO_2-free water to bring the solution to a final volume of 50 mL. Let stand several hours before use. Store in syringes at 4°C, tightly closed, and filter before use.

3. Methods

3.1. Induction of Developmental Cell Death: Stalk Cell Differentiation in Monolayer (see Notes 4–6)

1. Collect vegetative cells growing in HL-5 medium in late log phase (i.e., no more than 2×10^6 cells/mL if cells are grown without stirring), by shaking the flasks and centrifuging the resulting cell suspensions at 550g for 5 min.

2. Wash twice with SB buffer, count.

3. To each of two Lab-Tek chambers, add 3×10^5 cells in 1 mL of SB containing 3 mM cAMP.

4. Incubate for 8 h at 22°C. Most (>80%) cells should adhere firmly to the bottom slide. Because of the high concentration of cAMP, almost no aggregation should be seen at this stage, and the cells should be randomly scattered and isolated.

5. Carefully remove the liquid and wash once with 1 mL of SB.

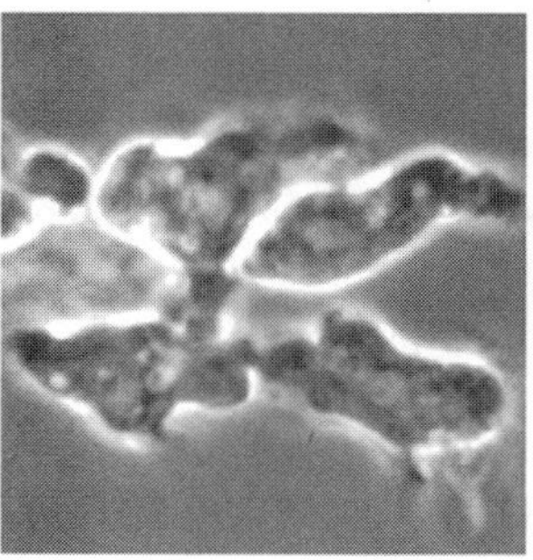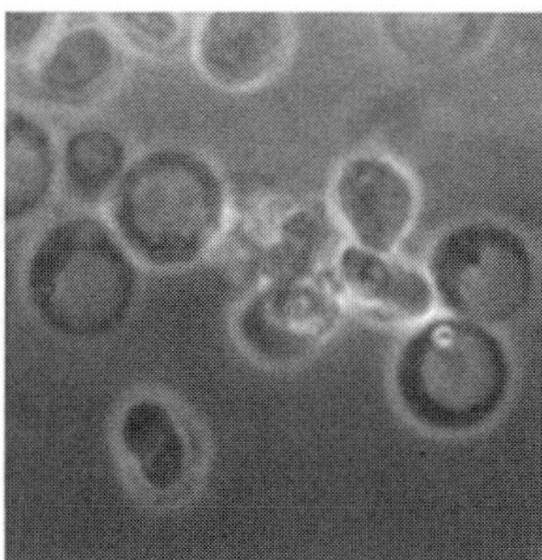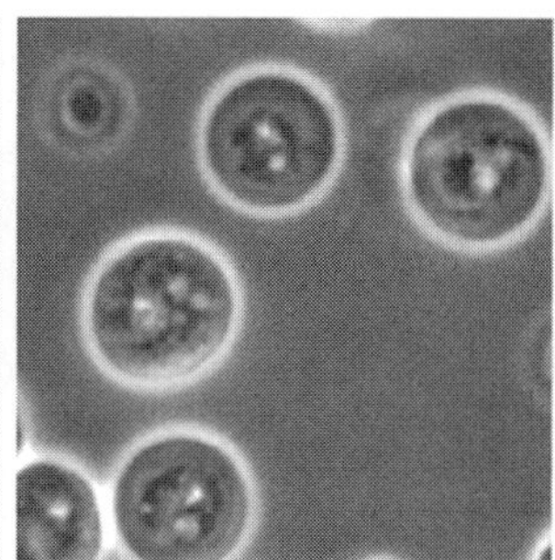

Fig. 1. Phase-contrast pictures of (left) vegetative HMX44A *Dictyostelium* cells, (middle) HMX44A cells subjected to starvation and differentiation-inducing factor (DIF) for 24 h showing vacuolated dying cells, and (right) HMX44A.*atg1⁻* cells subjected to starvation and DIF for 30 min, showing central condensation of cells doomed to die a nonvacuolar cell death.

6. Replace with:
   - 1 mL of SB containing 0.1 μ*M* DIF-1 in one chamber
   - 1 mL of SB with no DIF in the other chamber

7. Incubate for 24 h at 22°C.

At this stage, in the DIF-containing chamber, most cells should be differentiated to "stalk" cells: highly vacuolated, cellulose-encased, nonrefringent cells as seen by phase-contrast microscopy (*see* for instance **ref. 8**). Almost no vacuolated cells should be seen in the control chamber if a cell line producing little endogenous DIF (e.g., HMX44A) is used. Further incubation will not increase the proportion of dead cells much, but vacuolization will become progressively more prominent as the cytoplasm of dying/dead cells continues to shrink.

3.2. Qualitative Assessment of Cell Death

3.2.1. Mere Microscopic Examination Using Phase Contrast

Dictyostelium cells are small (around 10 μm across), thus a 100× objective is usually required. We use standard phase contrast (*see* **Fig. 1**). We feel that Nomarski differential interference contrast (DIC) shows fewer details. Time-lapse microscopy is often of tremendous help, as it particularly allows one to easily rank the various steps of a given process in time (see movies in **ref. 15**).

3.2.2. Propidium Iodide Staining

Propidium iodide (PI) is a DNA-intercalating dye that cannot cross cell membranes freely ; thus, cells will fluoresce only if membranes have become permeable, a late sign of cell death. This implies that cells must be stained fresh, e.g., they cannot be fixed for this test.

1. Add PI concentrated stock to cells in SB to reach a final concentration of 4 μM PI.
2. Incubate 10 min at room temperature away from light.
3. Wash twice carefully with SB.
4. View under the fluorescence microscope (Zeiss filter set 15: BP 546/12, FT 580, LP 590), observe red fluorescence.

3.2.3. Fluorescein Diacetate Staining

In contrast to PI, fluorescein diacetate (FDA) stains living cells. The nonfluorescent, hydrophobic compound freely enters the cell, where it is cleaved by cytoplasmic lipases of metabolically active cells into a green fluorophore that is unable to leave the cell if the membrane is intact. Again, cells must not have been fixed.

1. Wash cells once with SB.
2. Add FDA concentrated stock to cells in SB to reach a final concentration of 0.05 mg/mL.
3. Incubate 10 min at room temperature away from light.
4. Wash twice with SB.
5. View under the fluorescence microscope (Zeiss filter set 09: BP 450-490, FT 510, LP 515), observe green fluorescence.

FDA and PI can be used for double staining by directly mixing the two dyes.

3.2.4. Calcofluor Staining

As they differentiate, stalk cells encase themselves in a cellulose coat that may be labeled with calcofluor. Positive staining does not constitute evidence of cell death, but is nevertheless a useful differentiation marker. One should be aware that other cell types (e.g., spores and macrocysts) also secrete cellulose coats, but cell sizes are very different. A thin cellulose trail is also left on the substrate before the cell finally stops migrating and fully differentiates *(26)*. Cells may be fixed before staining.

1. Wash cells with SB.
2. Add calcofluor concentrated stock to cells in SB to reach a final concentration of 0.1%.
3. Incubate for 5 min at room temperature protected from light.
4. Wash twice with SB.
5. View under the fluorescence microscope (Zeiss filter set 01: BP 365/12, FT 395, LP 397), observe blue fluorescence.

3.2.5. Trypan Blue Staining

Trypan Blue is a dye that is commonly used to discriminate between live and dead mammalian cells, because it stains only cells with a compromised plasma membrane. Staining with Trypan Blue or PI thus provides comparable information. Trypan Blue-stained cells are less obvious than PI-labeled cells, but as fluorescence is not required, Trypan Blue is more convenient for routine quantification of live vs dead cells in a sample.

Trypan Blue is simply added to cells at a final concentration of 0.1–0.03% for at least 10 min, and the sample is directly examined using standard microscopy. The liquid layer must be thin; for cells in 25-cm^2 flasks, 1 mL of Trypan Blue solution is suitable. It is important not to allow cells to dry, as the liquid tends to accumulate as a meniscus at the corners of the flask. Results will be optically better if the cells are placed between a microscopic slide and a coverslip. Use of a blue filter enhances contrast.

Other stains may be used, as a function of the specific questions being asked, e.g., whether to stain cell organelles such as mitochondria or structures such as cytoskeleton components, or whether to detect metabolites such as reactive oxygen species. These other stains will not be described here.

3.2.6. Electron Microscopy

Electron microscopy is a valuable tool that allows the structure of organelles and cellular subsystems to be characterized with unprecedented detail and reliability. There are numerous protocols and staining techniques with which to study *Dictyostelium*. The protocol described here aims at the preservation of the ultrastructural morphology of *Dictyostelium* cells (*see* **Note 7** and **Figs. 2 and 3**). At the time of fixation, cells should be $1–2 \times 10^6$ per mL and the total volume should be at least 5 mL.

3.2.6.1. Fixation with Glutaraldehyde

1. Add an equal amount of double-strength fixative (2% glutaraldehyde in 200 m*M* HEPES buffer, pH 7.4) directly into the cell culture. Leave for 10 min at 23°C, followed by 30 min or longer at 4°C (here you can interrupt the procedure and leave the cells in the fixative or in the buffer at 4°C, overnight or for long-term storage or shipping).
2. Carefully scrape the cells off of the substrate using a rubber scraper or a piece of Teflon and transfer them into a tube.
3. Add bovine serum albumin (BSA) (1%) to prevent the cells from sticking to the sides of the tube and pellet scraped cells by centrifugation (3 min, 1500*g*).
4. Remove supernatant and add 1% glutaraldehyde in 200 m*M* HEPES buffer, pH 7.4. Leave for 1 h at 4°C.
5. If the pellet is tight, do not disturb it, but continue. If the pellet is loose or is falling apart, pellet the cells again in the presence of 1% BSA by centrifugation (3 min, 700*g*). Remove the excess of liquid and resuspend the cells in a drop of 2% agarose (LMP, prewarmed at 40°C) prepared in PBS. Quickly centrifuge the

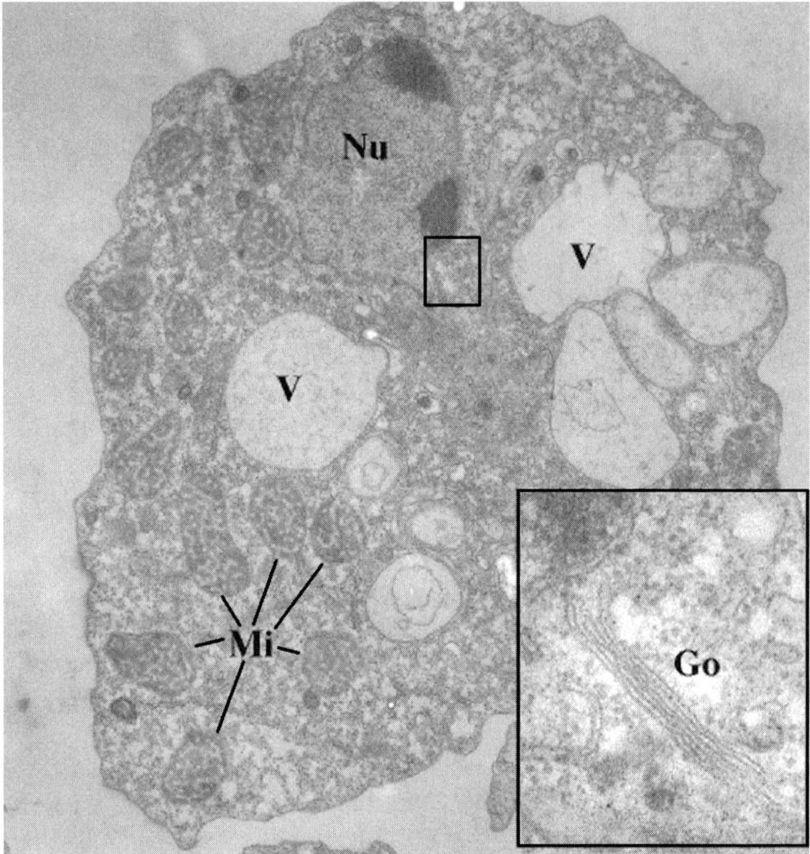

Fig. 2. An electron-microscope picture of a vegetative HMX44A wild-type cell showing numerous mitochondria, vacuoles, the nucleus, and a small Golgi apparatus. Original magnification: 8,000. Nu, nucleus; Mi, mitochondria; V, vacuoles. Inset: details of the Golgi apparatus (Go). Original magnification: 63,000.

cells in an Eppendorf tube (3 min, 700g) and cool the pellet on ice. Cut off the tip of the tube and remove the agarose-embedded pellet. On ice, slice the pellet into blocks of the desired size.

3.2.6.2. Washing in PBS, pH 7.4

Transfer the blocks into small glass vials and wash the cells with PBS, pH 7.4, four times for 5 min (HEPES buffer is very good for aldehyde fixation, but it reacts with OsO_4 and must be removed prior to osmification. The same is true for PIPES).

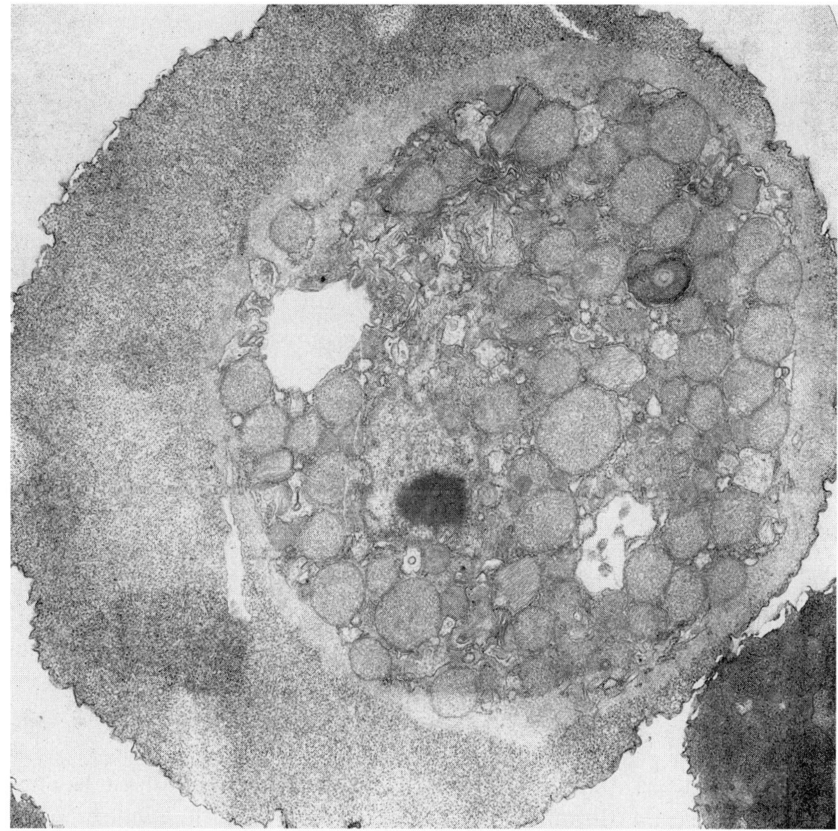

Fig. 3. An electron-microscope picture of an HMX44A. *atg1⁻* mutant cell, about 30 min after addition of differentiation-inducing factor. The cell is round shaped, showing perinuclear clustering of mitochondria and organelle-free periphery. Original magnification: 8000.

3.2.6.3. Postfixation and Staining with Osmium

1. Remove PBS and quickly add 1% OsO_4 in PBS, pH 7.4 for 1 h at 4°C.
2. Wash with distilled water, four times for 5 min (be careful to remove the last traces of phosphate from your samples before you incubate with uranyl acetate).

3.2.6.4. Postfixation and Staining with Uranyl Acetate

1. Add 1–2% uranyl acetate aqueous solution for 1 h, at 4°C in the dark (here you can interrupt the process and leave the cells in the dark at 4°C, continuing the next day).
2. Wash with distilled water, four times for 5 min.

3.2.6.5. Progressive Dehydration with an Acetone Series

Replace water with:

50% acetone in distilled water, three times for 5 min, at room temperature;
70% acetone in distilled water, three times for 5 min, at room temperature;
90% acetone in distilled water, three times for 5 min, at room temperature;
100% acetone, three times for 10 min (dried in molecular sieves), at room temperature.

3.2.6.6. Embedding and Polymerization in Resin

Immediately after the last dehydratation step, replace acetone with the following mixtures :

1. 1 part resin : 3 parts acetone (dried), 30 min, room temperature.
2. 1 part resin : 1 part acetone (dried), 30-60 min, room temperature.
3. 3 parts resin : 1 part acetone (dried), 30 min (or overnight), room temperature.
4. Pure resin, 2 h, room temperature.
5. Pure resin, 2 h, 37°C.
6. In the meantime, prepare paper labels (use a pencil, as printer ink or pen ink might smear) for the flat molds or capsule beams.
7. Transfer the cell pellets into the molds or capsules with fresh, pure resin, orient them for best sectioning position, remove bubbles, and polymerize at 60°C for 48 h.

3.2.6.7. Ultrathin Sections

Ultrathin sections (pale gold to silver: 90–60 nm) cut with a glass or diamond knife using an ultramicrotome are floated on distilled water and transferred to support grids (copper, EMS G200-Cu). These grids can be stored for months or years in indexed boxes (EMS 71140).

3.2.6.8. Staining of Ultrathin Sections

1. Put drops of filtered 1% uranyl acetate aqueous solution on a clean surface (parafilm) and float the grids on it (sections down) for 5 min, in the dark.
2. Wash on five drops of distilled water.
3. Dry the grids on filter paper before the next step. Sections should be up.
4. Float the grids on drops of filtered lead citrate for 1–2 min.
5. Wash on 10 drops of distilled water.
6. Dry the grids on filter paper.

3.3. Quantitative Assessment of Cell Death

3.3.1. Regrowth Assay

This assay provides a quantification of surviving cells (if surviving means ability to multiply).

1. Collect vegetative cells in late exponential growth phase.
2. Wash twice with SB buffer, count.
3. Plate two Lab-Tek chambers each with 3×10^5 cells (*see* **Note 8**) in 1 mL of SB containing 3 mM cAMP.
4. Incubate for 8 h at 22°C.
5. Carefully remove the liquid, wash once with 1 mL of SB, and replace with:
   - 1 mL of SB + DIF-1 0.1 µM in the first chamber;
   - 1 mL of SB in the second, control chamber (*see* **Note 9**).
6. Incubate for 24 h at 22°C. This leads, as described previously, to the death of most cells in the DIF-containing chamber (*see* **Note 10**).
7. From each chamber, carefully remove 0.5 mL of SB, and add 2 volumes (1 mL) of HL-5 to initiate regrowth of surviving cells.
8. Incubate at 22°C for 40 to 72 h.
9. Detach cells by vigorous flushes with a pipet. Under an inverted microscope, check that all vegetative cells are detached; many stalk cells will still adhere, which is not a problem because they are not to be counted.
10. Count amoeboid cells using a hemocytometer and phase-contrast optics. The rare heavily vacuolated, nonrefringent stalk cells are easily distinguished and excluded. Trypan Blue may be added to facilitate the identification of dead cells (*see* **Subheading 3.2.5.**).
11. Calculate the ratio of the number of regrowing cells in the DIF chamber to the number of cells in the control chamber (*see* **Note 11**). For HMX44A, this should be around 0.15. This ratio expresses the percentage of cells surviving after DIF-induced cell death. About 15% is the usual background of surviving HMX44A cells.

3.3.2. Flow Cytometry *(see **Note 12**)*

Cytometry provides an easy, precise, and objective quantification of death, which is particularly useful when testing the effect of modifiers (such as inhibitors) on cell death. Dead cells could be distinguished from live cells merely by morphological criteria. Using HMX44A.*atg1*⁻ mutant cells *(18)* 6 h after addition of DIF, a dot plot analysis with side scatter and forward scatter showed that presumably dead cells accumulated in a "cloud" (gated in R2) distinct from that of living cells (gated in R1) (*see* **Fig. 4**, left and middle). Only the R2 events, and almost all of them, stained positive with propidium iodide as shown by FL2 fluorescence intensity (*see* **Fig. 4**, right). Microscopy analysis after cell sorting of these two populations by flow cytometry confirmed that R2 events correspond to PI positive dead cells and that R1 events correspond to live cells.

1. Cell death is induced in monolayers by addition of DIF after a starvation period either in Lab-Tek culture chambers or in flasks.
2. Fifteen minutes before the time of analysis, the cells are resuspended, and 300-µL aliquots are transferred into RT15 (Dutscher) 0.8-mL tubes. The cell suspension

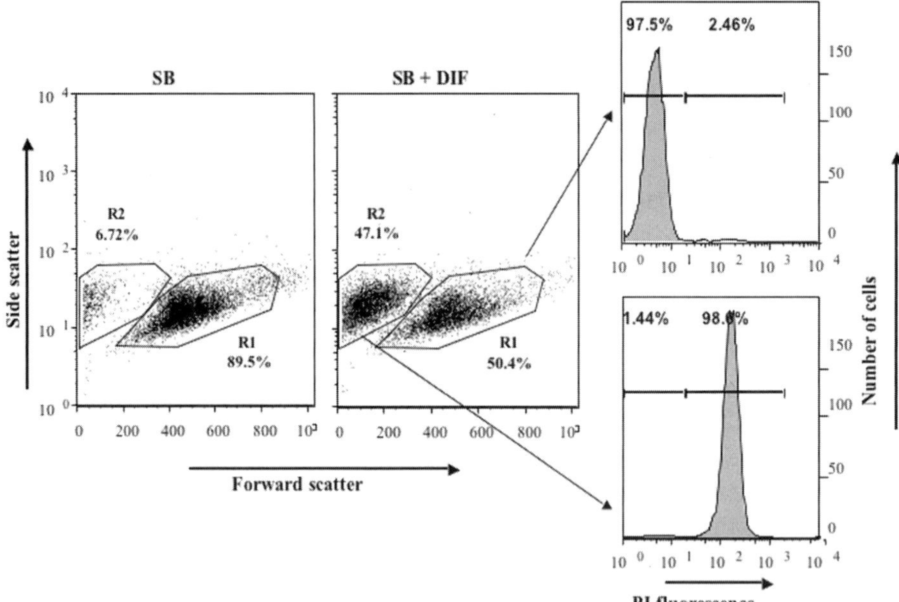

Fig. 4. Quantification of *Dictyostelium* cell death using flow cytometry analysis. To induce death, HMX44A. *atg1⁻* cells were subjected to starvation and cAMP, then incubated with or without differentiation-inducing factor (DIF) as described *(18)*. Cells were analyzed 6 h after DIF addition on morphological criteria. Left, dot plot of control cells without DIF. Middle, dot plot of cells with DIF. In these dot plots, the cells are distributed in two populations, gated in R1 and in R2, respectively. The presence of DIF induces a marked increase in the proportion of R2-gated cells. Right, distribution of cells after PI staining of each of the R1 and R2 cell populations. Most of the R1 cells are PI-negative, and most of the R2 cells are PI-positive, confirming that the latter population corresponds to dead cells.

should contain around 0.5×10^6 cells/mL. The cells can be analyzed directly. If one wants to use PI staining (which could be necessary if the treatment applied to cells affects their size or granularity), two tubes for each sample are prepared: a control (without any staining) and a test tube for PI staining. PI is used at 1 µg/mL and incubation is done with minimum light exposure for 10 min at 22°C, with no wash.

3. Cytometry analysis is then performed on a FACScan or a FACSCalibur cytometer from Becton Dickinson using CellQuest software. Just before analysis, the cells are resuspended by vortexing. The analysis is started by setting up the different parameters of the cytometer (Side scatter, Forward scatter, Threshold and fluorescence intensity) on control cells, without any staining. After set up the acquisition is made on 10^4 events per sample. Analysis of data is performed either with CELLQUEST or FlowJo softwares.

4. Notes

1. Use bacteriological peptone from OXOID; peptone from other sources can cause a dramatic decrease of *Dictyostelium* growth and impair development after starvation (our unpublished observations; *see also* **ref. 27**).
2. We have been routinely using Volvic source water instead of column-demineralized water. Source water contains a small amount of mineral salts and is therefore not strictly equivalent to fully demineralized water.
3. Although we routinely grow *Dictyostelium* in sterile conditions in the absence of antibiotics, accidental bacteriological contamination can usually be corrected by the addition of Penicillin-Streptomycin (100 U/mL and 100 µg/mL, respectively; Gibco-BRL, Grand Island, NY) to the culture medium. Fungal contaminations are more problematic because *Dictyostelium* is killed by fungizone; nystatin or gentamycin may be of help.
4. Programmed cell death in *Dictyostelium* is the outcome of terminal differentiation of stalk cells. This can be obtained in two different ways: either by inducing normal development at an air-wet solid interface (stalk cells then constitute the stalks and basal disks of the resulting fruiting bodies), or by in vitro stalk cell differentiation in monolayers under submerged conditions. The first method is the "natural" one, but is less convenient than the second one for most applications, because only approx 15% of the cells end up as stalk cells, the remainder differentiating into viable spores. Regrowth of stalk cells exclusively is thus difficult to score (although not impossible; *see* **ref. 7**). Also, microscopic observation of cells in a stalk is not easy; not only are further manipulations required in order to place the stalk on a microscope slide, but cell morphology is more difficult to assess because the cells are enclosed in the cellulose sheath tube (in addition to their own casing) and tightly packed. Furthermore, cells surviving in a stalk are difficult to isolate and manipulate.
5. The methods detailed here are meant for direct microscopic examination, usually of unfixed cells, mostly under the fluorescence microscope. A technical problem linked with such examination is the fact that when wild-type *Dictyostelium* cells differentiate in monolayers, some adhere very tightly to the substrate whereas others are found in suspension (most of them clustered). To get a representation of the total population, ideally both cell pools should be considered. If an inverted fluorescence microscope is available, differentiation may be conveniently carried out in plastic chambers on coverslips. These coverslips should be thin enough to allow microscopic examination using a 100× oil-immersion objective. Lab-Tek chambers have proven very useful for this purpose. Cell manipulations, staining, washes, and microscopic examination can be performed directly in these chambers.
6. This protocol is applicable to most of the usual strains. However, the percentage of cells differentiating into stalk cells is strongly strain-dependent: cells of V12M2 origin (such as HMX44) differentiate more efficiently than cells of NC4 origin (such as AX-2). This is largely because of a difference in sensitivity of inhibition of the DIF-dependent step by cAMP *(28)*, implying that an additional

washing step is recommended for some strains. The protocol detailed here, derived from the one described by Kay *(8)*, has been optimized for HMX44A.

7. There are many steps in this procedure in which things can go wrong, but unfortunately this can be determined only at the end of the procedure, by electron microscopy. Here are some tips that may help in avoiding frequent problems. **Always** use gloves and work under the hood, as most of the solutions are harmful, especially fixatives, uranyl, and resins. Dispose wastes (glutaraldehyde, osmium, uranyl, resins, lead) in appropriate containers. Handle the cells very carefully and avoid centrifugation before fixing. Never freeze the fixed cells. Never dry the cells at any stage. Molecular sieves (Fluka, ref 69828) are particles with a pore size of 0.3 nm that bind the water molecules. Put acetone and sieves in a bottle, shake it vigorously, and let it sit for 24 h. Avoid breathing on the sections while staining, especially while staining with lead citrate. If precipitates appear on the sections, make a "staining" chamber: put a piece of parafilm in a Petri dish, surround the parafilm with NaOH pellets, and replace the cover. Perform the lead citrate staining and the first wash in this chamber.

8. This test is sensitive to variations in initial density, so cells should be carefully counted before plating into flasks.

9. To be rigorous, a similar amount of absolute ethanol (the solvent of DIF) should be added to the control flask. However, we have never seen a significant effect of the addition of 0.1% ethanol in this assay.

10. If cells are incubated in starvation medium with DIF for longer than 24 h, vacuolization may seem more complete, but other phenomena may interfere with the results. In particular, and unexpectedly, differentiation into what appear to be macrocysts may occur with HMX44A cells maintained in SB without DIF (unpublished observation), lowering the frequency of regrowing cells.

11. Results of the test are collected after a period of exponential growth. Slight variations in culture conditions may thus significantly affect the results, although expression as a ratio prevents excessive departure from the usual values.

12. Flow cytometry is widely used to study the characteristics of cell death in mammalian cells. It allows quantitative measurement of fluorescence, size, and granularity of cells that can be applied to statistically significant numbers of cells. However, flow cytometry can only be employed with isolated cells in suspension. When they differentiate into stalk cells, wild-type *Dictyostelium* cells often adhere strongly to their substrate and form very tight cell clumps that are bound together with cellulose. This makes it difficult to analyze wild-type differentiating cells by flow cytometry. Vegetative cells, early wild-type differentiating cells, or round mutant *atg1⁻* cells that do not aggregate are, however, amenable to such analyses.

References

1. Raper, K. B. and Fennell, D. I. (1952) Stalk formation in *Dictyostelium*. *Bulletin of the Torrey Botanical Club* **79,** 25–51.
2. Maeda, Y. and Takeuchi, I. (1969) Cell differentiation and fine structures in the development of the cellular slime molds. *Devel. Growth Differ.* **11,** 232–245.

3. George, R. P., Hohl, H. R., and Raper, K. B. (1972) Ultrastructural development of stalk-producing cells in *Dictyostelium discoideum*, a cellular slime mould. *J. Gen. Microbiol.* **70,** 477–489.

4. de Chastellier, C. and Ryter, A. (1977) Changes of the cell surface and of the digestive apparatus of *Dictyostelium discoideum* during the starvation period triggering aggregation. *J. Cell Biol.* **75,** 218–236.

5. Quiviger, B., Benichou, J.-C., and Ryter, A. (1980) Comparative cytochemical localization of alkaline and acid phosphatases during starvation and differentiation of *Dictyostelium discoideum. Biol. Cellulaire.* **37,** 241–250.

6. Schaap, P., van der Molen, L., and Konijn, T. M. (1981) The vacuolar apparatus of the simple cellular slime mold *Dictyostelium minutum. Biol. Cell.* **41,** 133–142.

7. Whittingham, W. F. and Raper, K. B. (1960) Non-viability of stalk cells in *Dictyostelium. Proc. Natl Acad. Sci. USA* **46,** 642–649.

8. Kay, R. R. (1987) Cell differentiation in monolayers and the investigation of slime mold morphogens. *Methods in Cell Biology* **28,** 433–448.

9. Kopachik, W., Oohata, A., Dhokia, B., Brookman, J. J., and Kay, R. R. (1983) *Dictyostelium* mutants lacking DIF, a putative morphogen. *Cell* **33,** 397–403.

10. Town, C. D., Gross, J. D., and Kay, R. R. (1976) Cell differentiation without morphogenesis in *Dictyostelium discoideum. Nature* **262,** 717–719.

11. Town, C. and Stanford, E. (1979) An oligosaccharide-containing factor that induces cell differentiation in *Dictyostelium discoideum. Proc. Natl Acad. Sci. USA* **76,** 308–312.

12. Sobolewski, A., Neave, N., and Weeks, G. (1983) The induction of stalk cell differentiation in submerged monolayers of *Dictyostelium discoideum*. Characterization of the temporal sequence for the molecular requirements *Differentiation* **25,** 93–100.

13. Morris, H. R., Taylor, G. W., Masento, M. S., Jermyn, K. A., and Kay, R. R. (1987) Chemical structure of the morphogen differentiation inducing factor from *Dictyostelium discoideum. Nature* **328,** 811–814.

14. Cornillon, S., Foa, C., Davoust, J., Buonavista, N., Gross, J. D., and Golstein, P. (1994) Programmed cell death in *Dictyostelium. J. Cell Sci.* **107,** 2691–2704.

15. Levraud, J.-P., Adam, M., Luciani, M.-F., De Chastellier, C., Blanton, R. L., and Golstein, P. (2003) *Dictyostelium* cell death: early emergence and demise of highly polarized paddle cells. *J. Cell Biol.* **160,** 1105–1114.

16. Olie, R. A., Durrieu, F., Cornillon, S., Loughran, G., Gross, J., Earnshaw, W. C., and Golstein, P. (1998) Apparent caspase independence of programmed cell death in *Dictyostelium. Current Biol.* **8,** 955–958.

17. Roisin-Bouffay, C., Luciani, M. F., Klein, G., Levraud, J. P., Adam, M., and Golstein, P. (2004) Developmental cell death in *Dictyostelium* does not require paracaspase. *J. Biol Chem.* **279,** 11,489–11,494.

18. Kosta, A., Roisin-Bouffay, C., Luciani, M. F., Otto, G. P., Kessin, R. H., and Golstein, P. (2004) Autophagy gene disruption reveals a non-vacuolar cell death pathway in *Dictyostelium J. Biol.Chem.* **279,** 48,404–48,409.

19. Schaap, P., Nebl, T., and Fisher, P. R. (1996) A slow sustained increase in cytosolic Ca^{2+} levels mediates stalk gene induction by differentiation inducing factor in *Dictyostelium. EMBO J.* **15,** 5177–5183.

20. Li, G., Alexander, H., Schneider, N., and Alexander, S. (2000) Molecular basis for resistance to the anticancer drug cisplatin in *Dictyostelium*. *Microbiol. UK* **146,** 2219–2227.

21. Arnoult, D., Tatischeff, I., Estaquier, J., et al. (2001) On the evolutionary conservation of the cell death pathway: mitochondrial release of an apoptosis-inducing factor during *Dictyostelium discoideum* cell death. *Mol. Biol. Cell* **12,** 3016–3030.

22. Tatischeff, I., Petit, P. X., Grodet, A., Tissier, J. P., Duband-Goulet, I., and Ameisen, J. C. (2001) Inhibition of multicellular development switches cell death of *Dictyostelium discoideum* towards mammalian-like unicellular apoptosis. *Eur. J. Cell Biol.* **80,** 428–441.

23. Kawli, T., Venkatesh, B. R., Kennady, P. K., Pande, G., and Nanjundiah, V. (2002) Correlates of developmental cell death in *Dictyostelium discoideum*. *Differentiation* **70,** 272–281.

24. Katoch, B., and Begum, R. (2003) Biochemical basis of the high resistance to oxidative stress in *Dictyostelium discoideum*. *J. Biosci.* **28,** 581–588.

25. Levraud, J.-P., Adam, M., Cornillon, S., and Golstein, P. (2001) Methods to study cell death in *Dictyostelium discoideum in* "Cell Death. Methods in Cell Biology" (Schwartz, L. M., and Ashwell, J., Eds.), Vol. 66, pp. 469–97, Academic Press, San Diego.

26. Blanton, R. L. (1993) Prestalk cells in monolayer cultures exhibit two distinct modes of cellulose synthesis during stalk cell differentiation in *Dictyostelium*. *Development* **119,** 703–710.

27. Sussman, M. (1987) Cultivation and synchronous morphogenesis of *Dictyostelium* under controlled experimental conditions, in *Methods in Cell Biology* (Spudich, J. A., ed.). Harcourt Brace Jovanovich, New York: pp. 9–29.

28. Berks, M. and Kay, R. R. (1988) Cyclic AMP is an inhibitor of stalk cell differentiation in *Dictyostelium discoideum*. *Dev. Biol.* **126,** 108–114.

Index

A

Actin cytoskeleton,
    birefringence and polarized light
        scattering studies,
        macroscopic images of
          birefringence, 415
        materials, 411, 412, 418
        overview, 408, 409
        polarized light scattering, 415,
          416, 418, 419
        solution preparation, 413–415, 418
        theory, 409–411
    gel and bundle formation, 408
Adenylyl cyclase, *see* Cyclic AMP
Adhesion, *see* Cell–cell adhesion
Agglutination, *see* Cell–cell adhesion
Antisense-mediated gene silencing,
    applications, 211
    cotransformation with antisense
        constructs for RNA
        interference, 218, 225
    efficiency of silencing, 219, 220, 225
    essential gene silencing, 218, 219
    gene family silencing, 219
    stability of silencing, 221
Apoptosis, *see* Cell death
Axenic medium, *see* Culture,
    Dictyostelium; Parasexual
    cycle

B

Bidirectional orientation test, slug
    directional behavior analysis,
    159–161
Birefringence, *see* Actin cytoskeleton

**Bisulfite sequencing, *see* Chromatin
    remodeling**
BLAST, *see* dictyBase
Bonner, John, *Dictyostelium* research
    contributions, 7
Brefeld, Oskar, *Dictyostelium* research
    contributions, 4–6

C

Calcofluor, cell death staining, 540
cAMP, *see* Cyclic AMP
cAR1,
    chemotaxis signaling studies,
        cell culture, 284, 285, 292, 293
        cyclic AMP application and dye
        monitoring, 285, 293
        materials, 283, 284
        overview, 282
        simultaneous imaging of cyclic
        AMP stimulation and cell
        response,
        fluorescence resonance energy
          transfer imaging of protein–
          protein interactions, 288–290,
          292, 294
        gradient-exposed cells, 288, 294
        principles, 285–288
        uniformly stimulated cells,
        288, 293, 294
    signal transduction, 282
cDNA libraries,
    clone sequencing, 31, 32, 36, 37
    Dictyostelium resources, 32–34, 37
    Dicty_cDB database,
      access, 40

clone-based sequence and related
information, 40–42
expressed sequence tag clustering
and assembly, 42, 43
sequence catalogs and functional
classification, 43, 44
DNA microarray, *see* DNA
microarray
expressed sequence tag library
generation, 33, 35
full-length enriched library
generation, 35
functional genomics, 37–40, 47
gamete-enriched subtracted library
generation, 35, 36
promoter analysis *in silico*, 45–47
transcript and open reading frame
determination, 45
transcriptome analysis, 44, 45
Cell–cell adhesion,
assays,
agglutinometry,
development studies, 455, 456,
464, 465
instrumentation, 451, 452, 454,
455, 464
cell culture, 451
materials, 450, 451
modifying substance testing in
agglutinometer,
antibodies, 462
cell shape-modifying drugs and
chemicals, 463, 464
lectins, 462
overview, 460, 461
mutant analysis, 458–460, 466
single-cell assay in development,
457, 458, 465, 466
csA role, 449, 450
EDTA-resistant adhesion, 449, 450
glycoproteins in adhesion, 449, 450
Cell death,
pathways, 535, 536
assays,
calcofluor staining, 540
electron microscopy,
glutaraldehyde fixation, 541, 542
postfixation and osmium
staining, 543
postfixation and uranyl acetate
staining, 543
progressive dehydration with
acetone, 544
resin embedding and
polymerization, 544
sectioning and staining, 544
washing, 542
flow cytometry, 545, 546, 548
fluorescein diacetate staining, 540
materials, 537, 538, 547
overview, 536, 537
phase-contrast microscopy, 539
propidium iodide staining, 539, 540
regrowth assay, 544, 545, 548
stalk cell differentiation in monolayer
studies of developmental cell
death, 538, 539
Trypan Blue staining, 540, 541
Centrosome,
analysis,
immunofluorescence microscopy,
486–488
materials, 483, 484
protein yield measurement, 487, 488
two-dimensional gel electrophoresis
of components, 480, 481
function, 479
isolation,
cell preparation and lysis, 484, 488
DNase treatment and
sedimentation, 486, 488
materials, 482, 483, 487, 488
nuclei purification, 484, 488
overview, 480, 481, 485
pyrophosphate treatment, 484, 488
sucrose density gradient
centrifugation, 484–486
morphology, 479, 480

cGMP, *see* Cyclic GMP

Chemotaxis,
assay formats, 313
cAR1 signaling studies,
cell culture, 284, 285, 292, 293
cyclic AMP application and dye
monitoring, 285, 293
materials, 283, 284
overview, 282
simultaneous imaging of cyclic
AMP stimulation and cell
response,
fluorescence resonance energy
transfer imaging of protein–
protein interactions, 288–290,
292, 294
gradient-exposed cells, 288, 294
principles, 285–288
uniformly stimulated cells,
288, 293, 294
cyclic AMP as chemoattractant, 282,
311, 312, 394
developmental states, 311, 312
dynamic image analysis system for
motion analysis,
confocal microscopy,
fluorescence intensity analysis,
273
image acquisition, 273
outlining images, 273
sample preparation, 272, 273, 276
Dictyostelium preparation, 263
materials, 263, 265, 266, 273, 274
overview, 262
polymorphonuclear neutrophil
preparation, 263
three-dimensional analysis,
filopodia reconstruction, 270, 272
image capture, optical sectioning,
outlining, and reconstruction,
270
motility and dynamic
morphology parameters, 270
sample preparation, 268, 270
two-dimensional analysis,
computing parameters, 267, 268
image capture and outlining,
267, 275, 276
sample preparation, 266, 267, 274
folate as chemoattractant, 311, 312
green fluorescent protein assays,
cyclic AMP chemotaxis assays,
398–400, 402, 403
folate chemotaxis assays, 400–403
global chemoattractant
stimulation assays, 401–404
materials, 396–398
overview, 395, 396
random cell motility analysis,
398, 402
history of study, 7
mechanisms, 393
polymorphonuclear neutrophil, 262
translocation mechanisms, 261, 262
under-agarose assays,
cyclic AMP chemotaxis,
cell preparation and trough
filling, 317, 4318
gel preparation, 317
folate chemotaxis,
cell preparation and trough
filling, 316, 317, 322, 323
gel preparation, 315, 321, 322
trough formation, 315, 316, 322
individual cell analysis, 320, 321,
323
materials, 315, 321
overview, 313–315
population analysis, 318–320,
322, 323

Chromatin remodeling,
chromatin immunoprecipitation,
cell lysis and sonication, 496,
497, 504
elution, 497
materials, 494, 495, 504
precipitation, 497, 504
principles, 496

protein cross-linking, 496

developmental gene regulation, 491

DNA methylation analysis,
  bisulfite treatment, 499, 504
  genomic DNA preparation, 498, 499
  materials, 495
  primer design, 499, 500, 504
  principles, 498

DNA microarray studies, 497, 498

electrophoretic mobility shift assay,
  495, 502

immunoprecipitation,
  immunoprecipitation, 503, 504
  materials, 496
  nuclei purification, 503
  principles, 503

machinery, 492–494

pull-down of ectopically
    overexpressed proteins,
  cell extract preparation, 501
  materials, 495
  nickel-Sepharose bead
    preparation, 501
  overview, 500, 501
  pull-down, 502, 504

Confocal microscopy,
  cAR1 signaling studies,
    cell culture, 284, 285, 292, 293
    cyclic AMP application and dye
     monitoring, 285, 293
    materials, 283, 284
    overview, 282
    simultaneous imaging of cyclic
     AMP stimulation and cell
     response,
      fluorescence resonance energy
       transfer imaging of protein–
       protein interactions, 288–290,
       292, 294
      gradient-exposed cells, 288, 294
      principles, 285–288
      uniformly stimulated cells,
       288, 293, 294

dynamic image analysis system for
    chemotaxis motion analysis,
  fluorescence intensity analysis, 273
  image acquisition, 273
  outlining images, 273
  sample preparation, 272, 273, 276

green fluorescent protein-tagged host
    factors during *Legionella*
    invasion, 512–514

multicellular morphogenesis studies,
  303, 305

Cre-*lox*P system,
  electroporation of *Dictyostelium*,
    194, 195
  floxed *Bsr* cassette removal, 197–199
  gene mutation generation overview,
    187, 188
  knockout mutant validation, 195, 197
  materials, 189, 190
  targeting vector generation,
    competent *Escherichia coli*
     generation, 190, 191
    ligation and transformation, 193, 194
    polymerase chain reaction, 191, 198
    purification, 194, 198

Cryofixation, *Dictyostelium*,
  freeze-substitution,
    cleanup, 359, 360
    freeze-substitution module
     preparation, 358
    initial temperature series, 359, 363
    principles, 357, 358
    sample placement, 358, 359, 363,
     364
    temperature series continuation,
     immunolocalization studies, 359
     structural studies, 359, 363
  infiltration of specimens with
    embedment resin,
    acrylic resin infiltration, 360, 364
    embedment with flat-embedding
     techniques, 361, 362, 364
    epoxy resin infiltration, 360, 364
  materials, 341–345, 347–351, 362, 363

overview, 339–341
plunge-freezing,
    plunge-freezer preparation, 354, 355
    principles, 354
    sample preparation, 355, 356
    sample retrieval, 356, 357, 363
spray-freezing,
    airbrush assembly and testing, 352
    cryogen container assembly, 352
    principles, 351, 352
    sample preparation and spraying, 352–354, 363
    sprayer shutdown, 354
Culture, *Dictyostelium*,
advantages as model organism, 114
axenic media culture, 117
bacteria on agar surface culture, 116, 117, 121, 122
cloning of cells, 117, 122
fruiting body formation, 118, 119, 122
historical perspective, 6
macrocyst formation,
    maturation-inducing conditioning medium preparation, 120
    standard mated culture, 119, 120
    synchronous cell fusion, 120, 121, 123
    synchronous development of zygotes, 121
    variations, 120
materials, 114, 115, 121
nitrocellulose filter culture, 251
overview, 115, 116
parasexual cycle genetics, *see* Parasexual cycle
strain storage,
    amoebae, 118
    spores,
        freeze-dried spores, 118
        silica gel, 117, 118
Cyclic AMP (cAMP),
assays,
    binding assays,
        ammonium sulfate pellet assay, 379, 388
        phosphate buffer pellet assay, 378–380, 387, 388
        principles, 378, 387
        silicon oil assay, 379, 380, 387, 388
    cell culture, 374
    GTPase activity stimulating assay, 383, 384, 388
    GTPγS studies,
        adenylyl cyclase stimulation, 384, 385, 388
        cAMP stimulation of GTPγS binding, 382, 383
        inhibition of cAMP binding, 381, 382
    isotope dilution assay, 375–378, 387
    materials, 370–374
    membrane preparation, 374, 375
    second messenger responses, 380, 381, 388
chemoattractant, 282
chemotaxis studies, see Chemotaxis
ERK2 activation induction, *see* ERK2
receptor, *see* cAR1
signaling overview, 369, 370
Cyclic GMP (cGMP),
assays,
    cell culture, 374
    cyclic AMP response assay, 381
    GTPγS stimulation of guanylyl cyclase, 386–388
    isotope dilution assay, 375–378, 387
    materials, 370–374
    membrane preparation, 374, 375
signaling overview, 369, 370

D

Death, *see* Cell death
Developmental motility, *see* Multicellular morphogenesis

DIAS, *see* Dynamic image analysis system
dictyBase,
    BLAST server,
        databases and sequence download, 63, 64
        optimization, 64
    curated model, 52
    data and annotations, 52
    downloads, 53, 54
    functions, 51–53
    Gene Page,
        chromosomal coordinates and associated sequences, 67
        expression data, 68
        Gene Ontology, 67, 68
        names and identifiers, 67
        Navigation Bar, 66, 67
        overview, 64–66
        phenotype data, 68
        references and summary, 69
    Genome Browser,
        display configuration, 58, 59
        dump operations, 61
        Flip function, 61
        image output, 63
        Landmark or Region box, 60, 61
        restriction site annotation, 61
        Scroll/Zoom tool, 59
        sequence organization, 57
    origins, 51
    prospects, 73
    searching,
        expanded search, 56
        literature, 56
        search box output, 55
        searchable items, 54
        wildcard, 56
Dicty_cDB database,
    access, 40
    clone-based sequence and related information, 40–42
    expressed sequence tag clustering and assembly, 42, 43
    sequence catalogs and functional classification, 43, 44
Dictyostelium inverted repeat sequence (DIRS) elements, cluster mapping, 16, 17
Dicty Stock Center,
    functions, 69, 70
    nomenclature guidelines, 73
    ordering and depositing strains and plasmids, 72, 73
    searching, 71, 72
    strain and plasmid catalogs, 70
Directional statistics,
    acorn utilization, 163–165
    bidirectional orientation test, 159–161
    bimstat utilization, 165, 166, 169
    equality of concentration parameter tests, 158, 159
    graphical representation, 161, 163
    mean direction and concentration parameter estimation and confidence limits, 154, 156–158
    overview, 151, 152
    preferred direction and accuracy of orientation estimation, 161
    Rayleigh test for uniformity, 152, 154, 169
DIRS, *see* Dictyostelium inverted repeat sequence
DNA methylation, *see* Chromatin remodeling
DNA microarray,
    advantages, 75, 76
    cDNA library resources, 76, 77
    chromatin remodeling studies, 497, 498
    controls, 76, 77
    data analysis, 87–89, 91
    expression profiling applications in *Dictyostelium*, 37, 38, 78
    hybridization, 85, 90
    labeling with FairPlay kit,
        cDNA,
            generation, 82, 83, 89
            purification, 83, 89, 90

dye coupling reaction and
purification, 83, 84, 90
overview, 81, 82
materials, 78–80, 89
prehybridization, 84, 90
principles, 76, 77
RNA extraction, 80, 81, 89
scanning and quantification, 86, 91
spike mix addition and RNA
precipitation, 81, 82
washing, 85, 86, 90, 91
DNA sequence, *see* Genome sequence
Drug screening, *see* Pharmacogenetics
Dynamic image analysis system
(DIAS), *see* Chemotaxis

E
Electron microscopy (EM),
cell death analysis,
glutaraldehyde fixation, 541, 542
postfixation and osmium staining,
543
postfixation and uranyl acetate
staining, 543
progressive dehydration with
acetone, 544
resin embedding and
polymerization, 544
sectioning and staining, 544
washing, 542
cryofixation, *see* Cryofixation,
Dictyostelium
infiltration of specimens with
embedment resin,
acrylic resin infiltration, 360, 364
embedment with flat-embedding
techniques, 361, 362, 364
epoxy resin infiltration, 360, 364
Electrophoretic mobility shift assay
(EMSA), chromatin
remodeling studies, 495, 502
Electroporation, 194–195, 204–206
EM, *see* Electron microscopy

EMSA, *see* Electrophoretic mobility
shift assay
Endocytic pathway,
exocytosis versus recycling, 26
fluorescence assays,
endo-lysosomal pH measurements,
dual excitation ratio
fluorometry, 433, 436
dual fluorophores, 433, 434,
436, 437
fluorescent-labeled particles, 425,
434
fluorimetric analysis of
exocytosis, 431
fluorimetric analysis of
phagocytosis,
bacteria uptake assay, 429
latex bead assay, 428, 429, 436
overview of assays, 428, 436
yeast uptake assay, 429, 436, 437
fluorimetric analysis of pinocytosis,
FITC-dextran, 430, 436
TRIC-dextran, 430, 431, 436, 437
materials, 426–428, 436
microscopic analysis of
phagocytosis and fluid-phase
endocytosis, 435–437
mutant studies, 424, 425
particles for uptake, 425
presentation of fluorimetric
results, 431, 433
lysosomal enzyme recycling assay,
434, 435
overview, 423, 424
phagocytosis process, 439
Endolysosome purification,
cell breakage, 177, 182
magnetic purification, 179, 183
marker assays, 182
materials, 174, 175, 182
overview, 171, 172
Percoll gradient centrifugation for
endosome/lysosome
separation, 176, 177, 181, 183

probe labeling, 176, 181, 183
Epigenetics, *see* Chromatin remodeling
ERK2,
    cyclic AMP-induced activation,
      assays,
        affinity column preparation,
          472, 477
        antibody purification, 472,
          473, 477, 478
        glutathione *S*-transferase
          fusion protein preparation,
          471, 472, 477
        materials, 470, 471, 476, 477
        periodic change in level of
          phosphorylated protein, 475,
          476
        Western blot of activated
          protein, 473–475, 478
      overview, 469, 470
    developmental regulation, 469, 470
EST, *see* Expressed sequence tag
Exocytosis, *see* Endocytic pathway
Expressed sequence tag (EST),
    clustering and assembly, 42, 43
    library generation, 33, 35
Extracellular signal-regulated kinase,
      see ERK2

F

Fixation, *Dictyostelium*,
    coverslip preparation and cell
      plating, 332
    cryofixation for electron microscopy,
      see Cryofixation,
      Dictyostelium
    materials, 329, 330, 332
    methanol fixation, 335–337
    overview of strategies, 327–329
    picric acid-paraformaldehyde
      fixation, 336, 337
    rapid freezing in liquid ethane, 335
Flow cytometry,
    cell death assay, 545, 546, 548
    ploidy analysis, 132, 133

Fluorescein diacetate, cell death
      staining, 540
Fluorescence resonance energy transfer
      (FRET), cAR1 signaling and
      imaging of protein–protein
      interactions, 288–290, 292, 294
Fluorescence-activated cell sorting
      (FACS), *see* Flow cytometry
Folate chemotaxis, *see* Chemotaxis
FRET, *see* Fluorescence resonance
      energy transfer
Fruiting body, formation, 118, 119, 122

G

β-Galactosidase, *see* Reporter genes
Gene disruption, 187
Genome sequence, *Dictyostelium*,
    assembly, 16
    base content, 18
    cDNA libraries, *see* cDNA libraries
    chromosome features, 16–18, 51
    functional study prospects, 25, 26,
      37, 47
    genes,
      annotation, *see* dictyBase
      bacterial orthologs, 21
      duplications, 18
      prediction, 18–20
    phylogenetic analysis,
      comparison of amoebozoa, 24, 25
      tree construction, 22–24
    protein domain analysis, 20, 21
    sequencing, 16
GFP, *see* Green fluorescent protein
Green fluorescent protein (GFP),
    chemotaxis assays,
      cyclic AMP chemotaxis assays,
        398–400, 402, 403
      folate chemotaxis assays, 400–403
      global chemoattractant
        stimulation assays, 401–404
      materials, 396–398
      overview, 395, 396

random cell motility analysis, 398, 402

fluorescence resonance energy transfer imaging of protein–protein interactions, 288–290, 292, 294

fusion protein design, 240, 241, 244

host factor fusion protein confocal microscopy during *Legionella* invasion, 512–514

live cell imaging,
advantages and disadvantages of fluorescent proteins, 241, 242, 244

cameras and tracking software, 239, 240

cell culture and transformation, 237

chambers, 237, 238, 243

expression testing, 237, 243

filters, 239

fusion protein vectors,
construction, 242, 243
expression, 243

microscopes, 238

objective lenses, 239

overview, 229, 230

multicolor labeling, 235

photoactivatable proteins, 232

red fluorescent proteins, 232–234

variants, 230–232

vectors for expression, 236, 237

GTPase, cyclic AMP stimulating assay, 383, 384, 388

Guanylyl cyclase, *see* Cyclic GMP

H

HGT, *see* Horizontal gene transfer

Horizontal gene transfer (HGT), bacteria to *Dictyostelium*, 22

I

Immunoblot, *see* Western blot

Immunofluorescence microscopy, centrosomes, 486–488

fixation, *see* Fixation, *Dictyostelium*

staining, 336–338

Immunoprecipitation, *see* Chromatin remodeling

Infection, *see* *Legionella pneumophila*; Pathogens

Inositol trisphosphate,
assays,
cell culture, 374

cyclic AMP response assay, 381

GTPγS regulation of phospholipase C, 385, 386, 388

isotope dilution assay, 375–378, 387

materials, 370–374

membrane preparation, 374, 375

pharmacogenetic analysis,
drug treatment, 530, 533

extract preparation, 530, 533

radioassay, 530, 531, 533

signaling, 369, 370

In situ hybridization (ISH),
spatial expression profiling applications in *Dictyostelium*, 38–40

whole-mount *in situ* hybridization,
anti-DIG-alkaline phosphatase reaction, 253, 258

cell culture on nitrocellulose filters, 251

fixation, 252, 257

hybridization, 253, 257

materials, 229–251, 256

overview, 247, 248

probe labeling, 251, 252, 256, 257

ISH, *see* *In situ* hybridization

Isoelectric focusing, *see* Proteomics

K, L

Knockouts, *see* Cre-*lox*P system; Parasexual cycle

Legionella pneumophila,
Dictyostelium interaction studies,
cell culture, 510, 513, 514

confocal microscopy of green
    fluorescent protein-tagged host
    factors during invasion, 512–514
infection assay, 511, 512, 514
materials, 509, 510, 513
overview, 508, 509
phagocytosis assay, 510, 511, 514
safety, 510, 513
Dictyostelium uptake and infection,
    11, 508
Lithium, mechanism of action in
    bipolar disease, 518
Live cell imaging, *see* Green fluorescent
    protein; Chemotaxis;
    Multicellular morphogenesis
Lysosome, *see* Endolysosome
    purification

M

Macrocyst formation,
    maturation-inducing conditioning
        medium preparation, 120
    standard mated culture, 119, 120
    synchronous cell fusion, 120, 121, 123
    synchronous development of
        zygotes, 121
    variations on culture, 120
MAPK, *see* Mitogen-activated protein
    kinase
Mass spectrometry, *see* Proteomics
Micro-RNA (miRNA), gene expression
    regulation, 25, 26
miRNA, *see* Micro-RNA
Mitochondria purification,
    cell breakage, 177, 182
    centrifugation, 178, 182
    marker assays, 182
    materials, 174, 182
    overview, 171, 172
Mitogen-activated protein kinase
    (MAPK), *see* ERK2
Morphogenesis, *see* Multicellular
    morphogenesis

Motility, *see* Chemotaxis; Multicellular
    morphogenesis; Phototaxis;
    Spontaneous turning behavior;
    Thermotaxis
Multicellular morphogenesis,
    aggregation stream and mound
        preparation, 299
    materials for study, 298
    microscopy,
        confocal microscopy, 303, 305
        data analysis,
            cell tracks, 306
            fluorescence intensity, 306, 307
            image subtraction, 307
        fluorescence labeling for synergy
            experiments with mutant cells,
            298, 299, 307
        optical density wave
            visualization, 305
        sample preparation,
            cell movement studies, 300,
                302, 308
            fluorescent probe distribution
                during development studies,
                302, 308
    overview, 297, 298
    slug preparation, 299, 300

N

Necrosis, *see* Cell death
Neutrophil, *see* Polymorphonuclear
    neutrophil
Northern blot, small interfering RNA,
    222, 223, 225
Nuclei purification,
    cell breakage, 177, 182
    centrifugation, 178
    marker assays, 182
    materials, 174, 182
    overview, 171, 172

P

Parasexual cycle,

diploid selection from haploid
    parents, 127, 128, 131, 134
genetics applications, 126, 127
haploid segregation from diploid
    parents,
    axenic medium segregation, 132
    bacterial lawn segregation, 131, 132
    overview, 128–131
materials for genetic studies, 130
overview, 125, 126
ploidy analysis,
    cytological staining, 132
    fluorescence-activated cell
        sorting, 132, 133
Pathogens, *see also Legionella
    pneumophila,*
    Dictyostelium advantages as model
        system, 507, 508
    overview of *Dictyostelium* studies,
        507–509
PCR, *see* Polymerase chain reaction
Phagocytosis, *see* Endocytic pathway;
        Legionella pneumophila;
        Phagosome purification
Phagosome purification,
    cell lysis, homogenization, and
        centrifugation, 177, 182,
        441–443, 446, 447
    concentration measurement and
        yield, 444, 445
    latex beads and flotation, 179, 181, 183
    marker assays, 182
    materials, 175, 176, 182, 440, 441
    overview, 171, 172
    principles, 439, 440
    pulse-chase feeding, 445–447
Pharmacogenetics,
    drug target identification in
        Dictyostelium,
        drug-resistant clone selection,
            developmental selection, 520,
                521
            filter assay, 521, 531
            growth selection, 521

inositol trisphosphate analysis,
    drug treatment, 530, 533
    extract preparation, 530, 533
    radioassay, 530, 531, 533
inverse polymerase chain reaction
    rescue for gene identification,
        521–523, 532
materials, 518–520, 531
overview, 518
phenotypic effects, 520
recapitulation of gene knockout
    for validation,
    amplification of gene-controlling
        drug action, 524, 532
    DNA extraction, 528, 532
    overview, 523, 524
    polymerase chain reaction of
        transposase integrants, 527, 532
    polymerase chain reaction
        screening for homologous
        integrants, 528, 532
    transformant isolation, 527,
        528, 532
    transposon-targeted gene
        disruption, 525–527
Southern blot confirmation of
    gene ablation,
    blot transfer, 529
    DNA extraction, 529
    hybridization and
        autoradiography, 529, 530
overview, 517, 518
Phospholipase C (PLC), GTPγS
    regulation assay, 385, 386, 388
Phototaxis,
    assays,
        applications, 140, 141
        culture, incubation and analysis,
            146, 147, 167
        digitization of data, 149–151,
            167–169
        materials, 141, 142, 144
        qualitative tests, 144, 145, 166, 167
        quantitative tests, 145, 146, 167

spontaneous turning and average speed, 148, 149, 167

statistical analysis,

acorn utilization, 163–165

bidirectional orientation test, 159–161

bimstat utilization, 165, 166, 169

directional statistics overview, 151, 152

equality of concentration parameter tests, 158, 159

graphical representation, 161, 163

mean direction and concentration parameter estimation and confidence limits, 154, 156–158

preferred direction and accuracy of orientation estimation, 161

Rayleigh test for uniformity, 152, 154, 169

overview, 138

signaling pathways, 138–140

Phylogenetic analysis,

comparison of amoebozoa, 24, 25

tree construction, 22–24

Pinocytosis, *see* Endocytic pathway

Plasma membrane purification,

cell breakage, 177, 182

centrifugation, 177, 178, 182

marker assays, 182

materials, 172, 173, 182

overview, 171, 172

probe labeling, 176, 181, 183

Plasmids, *see* Dicty Stock Center

PLC, *see* Phospholipase C

Ploidy analysis,

cytological staining, 132

fluorescence-activated cell sorting, 132, 133

PMN, *see* Polymorphonuclear neutrophil

Polarized light scattering, *see* Actin cytoskeleton

Polyketide synthases, *Dictyostelium* genes, 9, 10

Polymerase chain reaction (PCR),

Cre-*lox*P system targeting vector generation, 191, 198

pharmacogenetics analysis,

inverse polymerase chain reaction rescue for gene identification, 521–523, 532

screening for homologous integrants, 528, 532

transposase integrants, 527, 532

Polymorphonuclear neutrophil (PMN),

dynamic image analysis system, *see* Chemotaxis

motility and chemotactic defects, 262

Programmed cell death, *see* Cell death

Promoters, analysis *in silico*, 45–47

Propidium iodide, cell death staining, 539, 540

Proteomics,

materials, 99, 100, 106

overview, 95, 96

protein domain analysis from sequences, 20, 21

sample preparation, 101, 106

two-dimensional polyacrylamide gel electrophoresis,

gel casting, 102

isoelectric focusing,

cup loading, 101, 102, 106, 107

equilibration, 102, 103

rehydration, 101, 106

principles, 96–99, 106

running conditions, 103

spot analysis,

database searching, 105, 108

digestion in-gel, 104, 107

mass spectrometry, 105, 108

picking, 103, 104, 107

staining, 103, 107

R

Raper, Kenneth, *Dictyostelium* research
contributions, 6
Rayleigh test for uniformity, slug
directional behavior analysis,
152, 154, 169
rDNA, *see* Ribosomal DNA
REMI, *see* Restriction enzyme-mediated
integration
Reporter genes,
applications, 248
detection, 248
β-galactosidase,
advantages, 248, 249, 255
cell culture on nitrocellulose
filters, 251
cell staining, 253–255, 258
materials, 251
spore staining, 255
vectors, 249, 255, 256
types, 248
Restriction enzyme-mediated integration
(REMI) mutagenesis,
applications, 201
blasticidin selection, 204, 205, 208
cloning of insertion site and flanking
DNA, 206–208
enzyme combinations, 202, 203
H50 method, 206
libraries for drug target identification
in *Dictyostelium*, 518, 520
materials, 203, 204, 207
principles, 201–203
transformant harvesting, 206
Ribosomal DNA (rDNA), palindromic
elements, 17, 18
RNA interference (RNAi),
applications, 211
construct preparation, 216, 217, 224
cotransformation with antisense
constructs, 218, 225

Dictyostelium genes in pathway,
212–216
efficiency of silencing, 219, 220, 225
essential gene silencing, 218, 219
failed approaches, 213, 223, 224
gene family silencing, 219
materials, 213
mechanism, 212, 221
resistant gene downregulation in
helF knockout mutants, 217,
218, 224, 225
small interfering RNA preparation,
Northern blot, 222, 223, 225
polyacrylamide gel
electrophoresis, 222
total RNA isolation, 221, 222
stability of silencing, 221
RNAi, *see* RNA interference

S

siRNA, *see* Small interfering RNA
Slug,
phototaxis, *see* Phototaxis
preparation of, 299–300
spontaneous turning behavior, *see*
Spontaneous turning behavior
thermotaxis, *see* Thermotaxis,
Small interfering RNA (siRNA), *see*
RNA interference
Southern blot, confirmation of gene
ablation,
blot transfer, 529
DNA extraction, 529
hybridization and autoradiography,
529, 530
Spontaneous turning behavior,
assays,
applications, 140, 141
digitization of data, 149–151,
167–169
materials, 141, 142, 144
qualitative tests, 144, 145, 166, 167

quantitative tests, 145, 146, 167
spontaneous turning and average
speed, 148, 149, 167
statistical analysis,
acorn utilization, 163–165
bidirectional orientation test,
159–161
bimstat utilization, 165, 166, 169
directional statistics overview,
151, 152
equality of concentration
parameter tests, 158, 159
graphical representation, 161, 163
mean direction and
concentration parameter
estimation and confidence
limits, 154, 156–158
preferred direction and
accuracy of orientation
estimation, 161
Rayleigh test for uniformity,
152, 154, 169
overview, 138
signaling pathways, 138–140
Spores,
β-galactosidase staining, 255
storage,
freeze-dried spores, 118
silica gel, 117, 118
Strains,
culture, *see* Culture, Dictyostelium
Dicty Stock Center, *see* Dicty Stock
Center
history of development, 8, 9
storage,
amoebae, 118
spores,
silica gel, 117, 118
freeze-dried spores, 118
Sussman, Maurice, *Dictyostelium*
research contributions, 7, 8
T
Thermotaxis,
assays,

applications, 140, 141
culture, incubation and analysis,
147, 148
digitization of data, 149–151,
167–169
materials, 141, 142, 144
qualitative tests, 144, 145, 166, 167
quantitative tests, 145, 146, 167
spontaneous turning and average
speed, 148, 149, 167
statistical analysis,
acorn utilization, 163–165
bidirectional orientation test,
159–161
bimstat utilization, 165, 166, 169
directional statistics overview,
151, 152
equality of concentration
parameter tests, 158, 159
graphical representation, 161, 163
mean direction and
concentration parameter
estimation and confidence
limits, 154, 156–158
preferred direction and accuracy
of orientation estimation,
161
Rayleigh test for uniformity,
152, 154, 169
overview, 138
signaling pathways, 138–140
Transposon, targeted gene disruption,
525–527
Trypan Blue, cell death staining, 540,
541
Two-dimensional polyacrylamide gel
electrophoresis, *see*
Centrosome; Proteomics

V, W

Valproic acid, mechanism of action in
bipolar disease, 518
Western blot, ERK2 activation studies,
473–475, 478